Die Grundlehren der mathematischen Wissenschaften

in Einzeldarstellung
mit besonderer Berücksichtigung
der Anwendungsgebiete

Band 205

S. M. Nikol'skiĭ

Approximation of Functions of Several Variables and Imbedding Theorems

Translated from the Russian by

J. M. Danskin

Springer-Verlag
Berlin Heidelberg New York 1975

Sergeĭ Mihaĭlovič Nikol'skiĭ
Steklov Mathematical Institute
Academy of Sciences, Moscow, U.S.S.R

Translator:
John M. Danskin (U.S.A.)

Title of the Russian Original Edition:
Približenie Funkciĭ Mnogih
Peremennyh i Teoremy Vloženiya
Publisher: "Nauka", Moscow 1969

AMS Subject Classifications (1970)
33 A 10, 35 C 10, 40 B 05, 41 A 50, 41 A 60, 42 A 24,
42 A 68, 44 A 40, 46 E 35, 46 E 25, 46 F 05, 46 E 30

ISBN 3-540-06442-7 Springer-Verlag Berlin Heidelberg New York
ISBN 0-387-06442-7 Springer-Verlag New York Heidelberg Berlin

Library of Congress Catalog Card Number 74-4652.

Printed in GDR.

Author's Preface to the English Edition

This English translation of my book "Približenie Funkciĭ Mnogih Peremennyh i Teoremy Vloženiya" is identical in content with the Russian original, published by "Nauka" in 1969. However, I have corrected a number of errors.

I am grateful to the publishing house Springer-Verlag for making my book available to mathematicians who do not know Russian.

I am also especially grateful to the translator, Professor John M. Danskin, who has fulfilled his task with painstaking care. In doing so he has showed high qualifications both as a mathematician and as a translator of Russian, which is considered by many to be a very difficult language.

The discussion in this book is restricted, for the most part, to functions everywhere defined in n-dimensional space. The study of these questions for functions given on bounded regions requires new methods. In connection with this I note that a new book, "Integral Representations of Functions and Imbedding Theorems", by O. V. Besov, V. P. Il'in, and myself, has just (May 1975) been published, by the publishing house "Nauka", in Moscow.

Moscow, U.S.S.R., May 1975

S. M. Nikol'skiĭ

Translator's Note

I am very grateful to Professor Nikol'skiĭ, whose knowledge of English, which is considered by many to be a very difficult language, is excellent, for much help in achieving a correct translation of his book. And I join Professor Nikol'skiĭ in thanking Springer-Verlag. The editing problem was considerable, and the typographical problem formidable.

Regensburg, West Germany
May 1975

John M. Danskin, jr.

Contents

Introduction

In the last two decades the theory of imbeddings of classes of differentiable functions of several variables, whose origins are to be found as far back as the work of Sobolev in the 1930's, has been considerably developed. At present a number of fundamental questions have been brought to completion, and the need for a compact exposition has become apparent. As to myself, I came to the questions of the theory of imbedding in connection with ideas long of interest to me connected with the classical theory of approximation of functions by polynomials, first of all by trigonometrical polynomials and their nonperiodic analogues, entire functions of exponential type.

These ideas, which I had to develop within the new context, served as a basis for me in the construction of the theory of imbeddings of H-classes. Here, already in questions on the traces of functions, there appeared not only direct theorems, but theorems completely inverting them. We may give the name "inverse theorems" to theorems on the extension of functions to a space from manifolds of lower dimensionality lying in it. Here we include not only the isotropic case of functions having differential properties which are the same in various directions, but also the anisotropic case.

Later on O. V. Besov constructed an analogous theory of imbedding of the B-classes introduced by him, also basing his work on the methods of the theory of approximation by trigonometric polynomials or by entire functions of exponential type. B-classes are significant because they, like the H-classes, are, as we say, closed in themselves relative to the imbedding theorems. We mean that the imbedding theorems of interest to us (we will not formulate them here) are expressed in terms of B-classes and in addition possess in a certain sense properties of transitivity and invertibility in the case of the problem of traces.

Sobolev proved his imbedding theorems for the classes $W_p^l = W_p^l(\Omega)$ of functions having, on a sufficiently general region Ω of n-dimensional space $\mathbb{R}_n$, derivatives to order l inclusive, which are integrable to the pth power ($1 \leqq p \leqq \infty$). The Sobolev classes may be called "discrete classes", because the parameter l expressing the differential properties of the functions entering into them runs through the discrete sequence $l = 0, 1, 2, \ldots$. In this sense the classes H and B are continuous. The reader on familiarizing himself with Chapter 9 will realize that those theorems of Sobolev, with the complements to them due to V. I. Kondrašov and V. P. Il'in which are accompanied by a change in the metric, are in a certain sense final, and even, to the extent that they make discreteness of classes possible, transitive.

As to imbedding theorems accompanied only by a change in the dimension without a change in the metric, theorems which we call theorems on traces, the matter is more complicated. Of course, the theorems of Sobolev gave a certain answer to the question as to what differential properties were possessed by the trace of a function of class $W_p^l(\Omega)$ on a manifold $\Gamma \subset \Omega$, but this answer was given in the terms of the classes W, and now we know that generally speaking if $p \neq 2$ the final answer to this question is not expressed in terms of the classes W.

The first final results on the problem of traces of W-classes were obtained for $p = 2$ by Aronszajn [1], and independently by V. M. Babič and L. N. Slobodeckiĭ [2]. In this case ($p = 2$) the fractional classes $W_2^l(\Omega)$ and $W_2^l(\Gamma)$, corresponding to any positive real parameter l, were introduced. In terms of these classes direct and completely inverse imbedding theorems were obtained. In the notation adopted in this book, $W_2^l = L_2^l = B_2^l$. The later investigations of Gagliardo [1], Besov [1, 2], P. I. Lizorkin [9] and S. V. Uspenskiĭ [1, 2], led to a complete solution of the problem of traces of functions of the classes W_p^l for any finite $p > 1$. The reader will find in the same Chapter 9 how this solution looks (putting $W_p^l = L_p^l$). Now we shall only say that the traces of functions f of class W_p^l for $p \neq 2$ belong generally speaking not to W-classes but to B-classes. This bears on the one hand on the fact that theorems on imbedding of different dimensions (theorems on traces) cease to be closed relative to the W-classes, and on the other hand on the fact that there is a close connection between the classes W and B. This connection was so close that in the days when not everything concerning these questions was clear, it was believed that the classes B_p^l for fractional l were the natural extensions of the integer classes of Sobolev, and they were denoted by W_p^l. In fact, the natural such extensions are the so-called Liouville classes L_p^r, to which Chapter 9 is devoted. Thus the classes W are treated there as well, since we take $W_p^l = L_p^l$ ($l = 0, 1, \ldots$). The reader should bear in mind that in this book the notation W_p^l is only used for $l = 0, 1, \ldots$. In this connection see 4.3.

Sobolev studied the functions of his classes using the integral representations introduced by himself. A considerable development of this work was carried out by V. P. Il'in, and then by O. V. Besov; in this connection see 6.10 below. Functions of the classes L_p^r are defined on the entire space, and in their integral representations it is extremely important to take care that the kernel of the representation decrease to zero sufficiently rapidly at infinity. The well-known kernels of Bessel-

MacDonald are of this sort. They are taken as basic for the repesentation of functions of the class L_p^r. We say "as basic", since we are in fact considering here anisotropic classes $L_p^{\mathbf{r}}$. The kernels of their integral representations are defined by complicating the MacDonald kernels. In writing Chapter 9 I have made essential use of materials given to me by my colleague P. I. Lizorkin, who quite recently found an entire system of imbedding theorems for general anisotropic classes $L_p^{\mathbf{r}}$, where $\mathbf{r}$ is any positive vector. His results are published for the present in the form of a short note.

In the one-dimensional case, where the problem of traces does not arise, theorems for imbedding different metrics for the classes L_p^r and, for noninteger r, for the classes H_p^r, were already obtained in the papers of Hardy and Littlewood.

The operations $I_{\mathbf{r}}$, defined by the Bessel-MacDonald kernels, are of universal character. In this book they are studied and applied in various situations. We use rather extensively the concept of generalized function. Hence the book contains a short section where there is presented, with complete proofs, only that information from the theory of generalized functions which the reader needs to know for the understanding of what follows. I introduce the concept of a generalized function regular in the sense of L_p, using the operation $I_{\mathbf{r}}$. For regular functions various proofs connected with multiplication by a generalized function are greatly simplified. I use this fact widely, since the generalized functions appearing in this book are regular.

The operation $I_{\mathbf{r}}$ has also been applied in an interesting way in Chapter 8. It realizes isomorphisms not only of the L-classes, but of the B- and H-classes as well, and it can serve as a means for integral representations of functions of these classes. These ideas, which in the periodic one-dimensional case go back to Hardy and Littlewood, were quite recently studied from various points of view in the papers of Aronszajn and Smith, Calderon, Taibleson, Lions, Lizorkin, the author and others.

It is natural that we have found in this book a place for the foundations of the theory of approximation of functions of several variables by trigonometric polynomials and entire functions of exponential type. They are of interest in themselves, but basically their role is subordinate. We further prove, by means of the methods of the theory of approximation, imbedding theorems for H- and B-classes, and give representations of functions of these classes in terms of series in entire functions of exponential type or in terms of trigonometric polynomials. Having these objectives in mind, along with the traditional inequalities we introduce and apply other inequalities, referring to different dimensions and metrics.

It should be noted that we give in this book complete proofs of the imbedding theorems for the above-mentioned classes of functions, defined on all of n-dimensional space $\mathbb{R}_n$. But these classes may be defined for regions $\Omega \subset \mathbb{R}_n$. Such definitions are given in the book. We also formulate (without proof) very general theorems on the extension of functions of these classes to the entire space (with the preservation of class). This makes it possible to extend theorems proved for the space $\mathbb{R}_n$ to the case of regions $\Omega \subset \mathbb{R}_n$.

Finally I note that recently investigations, begun by L. D. Kudrjavčev, have been carried out on the more general weight classes. In this book we restrict ourself to making only particular remarks concering the connections between the weight classes with the non-weight classes considered here.

Let me note further that over the last ten years and more there has been a constantly active seminar on the theory of differentiable functions of several variables at the Steklov Mathematical Institute, under the direction of V. I. Kondrašov, L. D. Kudrjavčev, and myself. O. V. Besov, Ja. S. Bugrov, V. I. Burenkov, A. A. Vašarin, P. I. Lizorkin, S. V. Uspenskiĭ, G. N. Jakovlev and other mathematicians took part in this seminar. Many of the results presented in this book are due to the participants and were discussed in the seminar from the moment of discovery.

In conclusion I think it appropriate here to express my deep gratitude to my colleagues O. V. Besov, who read the manuscript of the book, P. I. Lizorkin, who read Chapters 8 and 9, and S. A. Teljakovskiĭ, who read several chapters. They made many useful remarks, which I have taken into account one way or another.

I also thank T. A. Timan, who pointed out some errors in the manuscript.

Finally, I am deaply grateful to my young colleague V. I. Burenkov, the editor of the book. Much of his advice, relating not only to the form but also to the content, has been taken into account.

Chapter 1

Preparatory Information

1.1. The Spaces $C(\mathscr{E})$ and $L_p(\mathscr{E})$

In this book we shall consider functions depending, generally speaking, on several variables.

The symbol $\mathbb{R}_n (n = 1, 2, \ldots)$ will always denote n-dimensional Euclidean space of points $\boldsymbol{x} = (x_1, \ldots, x_n)$ with real coordinates. The length of the vector will be denoted as follows:

$$|\boldsymbol{x}| = \sqrt{\sum_1^n x_k^2}. \tag{1}$$

If $\mathscr{E}$ is a closed set, lying in $\mathbb{R}_n (\mathscr{E} \subset \mathbb{R}_n)$, then $C(\mathscr{E})$ will denote the collection of all functions (real or complex) $f = f(\boldsymbol{x})$ which are uniformly continuous on $\mathscr{E}$.

To each function $f \in C(\mathscr{E})$ we attach a norm (in the sense of $C(\mathscr{E})$)

$$\|f\|_{C(\mathscr{E})} = \sup_{x \in \mathscr{E}} |f(\boldsymbol{x})|. \tag{2}$$

In the case of a bounded (closed) set $\mathscr{E}$ the sup will be replaced by max.

If p is a real number satisfying the inequalities $1 \leqq p < \infty$, and if on some measurable set $\mathscr{E} \subset \mathbb{R}_n$, not necessarily bounded, there is given a real or complex function f such that the function $|f|^p$ is integrable (summable) in the sense of Lebesgue on $\mathscr{E}$, then we put

$$\|f\|_{L_p(\mathscr{E})} = \left(\int_{\mathscr{E}} |f|^p \, d\mathscr{E} \right)^{1/p}. \tag{3}$$

The quantity (3) is called the *norm* of f in the sense of $L_p(\mathscr{E})$. We denote by $L_p(\mathscr{E})$ the collection of all functions having a finite norm (3). If $p = 1$ we frequently drop the subscript p. Thus we write $L_1(\mathscr{E}) = L(\mathscr{E})$.

We will not distinguish two *equivalent* functions $f_1, f_2 \in L_p(\mathscr{E})$, i.e. functions differing on a set of measure zero. We shall take them to be equal to one and the same element of the functional space $L_p(\mathscr{E})$ and write $f_1 = f_2$. In particular, if $f \in L_p(\mathscr{E})$ is equal to zero for almost all $\boldsymbol{x} \in \mathscr{E}$, we will write $f = 0$, in this way identifying this function to the function identically zero on $\mathcal{E}$. Thus the equation $\|f_1 - f_2\|_{L_p(\mathcal{E})} = 0$ implies that $f_1 - f_2 = 0$ and $f_1 = f_2$.

The set $\mathcal{E}$ may have a dimension m less than n, and then the integral entering into equation (3) will be understood in the sense of the natural (m-dimensional) Lebesgue measure defined on the set $\mathcal{E}$. We will not be concerned with sets $\mathcal{E}$ of complicated structure. Often $\mathcal{E}$ will coincide with the entire space $\mathbb{R}_n$ or with an m-dimensional subspace of $\mathbb{R}_n$, or an m-dimensional cube or ball lying in $\mathbb{R}_n$. Finally, $\mathcal{E}$ might be a smooth or piecewise smooth hypersurface consisting of sufficiently smooth pieces, and then the measure of the measurable subset $\mathcal{E}$ on the basis of which the integral appearing on the right side of (3) is defined is a generalization (extension) of the usual concept of area of a hypersurface.

It is natural to extend the definition (3) to the case $p = \infty$. Indeed, if the function $f(\boldsymbol{x})$ is measurable and essentially bounded on the bounded set $\mathcal{E}$, i.e. if the quantity

$$\sup_{x\in\mathcal{E}} \operatorname{vrai} |f(\boldsymbol{x})| = M_f,$$

which we call the *essential maximum*[1] of $|f(\boldsymbol{x})|$ on $\mathcal{E}$, exists, then the equation

$$\lim_{p\to\infty} \|f\|_{L_p(\mathcal{E})} = M_f \tag{4}$$

holds.

This is proved as follows. Let μe denote the measure of e. If $M_f = 0$ or $\mu\mathcal{E} = 0$, then (4) is obvious. Suppose that $0 < M_f < \infty$. If $\mathcal{E}$ is a bounded measurable set, then

$$\left(\int_{\mathcal{E}} |f(\boldsymbol{x})|^p \, d\boldsymbol{x}\right)^{1/p} \leqq M_f(\mu\mathcal{E})^{1/p}.$$

Accordingly,

$$\overline{\lim_{p\to\infty}} \|f\|_{L_p(\mathcal{E})} \leqq M_f. \tag{5}$$

If $\mathcal{E}$ is an unbounded measurable set, then equation (5) is in general not satisfied, as shown by the example $\mathcal{E} = \mathbb{R}_n$, $f(\boldsymbol{x}) = 1$. However one can prove this inequality under the hypotheses that $f(\boldsymbol{x}) \in L_p(\mathcal{E})$ for all sufficiently large p and that $\overline{\lim\limits_{p\to\infty}} \|f\|_{L_p(\mathcal{E})} < \infty$. In this case

$$\left(\int_{\mathcal{E}} |f(\boldsymbol{x})|^p \, d\boldsymbol{x}\right)^{1/p} \leqq M_f^{1/2} \left(\int_{\mathcal{E}} |f(\boldsymbol{x})|^{p/2} \, d\boldsymbol{x}\right)^{1/p},$$

[1] M_f is the smallest of the numbers M having the property that the set of all $\boldsymbol{x} \in \mathcal{E}$ for which $|f(\boldsymbol{x})| > M$ is of measure zero. It is easy to see that such a number exists.

so that

$$\overline{\lim_{p\to\infty}} \left(\int_{\mathscr{E}} |f(\boldsymbol{x})|^p \, d\boldsymbol{x} \right)^{1/p} \leqq M_f^{1/2} \left[\overline{\lim_{p\to\infty}} \left(\int_{\mathscr{E}} |f(\boldsymbol{x})|^p \, d\boldsymbol{x} \right)^{1/p} \right]^{1/2},$$

which implies inequality (5).

On the other hand, from the definition of essential maximum of a function it follows that there exists a bounded set $\mathscr{E}_1$ of positive measure such that for all of its points the inequality

$$|f(\boldsymbol{x})| > M_f - \varepsilon$$

is satisfied, where $0 < \varepsilon \leqq M_f$. Hence

$$\|f\|_{L_p(\mathscr{E})} \geqq \left(\int_{\mathscr{E}} (M_f - \varepsilon)^p \, d\mathscr{E}_1 \right)^{1/p} = (M_f - \varepsilon) \, (\mu\mathscr{E}_1)^{1/p},$$

so that

$$\underline{\lim}_{p\to\infty} \|f\|_{L_p(\mathscr{E})} \geqq M_f - \varepsilon .$$

Since ε is arbitrary, then

$$\underline{\lim}_{p\to\infty} \|f\|_{L_p(\mathscr{E})} \geqq M_f . \tag{6}$$

We note that inequality (6) is valid for any measurable set $\mathscr{E}$.

(4) follows from (5) and (6).

Thus we have proved that if the function $f(\boldsymbol{x})$ is essentially bounded on a bounded measurable set $\mathscr{E}$, then the limit

$$\lim_{p\to\infty} \|f\|_{L_p(\mathscr{E})} \tag{7}$$

exists and is finite and equals the essential maximum of $f(\boldsymbol{x})$ on $\mathscr{E}$.

On the other hand, from the existence of the limit (7) it follows that $f(\boldsymbol{x})$ is bounded on $\mathscr{E}$. Indeed, if this were not so, then for any sufficiently large N there would exist a bounded measurable subset $\mathscr{E}'$ of $\mathscr{E}$, of positive measure, on which

$$|f(\boldsymbol{x})| > N .$$

Then for any $p \geqq 1$

$$\|f\|_{L_p(\mathscr{E})} \geqq N(\mu\mathscr{E}')^{1/p},$$

so that

$$\lim_{p\to\infty} \|f\|_{L_p(\mathscr{E})} \geqq N .$$

Since N may be as large as desired, then the limit (7) cannot be finite, and we have arrived at a contradiction.

The foregoing considerations show the appropriateness of the notation

(8) $$\|f\|_{L_\infty(\mathscr{E})} = \sup_{x\in\mathscr{E}} \text{vrai}\, |f(x)|,$$

complementing the notation (3) to the case $p = \infty$. In functional analysis it is customary to denote the norm (8) in the following way:

(9) $$\|f\|_{M(\mathscr{E})} = \sup_{x\in\mathscr{E}} \text{vrai}\, |f(x)|.$$

We shall also sometimes use this notation, thus taking

(10) $$\|f\|_{M(\mathscr{E})} = \|f\|_{L_\infty(\mathscr{E})}.$$

The symbol $M(\mathscr{E})$ will denote the collection of all functions f having a finite norm in the sense of $M(\mathscr{E})$.

If $\mathscr{E}$ is a bounded closed set and the function $f(x)$ is continuous on $\mathscr{E}$, the quantity (8) is equal to the ordinary maximum of the function $|f(x)|$ on $\mathscr{E}$. In this case

(11) $$\|f\|_{L_\infty(\mathscr{E})} = \|f\|_{C(\mathscr{E})}.$$

1.1.1. In the case when the function $f(x) = f(x_1, \ldots, x_n)$ is a periodic function of period 2π with respect to all its variables, i.e. if it satisfies the identity

(1) $$f(x_1, \ldots, x_{l-1}, x_l + 2\pi, x_{l+1}, \ldots, x_n) = f(x_1, \ldots, x_n)$$

for all or almost all x and $l = 1, \ldots, n$, then, as the set $\mathscr{E}$ with respect to which the norm is taken, we will consider the n-dimensional cube

$$\Delta^{(n)} = \{0 \leqq x_i \leqq 2\pi;\, i = 1, \ldots, n\}$$

in the space $\mathbb{R}_n$. We will denote the corresponding norm as follows:

(2) $$\|f\|_{L_p(\Delta^{(n)})} = \|f\|_{L_p^*}^{(n)}, \quad \|f\|_{M(\Delta^{(n)})} = \|f\|_{M^*}^{(n)}, \quad \|f\|_{C(\Delta^{(n)})} = \|f\|_{C^*}^{(n)}.$$

The presence of the asterisk will always indicate that the function f is periodic and that its norm is computed relative to the cube defined by the period of the function.

For $n = 1$ we will write as a rule $\|f\|_{L_p^*}$, $\|f\|_{M^*}$, in place of $\|f\|_{L_p^*}^{(1)}$, $\|f\|_{M^*}^{(1)}$, $\|f\|_{C^*}^{(1)}$ respectively.

The collection of all periodic functions of period 2π defined on $\mathbb{R}_n$ and having finite norm $\|f\|_{L_p^*}^{(n)}$ will be denoted by $L_{p}^{*(n)}$. The collection of all continuous functions of period 2π defined on $\mathbb{R}_n$ will be denoted by the symbol $C_*^{(n)}$.

However, when possible we shall omit the sign (n) in these notations.

Very frequently we shall consider a measurable set $\mathscr{E} = \mathbb{R}_m \times \mathscr{E}' \subset \mathbb{R}_n$, the topological product of the m-dimensional subspace $\mathbb{R}_m (m < n)$ of points $(x_1, \ldots, x_m)$ and a measurable set $\mathscr{E}' \subset \mathbb{R}_{n-m}$, where $\mathbb{R}_{n-m}$ is the set of points $(x_{m+1}, \ldots, x_n)$.

Here the functional space consisting of functions $f(\boldsymbol{x})$ measurable on $\mathscr{E}$, periodic of period 2π relative to the variables $x_1, \ldots, x_m$ and summable to the pth power on the set $\Delta_m \times \mathscr{E}'$, where

$$\Delta_m = \{0 \leqq x_k \leqq 2\pi, k = 1, \ldots, m\},$$

will be denoted by $L_p^*(\mathscr{E})$. The asterisk calls attention to the presence of periodicity (relative to Δ_m) of the function $f \in L_p^*(\mathscr{E})$ and to the fact that the norm of the function $f \in L_p^*(\mathscr{E})$ is defined by the equation

$$\|f\|_{L_p^*(\mathscr{E})} = \left(\int_0^{2\pi} \cdots \int_0^{2\pi} \int_{\mathscr{E}'} |f(x_1, \ldots, x_m, x_{m+1}, \ldots, x_n)|^p \times dx_1 \ldots dx_m dx_{m+1} \ldots dx_n \right)^{1/p}.$$

1.1.2. We shall make extensive use of the fact that for a summable periodic function φ, i.e. one lying in $\overset{(n)}{L_p^*}$, the equation

$$\|\varphi(\boldsymbol{x} + \boldsymbol{a})\|_{L_p^*} = \|\varphi(\boldsymbol{x})\|_{L_p^*} \tag{1}$$

holds for any $\boldsymbol{a} \in \mathbb{R}_n$, and for functions $\varphi(\boldsymbol{x}) \in L_p(\mathbb{R}_n)$ the equation

$$\|\varphi(\boldsymbol{x} + \boldsymbol{a})\|_{L_p(\mathbb{R}_n)} = \|\varphi(\boldsymbol{x})\|_{L_p(\mathbb{R}_n)} \tag{2}$$

holds.

1.2. Normed Linear Spaces

1.2.1. Linear sets. A set G of elements $\boldsymbol{x}, \boldsymbol{y}, \boldsymbol{z}, \ldots$ is said to be a *linear set*, if under some rule there corresponds to each pair $\boldsymbol{x}, \boldsymbol{y}$ of its elements an element $\boldsymbol{z} = \boldsymbol{x} + \boldsymbol{y}$, lying in G and called the *sum* of $\boldsymbol{x}$ and $\boldsymbol{y}$, and if to each real (complex) number α and element $\boldsymbol{x} \in G$ there corresponds an element $\alpha\boldsymbol{x} \in G$, called the *product* of the number α and the element $\boldsymbol{x}$, and in addition the operations of addition and multiplication are subjected to the following axioms:

1) $\boldsymbol{x} + \boldsymbol{y} = \boldsymbol{y} + \boldsymbol{x}$,
2) $(\boldsymbol{x} + \boldsymbol{y}) + \boldsymbol{z} = \boldsymbol{x} + (\boldsymbol{y} + \boldsymbol{z})$,
3) $\boldsymbol{x} + \boldsymbol{y} = \boldsymbol{x} + \boldsymbol{z}$ implies $\boldsymbol{y} = \boldsymbol{z}$,
4) $\alpha\boldsymbol{x} + \alpha\boldsymbol{y} = \alpha(\boldsymbol{x} + \boldsymbol{y})$,
5) $\alpha\boldsymbol{x} + \beta\boldsymbol{x} = (\alpha + \beta)\boldsymbol{x}$,
6) $\alpha(\beta\boldsymbol{x}) = (\alpha\beta)\boldsymbol{x}$,
7) $1 \cdot \boldsymbol{x} = \boldsymbol{x}$.

The set G is a *real* or *complex* linear set according as the numbers α, β entering into its definition are real or complex.

It follows from the definition of a linear space that there exists in it a unique element θ—the *null*, or *zero, element* such that for all $\boldsymbol{x} \in G$ the following equations hold:

$$\boldsymbol{x} + \theta = \boldsymbol{x}, \quad 0 \cdot \boldsymbol{x} = \theta.$$

Indeed, suppose that the elements $\boldsymbol{x}$ and $\boldsymbol{y}$ lie in G. Put $\theta = \theta_x = 0 \cdot x$ and $\theta_y = 0 \cdot y$. Then

$$\boldsymbol{x} + \theta_x = 1 \cdot \boldsymbol{x} + 0 \cdot \boldsymbol{x} = 1 \cdot \boldsymbol{x} = \boldsymbol{x}$$

and analogously

$$\boldsymbol{y} + \theta_y = \boldsymbol{y}.$$

From the equations just obtained, we find on the basis of the axioms that

$$\boldsymbol{x} + \boldsymbol{y} + \theta_x = \boldsymbol{x} + \boldsymbol{y} + \theta_y,$$

so that

$$\theta_x = \theta_y = \theta.$$

Now put $-1 \cdot \boldsymbol{x} = -\boldsymbol{x}$. Then $\boldsymbol{x} + (-\boldsymbol{x}) = \theta$. If $\boldsymbol{x}$ and $\boldsymbol{y}$ are arbitrary elements of G, then the equation $\boldsymbol{x} + \boldsymbol{z} = \boldsymbol{y}$ has the solution $\boldsymbol{z} = \boldsymbol{y} + (-\boldsymbol{x})$, uniquely according to Axiom 3). It is natural to call this solution the *difference* between $\boldsymbol{y}$ and $\boldsymbol{x}$ and to write $\boldsymbol{z} = \boldsymbol{y} - \boldsymbol{x}$. Thus, besides addition, an operation of subtraction is defined in G.

The axioms of a linear set give us the right, using the operations of addition, subtraction, and multiplication by a number, to transform finite sums of the form

$$\alpha\boldsymbol{x} + \beta\boldsymbol{y} + \cdots + \gamma\boldsymbol{z},$$

in a way similar to that which is done with elementary algebraic expressions.

Every set $G_1 \subset G$, containing along with the elements $\boldsymbol{x}$ and $\boldsymbol{y}$ the element $\alpha\boldsymbol{x} + \beta\boldsymbol{y}$, where α and β are real (complex) numbers, is obviously in its turn a linear set.

A finite system of elements $\boldsymbol{x}_1, \ldots, \boldsymbol{x}_n$ of G is said to be *linearly independent*, if from the equation

$$\alpha_1\boldsymbol{x}_1 + \cdots + \alpha_n\boldsymbol{x}_n = \theta$$

it follows that $\alpha_1 = \cdots = \alpha_n = 0$. In the contrary case that system is said to be *linearly dependent*.

The set of functions $C(\mathscr{E})$ defined in §1.1 is obviously a linear set. The null element in $C(\mathscr{E})$ is the function identically equal to zero.

The set $L_p(\mathscr{E})$ of functions f integrable to the pth power on the measurable set $\mathscr{E}$ is also a linear set, with a null element which is a function almost everywhere equal to zero on $\mathscr{E}$ (equivalent to zero).

1.2.2. Banach spaces. The spaces $L_p(\mathscr{E})$, $C(\mathscr{E})$. A linear (real or complex) set E is said to be a *normed space* if to each element $x \in E$ there can be assigned a nonnegative number $\|x\|$, called the *norm* of the element x (in the space E or in the sense of the space E), satisfying the following conditions:

1) if $\|x\| = 0$, then $x = \theta$;

2) $\|x + y\| \leqq \|x\| + \|y\|$ $(x, y \in E)$;

3) $\|\alpha\| = |\alpha|\,\|x\|$,

where $x \in E$ and α is any (real or complex) number.

Condition (2) implies the inequality

$$| \|x\| - \|y\| | \leqq \|x - y\| \quad (x, y \in E). \tag{1}$$

The normed space E is said to be *complete* if, from the fact that the sequence $x_n \in E$ satisfies the Cauchy condition

$$\lim_{n,m\to\infty} \|x_n - x_m\| = 0,$$

it follows that there exists in E an element x_0 for which the equation

$$\lim_{n\to\infty} \|x_n - x_0\| = 0 \tag{2}$$

holds.

The fact that condition (2) holds is also written as follows:

$$\lim_{n\to\infty} x_n = x_0,$$

and it is spoken as follows: x_n tends to x_0 in the norm of the space E or in the sense of the metric of E.

A complete normed linear space is said also to be a *Banach space*.

The functional set $C(\mathscr{E})$ defined in 1.1. is obviously a Banach space. It is also well-known that the set $L_p(\mathscr{E})$ of functions $(1 \leqq p \leqq \infty)$ also forms a Banach space. Here $C(\mathscr{E})$ or $L_p(\mathscr{E})$ are said to be real or complex spaces depending on whether they consist of real or complex functions f. In the first case it is permissible to multiply f by real numbers, and in the second case by complex numbers.

1.2.3. Finite-dimensional space. The set $\mathfrak{M} \subset E$ is said to be a subspace of the Banach space E if it is a linear set closed relative to the norm $\|\cdot\|$.

Suppose given a linearly independent system $x_1, \ldots, x_n$ of elements of E. The set $\mathfrak{M}_n$ of elements of the form

$$y = \sum_1^n c_k x_k, \tag{1}$$

where $c = (c_1, \ldots, c_n)$ is any set of real (complex) numbers, is said to be *an n-dimensional* (*finite-dimensional*) space. If $\mathfrak{M}_n$ is a subset of E, then $\mathfrak{M}_n$ is said further to be an *n-dimensional subspace* of E, and the system of elements $x_1, \ldots, x_n$ its *basis*. In order to justify this definition, we must verify that $\mathfrak{M}_n$ is a closed linear set. The linearity of $\mathfrak{M}_n$ is obvious. We shall establish below that it is closed.

If along with an element y given by equation (1) we are given another element

$$y' = \sum_1^n c_k' x_k,$$

defined by a system $c' = (c_1', \ldots, c_n')$, then obviously

$$\|y - y'\| \leqq \sum_1^n |c_k - c_k'| \, \|x_k\|.$$

Hence it follows that

$$\lim_{|c-c'|\to 0} \|y - y'\| = 0. \tag{2}$$

Property (2) means that the element y depends continuously, relative to the norm, on the coefficients c_k defining it. In view of the inequality

$$|\, \|y\| - \|y'\| \,| \leqq \|y - y'\|$$

it follows also from (2) that

$$\lim_{|c-c'|\to 0} \|y'\| = \|y\|. \tag{3}$$

Thus the norm

$$\|y\| = \Phi(c_1, \ldots, c_n) = \Phi(c)$$

is a continuous function of $c = (c_1, \ldots, c_n)$.

The minimum of this function on the closed and bounded set defined by the equation

$$|c| = \sqrt{\sum_1^n c_k^2} = 1,$$

is achieved for some system of coefficients $\boldsymbol{c} = (c_1, \ldots, c_n)$, and is equal to the number

$$\frac{1}{\lambda} = \Phi(\boldsymbol{c}^{(0)}) > 0,$$

a fortiori positive, since the system $\boldsymbol{x}_1, \ldots, \boldsymbol{x}_n$ is linearly independent.

Suppose we are given any system of numbers $\boldsymbol{c} = (c_1, \ldots, c_n)$, with $|\boldsymbol{c}| > 0$. Put

$$\boldsymbol{c}' = \frac{\boldsymbol{c}}{|\boldsymbol{c}|}.$$

Then in view of the fact that $|\boldsymbol{c}'| = 1$, we will have the inequality

$$\frac{1}{\lambda} \leqq \left\| \sum_1^n c_k' \boldsymbol{x}_k \right\|$$

which, after multiplication on left and right by $\lambda|\boldsymbol{c}|$ turns into the inequality

(4) $$|\boldsymbol{c}| \leqq \lambda \left\| \sum_1^n c_k \boldsymbol{x}_k \right\|.$$

Now it is already easy to show that the linear set $\mathfrak{M}_n$ is closed and thus is a space.

In fact, since

(5) $$\boldsymbol{y}_l = \sum_1^n c_k^{(l)} \boldsymbol{x}_k \quad (l = 1, 2, \ldots)$$

and

$$\|\boldsymbol{y}_l - \boldsymbol{y}_m\| \to 0 \quad (l, m \to \infty),$$

it follows in view of (4) that

$$|\boldsymbol{c}^{(l)} - \boldsymbol{c}^{(m)}| \leqq \lambda \|\boldsymbol{y}_l - \boldsymbol{y}_m\| \to 0 \quad (l, m \to \infty),$$

where $\boldsymbol{c}^{(l)} = (c_1^{(l)}, \ldots, c_n^{(l)})$ $(l = 1, 2, \ldots)$. Accordingly, the limit

(6) $$\lim_{l \to \infty} \boldsymbol{c}^{(l)} = \boldsymbol{c}^{(0)}$$

exists, so that

(7) $$\|\boldsymbol{y}_l - \boldsymbol{y}_0\| \to 0 \ (l \to \infty),$$

where

(8) $$\boldsymbol{y}_0 = \sum_1^n c_k^{(0)} \boldsymbol{x}_k \in \mathfrak{M}_n.$$

Now we note a further important property of a finite-dimensional space $\mathfrak{M}_n$, following directly from inequality (4). It consists in that every bounded (relative to the norm) set $\Omega \subset \mathfrak{M}_n$ is compact in $\mathfrak{M}_n$, i.e. from any sequence of elements $\boldsymbol{y}_l \in \Omega$ $(l = 1, 2, \ldots)$ it is possible to select a subsequence converging in the norm to some element of $\mathfrak{M}_n$. Indeed, from the fact that the elements $\boldsymbol{y}_l$ defined by the equations (5) form a bounded set, it follows in view of (4) that the vectors $\boldsymbol{c}^{(l)}$ are also uniformly bounded. But then for some subsequence of integers l, equation (6) will be satisfied for some vector $\boldsymbol{c}^{(0)}$, and accordingly relations (7) and (8) as well.

Remark. In special courses in functional analysis it is proved conversely that if every bounded set lying in a given Banach space $\mathfrak{M}$ is compact, then $\mathfrak{M}$ is a finite-dimensional space, i.e. all of its elements may be described in the form of a finite sum (1), where the elements $\boldsymbol{x}_1, \ldots, \boldsymbol{x}_n$ form a linearly independent system.

Since

$$|c_k| \leqq \sqrt{\sum_1^n c_k^2},$$

then

$$\sum_1^n |c_k| \leqq n\, |\boldsymbol{c}|,$$

so that, if one supposes that

$$M \geqq \|\boldsymbol{x}_k\| \ (k = 1, \ldots, n),$$

then, once again taking (4) into account, we get

$$\left\| \sum_1^n c_k \boldsymbol{x}_k \right\| \leqq M \sum_1^n |c_k| \leqq Mn\, |\boldsymbol{c}| \leqq Mn\lambda \left\| \sum_1^n c_k \boldsymbol{x}_k \right\|.$$

Thus we have proved that for any system c_k $(k = 1, \ldots, n)$ we have the inequalities

$$\frac{1}{\lambda} \left(\sum_1^n c_k^2 \right)^{1/2} \leqq \left\| \sum_1^n c_k \boldsymbol{x}_k \right\| \leqq Mn \left(\sum_1^n c_k^2 \right)^{1/2}, \tag{9}$$

where λ and M are positive numbers depending on the properties of the norm defined in $\mathfrak{M}_n$.

If in the n-dimensional set in question we introduce another norm $\|\cdot\|'$, thus defining another space $\mathfrak{M}'_n$, then we obtain new inequalities

$$\frac{1}{\lambda'} \left(\sum_1^n c_k^2 \right)^{1/2} \leqq \left\| \sum_1^n c_k \boldsymbol{x}_k \right\|' \leqq M'n \left(\sum_1^n c_k^2 \right)^{1/2}, \tag{10}$$

where λ' and M' are positive numbers in general different from λ and M. It follows from (9) and (10) that

$$(11)\qquad \frac{1}{\lambda M' n}\left\|\sum_1^n c_k \boldsymbol{x}_k\right\|' \leqq \left\|\sum_1^n c_k \boldsymbol{x}_k\right\| \leqq \lambda' M n \left\|\sum_1^n c_k \boldsymbol{x}_k\right\|'.$$

1.2.4. Equivalent normed spaces. If a linear set is normed in two ways, leading to two normed spaces E_1 and E_2, and if there exist two positive numbers c_1, c_2, not depending on $\boldsymbol{x} \in E_1, E_2$, such that

$$(1)\qquad c_1 \|\boldsymbol{x}\|_{E_1} \leqq \|\boldsymbol{x}\|_{E_2} \leqq c_2 \|\boldsymbol{x}\|_{E_1}$$

for all $\boldsymbol{x} \in E_1, E_2$, then the spaces E_1 and E_2 are said to be *equivalent*.

As a rule, we will not distinguish equivalent norms, i.e. we will use one and the same notation for equivalent norms.

It follows from the inequality 1.2.3. (11) that any two normalizations of an n-dimensional linear set lead to equivalent normed spaces.

In our later considerations, the finite-dimensional subspaces with which we shall be dealing will usually be sets of trigonometric or algebraic polynomials of one variable of a given degree ν, or of n variables of given degrees $\nu_1, \ldots, \nu_n$, or simply systems $\xi = (\xi_1, \ldots, \xi_n)$ of n numbers normalized in one or another way.

1.2.5. Real Hilbert space. Suppose that H is a real linear set and that to each two of its elements f, φ it is possible to assign a real number (f, φ), called the *scalar product of f and φ*, having the following properties:

1) $(f, f) \geqq 0$; if $(f, f) = 0$ it follows that $f = \theta$, the null element of H;
2) $(f, \varphi) = (\varphi, f)$;
3) $(c_1 f_1 + c_2 f_2, \varphi) = c_1(f_1, \varphi) + c_2(f_2, \varphi)$ for any real numbers c_1, c_2 and elements $f, \varphi, f_1, f_2 \in H$.

One introduces into H the norm

$$\|f\| = (f, f)^{1/2}.$$

It is not hard to verify that this expression in fact defines a norm. With this norm H is a normed space. If H is a complete space, then H is said to be a *Hilbert space (real)*.

We note that for arbitrary real λ and $f, \varphi \in H$

$$0 \leqq (\lambda f + \varphi, \lambda f + \varphi) = \lambda^2 (f, f) + 2\lambda (f, \varphi) + (\varphi, \varphi),$$

so that

$$|(f, \varphi)| \leqq (f, f)^{1/2} (\varphi, \varphi)^{1/2} = \|f\| \, \|\varphi\|.$$

The space $L_2(\Omega)$ of real functions measurable on Ω and having integrable squares on Ω, taken with the scalar product

$$(f, \varphi) = \int_\Omega f(x)\, \varphi(x)\, dx \quad \big(f, \varphi \in L_2(\Omega)\big),$$

serves as an important example of a real Hilbert space. We shall encounter other examples (see, for example, 4.3.1. (4)).

It is easy to see that for any $f, \varphi \in H$ the equation

$$(1) \qquad \|f + \varphi\|^2 + \|f - \varphi\|^2 = 2(\|f\|^2 + \|\varphi\|^2)$$

holds. This recalls a well-known fact from geometry: the sum of the diagonals of a parallelogram is equal to the sum of the squares of its sides.

The space $L_p(\Omega)$ for $p \neq 2$ is not Hilbertian, since it is possible to indicate functions f, φ lying in it for which equation (1) is not satisfied.

1.2.6. Distance of an element from a subspace. Best approximation. Suppose that $\mathfrak{M}$ is a subspace of the Banach space E, and suppose that $\boldsymbol{y} \in E$. The *distance from* $\boldsymbol{y}$ *to* $\mathfrak{M}$ is the lower bound

$$(1) \qquad E(\boldsymbol{y}) = \inf_{x \in \mathfrak{M}} \|\boldsymbol{y} - \boldsymbol{x}\|,$$

extended over all elements $\boldsymbol{x} \in \mathfrak{M}$. Following the tradition which has arisen in the theory of the approximation of functions, we shall frequently call the number $E(\boldsymbol{y})$ the *best approximation* of the element $\boldsymbol{y}$ using elements $\boldsymbol{x} \in \mathfrak{M}$.

It may turn out that there is in $\mathfrak{M}$ an element $\boldsymbol{x}_*$ such that the lower bound in (1) is taken on, i.e.

$$(2) \qquad E(\boldsymbol{y}) = \min_{x \in \mathfrak{M}} \|\boldsymbol{y} - \boldsymbol{x}\| = \|\boldsymbol{y} - \boldsymbol{x}_*\|.$$

In this case the element $\boldsymbol{x}_*$ is said to be the *best element approximating to* $\boldsymbol{y}$ among the elements $\boldsymbol{x} \in \mathfrak{M}$.

It is important to note some quite general cases when it is possible to assert in advance that a best element exists. Moreover, another question, as to whether the best element for a given problem is unique, is of interest.

It is not hard to see that if $\mathfrak{M} = \mathfrak{M}_n$ is a finite-dimensional subspace of any normed space E, then for each element $\boldsymbol{y} \in E$ a best element approximating to $\boldsymbol{y}$ among the elements $\boldsymbol{x} \in \mathfrak{M}$ always exists. Indeed, suppose that

$$E(\boldsymbol{y}) = \inf_{x \in \mathfrak{M}_n} \|\boldsymbol{y} - \boldsymbol{x}\|.$$

Then there exists a *minimizing* sequence $\{x^{(l)}\}$ of elements of $\mathfrak{M}$ such that

$$\|y - x^{(l)}\| = E(y) + \varepsilon_l \left(\varepsilon_l \geqq 0,\ \varepsilon_l \underset{l\to\infty}{\to} 0\right).$$

This sequence is bounded and accordingly compact, and thus one of its subsequences converges in norm to some element $x_* \in \mathfrak{M}_n$. It is not hard to see that x_* is a best element, approximating to y among $x \in \mathfrak{M}_n$. In general it is not unique.

If $\mathfrak{M}$ is an infinite-dimensional (not finite-dimensional) subspace of E, then in problem (1) there may not in general exist a best element. Such a phenomenon is observed, for example, in the spaces $L_\infty(\mathscr{E})$ and $L_1(\mathscr{E})$. However, for p satisfying $1 < p < \infty$ the existence of a best element holds for any function $f \in L_p(\mathscr{E})$ and any subspace $\mathfrak{M} \subset L_p(\mathscr{E})$. Moreover, in this case the best function is always unique; these facts are proved below in 1.3.6. In the spaces $L_1(\mathscr{E})$ and $L_\infty(\mathscr{E})$, if the best element exists, then it is not always unique (see 1.2.7., Examples 1) and 2)). However, in the spaces L, L_∞ and C there exist cases of uniqueness of the best function, important for analysis. But these cases depend on the special properties of the subspace $\mathfrak{M}$ and of the functions f being approximated. Such problems will not be considered in this book.

1.2.7. *Example 1.* Suppose that $f(x) = \operatorname{sign} x$. We will approximate it in the metric of $L(-1,1)$[1] by means of constant functions $\varphi(x) = c$, i.e. we seek a constant λ for which the minimum

$$\min_c \|f - c\|_{L(-1,+1)} = \min_c \int_{-1}^{1} |f(x) - c|\, dx = \int_{-1}^{1} |f(x) - \lambda|\, dx$$

is achieved. It is not hard to see that this minimum is achieved for any constant λ satisfying $-1 \leqq \lambda \leqq 1$.

As to the notations occurring in the preceding paragraph, we should say that we are approximating the function $f \in L(-1, +1)$ by means of functions $\varphi(x) = c$ constant on $(-1, +1)$. These constant functions form a one-dimensional subspace of the space $L(-1, +1)$. The best function turns out to be not unique.

Example 2. The function $f(x) = \operatorname{sign} x$ is to be approximated now in the metric of $L_\infty(-1, +1) = M(-1, +1)$ by means of linear functions

$$\varphi(x) = Ax + B,$$

where A and B are arbitrary real numbers.

---

[1] $L_p(a, b)$ denotes $L_p(\mathscr{E})$, where $\mathscr{E}$ is the segment $[a, b]$.

It is not hard to see that

$$\min_{A,B} \|f(x) - Ax - B\|_{M(-1,+1)} = \min_{A,B} \max_{-1 \leq x \leq 1} |f(x) - Ax - B|$$
$$= \|f(x) - \lambda x\|_{M(-1,+1)}$$

where λ may be any number satisfying $|\lambda| \leq 1$.

Thus, in this example as well, the best function is not unique.

1.2.8. Linear operators. If E and E' are Banach spaces and if to each element $\boldsymbol{x} \in E$ there corresponds under some rule a definite element

$$\boldsymbol{y} = A(\boldsymbol{x}),$$

lying in E', then we say that A is an *operator* mapping E into E'. The operator A is *linear* if for any two elements $\boldsymbol{x}_1, \boldsymbol{x}_2 \in E$ and numbers c_1, c_2 (real or complex in accordance with the reality or complexity of E and E') the equation

$$A(c_1\boldsymbol{x}_1 + c_2\boldsymbol{x}_2) = c_1 A(\boldsymbol{x}_1) + c_2 A(\boldsymbol{x}_2)$$

holds.

The linear operator A is said to be *bounded*, if there exists a positive constant M such that the inequality

$$(1) \qquad \|A(\boldsymbol{x})\|_{E'} \leq M\|\boldsymbol{x}\|_E$$

holds for all $\boldsymbol{x} \in E$. The smallest constant for which this inequality holds for all $\boldsymbol{x} \in E$ is said to be the *norm* of the linear operator A and is denoted by the symbol $\|A\|$. The norm of an operator may also be defined as one of the upper bounds:

$$\|A\| = \sup_{0 \neq \boldsymbol{x} \in E} \frac{\|A(\boldsymbol{x})\|_{E'}}{\|\boldsymbol{x}\|_E} = \sup_{\|\boldsymbol{x}\|_E \leq 1} \|A(\boldsymbol{x})\|_{E'}.$$

The operator A is said to be *completely continuous* if it maps every bounded set $\mathscr{E} \subset E$ into a compact set lying in E'. In other words, for any bounded sequence $\{\boldsymbol{x}_l\}$ of elements of E, it is possible to select from it a subsequence $\{\boldsymbol{x}_{l_k}\}$ and an element $\boldsymbol{y}_0 \in E'$ such that

$$\lim_{k \to \infty} A(\boldsymbol{x}_{l_k}) = \boldsymbol{y}_0.$$

If the space E' is finite-dimensional, then every bounded linear operator mapping E into E' is a completely continuous operator, since A maps each bounded set of E into a bounded set of E', and the latter, because of the finite-dimensional character of E', is compact.

Let us consider an example. Suppose that E as before denotes a Banach space, and suppose that $\mathfrak{M}$ is a finite-dimensional subspace of it.

Suppose further that to each element $\boldsymbol{x} \in E$ there corresponds only one element $\boldsymbol{x}_* = A(\boldsymbol{x})$, best approximating to $\boldsymbol{x}$ among the elements $\boldsymbol{u} \in \mathfrak{M}$. In other words, suppose that $A(\boldsymbol{x})$ is the unique element of $\mathfrak{M}$ for which the equation

$$\min_{\boldsymbol{u}\in\mathfrak{M}} \|\boldsymbol{x} - \boldsymbol{u}\|_E = \|\boldsymbol{x} - A(\boldsymbol{x})\|_E$$

is satisfied.

Then $A(\boldsymbol{x})$ is an operator, mapping E into $\mathfrak{M}$. This operator is in general nonlinear (it is linear if E is a Hilbert space), but completely continuous, as is clear from the following considerations. It follows from the inequality

$$\|A(\boldsymbol{x})\|_E - \|\boldsymbol{x}\|_E \leqq \| \boldsymbol{x} - A(\boldsymbol{x}) \|_E \leqq \|\boldsymbol{x}\|_E$$

that

$$\|A(\boldsymbol{x})\|_E \leqq 2\|\boldsymbol{x}\|_E .$$

Hence it results that the operator A maps a bounded set of elements of E into a bounded set of elements of $\mathfrak{M}$. But since $\mathfrak{M}$ is finite-dimensional, that bounded set is compact.

Remark. The definition of a completely continuous operator may be extended also to many-valued operators mapping E into E', i.e. to operators such that to each $\boldsymbol{x} \in E$ there corresponds generally speaking not just one element $\boldsymbol{y} = A(\boldsymbol{x})$. The many-valued operator A is said to be completely continuous if from each bounded sequence of elements $\boldsymbol{x}_l \in E$ it is possible to select a subsequence $\{\boldsymbol{x}_{l_k}\}$, and such definite values of the operator $A(\boldsymbol{x}_{l_k})$, that the sequence $\{A(\boldsymbol{x}_{l_k})\}$ converges in E'.

The example presented above of an operator $A(\boldsymbol{x})$ of best approximation of an element $\boldsymbol{x}$ by means of elements of a finite-dimensional subspace $\mathfrak{M}$ yields in the general case a many-valued operator, which is a completely continuous operator in the sense indicated above.

1.3. Properties of the Space $L_p(\mathscr{E})$

Here we only formulate and explain certain properties of the space $L_p(\mathscr{E})$, referring the reader to their proofs in other places. In this connection see the Remark to Chapter 1 at the end of the book.

1.3.1. It was already noted in 1.2.2. that $L_p(\mathscr{E})$ is a (real or complex) Banach space. Thus, for the elements of $L_p(\mathscr{E})$ the following properties are fulfilled:

1) the norm

$$\|f\|_{L_p(\mathscr{E})} = \left(\int_{\mathscr{E}} |f|^p \, d\mathscr{E}\right)^{1/p}$$

of each function $f \in L_p(\mathscr{E})$ is nonnegative and equal to zero only for a function f_0 equivalent to zero ($f_0 = 0$);

2) $\|f_1 + f_2\|_{L_p(\mathscr{E})} \leqq \|f_1\|_{L_p(\mathscr{E})} + \|f_2\|_{L_p(\mathscr{E})}$;

3) $\|cf\|_{L_p(\mathscr{E})} = |c| \, \|f\|_{L_p(\mathscr{E})}$,

where c is any (real or complex) number;

4) if $f_k \in L_p(\mathscr{E})$ and

$$\|f_k - f_l\|_{L_p(\mathscr{E})} \to 0 \quad (k, l \to \infty),$$

then there exists a function $f \in L_p(\mathscr{E})$ for which

$$\lim_{k\to\infty} \|f_k - f_*\|_{L_p(\mathscr{E})} = 0. \tag{1}$$

Properties 1) and 2) are obvious. Inequality 2) is called the Minkowski inequality. It can become an equation if and only if the functions f_1 and f_2 are linearly dependent as elements of L_p, i.e. $c_1 f_1 + c_2 f_2 = 0$ for some (real or complex) c_1, c_2 ($|c_1| + |c_2| > 0$). Property 4) is the theorem on the completeness of the space L_p.

We shall write

$$\Psi(x) = u_0(x) + u_1(x) + \cdots \quad (x \in \mathscr{E})$$

and say that the series on the right side of this equation converges in the sense of $L_p(\mathscr{E})$ to its sum $\Psi(x)$, if

$$\lim_{N\to\infty} \left\| \Psi - \sum_0^N u_k \right\|_{L_p(\mathscr{E})} = 0.$$

The Minkowski inequality extends by induction to the case of N functions, and then it has the form

$$\left\| \sum_1^N f_k \right\|_{L_p(\mathscr{E})} \leqq \sum_1^N \|f_k\|_{L_p(\mathscr{E})}. \tag{2}$$

From this there easily follows the inequality

$$\left\| \sum_1^\infty f_k \right\|_{L_p(\mathscr{E})} \leqq \sum_1^\infty \|f_k\|_{L_p(\mathscr{E})}, \tag{3}$$

corresponding to the case $N = \infty$. It may be read as follows. If the functions $f_k \in L_p(\mathscr{E})$ ($k = 1, 2 \ldots$), and if the series of numbers on the right side of (3) converges, then the series $f_1 + f_2 + \ldots$ converges in the sense of $L_p(\mathscr{E})$ to some function, lying in $L_p(\mathscr{E})$, which we shall denote by $\sum_1^\infty f_k$, and inequality (3) holds for this function.

We note another fact which we shall need. If the series

$$f(\boldsymbol{x}) = f_1(\boldsymbol{x}) + f_2(\boldsymbol{x}) + \cdots$$

converges in the ordinary sense almost everywhere on $\mathscr{E}$ to a function f, and if moreover it converges to an f_* in the sense of $L_p(\mathscr{E})$, then $f(\boldsymbol{x}) = f_*(\boldsymbol{x})$ almost everywhere on $\mathscr{E}$. Indeed, by hypothesis the sum $S_n(\boldsymbol{x})$ of the first n terms of our series converges in the metric of $L_p(\mathscr{E})$ to f_*. But then, as is known from the theory of functions of a real variable, there exists a subsequence of indices $n_1, n_2, \ldots$ such that $S_{n_k}(\boldsymbol{x})$ converges in the ordinary sense almost everywhere to $f_*(\boldsymbol{x})$ on $\mathscr{E}$, and since also $S_{n_k}(\boldsymbol{x})$ converges amost everywhere to $f(\boldsymbol{x})$, then almost everywhere $f(\boldsymbol{x}) = f_*(\boldsymbol{x})$.

In the left side of (2) one first carries out the operation of summing relative to the index k, and then one applies to the result the operation of taking the norm. On the right side these two operations are reversed. In what follows we shall present a similar inequality, in which the operation of summing relative to k is replaced by the operation of integrating relative to a variable k.

1.3.2. Generalized Minkowski inequality. For a function $k(\boldsymbol{u}, \boldsymbol{y})$, given on a measurable set $\mathscr{E} = \mathscr{E}_1 \times \mathscr{E}_2 \subset \mathbb{R}_n$, where $\boldsymbol{x} = (\boldsymbol{u}, \boldsymbol{y})$, $\boldsymbol{u} = (x_1, \ldots, x_m)$, $\boldsymbol{y} = (x_{m+1}, \ldots, x_n)$, the following inequality holds:

$$\text{(1)}\qquad \Big(\int_{\mathscr{E}} \Big|\int_{\mathscr{E}} k(\boldsymbol{u}, \boldsymbol{y})\, d\boldsymbol{y}\Big|^p d\boldsymbol{u}\Big)^{1/p} \leq \int_{\mathscr{E}_2} \Big(\int_{\mathscr{E}_1} |k(\boldsymbol{u}, \boldsymbol{y})|^p\, d\boldsymbol{u}\Big)^{1/p} d\boldsymbol{y},$$

$$1 \leqq p \leqq \infty.$$

This must be understood in the sense that if the right side has a meaning, i.e. if for almost all $\boldsymbol{y} \in \mathscr{E}_2$ the inside integral with respect to $\mathscr{E}_1$ exists, and the outside integral with respect to $\mathscr{E}_2$ then exists, then the left side has a sense as well and does not exceed the right.

1.3.3. Inequality 1.3.2. (1) will be frequently applied in particular in the following circumstances:

$$\begin{aligned}\Big(\int \Big|\int K(\boldsymbol{t} - \boldsymbol{x})\, f(\boldsymbol{t})\, d\boldsymbol{t}\Big|^p d\boldsymbol{x}\Big)^{1/p} &= \Big(\int \Big|\int K(\boldsymbol{t})\, f(\boldsymbol{t} + \boldsymbol{x})\, d\boldsymbol{t}\Big|^p d\boldsymbol{x}\Big)^{1/p}\\ &\leqq \int |K(\boldsymbol{t})| \Big(\int |f(\boldsymbol{t} + \boldsymbol{x})|^p\, d\boldsymbol{x}\Big)^{1/p} d\boldsymbol{t}\\ &= \int |K(\boldsymbol{t})|\, d\boldsymbol{t} \Big(\int |f(\boldsymbol{u})|^p\, d\boldsymbol{u}\Big)^{1/p} = \|K\|_{L(\mathbb{R})} \|f\|_{L_p(\mathbb{R})},\end{aligned}$$

where $1 \leqq p \leqq \infty$, $K \in L(\mathbb{R})$, $f \in L_p(\mathbb{R})$[1].

[1] Here and in what follows $\int \equiv \int_{\mathbb{R}}$, $\mathbb{R} = \mathbb{R}_n$.

If the functions $K(t)$ and $f(t)$ are periodic of period 2π and if $K \in L(0,2\pi)$, $f \in L_p(0,2\pi)$, then the analogous inequality

$$\left(\int_0^{2\pi} \left| \int_0^{2\pi} K(t-x)\, f(t)\, dt \right|^p dx\right)^{1/p} \leqq \|K\|_{L(0,2\pi)} \|f\|_{L_p(0,2\pi)}$$

holds, or the similar inequality for periodic functions of n variables.

1.3.4. Hőlder's inequality. If $f_1 \in L_p(\mathscr{E})$, $f_2 \in L_q(\mathscr{E})$ and $1/p + 1/q = 1$, where $1 \leqq p < \infty$, the *Hőlder's inequality*

$$\int_{\mathscr{E}} |f_1 f_2|\, d\mathscr{E} \leqq \|f_1\|_{L_p(\mathscr{E})} \|f_2\|_{L_q(\mathscr{E})} \tag{1}$$

holds, where we understand that $q = \infty$ if $p = 1$. Equality can hold if and only if there exists a constant c such that $|f_2(\boldsymbol{x})| = c|f_1(\boldsymbol{x})|^{p/q}$ almost everywhere.

1.3.5. Clarkson's inequalities, uniform convexity. For Clarkson's inequalities see, for example, the book [4] of S. L. Sobolev. These inequalities are as follows: suppose that $f_1, f_2 \in L_p(\mathscr{E})$. If $2 \leqq p \leqq \infty$, then

$$\left\|\frac{f_1 + f_2}{2}\right\|^p_{L_p(\mathscr{E})} + \left\|\frac{f_1 - f_2}{2}\right\|^p_{L_p(\mathscr{E})} \leqq \frac{1}{2}\|f_1\|^p_{L_p(\mathscr{E})} + \frac{1}{2}\|f_2\|^p_{L_p(\mathscr{E})}. \tag{1}$$

If on the other hand $1 < p \leqq 2$, $\dfrac{1}{p} + \dfrac{1}{q} = 1$, then

$$\left\|\frac{f_1 + f_2}{2}\right\|^q_{L_p(\mathscr{E})} + \left\|\frac{f_1 - f_2}{2}\right\|^q_{L_p(\mathscr{E})} \leqq \left(\frac{1}{2}\|f_1\|^p_{L_p(\mathscr{E})} + \frac{1}{2}\|f_2\|^p_{L_p(\mathscr{E})}\right)^{\frac{1}{p-1}}. \tag{2}$$

For $p = 2$ inequalities (1) and (2) become equations, the parallelogram equation.

One says that the Banach space E is *uniformly convex* if it follows from the fact that

$$\max_{0 \leqq \alpha \leqq 1} \left(1 - \|\alpha \boldsymbol{x}_1^{(n)} + (1-\alpha)\, \boldsymbol{x}_2^{(n)}\|\right) \xrightarrow[n\to\infty]{} 0,$$

where $\|\boldsymbol{x}_1^{(n)}\| = \|\boldsymbol{x}_2^{(n)}\| = 1$, that

$$\|\boldsymbol{x}_1^{(n)} - \boldsymbol{x}_2^{(n)}\| \xrightarrow[n\to\infty]{} 0.$$

It follows from the Clarkson inequalities (1), (2) that the spaces $L_p(\mathscr{E})$ are uniformly convex. Indeed, suppose that

$$\|\cdot\|_{L_p(\mathscr{E})} = \|\cdot\| \quad \text{and} \quad \|f_1^{(n)}\| = \|f_2^{(n)}\| = 1.$$

Then it follows from (1) and (2) that

$$(3)\qquad \left\|\frac{f_1^{(n)} - f_2^{(n)}}{2}\right\|^\lambda \leqq 1 - \left\|\frac{f_1^{(n)} + f_2^{(n)}}{2}\right\|^\lambda,$$

where $\lambda = p$ in case (1) and $\lambda = q$ in case (2). If now

$$\max_{0\leqq\alpha\leqq 1} (1 - \|\alpha f_1^{(n)} + (1-\alpha) f_2^{(n)}\|) \xrightarrow[n\to\infty]{} 0,$$

then

$$1 - \left\|\frac{f_1^{(n)} + f_2^{(n)}}{2}\right\| \to 0.$$

But then the right side of (3) tends to zero, and the left side along with it. Hence

$$\|f_1^{(n)} - f_2^{(n)}\| \xrightarrow[n\to\infty]{} 0.$$

1.3.6. Theorem. *Suppose that E (in particular, $L_p(\mathscr{E})$) is a uniformly convex Banach space, $\mathfrak{M}$ a subspace of E and $\boldsymbol{y} \in E - \mathfrak{M}$.*

Then there exists a unique element $\boldsymbol{u} \in \mathfrak{M}$ yielding the best approximation to y among the elements of $\mathfrak{M}$:

$$(1)\qquad \|\boldsymbol{y} - \boldsymbol{u}\| = \inf_{\boldsymbol{x}\in\mathfrak{M}} \|\boldsymbol{y} - \boldsymbol{x}\|.$$

Proof. Suppose that

$$\inf_{\boldsymbol{x}\in\mathfrak{M}} \|\boldsymbol{y} - \boldsymbol{x}\| = d \quad (d > 0).$$

Then there exists a minimizing sequence of elements $\boldsymbol{x}_n \in \mathfrak{M}$, for which

$$\|\boldsymbol{y} - \boldsymbol{x}_n\| = d + \varepsilon_n \ (\varepsilon_n \to 0,\ \varepsilon_n \geqq 0).$$

We will suppose that $\boldsymbol{x}, \boldsymbol{x}'$ denote any elements of $\mathfrak{M}$. Obviously the elements

$$\boldsymbol{w}_n = \frac{\boldsymbol{y} - \boldsymbol{x}_n}{d + \varepsilon_n}$$

have unit norms, and for any $\alpha, \beta \geqq 0$ with $\alpha + \beta = 1$,

$$0 \leqq 1 - \|\alpha \boldsymbol{w}_n + \beta \boldsymbol{w}_m\| = 1 - \left\|\left(\frac{\alpha}{d+\varepsilon_n} + \frac{\beta}{d+\varepsilon_m}\right)\boldsymbol{y} - \boldsymbol{x}\right\| =$$

$$= 1 - \left(\frac{\alpha}{d+\varepsilon_n} + \frac{\beta}{d+\varepsilon_m}\right)\|\boldsymbol{y} - \boldsymbol{x}'\| \leqq$$

$$\leqq 1 - \left(\frac{\alpha}{d+\varepsilon_n} + \frac{\beta}{d+\varepsilon_m}\right) d = \eta_{nm} \xrightarrow[n,m\to\infty]{} 0,$$

uniformly for the α, β in question.

In such a case, from the definition of a uniformly convex space,

$$\|\boldsymbol{w}_n - \boldsymbol{w}_m\| \xrightarrow[n,m\to\infty]{} 0.$$

But

$$\|\boldsymbol{w}_n - \boldsymbol{w}_m\| = \left\| y\left(\frac{1}{d+\varepsilon_n} - \frac{1}{d+\varepsilon_m}\right) - \left(\frac{\boldsymbol{x}_n}{d+\varepsilon_n} - \frac{\boldsymbol{x}_m}{d+\varepsilon_m}\right)\right\|$$

$$= \left\|\frac{\boldsymbol{x}_n}{d+\varepsilon_n} - \frac{\boldsymbol{x}_m}{d+\varepsilon_m}\right\| + o(1) = \frac{1}{d}\|\boldsymbol{x}_n - \boldsymbol{x}_m\| + o(1)\ (n, m \to \infty),$$

since the elements $\boldsymbol{x}_n$, $\boldsymbol{x}_m$ are bounded in norm. We have proved that

$$\|\boldsymbol{x}_n - \boldsymbol{x}_m\| \xrightarrow[n,m\to\infty]{} 0.$$

Because of the completeness of E and the fact that $\mathfrak{M}$ is closed, there exists an element $\boldsymbol{u} \in \mathfrak{M}$ such that $\boldsymbol{x}_n \to \boldsymbol{u}$, and (1) is obviously fulfilled.

Now suppose that there exists a further element $\boldsymbol{u}'$ for which (1) is satisfied. For $0 \leqq \alpha \leqq 1$ we have

$$d \leqq \|a\boldsymbol{u} + (1-\alpha)\,\boldsymbol{u}' - \boldsymbol{y}\| \leqq \alpha\,\|\boldsymbol{u} - \boldsymbol{y}\|$$
$$+ (1-\alpha)\,\|\boldsymbol{u}' - \boldsymbol{y}\| = \alpha d + (1-\alpha)\,d = d.$$

Accordingly, $\|\alpha\boldsymbol{u} + (1-\alpha)\,\boldsymbol{u}' - \boldsymbol{y}\| = d$. Thus

$$\left\|\alpha\,\frac{\boldsymbol{u}-\boldsymbol{y}}{d} + (1-\alpha)\,\frac{\boldsymbol{u}'-\boldsymbol{y}}{d}\right\| = 1,$$

while $\left\|\dfrac{\boldsymbol{u}-\boldsymbol{y}}{d}\right\| = \left\|\dfrac{\boldsymbol{u}'-\boldsymbol{y}}{d}\right\| = 1$. Because of the uniform convexity of the space E $\left(\boldsymbol{x}_1^{(n)} = \dfrac{\boldsymbol{u}-\boldsymbol{y}}{d},\ \boldsymbol{x}_2^{(n)} = \dfrac{\boldsymbol{u}'-\boldsymbol{y}}{d}\right)$, we get

$$\frac{\boldsymbol{u}-\boldsymbol{y}}{d} = \frac{\boldsymbol{u}'-\boldsymbol{y}}{d},$$

i.e. $\boldsymbol{u} = \boldsymbol{u}'$.

1.3.7. It is frequently necessary for us to make use of the following fact, related to the theory of functions of a real variable.

Suppose that $f, f_k \in f_p(\mathscr{E})$ $(k = 1, 2, \ldots)$ and

(1) $$\lim_{k\to\infty} \|f - f_k\|_{L_p(\mathscr{E})} = 0\ (k \to \infty),\ (1 \leqq p \leqq \infty).$$

Then there exists a subsequence $\{k_l\}$ of the natural numbers such that

(2) $$\lim_{k_l\to\infty} f_{k_l}(\boldsymbol{x}) = f(\boldsymbol{x}) \text{ for almost all } \boldsymbol{x} \in \mathscr{E}.$$

Thus there exists a set $\mathscr{E}' \subset \mathscr{E}$, distinct from $\mathscr{E}$ by a set of measure zero, such that f and $f_{k_l}(l = 1, 2, \ldots)$ are finite on $\mathscr{E}'$ and equation (2) is satisfied for all $x \in \mathscr{E}$. Hence it easily follows that if, along with (1), for some p_* one has $\lim \|f_* - f_k\|_{L_{p^*}(\mathscr{E})} = 0$, then $f_* = f$ on $\mathscr{E}$, i.e. the functions f and f_* are equivalent on $\mathscr{E}$.

If the measurable set $\mathscr{E} = \mathscr{E}_1 \times \mathscr{E}_2$ is the topological product of two measurable sets $\mathscr{E}_1$, $\mathscr{E}_2$, so that each point $x \in \mathscr{E}$ may be represented in the form of a pair $x = (y, z)$, where $y \in \mathscr{E}_1$, $z \in \mathscr{E}_2$, then we may suppose that

$$f(x) = f(y, z), \quad f_k(x) = f_k(y, z) \quad (k = 1, 2, \ldots).$$

In this case

$$\|f - f_k\|_{L_p(\mathscr{E})} = \Big(\int_{\mathscr{E}_1} \Psi_k(y)\, dy\Big)^{1/p} \quad (k = 1, 2, \ldots),$$

where

$$(3) \quad \Psi_k(y) = \int_{\mathscr{E}_2} |f(y, z) - f_k(y, z)|^p\, dz = \|f(y, z) - f_k(y, z)\|^p_{L_p(\mathscr{E}_2)}$$

is a nonnegative function summable on $\mathscr{E}_1$, i.e. $\Psi_k \in L_1(\mathscr{E}_1)$.

It follows from equation (1) that

$$\lim_{k\to\infty} \int_{\mathscr{E}_1} \Psi_k(y)\, dy = 0,$$

and thus, applying to Ψ_k the property mentioned above (where we have to take $p = 1$ and replace $\mathscr{E}$ by $\mathscr{E}_1$), we arrive at the following lemma.

1.3.8. Lemma. *It follows from equation 1.3.7. (1), where the measurable set $\mathscr{E} = \mathscr{E}_1 \times \mathscr{E}_2$ is the topological product of the measurable sets $\mathscr{E}_1$, $\mathscr{E}_2$, that for some subsequence $\{k_l\}$ of the natural numbers the equation*

$$(1) \qquad \lim_{k_l\to\infty} \|f(y, z) - f_{k_l}(y, z)\|_{L_p(\mathscr{E})} = 0$$

is satisfied for almost all $y \in \mathscr{E}_1$.

From this lemma and the remark at the beginning of subsection 1.3.7. the following lemma follows easily.

1.3.9. Lemma. *Suppose that $\mathscr{E} = \mathscr{E}_1 \times \mathscr{E}_2$ is defined as in the preceding lemma, and suppose that for the sequence of functions $f_k \in L_p(\mathscr{E})$ $(k = 1, 2, \ldots)$ and the function $f \in L_p(\mathscr{E})$ the equation*

$$(1) \qquad \lim \|f - f_k\|_{L_p(\mathscr{E})} = 0 \quad (1 \leqq p \leqq \infty).$$

is satisfied.

Suppose moreover that for some number p' $(1 \leqq p' \leqq \infty)$, *in general distinct from* p, *and function* f_*, *the equation*

$$\lim_{k\to\infty} \|f_*(\boldsymbol{y}, \boldsymbol{z}) - f_k(\boldsymbol{y}, \boldsymbol{z})\|_{L_{p'}(\mathscr{E}_2)} = 0 \tag{2}$$

is satisfied for almost all $\boldsymbol{y} \in \mathscr{E}_1$.

Then $f = f_*$, *i.e. the functions* $f(\boldsymbol{x})$ *and* $f_*(\boldsymbol{x})$ *are equivalent on* $\mathscr{E}$.

Proof. In view of the preceding lemma, for some subsequence $\{k_l\}$ of the integers, and for some set $\mathscr{E}_1' \subset \mathscr{E}_1$ differing from $\mathscr{E}_1$ by a set of measure zero (in the sense of $\mathscr{E}_1$), equation 1.3.8. (1) holds for all $\boldsymbol{y} \in \mathscr{E}_1'$. We may suppose that equation (2) also holds for almost all $\boldsymbol{y} \in \mathscr{E}_1'$. Thus, if $\boldsymbol{y} \in \mathscr{E}_1'$, (1) and (2) hold simultaneously for it.

But then the equation $f(\boldsymbol{y}, \boldsymbol{z}) = f_*(\boldsymbol{y}, \boldsymbol{z})$ holds for almost all $\boldsymbol{z} \in \mathscr{E}_2$, i.e. almost everywhere on the measurable set $\mathscr{E} = \mathscr{E}_1 \times \mathscr{E}_2$.

We note further the following theorems.

1.3.10. Theorem (P. Fatou)[1]. *If the sequence* $\{f_n\}$ *of measurable and nonnegative functions converges almost everywhere to the function* $F(x)$ *on a measurable set* $\mathscr{E} \subset \mathbb{R}_n$, *then*

$$\int_{\mathscr{E}} F d\boldsymbol{x} \leqq \sup \left\{ \int_{\mathscr{E}} f_n \, d\boldsymbol{x} \right\}.$$

1.3.11. Theorem[2]. *Suppose given a sequence* $\{f_k\}$ *of functions, bounded in the sense of* $L_p(\mathscr{E})$ $(1 < p < \infty)$:

$$\|f_k\|_{L_p(\mathscr{E})} \leqq M.$$

Then it is possible to select a subsequence $\{f_{k_l}\}$ *which converges weakly to some function* $f \in L_p(\mathscr{E})$ *with* $\|f\|_{L_p(\mathscr{E})} \leqq M$. *This means that the equation*

$$\lim_{k\to\infty} \int_{\mathscr{E}} f_k \varphi \, d\boldsymbol{x} = \int_{\mathscr{E}} f \varphi \, d\boldsymbol{x}$$

holds for any function $\varphi \in L_q(\mathscr{E})$ $\left(\frac{1}{p} + \frac{1}{q} = 1\right)$.

1.3.12. The function $f \in L_p(\mathscr{E})$ is said to be *continuous in the large* in $L_p(\mathscr{E})$, it for any $\varepsilon > 0$ there exists a $\delta(\varepsilon) > 0$ such that

$$\|f(\boldsymbol{x} + \boldsymbol{y}) - f(\boldsymbol{x})\|_{L_p(\mathscr{E}_\delta)} < \varepsilon$$

---

[1] See, for example, the book of I. P. Natanson [1] or any other standard text on measure theory.

[2] See, for example, the book of V. I. Smirnov [1] or any book on functional analysis, e.g. Banach [1].

for any $|\boldsymbol{y}| < \delta$. Here $\mathscr{E}_\delta$ is the set of those $\boldsymbol{x} \in \mathscr{E}$ such that $\boldsymbol{x} + \boldsymbol{y} \in \mathscr{E}$ for any $\boldsymbol{y}$ satisfying the inequality $|\boldsymbol{y}| < \delta$.

Theorem. *Every function $f(\boldsymbol{x}) \in L_p(\mathscr{E})$, $1 \leqq p < \infty$ is continuous in the large in $L_p(\mathscr{E})$.*

1.3.13. We shall use extensively also the following inequalities for $1 \leqq p \leqq \infty$:

$$\left(\sum_1^\infty |a_k + b_k|^p\right)^{1/p} \leqq \left(\sum_1^\infty |a_k|^p\right)^{1/p} + \left(\sum_1^\infty |b_k|^p\right)^{1/p}, \tag{1}$$

$$\sum_1^\infty |a_k b_k| \leqq \left(\sum_1^\infty |a_k|^p\right)^{1/p} \left(\sum_1^\infty |b_k|^q\right)^{1/q}, \quad \text{if} \quad \frac{1}{p} + \frac{1}{q} = 1, \tag{2}$$

where a_k and b_k are arbitrary numbers, and q is taken to be ∞ if $p = 1$. They are called respectively the Minkowski and Hölder inequalities for sums.

Using (1) one may establish that the linear n-dimensional set of vectors $\xi = \{\xi_1, \ldots, \xi_n\}$ with the norm

$$|\xi|_{l_p^n} = \left(\sum_1^n |\xi_k|^p\right)^{1/p}, \quad 1 \leqq p \leqq \infty,$$

is a normed space. In particular, it follows from 1.2.4. that for any p and p' with $1 \leqq p < p' \leqq \infty$

$$c_1 \|\xi\|_{l_p^n} \leqq \|\xi\|_{l_{p'}^n} \leqq c_2 \|\xi\|_{l_p^n},$$

where c_1, c_2 are positive constants not depending on ξ. Of course, these inequalities may be obtained directly and the exact values of c_1 and c_2 established.

1.4. Averaging of Functions According to Sobolev

This process is due to Sobolev [4].

Denote by

$$\sigma_\varepsilon = \{|\boldsymbol{x}| \leqq \varepsilon\}, \; \sigma_1 = \sigma,$$

the ball in $\mathbb{R} = \mathbb{R}_n$ of radius ε and with center at the origin.

Suppose that $\psi(t)$ is an infinitely differentiable even nonnegative function of one variable $t (-\infty < t < \infty)$, equal to zero for $|t| \geqq 1$ and such that

$$\int_{|\boldsymbol{x}| \leqq 1} \psi(|\boldsymbol{x}|)\, d\boldsymbol{x} = 1, \tag{1}$$

where $\varkappa_n$ is the volume of the unit $(n-1)$-dimensional sphere in $\mathbb{R}_n$.

As ψ we may choose the function

$$\psi(t) = \begin{cases} \frac{1}{\lambda_n} e^{\frac{t^2}{t^2-1}}, & 0 \leqq |t| < 1, \\ 0, & 1 \leqq |t|, \end{cases}$$

where the constant λ_n is chosen so that condition (1) is satisfied.

The function

$$\varphi_\varepsilon(\boldsymbol{x}) = \frac{1}{\varepsilon^n} \varphi\left(\frac{\boldsymbol{x}}{\varepsilon}\right), \quad \varphi(\boldsymbol{x}) = \psi(|\boldsymbol{x}|), \quad \varepsilon > 0, \tag{2}$$

infinitely differentiable on $\mathbb{R}$ (taking account of the evenness of ψ), has a support on σ_ε and satisfies the condition

$$\int \varphi_\varepsilon(\boldsymbol{x})\, d\boldsymbol{x} = \frac{1}{\varepsilon^n} \int \varphi\left(\frac{\boldsymbol{x}}{\varepsilon}\right) d\boldsymbol{x} = 1. \tag{3}$$

Suppose that $g \subset \mathbb{R}_n = \mathbb{R}$ is an open set and $f \in L_p(g)$ $(1 \leqq p \leqq \infty)$. Put $f = 0$ on $\mathbb{R} - g$. The function

$$f_\varepsilon(\boldsymbol{x}) = f_{g,\varepsilon}(\boldsymbol{x}) = \frac{1}{\varepsilon^n} \int \varphi\left(\frac{\boldsymbol{x} - \boldsymbol{u}}{\varepsilon}\right) f(\boldsymbol{u})\, d\boldsymbol{u} = \frac{1}{\varepsilon^n} \int \varphi\left(\frac{\boldsymbol{u}}{\varepsilon}\right) f(\boldsymbol{x} - \boldsymbol{u})\, d\boldsymbol{u} \tag{4}$$

is called the *Sobolev ε-average*. This is evidently an infinitely differentiable function on $\mathbb{R}$.

For the moment let us focus our attention on the following important property of f_ε:

$$\|f_\varepsilon - f\|_p \to 0 \ (\varepsilon \to 0,\ \|\cdot\|_p = \|\cdot\|_{L_p(\mathbb{R})},\ 1 \leqq p < \infty). \tag{5}$$

It shows that for finite $p (1 \leqq p < \infty)$ the set of infinitely differentiable functions on $\mathbb{R}$ is everywhere dense in $L_p(g)$, i.e. no matter how the open set g might be constructed, for each $f \in L_p(g)$ it is possible to find a family of infinitely differentiable functions f_ε on $\mathbb{R}$, i.e. the Sobolev averages, such that (5) is satisfied.

Indeed, in view of (3)

$$f_\varepsilon(\boldsymbol{x}) - f(\boldsymbol{x}) = \frac{1}{\varepsilon^n} \int \varphi\left(\frac{\boldsymbol{x} - \boldsymbol{u}}{\varepsilon}\right) [f(\boldsymbol{u}) - f(\boldsymbol{x})]\, d\boldsymbol{u} =$$

$$= - \int \varphi(\boldsymbol{v})\, [f(\boldsymbol{x} - \varepsilon \boldsymbol{v}) - f(\boldsymbol{x})]\, d\boldsymbol{v},$$

so that, applying the generalized Minkowski inequality and taking account of the fact that φ has its support on σ, we obtain

$$(6) \qquad \|f_\varepsilon - f\|_p \leqq \int \varphi(\boldsymbol{v}) \, \|f(\boldsymbol{x} - \varepsilon\boldsymbol{v}) - f(\boldsymbol{x})\|_p \, d\boldsymbol{v}$$
$$\leqq \sup_{|\boldsymbol{v}|<\varepsilon} \|f(\boldsymbol{x} - \boldsymbol{v}) - f(\boldsymbol{x})\|_p \to 0 \quad (\varepsilon \to 0).$$

In the case $p = \infty$ property (5) is not satisfied. However, if we suppose that $g = \mathbb{R}$ and $f(\boldsymbol{x})$ is uniformly continuous on $\mathbb{R}(f \in C(\mathbb{R}))$, then (6) may be written in the form

$$\|f_\varepsilon - f\|_\infty \leqq \sup_{|\boldsymbol{v}|<\varepsilon} |f(\boldsymbol{x} - \boldsymbol{v}) - f(\boldsymbol{x})| \to 0 \quad (\varepsilon \to 0).$$

We note also the inequality

$$(7) \qquad \|f_\varepsilon\|_p \leqq \frac{1}{\varepsilon^n} \int \varphi\left(\frac{\boldsymbol{u}}{\varepsilon}\right) \|f(\boldsymbol{x} - \boldsymbol{u})\|_p \, d\boldsymbol{u} = \|f\|_p \quad (1 \leqq p \leqq \infty).$$

1.4.1. Finite functions. Suppose that $g \subset \mathbb{R}$ is an open set. The function $\varphi(\boldsymbol{x})$ is said to be *finite* in g if it is defined there and has a compact support lying inside g. The *support* of a function is the closure of the set of all points where it is not equal to zero.

Lemma. *If $f \in L_p(g)$ $(1 \leqq p < \infty)$, then there exists a sequence of infinitely differentiable functions φ_l, finite in g and having the properties*

$$\|f - \varphi_l\|_p \to 0 \ (l \to \infty),$$
$$|\varphi_l(\boldsymbol{x})| \leqq \sup_{\boldsymbol{x} \in g} \operatorname{vrai} |f(\boldsymbol{x})|.$$

If f lies both in L_p and $L_{p'}$ with $1 \leqq p, p' < \infty$, then *the sequence $\{\varphi_l\}$ may be taken to be the same for both.*

Proof. Suppose given $\eta > 0$, and select an open bounded set $\Omega \subset \bar{\Omega} \subset g$ such that

$$\|f\|_{L_p(g-\Omega)} < \frac{\eta}{2}.$$

Denote by d the distance from Ω to the boundary of g (evidently $d > 0$; if g is unbounded, then $d = \infty$). Introduce further the function

$$f_\Omega(\boldsymbol{x}) = \begin{cases} f(\boldsymbol{x}), & \boldsymbol{x} \in \Omega, \\ 0, & \boldsymbol{x} \notin \Omega. \end{cases}$$

Its ε-average $f_{\Omega,\varepsilon} = \varphi$ for $\varepsilon < d$ is an infinitely differentiable finite function in g, for which the inequalities

$$\|f - f_{\Omega,\varepsilon}\|_{L_p(g)} \leqq \|f - f_\Omega\|_{L_p(g)} + \|f_\Omega - f_{\Omega,\varepsilon}\|_{L_p(g)} =$$

$$= \|f\|_{L_p(g-\Omega)} + \|f_\Omega - f_{\Omega,\varepsilon}\|_{L_p(g)} < \frac{\eta}{2} + \frac{\eta}{2} = \eta,$$

hold whenever ε is sufficiently small.

Further (see 1.4. (7))

$$|f_{\Omega,\varepsilon}(\boldsymbol{x})| \leqq \sup_{\boldsymbol{x}\in\mathbb{R}} \operatorname{vrai} |f_\Omega(\boldsymbol{x}) \leqq \sup_{\boldsymbol{x}\in\mathbb{R}} \operatorname{vrai} |f(\boldsymbol{x})|.$$

Therefore, if $\eta = \eta_l \to 0$, then, taking $\varepsilon = \varepsilon_l$, $\Omega = \Omega_l$, we find that the functions $\varphi_l = f_{\Omega_l,\varepsilon_l}$ satisfy the requirements of the lemma. In addition if at the same time $f \in L_p, L_{p'}$, then for both p and p' we may choose the same Ω_l and ε_l, and hence the same φ_l.

1.4.2. Lemma. *If $f \in L_\infty(g)$, i.e. if it is a measurable and essentially bounded function on the open set $g \subset \mathbb{R}$, then there exists a sequence of infinitely differentiable functions φ_l, finite in g and satisfying the conditions*

(1) $$\lim_{l\to\infty} \varphi_l(\boldsymbol{x}) = f(\boldsymbol{x}) \qquad \textit{almost everywhere in } g,$$

(2) $$|\varphi_l(\boldsymbol{x})| \leqq \sup_{\boldsymbol{x}\in g} \operatorname{vrai} |f(\boldsymbol{x})|.$$

Proof. Denote by g_N the intersection of g with the ball $|x| < N$. Let η_N tend monotonically to zero, $N = 1, 2, \ldots$. Since $f \in L(g_N)$, then one can find a function f_N finite in g_N and accordingly in g such that

(3) $$\|f - f_N\|_{L(g_N)} < \eta_N$$

and

(4) $$|f_N(\boldsymbol{x})| \leqq \sup_{\boldsymbol{x}\in g} \operatorname{vrai} |f(\boldsymbol{x})|.$$

It follows from (3) and (4) that it is possible to select from the sequence $\{f_N\}$ a subsequence $\{\varphi_l\}$ satisfying the requirements of the lemma.

1.5. Generalized Functions

We introduce the class S (L. Schwartz [1]) of generalized functions $\varphi = \varphi(\boldsymbol{x})$. The function φ is defined on $\mathbb{R}$, it is complex-valued ($\varphi = \varphi_1 + i\varphi_2$, φ_1, φ_2 real), infinitely differentiable on $\mathbb{R}$, and it has the property that for any nonnegative number l (it is sufficient that l be an integer) and nonnegative integer vector $\boldsymbol{k} = (k_1, \ldots, k_n)$

$$(1) \qquad \sup_{\boldsymbol{x}} (1 + |\boldsymbol{x}|^l)\, |\varphi^{(\boldsymbol{k})}(\boldsymbol{x})| = \varkappa(l, \boldsymbol{k}, \varphi) < \infty,$$

where

$$\varphi^{(\boldsymbol{k})} = \frac{\partial^{|\boldsymbol{k}|}\varphi}{\partial x_1^{k_1} \cdots \partial x_n^{k_n}}, \quad |\boldsymbol{k}| = \sum_{j=1}^{n} k_j.$$

We put $L_p = L_p(\mathbb{R})$. It follows from (1) in particular that

$$|\varphi^{(\boldsymbol{k})}(\boldsymbol{x})| \leqq \frac{1}{2}\, \varkappa(0, \boldsymbol{k}, \varphi) < \infty,$$

i.e. the function $\varphi^{(\boldsymbol{k})} \in S$ is bounded ($\varphi^{(\boldsymbol{k})} \in L^\infty$). Further $\varphi^{(\boldsymbol{k})} \in L_p$ ($1 \leqq p < \infty$), since

$$\int |\varphi^{(\boldsymbol{k})}(\boldsymbol{x})|^p\, d\boldsymbol{x} \leqq c_1 \int \frac{\left|\varphi^{(\boldsymbol{k})}(\boldsymbol{x}) \left(1 + |\boldsymbol{x}|^{\frac{n+1}{p}}\right)\right|^p d\boldsymbol{x}}{(1 + |\boldsymbol{x}|)^{n+1}} \leqq$$

$$\leqq c_1 \varkappa^p \left(\frac{n+1}{p}, \boldsymbol{k}, \varphi\right) \int \frac{d\boldsymbol{x}}{(1 + |\boldsymbol{x}|)^{n+1}} = c_2 \varkappa^p \left(\frac{n+1}{p}, \boldsymbol{k}, \varphi\right) < \infty,$$

where n is the dimension of $\mathbb{R}$. Thus for any $\boldsymbol{k}$

$$(2) \qquad \varphi^{(\boldsymbol{k})} \in L_p \text{ and } \|\varphi^{(\boldsymbol{k})}\|_{L_p} \leqq c\varkappa \left(\frac{n+1}{p}, \boldsymbol{k}, \varphi\right).$$

Moreover,

$$(3) \qquad |\varphi^{(\boldsymbol{k})}(\boldsymbol{x})| \leqq \frac{\varkappa(1, \boldsymbol{k}, \varphi)}{1 + |\boldsymbol{x}|} \to 0 \quad (|\boldsymbol{x}| \to \infty).$$

If $\varphi_m, \varphi \in S$ ($m = 1, 2, \ldots$) and if for any nonnegative integer l and integer vector $\boldsymbol{k}$

$$\varkappa(l, \boldsymbol{k}, \varphi_m - \varphi) \to 0 \quad (m \to \infty),$$

then we will write

$$\varphi_m \to \varphi(S).$$

Concerning a function ψ, infinitely differentiable on $\mathbb{R}$, we shall say that it *has polynomial growth*, if for any nonnegative vector $\boldsymbol{k}$ there exists an $l = l(\boldsymbol{k})$ such that

$$|\psi^{(\boldsymbol{k})}(\boldsymbol{x})| < c(1 + |\boldsymbol{x}|^l),$$

where c does not depend on $\boldsymbol{x}$.

If $\varphi \in S$, then $\psi\varphi \in S$, since

$$(\psi\varphi)^{(\boldsymbol{k})} = \sum_{|\boldsymbol{s}| \leqq |\boldsymbol{k}|} C_{\boldsymbol{k}}^{\boldsymbol{s}} \psi^{(\boldsymbol{s})} \varphi^{(\boldsymbol{k}-\boldsymbol{s})}$$

$$(\boldsymbol{k} = (k_1, \ldots, k_n),\ \boldsymbol{s} = (s_1, \ldots, s_n),$$

$$C_{\boldsymbol{k}}^{\boldsymbol{s}} = \frac{\boldsymbol{k}!}{\boldsymbol{s}!(\boldsymbol{k}-\boldsymbol{s})!},\ \boldsymbol{k}! = \prod_{j=1}^{n} k_j!),$$

and if m is an integer, then

$$|(1 + |\boldsymbol{x}|^m)\, \psi^{(\boldsymbol{s})} \varphi^{(\boldsymbol{k}-\boldsymbol{s})}| \leqq c_1 |\, (1 + |\boldsymbol{x}|^m)\, (1 + |\boldsymbol{x}|^{l(\boldsymbol{s})})\, \varphi^{(\boldsymbol{k}-\boldsymbol{s})}|$$
$$\leqq c_2\, |(1 + |\boldsymbol{x}|^{m+l(\boldsymbol{s})})\, \varphi^{(\boldsymbol{k}-\boldsymbol{s})}| \leqq c_2 \varkappa(m + l(\boldsymbol{s}), \boldsymbol{k} - \boldsymbol{s}, \varphi).$$

Moreover these inequalities show that if

$$\varphi_m, \varphi \in S \quad \text{and} \quad \varphi_m \to \varphi(S), \quad \text{then } \psi\varphi_m \to \psi\varphi(S).$$

The Fourier transform of the function φ will be denoted as follows:

$$\tilde{\varphi}(\boldsymbol{x}) = \frac{1}{(2\pi)^{n/2}} \int \varphi(\boldsymbol{\lambda})\, e^{-i\boldsymbol{\lambda}\boldsymbol{x}}\, d\boldsymbol{\lambda}, \quad \boldsymbol{\lambda}\boldsymbol{x} = \sum_{1}^{n} \lambda_j x_j,$$

and the transformation inverse to it as follows:

$$\hat{\varphi}(\boldsymbol{x}) = \frac{1}{(2\pi)^{n/2}} \int \varphi(\boldsymbol{\lambda})\, e^{i\boldsymbol{\lambda}\boldsymbol{x}} d\boldsymbol{\lambda}.$$

We shall show that if $\varphi \in S$, then $\tilde{\varphi}, \hat{\varphi} \in S$, and, for any positive number and integer vector $\boldsymbol{k}$,

(4) $$(1 + |\boldsymbol{x}|^l)\, |\tilde{\varphi}^{(\boldsymbol{k})}(\boldsymbol{x})| \leqq c_{l,\boldsymbol{k}} \sum_{(l',\boldsymbol{k}') \in \mathscr{E}_{l\boldsymbol{k}}} \varkappa(l', \boldsymbol{k}', \varphi),$$

where $c_{l,\boldsymbol{k}}$ is a constant depending on $(l, \boldsymbol{k})$ and $\mathscr{E}_{l\boldsymbol{k}}$ is a finite set of pairs $(l', \boldsymbol{k})$ depending on $(l, \boldsymbol{k})$. Hence it follows in particular that if φ_m, $\varphi \in S$, $\varphi_m \to \varphi(S)$, then $\tilde{\varphi}_m \to \tilde{\varphi}(S)$ and $\hat{\varphi}_m \to \hat{\varphi}(S)$.

Indeed,

$$\tilde{\varphi}^{(\boldsymbol{k})}(\boldsymbol{x}) = \int \psi(\boldsymbol{\lambda})\, e^{-i\boldsymbol{\lambda}\boldsymbol{x}}\, f\boldsymbol{\lambda},$$

where

$$\psi(\boldsymbol{\lambda}) = \frac{(-i\boldsymbol{\lambda})^{\boldsymbol{k}}}{(2\pi)^{n/2}} \varphi(\boldsymbol{\lambda}) \quad (\boldsymbol{\lambda}^{\boldsymbol{k}} = \lambda_1^{k_1} \cdots \lambda_n^{k_n}).$$

Obviously $\psi(\boldsymbol{\lambda}) \in S$,

(5) $$(1 + |\boldsymbol{x}|) \leqq (1 + |x_1| + \cdots + |x_n|)$$

and

$$(6)\qquad |\tilde{\varphi}^{(\boldsymbol{k})}(\boldsymbol{x})| \leqq c \int |\boldsymbol{\lambda}|^{|\boldsymbol{k}|}\, |\varphi(\boldsymbol{\lambda})|\, d\boldsymbol{\lambda}$$

$$\leqq c \int \frac{|\boldsymbol{\lambda}|^{|\boldsymbol{k}|} d\boldsymbol{\lambda}}{1 + |\boldsymbol{\lambda}|^{|\boldsymbol{k}|+n+2}}\, \varkappa(|\boldsymbol{k}| + n + 2, \boldsymbol{0}, \varphi) \leqq c_1 \varkappa(|\boldsymbol{k}| + n + 2, \boldsymbol{0}, \varphi).$$

Further for $|x_j| \leqq 1$

$$|\boldsymbol{x}_j \tilde{\varphi}^{(\boldsymbol{k})}(\boldsymbol{x})| \leqq |\tilde{\varphi}^{(\boldsymbol{k})}(\boldsymbol{x})| \leqq c_1 \varkappa(|\boldsymbol{k}| + n + 2, \boldsymbol{0}, \varphi),$$

and for $|x_j| \geqq 1$, taking account of the fact that Δ_N is the portion of $\mathbb{R}$ where $|\lambda_j| < N$ and recalling (see (3)) that $\psi \to 0$ for $\lambda_j = \pm N \to \infty$, we obtain

$$(7)\qquad \tilde{\varphi}^{(\boldsymbol{k})}(\boldsymbol{x}) = \lim_{N\to\infty} \left\{ \psi(\boldsymbol{\lambda}) \frac{e^{-i\boldsymbol{\lambda}\boldsymbol{x}}}{-ix_j}\Bigg|_{\lambda_j=-N}^{\lambda_j=N} + \frac{1}{ix_j} \int_{\Delta_N} \frac{\partial \psi}{\partial \lambda_j} e^{-i\boldsymbol{\lambda}\boldsymbol{x}}\, d\boldsymbol{\lambda} \right\} =$$

$$= \frac{1}{ix_j} \int \frac{\partial \psi}{\partial \lambda_j} e^{-i\boldsymbol{\lambda}\boldsymbol{x}}\, d\boldsymbol{\lambda} = \frac{c}{x_j} \int \left(\frac{\partial}{\partial \lambda_j} \left(\prod_{s=1}^{n} \lambda_s^{k_s} \right) \varphi(\boldsymbol{\lambda}) + \prod_{s=1}^{n} \lambda_s^{k_s} \frac{\partial \varphi}{\partial \lambda_j} \right) e^{-i\boldsymbol{\lambda}\boldsymbol{x}}\, d\boldsymbol{\lambda}.$$

Since $\left| \frac{\partial}{\partial \lambda_j} \left(\prod_{s=1}^{n} \lambda_s^{k_s} \right) \right| \leqq c_1 |\boldsymbol{\lambda}|^{|\boldsymbol{k}|-1}$ (for $\boldsymbol{k} = \boldsymbol{0}$ we need to suppose that $|\boldsymbol{k}| - 1 = 0$), then

$$(8)\qquad |x_j \tilde{\varphi}^{(\boldsymbol{k})}(\boldsymbol{x})| \leqq c_2 \int \frac{|\boldsymbol{\lambda}|^{|\boldsymbol{k}|-1} + |\boldsymbol{\lambda}|^{|\boldsymbol{k}|}}{1 + |\boldsymbol{\lambda}|^{n+|\boldsymbol{k}|+2}} \big(\varkappa(n + |k| + 2, \boldsymbol{0}, \varphi)$$

$$+ \varkappa(n + |\boldsymbol{k}| + 2, \boldsymbol{e}_j, \varphi)\big)\, d\boldsymbol{\lambda},$$

where $\boldsymbol{e}_j$ is the unit vector directed along the x_j axis.

It follows from (5), (6) and (8) that

$$(1 + |\boldsymbol{x}|)\, |\tilde{\varphi}^{(\boldsymbol{k})}(\boldsymbol{x})| \leqq c_{1\boldsymbol{k}} \left(\varkappa(n + |\boldsymbol{k}| + 2, \boldsymbol{0}, \varphi) + \sum_{j=1}^{n} \varkappa(n + |\boldsymbol{k}| + 2, \boldsymbol{e}_j, \varphi) \right),$$

and we have proved inequality (4) for any $\boldsymbol{k}$ and $l = 1$. For an arbitrary l the proof is analogous; we need only integrate by parts in equation (7) l times instead of once.

For $\varphi, \psi \in S$ put[1]

$$(\varphi, \psi) = \int \varphi(\boldsymbol{x})\, \psi(\boldsymbol{x})\, d\boldsymbol{x}.$$

[1] We direct the reader's attention to the fact that the ψ in the integrand is taken without the complex conjugation sign (see V. S. Vladimirov [1]).

From the theory of Fourier integrals it is known that

$$(\tilde{\varphi}, \psi) = (\varphi, \tilde{\psi}), \ (\hat{\varphi}, \psi) = (\varphi, \hat{\psi}).$$

The continuous linear functional (f, φ) on S is called a *generalized function* (*over* S).

Thus, if $\varphi_1, \varphi_2, \varphi_m, \varphi \in S$, c_1, c_2 are complex numbers and $\varphi_m \to \varphi(S)$, then

$$(f, c_1\varphi_1 + c_2\varphi_2) = c_1(f, \varphi_1) + c_2(f, \varphi_2)$$

$$(f, \varphi_m) \to (f, \varphi) \ (m \to \infty).$$

The set of all generalized (over S) functionals is denoted by S'.

The *derivative* of $f \in S'$ relative to the variable x_j is defined as a linear functional:

$$(f'_{x_j}, \varphi) = -(f, \varphi'_{x_j}).$$

If $f(\boldsymbol{x})$ is an ordinary measurable function defined on $\mathbb{R}$ and such that the integral

$$(f, \varphi) = \int f(\boldsymbol{x}) \varphi(\boldsymbol{x}) \, d\boldsymbol{x} \tag{9}$$

exists for all $\varphi \in S$, which is in fact a linear functional over S, then the generalized function defined by the equation (9) is *identified* with $f(\boldsymbol{x})$. For example, if $f \in L_p(1 \leqq p \leqq \infty)$, then the integral (9) is a linear functional over S. Indeed,

$$\int |f(\boldsymbol{x}) \varphi(\boldsymbol{x})| \, d\boldsymbol{x} \leqq \left(\int |f|^p \, d\boldsymbol{x}\right)^{1/p} \left(\int |\varphi|^q \, d\boldsymbol{x}\right)^{1/q}$$

$$\leqq c\varkappa\left(\frac{n+1}{q}, \ 0, \varphi\right)\left(\frac{1}{p} + \frac{1}{q} = 1\right),$$

so that the integral (9) is finite for all $\varphi \in S$ and continuous in S. The linearity of (9) is obvious. If $f(\boldsymbol{x}) \in S$, $\boldsymbol{a} \in \mathbb{R}$ and $c \neq 0$ is a real number, the $f(\boldsymbol{x} + \boldsymbol{a})$, $f(c\boldsymbol{x}) \in S'$ and they are defined respectively as functionals

$$\big(f(\boldsymbol{x} + a), \varphi(\boldsymbol{x})\big) = \big(f(\boldsymbol{x}), \varphi(\boldsymbol{x} - \boldsymbol{a})\big),$$

$$f(c\boldsymbol{x}) = \frac{1}{|c|}\left(f(\boldsymbol{x}), \varphi\left(\frac{\boldsymbol{x}}{c}\right)\right).$$

If f is a generalized function and ψ an infinitely differentiable function of polynomial growth, then the functional over S, defined by the equation

$$(f\psi, \varphi) = (f, \psi\varphi),$$

is obviously also a generalized function. We will denote it by $f\psi$ or $\psi f (f\psi = \psi f)$.

If ψ_1 and ψ_2 are two infinitely differentiable functions with polynomial growth, then their product has the same property. In this connection it is easy to see that if $f \in S'$ then

$$(\psi_1\psi_2)f = \psi_1(\psi_2 f).$$

It is clear that if $f(x)$ is an ordinary function lying in L and $\psi(\boldsymbol{x})$ is an infinitely differentiable function with polynomial growth, then the ordinary product $f(\boldsymbol{x})\,\psi(\boldsymbol{x})$ corresponds according to the rule of identification of a generalized function to the product of the generalized functions f and ψ.

The *Fourier transforms* (direct and inverse) for $f \in S'$ are defined respectively by the equations

$$(\tilde{f}, \varphi) = (f, \tilde{\varphi}),\ (\hat{f}, \varphi) = (f, \hat{\varphi})\ (\varphi \in S).$$

Since $\varphi_m \to \varphi(S)$ $(\varphi_m, \varphi \in S)$ implies $\tilde{\varphi}_m \to \tilde{\varphi}$, $\hat{\varphi}_m \to \hat{\varphi}(S)$, then $\tilde{f}, \hat{f} \in S'$. If $f(\boldsymbol{x}) \in L_p$ $(1 \leqq p \leqq \infty)$ is an ordinary L_p function on $\mathbb{R}$, then, as we know, it is a generalized function and has a Fourier transform $\tilde{f}$, which is generally speaking a generalized function. If $f \in L_2$, then by the Plancherel theorem (see the book of N. I. Ahiezer [1])

$$\tilde{f}(\boldsymbol{x}) = \frac{1}{(2\pi)^{n/2}} \lim_{N\to\infty} \int_{\Delta_N} f(\boldsymbol{t})\, e^{-i\boldsymbol{x}\boldsymbol{t}}\, d\boldsymbol{t},$$

$$\Delta_N = \{|x_j| < N;\ j = 1, \ldots, n\},$$

and the convergence is understood in the sense of L_2. Here $\int \tilde{f}\varphi d\boldsymbol{x} = \int f\tilde{\varphi} d\boldsymbol{x}$ (for all $\varphi \in S$), which shows that in the case at hand the ordinary Fourier transform coincides with (is identified with) the generalized transform.

Suppose that $\varphi \in S$. Then

$$\varphi^{(\boldsymbol{k})}(\boldsymbol{x}) = \frac{1}{(2\pi)^{n/2}} \int (i\boldsymbol{u})^{\boldsymbol{k}}\, \tilde{\varphi}(\boldsymbol{u})\, e^{i\boldsymbol{x}\boldsymbol{u}}\, d\boldsymbol{u}$$

and

$$\widetilde{\varphi^{(\boldsymbol{k})}} = (i\boldsymbol{u})^{\boldsymbol{k}}\, \tilde{\varphi}(\boldsymbol{u})\ (\boldsymbol{u}^{\boldsymbol{k}} = u_1^{k_1}, \ldots, u_n^{k_n}).$$

Further

$$\tilde{\varphi}^{(\boldsymbol{k})}(\boldsymbol{x}) = \frac{1}{(2\pi)^{n/2}} \int (-\,i\boldsymbol{u})^{\boldsymbol{k}}\, \varphi(\boldsymbol{u})\, e^{-i\boldsymbol{x}\boldsymbol{u}}\, d\boldsymbol{u} = \widetilde{(-\,i\boldsymbol{u})^{\boldsymbol{k}}\, \varphi(\boldsymbol{u})}.$$

For generalized functions $f \in S'$ the analogous equations hold

$$(10) \qquad \widetilde{f^{(\boldsymbol{k})}} = (i\boldsymbol{u})^{\boldsymbol{k}}\, \tilde{f},\ \tilde{f}^{(\boldsymbol{k})} = \widetilde{(-\,i\boldsymbol{u})^{\boldsymbol{k}}\, f}.$$

Indeed, if $f \in S'$, $\varphi \in S$, then

$$(\widetilde{f^{(k)}}, \varphi) = (-1)^{|k|} (f, \tilde{\varphi}^{(k)}) = (-1)^{|k|} (f, \widetilde{(-iu)^k \varphi}) = ((iu)^k \tilde{f}, \varphi),$$

$$(\tilde{f}^{(k)}, \varphi) = (-1)^{|k|} (f, \widetilde{\varphi^{(k)}}) = (-1)^{|k|} (f, (iu)^k \tilde{\varphi}) = (\widetilde{(-iu)^k f}, \varphi).$$

Suppose first that $\varphi \in S$ and

$$\Delta_N = \{|x_j| < N; j = 1, \dots, n\} \subset \mathbb{R}.$$

Then

$$(\tilde{1}, \varphi) = (1, \tilde{\varphi}) = \frac{1}{(2\pi)^{n/2}} \int dx \int \varphi(t)\, e^{-ixt}\, dt =$$

$$= \frac{1}{(2\pi)^{n/2}} \lim_{N\to\infty} \int \varphi(t)\, dt \int_{\Delta_N} e^{-ixt}\, dx =$$

$$= (2\pi)^{n/2} \lim_{N\to\infty} \frac{1}{\pi^n} \int \varphi(t) \prod_{j=1}^{n} \frac{\sin N t_j}{t_j}\, dt = (2\alpha)^{n/2} \varphi(0).$$

The last equation follows from the ordinary theory of the Fourier integral.

Thus

$$\tilde{1} = (2\pi)^{n/2} \delta(x),$$

where $\delta(x)$ is the well-known delta function, i.e. the generalized function defined by the equation

$$(\delta, \varphi) = \varphi(0) \quad (\varphi \in S).$$

Hence, if $k = (k_1, \dots, k_n)$ is a vector with nonnegative integer components, then

$$\widetilde{x^k} = i^k \widetilde{(-ix)^k \cdot 1} = i^k (2\pi)^{n/2} \delta^{(k)}(x). \tag{11}$$

Further

$$(\tilde{\delta}, \varphi) = (\delta, \tilde{\varphi}) = \frac{1}{(2\pi)^{n/2}} \int \varphi(t)\, dt,$$

i.e.

$$\tilde{\delta} = \frac{1}{(2\pi)^{n/2}}. \tag{12}$$

For functions $f, f_l \in S'$ $(l = 1, 2, \dots)$ one writes

$$f_l \to f(S'), \text{ if } (f_l, \varphi) \to (f, \varphi) \tag{13}$$

for all $\varphi \in S$, and one says that f_l tends to f in the sense of S', or *weakly*.

If f_l and f are ordinary integrable functions such that almost everywhere

$$f_l(\boldsymbol{x}) \to f(\boldsymbol{x}) \ (l \to \infty)$$

and

$$|f_l(\boldsymbol{x})| \leqq \Phi(\boldsymbol{x}) \in L \ (l = 1, 2, \ldots),$$

where Φ does not depend on l, then evidently $f_l, f \in S'$ and, according to a theorem of Lebesgue, $f_l \to f$ weakly.

It follows obviously from (13) that if $f_l \to f(S')$, then

$$\tilde{f}_l \to \tilde{f}, \hat{f}_l \to \hat{f}(S'), \tag{14}$$

$$\lambda f_l \to \lambda f(S'), \tag{15}$$

$$f_l^{(s)} \to f^{(s)}(S'), \tag{16}$$

where λ is an infinitely differentiable function of polynomial growth. Suppose that $\varphi \in S$, and $\boldsymbol{\mu} = (\mu_1, \ldots, \mu_n)$, $\boldsymbol{t} = (t_1, \ldots, t_n)$ are real vectors. Then

$$\widehat{e^{i\mu t}\tilde{\varphi}} = \frac{1}{(2\pi)^{n/2}} \int e^{i\mu t} \frac{1}{(2\pi)^{n/2}} \int \varphi(\boldsymbol{u})\, e^{-iut}\, d\boldsymbol{u} e^{ixt}\, d\boldsymbol{t} = \tag{17}$$

$$= \frac{1}{(2\pi)^n} \int e^{ixt}\, d\boldsymbol{t} \int \varphi(\boldsymbol{\mu} + \boldsymbol{v})\, e^{-ivt}\, d\boldsymbol{v} = \varphi(\boldsymbol{\mu} + \boldsymbol{x}).$$

If the vector $f \in S'$, then, taking account of the fact that the function $e^{i\mu t}$ is infinitely differentiable and bounded along with its derivatives (of polynomial growth), we find for $\varphi \in S$

$$\left(\overline{e^{i\mu t} f}, \varphi\right) = \left(f, \widehat{e^{i\mu t}\tilde{\varphi}}\right) = \left(f, \varphi(\boldsymbol{\mu} + \boldsymbol{x})\right) = \left(f(\boldsymbol{x} - \boldsymbol{\mu}), \varphi(\boldsymbol{x})\right),$$

i.e.

$$\overline{e^{i\mu t}\tilde{f}} = f(\boldsymbol{x} - \boldsymbol{\mu}) \ (f \in S'). \tag{18}$$

Further

$$\left(\widehat{e^{i\mu t}\tilde{f}}, \varphi\right) = \left(f, \overline{e^{i\mu t}\hat{\varphi}}\right) = \left(f(\boldsymbol{x}), \varphi(\boldsymbol{x} - \boldsymbol{\mu})\right),$$

i.e.

$$\widehat{e^{i\mu t}\tilde{f}} = f(\boldsymbol{x} + \boldsymbol{\mu}) \ (f \in S'). \tag{19}$$

1.5.1. Convolution. Multiplicator. We will frequently have to deal with the situation when some measurable function $\mu(\boldsymbol{x})$ is multiplied by $\tilde{f}(\boldsymbol{x})$, where $f \in L_p = L_p(\mathbb{R})$ $(\mathbb{R} = \mathbb{R}_n)$. If $\tilde{f}(\boldsymbol{x})$ is an ordinary function then it is natural to suppose that

$$\mu\tilde{f} = \mu(\boldsymbol{x})\, \tilde{f}(\boldsymbol{x}). \tag{1}$$

However, even in this apparently simple case there can arise difficulties, consisting in that along with the definition of $\mu\tilde{f}$ by means of equation (1) there may be another definition of μf as some generalized function (belonging to S'), and then the question as to the identification of these two definitions arises.

In the case when $\tilde{f} \in S'$ is a generalized function, there is until now only a single definition of $\mu\tilde{f}$ at our disposal, under the hypothesis that μ is an infinitely differentiable function of polynomial growth. Indeed, $\mu\tilde{f}$ is defined as a functional

$$(\mu\tilde{f}, \varphi) = (\tilde{f}, \mu\varphi). \tag{2}$$

If $\tilde{f} \in L_p$, then this definition is consistent with formula (1), since in this case

$$(\mu\tilde{f}, \varphi) = \int [\mu(\boldsymbol{x})\, \tilde{f}(\boldsymbol{x})]\, \varphi(\boldsymbol{x})\, d\boldsymbol{x} = \int \tilde{f}(\boldsymbol{x})\, [\mu(\boldsymbol{x})\, \varphi(\boldsymbol{x})]\, d\boldsymbol{x} = (\tilde{f}, \mu\varphi).$$

Below we shall introduce another definition of $\mu\tilde{f}$, where μ belongs to some class of measurable functions bounded on $\mathbb{R} = \mathbb{R}_n$ and $f \in L_p$. This section is devoted to the important case when $\hat{\mu} = K \in L$. In this case the function

$$\mu(\boldsymbol{x}) = \frac{1}{(2\pi)^{n/2}} \int \hat{\mu}(\boldsymbol{u})\, e^{-i\boldsymbol{x}\boldsymbol{u}}\, d\boldsymbol{u}$$

is bounded and continuous on $\mathbb{R}$, but, of course, it is not an arbitrary continuous function on $\mathbb{R}$.

If $f \in S$, then the functions $\tilde{f}, \mu\tilde{f}, \widehat{\mu\tilde{f}} \in S'$. But they may also be calculated by means of ordinary analysis (explanation below):

$$\widehat{\mu f} = \widehat{\tilde{k}\tilde{f}} = \frac{1}{(2\pi)^{3n/2}} \int e^{i\boldsymbol{x}\boldsymbol{u}}\, d\boldsymbol{u} \int K(\xi)\, e^{-iu\xi}\, d\xi \int f(\eta)\, e^{-iu\eta} = d\eta \tag{3}$$

$$= \frac{1}{(2\pi)^{3n/2}} \int e^{i\boldsymbol{x}\boldsymbol{u}}\, d\boldsymbol{u} \int K(\xi)\, d\xi \int f(\lambda - \xi)\, e^{-iu\lambda}\, d\lambda =$$

$$= \frac{1}{(2\pi)^{3n/2}} \int e^{i\boldsymbol{x}\boldsymbol{u}}\, d\boldsymbol{u} \int e^{-iu\lambda}\, d\lambda \int K(\xi)\, f(\lambda - \xi)\, d\xi =$$

$$= \frac{1}{(2\pi)^{n/2}} \int K(\xi)\, f(\lambda - \xi)\, d\xi = K * f.$$

Since $f \in S$, then the integral in η in the top line is a function of $\boldsymbol{u}$, belonging to $S \subset L$. It is multiplied by the integral with respect to ξ, which is a continuous bounded function. The product belongs to L, so that after multiplication by $e^{i\boldsymbol{x}\boldsymbol{u}}$ and integration with respect to $\boldsymbol{u}$ we obtain a continuous function $\widehat{\tilde{k}f}$ of $\boldsymbol{x}$. The change of the order of integration relative to ξ and λ is legitimate, since $K, f \in L$. Here we use Fubini's theorem.

The integral in the next-to last term of these relations is called the *convolution* of K and f. In considerations when K is fixed and $f \in L_p$ an arbitrary function, K is called the *kernel of the convolution*, and the function μ the *multiplicator*, in L_p.

The right side of (3) has meaning not only for $f \in S$ but also for $f \in L_p$. For any function $K \in L$ and $f \in L_p$ $(1 \leqq p \leqq \infty)$, their convolution has meaning:

$$(4) \quad K * f = \frac{1}{(2\pi)^{n/2}} \int K(x - u)\, f(u)\, du = \frac{1}{(2\pi)^{n/2}} \int K(u)\, f(x - u)\, du,$$

satisfying furthermore the important inequality (see 1.3.3. (1))

$$(5) \quad \|K * f\|_p \leqq \frac{1}{(2\pi)^{n/2}} \|K\|_L \, \|f\|_p \; (\|\cdot\|_p = \|\cdot\|_{L_p(\mathbb{R})}, \; \|\cdot\|_L = \|\cdot\|_{L(\mathbb{R})}).$$

Equation (3), valid for functions $f \in S$, yields a basis for putting by definition

$$(6) \quad \mu\tilde{f} = \widetilde{K * f} \; (\hat{\mu} = K \in L, \, f \in L_p).$$

Since, by (5), $K * f \in L_p \subset S'$, then $\widetilde{K * f} \in S'$, and we agree to denote this last generalized function by $\mu\tilde{f}$.

We shall show that for any function $f \in L_p (1 \leqq p \leqq \infty)$ there exists a sequence of infinitely differentiable finite functions f_l such that as $l \to \infty$

$$(7) \quad \tilde{f}_l \to f(S') \text{ and } \mu\tilde{f}_l \to \mu\tilde{f}(S').$$

If p is finite, then we define (see subsection 1.4.1.) a sequence of finite functions f_l such that

$$\|f - f_l\|_p \to 0 \quad (l \to \infty),$$

$$\|(\hat{\mu} * f) - (\hat{\mu} * f_l)\|_p \leqq \|\hat{\mu}\|_L \, \|f - f_l\|_p \to 0.$$

Accordingly, in the weak sense as well $f_l \to f$ and $\mu\tilde{f}_l \to \mu\tilde{f}$. Now if $p = \infty$, then we define (see 1.4.2.) a sequence of infinitely differentiable finite functions f_l, boundedly converging almost everywhere and thus converging weakly to f. In view of the fact that $\hat{\mu} \in L$ and $f - f_l \in L_\infty$, the function

$$(\hat{\mu} * f) - (\hat{\mu} * f_l) = \frac{1}{(2\pi)^{n/2}} \int \hat{\mu}(x - t)\, [f(t) - f_l(t)]\, dt$$

of x is continuous and bounded on $\mathbb{R}$. On the basis of the theorem of Lebesgue on the limit under the integral sign it converges boundedly for all x to zero. Accordingly, as a generalized function it tends weakly to zero, i.e. (7) holds.

If $\hat{\mu} = K \in L$ and at the same time μ is infinitely differentiable and of polynomial growth, then we have at our disposal two definitions of the product $\mu\tilde{f}(f \in L_p)$. On the one hand, one has the functional

$$(\mu\tilde{f}, \varphi) = (\tilde{f}, \varphi\mu)\ (\varphi \in S),$$

and on the other hand one has the functional (6). We shall show that these functionals are equal.

Suppose that $\{f_l\}$ is a sequence of infinitely differentiable finite functions for which $\mu\tilde{f}_l \to \mu\tilde{f}$ weakly. Then, if not only $\hat{\mu} \in L$, but also μ, are infinitely differentiable and of polynomial growth, then

$$(\mu\tilde{f}, \varphi) = \lim_{l\to\infty} (\mu\tilde{f}_l, \varphi) = \lim_{l\to\infty} (\tilde{f}, \mu\varphi) = (\tilde{f}, \mu\varphi), \tag{8}$$

and we have proved the equality of the functionals in question.

Thus, the definition (2) for an infinitely differentiable function μ of polynomial growth and the definition (6), where $\hat{\mu} \in L$, do not contradict one another for any function $f \in L_p (1 \leqq p \leqq \infty)$.

If λ and μ are differentiable functions of polynomial growth and $f \in S$, then we know (see 1.5.) that

$$\lambda(\mu f) = \mu(\lambda f) = (\lambda\mu)\, f. \tag{9}$$

If now $\hat{\lambda} = K_1 \in L$, $\hat{\mu} = K_2 \in L$, then also $\widehat{\lambda\mu} \in L$ and for all $f \in L_p$ $(1 \leqq p \leqq \infty)$

$$\lambda(\mu\tilde{f}) = \mu(\lambda\tilde{f}) = (\lambda\mu)\, \tilde{f}. \tag{10}$$

Indeed, it is easily verified by the methods of ordinary analysis that under the indicated conditions the function

$$K = K_1 * K_2 = \int K_1(\boldsymbol{x} - \boldsymbol{u})\, K_2(\boldsymbol{u})\, d\boldsymbol{u}$$

belongs to L and that

$$K_1 * (K_2 * f) = K_2 * (K_1 * f) = (K_1 * K_2) * f, \tag{11}$$

which, in view of (6), is equivalent to (10). We do not intend to consider in all generality the case when the multiplicator is a product $\lambda\mu$, where $\hat{\lambda} \in L$ and μ is an infinitely differentiable function of polynomial growth. We will not need this in what follows. But there is one case which we shall need—the case of the factor $V^{-1}\mu V$, where $\hat{V}$, $\hat{\mu} \subset L$ and V moreover is a positive infinitely differentiable function of polynomial growth. If $f \in L_p$, then the operation

$$V^{-1}\mu V\tilde{f} = V^{-1}(\mu(V\tilde{f}))$$

has meaning. Indeed, $V\tilde{f}$ may be understood in the sense (1) or (6). This leads to one and the same result. The operation $\mu(V\tilde{f})$ (over $V\tilde{f}$) may in every case be understood in the sense (2). For this we need only remark that $\mu(V\tilde{f}) \in S'$, because $\widehat{\mu(V\tilde{f})} \in L_p$.

It is important that the equation

$$V^{-1}\mu V\tilde{f} = (V^{-1}\mu V)\tilde{f} = \mu\tilde{f} \tag{12}$$

holds for every $f \in L_p$. Indeed, if f is a finite function, then it reduces to the corresponding obvious equation between ordinary functions. If now $f \in L_p$, then, as we know, it is possible to select a sequence of infinitely differentiable finite functions f_l such that at the same time $\mu\tilde{f}_l \to \mu\tilde{f}$ and $(\mu V)\tilde{f}_l \to (\mu V)\tilde{f}$ weakly ($\widehat{\mu V} \in L!$). But then, taking account of the fact that V^{-1} is an infinitely differentiable function of polynomial growth, and $V^{-1}\mu V\tilde{f}_l \to V^{-1}\mu V\tilde{f}$ weakly. Therefore equation (12) may be obtained by a passage to the limit as $l \to \infty$ of the already established equation

$$(V^{-1}\mu V\tilde{f}_l, \varphi) = (\mu\tilde{f}_l, \varphi) \ (\varphi \in S).$$

1.5.1.1. General definition of a multiplicator in $L_p(1 \leqq p \leqq \infty)$. Suppose that $\mu = \mu(\boldsymbol{x})$ is a bounded function measurable on $\mathbb{R} = \mathbb{R}_n$, so that $\mu \in S'$.

We emphasize that if $f \in S$, then $\tilde{f} \in S$ is an infinitely differentiable function of polynomial growth. Accordingly, the product $\mu\tilde{f} \in S'$ is defined:

$$(\mu\tilde{f}, \varphi) = (\mu, \tilde{f}\varphi), \tag{1}$$

which is represented by the measurable function

$$\mu\tilde{f} = \mu(\boldsymbol{x})\,\tilde{f}(\boldsymbol{x}).$$

By definition the function μ is said to be a *multiplicator in* $L_p(1 \leqq p \leqq \infty)$, if it is measurable and bounded on $\mathbb{R}$ and if for any infinitely differentiable function (or, what is the same thing, for any function $f \in S$) the equation

$$\|\widehat{\mu\tilde{f}}\|_p \leqq c_p\|f\|_p \tag{2}$$

is satisfied, where the constant c_p does not depend on f.

If now $f \in L_p$ and f_l is an infinitely differentiable finite function for which $\|f - f_l\| \to 0$ $(l \to \infty)$, then it follows from (2) that

$$\|\widehat{\mu\tilde{f}_k} - \widehat{\mu\tilde{f}_l}\|_p \leqq c_p\|f_k - f_l\|_p \to 0 \ (k, l \to \infty).$$

Accordingly, there exists a function $F \in L_p$ to which, as $l \to \infty$, the series $\widehat{\mu\tilde{f}_l}$ tends. It is natural to denote it by

$$F = \widehat{\mu\tilde{f}} = \hat{\mu} * f, \tag{3}$$

calling $\hat{\mu} * f$ the convolution of the function $\hat{\mu}$ (generally speaking generalized) with f. The second term in (3) means that we define $\mu\tilde{f}$ by means of the equation

$$\mu\tilde{f} = \widetilde{\hat{\mu} * f}, \tag{4}$$

where $\hat{\mu} * f$ is understood in the way indicated above. Thus we have defined the product $\mu\tilde{f}$ for functions $f \in L_p$ $(1 \leqq p \leqq \infty)$. For $p = \infty$ this definition already does not go through, since a function bounded on $\mathbb{R}$ cannot be approximated arbitrarily well in the metric of L_∞ by finite functions. But for our needs the definition of $\hat{\mu}\tilde{f}$ introduced in the preceding section will be fully satisfactory in the case $p = \infty$, when $\hat{\mu} = K \in L$.

A multiplicator μ (satisfying property (2)) will be called a *Marcinkiewicz multiplier* (see further 1.5.3).

Obviously,

$$\|\widehat{\mu\tilde{f}}\|_p \leqq c_p \|f\|_p \tag{5}$$

for all $f \in L_p (1 \leqq p < \infty)$, where the c_p is the same constant as in the corresponding inequality for $f \in S$.

A function μ for which $\hat{\mu} \in L$ is obviously a multiplicator in the sense of the definition just now indicated, because (see 1.5.1.) for infinitely differentiable finite functions f

$$\widehat{\mu\tilde{f}} = \frac{1}{(2\pi)^{n/2}} \int \hat{\mu}(\boldsymbol{x} - \boldsymbol{u})\, f(\boldsymbol{u})\, d\boldsymbol{u}, \tag{6}$$

from which (2) immediately follows, where

$$c_p = \frac{1}{(2\pi)^{n/2}} \|\hat{\mu}\|_L .$$

This definition for $\hat{\mu} \in L$ is equivalent to the corresponding definition of multiplicator introduced in 1.5.1. To say that the function $\widehat{\mu\tilde{f}} = \hat{\mu} * f$ is defined (for $f \in L_p$, $1 \leqq p < \infty$, and $\hat{\mu} \in L$) as an integral (6), or as a limit in the metric of L_p of such an integral, calculated for an infinitely differentiable finite function f_l, when $\|f - f_l\| \to 0$, is, obviously, all the same.

But we have generalized here the concept of multiplicator and convolution, since $\hat{\mu}$ may fail to belong to L and be a generalized (not ordinary) function.

We note that if $f \in L_p (1 \leqq p < \infty)$ and f_l are infinitely differentiable finite functions for which $\|f_l - f\|_p \to 0$ $(l \to \infty)$, and μ a multiplicator, then

$$\|\widehat{\mu\tilde{f}} - \widehat{\mu\tilde{f}_l}\|_p \leqq c_p \|f - f_l\|_p \to 0,$$

from which it follows that

(7) $$\mu\tilde{f}_l \to \mu\tilde{f}(S').$$

If μ is a Marcinkiewicz multiplier and is at the same time an infinitely differentiable function of polynomial growth, then for $f \in L_p$ and a sequence $\{f_l\}$ of infinitely differentiable finite functions for which (7) is satisfied, we will have

(8) $$(\mu\tilde{f}, \varphi) = \lim_{l\to\infty} (\mu\tilde{f}_l, \varphi) = \lim_{l\to\infty} (\tilde{f}_l, \mu\varphi) = (\tilde{f}, \mu\varphi).$$

In the first term of (8) $\mu\tilde{f}$ is understood in the sense (4). In the second equation (8) the transfer of μ across the comma is legitimate, since μ is an infinitely differentiable function of polynomial growth. The last equation is based on the fact that $f_l \to f(S)$ and $\mu\varphi \in S$.

Equation (8) shows that if μ is a Marcinkiewicz multiplier and at the same time is an infinitely differentiable function of polynomial growth, then the definitions of $\mu\tilde{f}$ corresponding to these facts for $f \in L_p$ $(1 \leq p < \infty)$ do not contradict one another.

We shall show that along with λ and μ the product $\lambda\mu$ is a multiplicator in L_p and that the following equation holds:

(9) $$\lambda\mu\tilde{f} = \mu\lambda\tilde{f} = (\lambda\mu)\tilde{f} \; (f \in L_p, 1 \leq p < \infty).$$

Indeed, we begin with the assertion, interesting in itself, that if the function $\lambda(\boldsymbol{x})$ is measurable and bounded, and the function F belongs not only to L_p but also to L_2, then

(10) $$\lambda\tilde{F} = \lambda(\boldsymbol{x})\,\tilde{F}(\boldsymbol{x}),$$

i.e. the product $\lambda\tilde{F}$, understood in the sense (4), represents the ordinary product of the functions $\lambda(\boldsymbol{x})$ and $\tilde{F}(\boldsymbol{x})$. Indeed, since $F \in L_2$, then there exists a sequence of infinitely differentiable finite functions $F_k (k = 1, 2, \ldots)$ such that (see subsection 1.4.1)

$$\|F - F_k\|_p \to 0,$$

$$\|F - F_k\|_2 \to 0.$$

For the infinitely differentiable finite functions F_k we have

(11) $$\lambda\tilde{F}_k = \lambda(\boldsymbol{x})\,\tilde{F}_k(\boldsymbol{x})$$

by definition. On the other hand,

$$\widehat{\lambda\tilde{F}_k} \to \widehat{\lambda\tilde{F}}(L_p),$$

and accordingly in S', which means that

(12) $$\lambda\tilde{F}_k \to \lambda\tilde{F}(S').$$

As a consequence of the boundedness of λ and on the basis of the Parseval equation

$$\|\lambda(\boldsymbol{x})\, \tilde{F}_k(\boldsymbol{x}) - \lambda(\boldsymbol{x})\, \tilde{F}(\boldsymbol{x})\|_2 \leqq c\|\tilde{F}_k - \tilde{F}\|_2 = c\|F_k - F\|_2 \to 0,$$

from which it follows that

$$\lambda(\boldsymbol{x})\, \tilde{F}_k(\boldsymbol{x}) \to \lambda(\boldsymbol{x})\, \tilde{F}(\boldsymbol{x})\ (S'). \tag{13}$$

It follows from (11)—(13) obviously that our assertion (10) holds.

Now suppose we are given an arbitrary finite function $f \in S$. We put

$$F = \widehat{\mu \tilde{f}}.$$

Since $\tilde{f} \in L_2$, then in view of the boundedness of μ, also $\mu\tilde{f} \in L_2$ and $F \in L_2$. Accordingly, in view of (10)

$$\lambda\mu\tilde{f} = \lambda(\boldsymbol{x})\, (\mu\tilde{f})\, (\boldsymbol{x}) = \lambda(\boldsymbol{x})\, \mu(\boldsymbol{x})\, \tilde{f}(\boldsymbol{x}) = \tag{14}$$

$$= \mu(\boldsymbol{x})\, \lambda(\boldsymbol{x})\, \tilde{f}(\boldsymbol{x}) = \big(\lambda(\boldsymbol{x})\, \mu(\boldsymbol{x})\big)\, \tilde{f}(\boldsymbol{x}) = (\lambda\mu)\, \tilde{f}.$$

In this way equation (9) is proved to this point for $f \in S$. Therefore for $f \in S$, taking account of the fact that λ and μ are multiplicators,

$$\|\widehat{(\lambda\mu)\, \tilde{f}}\|_p = \|\widehat{\lambda(\mu\tilde{f})}\|_p \leqq c\|\widehat{\mu\tilde{f}}\|_p \leqq cc'\, \|f\|_p,$$

and we have proved that $\lambda\mu$ is also a multiplicator. It remains to prove equation (9) for arbitrary functions $f \in L_p$ $(1 \leqq p < \infty)$. To this end we need to take a sequence of infinitely differentiable finite functions f_l, converging to f in the metric of L_p, substitute f_l in the place of f in (14), apply the operation to all the terms of (14), and pass to the limit as $l \to \infty$ in the sense of L_p.

1.5.1.2. **Lemma.** *Suppose that $\boldsymbol{a} \in \mathbb{R}_n$ is a fixed point. Then along with $\mu(\boldsymbol{x})$ the function $\mu(\boldsymbol{x} - \boldsymbol{a})$ is a Marcinkiewicz multiplier, and the following equation holds for all $f \in L_p$:*

$$\widehat{e^{iat}\mu(t)\, \widetilde{e^{-iat}f}} = \widehat{\mu(\boldsymbol{x} - \boldsymbol{a})\, \tilde{f}}\ (\text{for all } f \in L_p). \tag{1}$$

From this it follows that

$$\|\widehat{\mu(\boldsymbol{x} - \boldsymbol{a})\, \tilde{f}}\|_p = \|\widehat{\mu \widetilde{e^{-iat}f}}\|_p \leqq c_p\|e^{-iat}f\|_p = c_p\|f\|_p. \tag{2}$$

Thus, the constant c_p in this inequality is the same as in the corresponding inequality for $\mu(\boldsymbol{x})$.

Proof. Put

(3) $$f_\beta = e^{i\beta t} f(t) \quad (\beta \in \mathbb{R}_n).$$

Then (see 1.5. (18))

$$\tilde{f} = \widetilde{e^{-i\beta t} f_\beta} = \tilde{f}_\beta(x + \beta).$$

Therefore (see again 1.5. (18))

$$\widetilde{\mu(x-a)\tilde{f}} = \widetilde{\mu(x-a)\tilde{f}_{-a}(x-a)} = e^{iat}\widetilde{\mu(x)\tilde{f}_{-a}(x)},$$

from which we get (1) in view of (3).

1.5.2. Periodic functions from L_p^*. The functions

$$\omega_n(x) = \operatorname{sign} \sin(2^{n+1}\pi x) \ (0 \leqq x \leqq 1),$$

$n = 0, 1, \ldots,$ form an orthogonal and normal Rademacher system on $[0, 1]$.

Here and in what follows we will frequently write $A \ll B$ in place of $A \leqq cB$, where c is a constant.

For any double sequence $\{a_{mn}\}$ of complex numbers and $p > 0$ the inequalities

(1) $$\left(\sum |a_{mn}|^2\right)^{p/2} \ll \int_0^1\int_0^1 \left|\sum a_{mn}\omega_m(\theta)\,\omega_n(\theta')\right|^p d\theta\, d\theta' \ll \left(\sum |a_{mn}|^2\right)^{p/2}$$

hold, with constants not depending on the a_{mn}.

Indeed, if s is a natural number, then, making use of the polynomial formula of Newton and the fact that $[\omega_n(\theta)]^l = \omega_n(\theta)$ for odd l, we get, at present for real $a_{m,n}$,

$$\int_0^1\int_0^1 \left|\sum a_{mn}\omega_m(\theta)\,\omega_n(\theta)\right|^{2s} d\theta\, d\theta' = \sum \frac{(2s)!}{(2\alpha_1)!\cdots(2\alpha_{2s})!} a_{m_1n_1}^{2\alpha_1}\cdots a_{m_{2s}n_{2s}}^{2\alpha_{2s}}$$

$$\leqq \frac{(2s)!}{s!\,2^s}\sum \frac{s!}{\alpha_1!\cdots\alpha_{2s}!} a_{m_1n_1}^{2\alpha_1}\cdots a_{m_{2s}n_{2s}}^{2\alpha_{2s}}$$

$$= \frac{(2s)!}{s!\,2^s}\left(\sum a_{mn}^2\right)^s \quad (\alpha_1 + \cdots + \alpha_{2s} = s).$$

But then, if the $a_{mn} = d_{mn} + i\beta_{mn}$ are complex,

(2) $$\int_0^1\int_0^1 |a_{mn}\omega_m(\theta)\,\omega_n(\theta')|^{2s} d\theta\, d\theta'$$

$$\ll \int_0^1\int_0^1 |a_{mn}\omega_n(\theta)\,\omega_n(\theta')|^{2s} d\theta\, d\theta' + \int_0^1\int_0^1 |\beta_{mn}\omega_n(\theta)\,\omega_n(\theta')|^{2s} d\theta\, d\theta'$$

$$\ll \left(\left(\sum d_{mn}^2\right)^{1/2} + \left(\sum \beta_{mn}^2\right)^{1/2}\right) \ll \left(\sum |a_{mn}|^2\right).$$

Therefore for any $p > 0$, if we chose the natural number s such that $2s \geqq p$, we will find using the Hölder inequality that

$$\left(\int_0^1\int_0^1 |\sum a_{mn}\omega_m(\theta)\ \omega_n(\theta')|^p\, d\theta\, d\theta'\right)^{1/p}$$

$$\leqq \left(\int_0^1\int_0^1 |\sum a_{mn}\omega_m(\theta)\ \omega_n(\theta')|^{2s}\, d\theta\, d\theta'\right)^{2s} \ll (\sum |a_{mn}|^2)^{1/2},$$

which proves the second inequality (1). Further, if $p \geqq 2$, then

$$(\sum |a_{mn}|^2)^{1/2} = \left(\int_0^1\int_0^1 |\sum a_{mn}\omega(\theta)\ \omega_n(\theta')|^2\, d\theta\, d\theta'\right)^{1/2},$$

$$\ll \left(\int_0^1\int_0^1 |\sum a_{mn}\omega_m(\theta)\ \omega_n(\theta')|^p\, d\theta\, d\theta'\right)^{1/p},$$

which proves the first inequality (1). It remains to prove it for $p < 2$. In view of (2)

$$\int_0^1\int_0^1 |\sum a_{mn}\omega_m(\theta)\ \omega_n(\theta')|^4\, d\theta\, d\theta' \leqq 3\ (\sum |a_{mn}|^2)^{1/2}.$$

Suppose $s^2 = \sum |a_{mn}|^2$, $S = \sum a_{mn}\omega_m(\theta)\ \omega_n(\theta')$. Let A be a set of points (θ, θ') such that $|S(\theta, \theta')| > \frac{s}{2}$, CA the complement to A in the unit square and $|A|$ and $|CA|$ their measures. Then

$$s^2 = \int_0^1\int_0^1 S^2\, d\theta\, d\theta' = \int\int_A + \int\int_{CA}$$

$$\leqq \frac{1}{4}\, s^2\, |CA| + \sqrt{|A|}\left(\int_0^1\int_0^1 S^4\, d\theta\, d\theta'\right)^{1/2} \leqq \left(\frac{1}{4} + \sqrt{3}\ \sqrt{|A|}\right) s^2.$$

This means that

$$1 < \frac{1}{4} + 2\sqrt{|A|} \quad \text{or} \quad |A| > \frac{1}{8}.$$

Accordingly,

$$\int_0^1\int_0^1 |S|^p\, d\theta\, d\theta' \geqq \frac{1}{8}\cdot\frac{1}{2^p}\, s^p \geqq \frac{1}{32}\, s^p,$$

which proves the first inequality (1) for $p < 2$. This proves inequalities (1) completely.

Analogously one proves inequalities corresponding to (2) in the n-dimensional case:

$$(3) \qquad (\sum |a_{\boldsymbol{k}}|^2)^{1/2} \ll \left(\int_0^1 \cdots \int_0^1 |\sum a_{\boldsymbol{k}} \omega_{\boldsymbol{k}}(\theta)|^p \, d\theta \right)^{1/p} \ll (\sum |a_{\boldsymbol{k}}|^2)^{1/2},$$

where the $\boldsymbol{k} = (k_1, \ldots, k_n)$ are all-possible integer-termed nonnegative vectors and

$$(4) \qquad \omega_{\boldsymbol{k}}(\boldsymbol{\theta}) = \omega_{k_1}(\theta_1), \ldots, \omega_{k_n}(\theta_n).$$

Suppose that $f(t) \in L_p^* = L_p(0, 2\pi)$ $(1 < p < \infty)$ is a function of one variable t of period 2π, decomposed into a Fourier series of the form

$$f(t) = \sum_0^\infty c_k e^{ikt}.$$

It is known (see Zygmund [1], Ch. VII) that for $1 < p < \infty$ this series converges to f in the sense of L_p^*.

Suppose we are given an increasing sequence of natural numbers

$$0 = n_0 < 1 = n_1 < n_2 < \ldots,$$

satisfying the condition

$$(5) \qquad \frac{n_{k+1}}{n_k} > \alpha > 1 \quad (k = 1, 2, \ldots).$$

Introduce the functions

$$\delta(f) = c_0, \ \delta_k(f) = \sum_{n_{k-1}+1}^{n_k} c_\nu e^{i\nu t} \quad (k = 1, 2, \ldots).$$

Then the series

$$f(t) = \sum_0^\infty \delta_k(f)$$

converges to f in the sense of L_p^*. Put further

$$f_1(t) = \sum_0^\infty \varepsilon_k \delta_k(f) \quad (\varepsilon_k = \pm 1; \, k = 0, 1, \ldots),$$

where the numbers $\varepsilon_k = \pm 1$ depend in any way on k. Then, from Littlewood and Paley [4] (see also Zygmund [2], II, XV) we have the inequality

$$(6) \qquad \|f\|_p \ll \|f_1\|_p \ll \|f\|_p$$

with constants depending on α but not on f and the distribution of the ε_k, and with norms taken relative to the period. These assertions easily extend to functions of several variables

$$f(\boldsymbol{x}) = \sum_{\boldsymbol{\nu} \geqq 0} c_{\boldsymbol{\nu}} e^{i\boldsymbol{\nu x}} = \sum_{\boldsymbol{k}} \delta_{\boldsymbol{k}}(f) \in L_p^* \tag{7}$$

$$\left(\boldsymbol{\nu} = (\nu_1, \ldots, \nu_n); \boldsymbol{k} = (k_1, \ldots, k_n)\right),$$

where

$$\delta_{\boldsymbol{k}}(f) = \delta_{k_1 x_1} \ldots \delta_{k_n x_n}(f)$$

and

$$\varepsilon_{\boldsymbol{k}} = \varepsilon_{k_1} \ldots \varepsilon_{k_n}, f_1 = \sum \varepsilon_{\boldsymbol{k}} \delta_{\boldsymbol{k}}(f).$$

Here we are supposing that the ε_s, $s = 1, \ldots, n$, can take on only the values ± 1. For such f and f_1 we also have the inequalities

$$\|f\|_p \ll \|f_1\|_p \ll \|f\|_p \ (1 < p < \infty), \tag{8}$$

where the norms are now taken relative to the n-dimensional period $\{0 < x_j < 2\pi; j = 1, \ldots, n)$. We will suppose that

$$\delta_{\boldsymbol{k}'}(f) = \delta_{k_1 x_1} \ldots \delta_{k_{n-1} x_{n-1}}(f), d\boldsymbol{x}' = dx_1, \ldots, dx_{n-1},$$

$$\varepsilon_{\boldsymbol{k}'} = \varepsilon_{k_1} \ldots \varepsilon_{k_{n-1}}.$$

Therefore, if we accept that inequalities (8) are true for $n - 1$, and suppose that the integrals are taken relative to the corresponding periods, then we obtain

$$\|f\|_p^p = \int dx_n \int |f|^p \, d\boldsymbol{x}' \ll \int dx_n \int |\sum \varepsilon_{\boldsymbol{k}'} \delta_{\boldsymbol{k}'}(f)|^p \, d\boldsymbol{x}' =$$

$$= \int d\boldsymbol{x}' \int |\sum \varepsilon_{\boldsymbol{k}'} \delta_{\boldsymbol{k}'}(f)|^p \, dx_n$$

$$\ll \int d\boldsymbol{x}' \int |\sum \varepsilon_{\boldsymbol{k}} \delta_{\boldsymbol{k}}(f)|^p \, dx_n \ll \int dx_n \int |f|^p \, d\boldsymbol{x}' = \|f\|_p^p,$$

i.e. (8), if we take into account the fact that

$$\sum \varepsilon_{k_n} \delta_{k_n x_n} \sum \varepsilon_{\boldsymbol{k}'} \delta_{\boldsymbol{k}'}(f) = \sum \varepsilon_{\boldsymbol{k}} \delta_{\boldsymbol{k}}(f).$$

Finally, for the functions (7) we may write the inequalities

$$\|f\|_p \ll \|\sum |\delta_{\boldsymbol{k}}^2(f)|^{1/2}\|_p \ll \|f\|_p \ (1 < p < \infty) \tag{9}$$

with constants not depending on f.

Indeed, putting $\Omega = \{0 \leqq \theta_i \leqq 1\}$, we get

$$(10)\qquad \|f\|_p^p = \int_\Omega \|f\|_p^p \, d\theta \ll \int_\Omega \|\sum \omega_{\boldsymbol{k}}(\theta)\, \delta_{\boldsymbol{k}}(f)\|_p^p \, d\theta =$$

$$= \left\| \int_\Omega |\sum \omega_{\boldsymbol{k}}(\theta)\, \delta_{\boldsymbol{k}}(f)|^p \, d\theta \right\|_p^p \ll \|(\sum |\delta_{\boldsymbol{k}}(f)|^2)^{1/2}\|_p^p$$

$$\ll \left\| \int_\Omega |\sum \omega_{\boldsymbol{k}}(\theta)\, \delta_{\boldsymbol{k}}(f)|^p \, d\theta \right\|_p^p = \int_\Omega \|\sum \omega_{\boldsymbol{k}}(\theta)\, \delta_{\boldsymbol{k}}(f)\|_p^p \, d\theta$$

$$\ll \int_\Omega \|f\|_p^p \, d\theta = \|f\|_p^p.$$

The transition from the 2nd to the 3rd and from the 6th to the 7th terms are carried out on the basis of inequality (8) with $\varepsilon_k = \omega_k(\theta)$, and those from the 4th to the 5th and thereupon to the 6th terms on the basis of inequality (3).

It follows from (9) that if $f \in L_p^*$ is a function whose Fourier coefficients $c_{\boldsymbol{k}}$ are not equal to zero unless $\boldsymbol{k} \geqq 0$, and $\mathscr{E}$ is any set of vectors $\boldsymbol{k}$, then the function

$$\sum_{\boldsymbol{k} \in \mathscr{E}} \delta_{\boldsymbol{k}}(f) = \varphi = \sum_{\boldsymbol{k}} \delta_{\boldsymbol{k}}(\varphi)$$

generates the Fourier series of some function $\varphi \in L_p^*$, for which the following inequalities are satisfied:

$$\|\varphi\|_p \ll \left\| \Big(\sum_{\boldsymbol{k} \in \mathscr{E}} |\delta_{\boldsymbol{k}}(f)|^2\Big)^{1/2} \right\|_p \leqq \left\| \Big(\sum_{\boldsymbol{k}} |\delta_{\boldsymbol{k}}^2(f)|^2\Big)^{1/2} \right\|_p \ll \|f\|_p.$$

We note that if an arbitrary periodic function of one variable

$$f(t) = \sum_{-\infty}^{\infty} c_k e^{ikt} = \sum_{k \geqq 0} + \sum_{k<0} = f_+ + f_-$$

lies in L_p^*, then its conjugate (trigonometric) function (for $1 < p < \infty$)

$$(11)\qquad f_*(t) = -i \sum_{-\infty}^{\infty} \operatorname{sign} k c_k e^{ikt}$$

(here sign $0 = 1$), and along with it $if_* = f_+ - f_-$, also belong to L_p^* (Zygmund [1], Ch. VII, 2.5.). Accordingly, $f_+ = \frac{1}{2}(f + if_*) \in L^*$ and $\|f_+\|_p \ll \|f\|_p$. Hence it follows by induction that also for the functions

$$f(\boldsymbol{x}) = \sum c_{\boldsymbol{k}} e^{i\boldsymbol{k}\boldsymbol{x}}$$

of several variables, if one puts

$$f_+ = \sum_{k \geqq 0} c_k e^{ikx},$$

then from the fact that $f \in L_p^* = \overset{(n)}{L_p^*}$ it follows that $\|f_+\|_p \ll \|f\|_p$.

Suppose that $\boldsymbol{k} = (k_1, \ldots, k_n)$ is an arbitrary integer-termed vector, not necessarily nonnegative. Put

$$\text{(12)} \qquad \delta_{\boldsymbol{k}}(f) = \sum_{\pm n_{|k_1|-1}+1}^{\pm n_{|k_1|}} \cdots \sum_{\pm n_{|k_n|-1}+1}^{\pm n_{|k_n|}} c_m e^{imx},$$

where at the jth place one puts in $+$ or $-$ according as to whether $k_j > 0$ or $k_j < 0$. For $k_j = 0$ we need to take $n_{|k_j|-1} + 1 = 0$. On the basis of what was said above it is obvious that along with inequalities (9) we also have the inequalities

$$\text{(13)} \qquad \|f\|_p \ll \left\| \left(\sum |\delta_{\boldsymbol{k}}(f)|^2\right)^{1/2} \right\|_p \ll \|f\|_p$$

analogous to them, for any periodic function $f(\boldsymbol{x})$, not necessarily just one for which $c_{\boldsymbol{k}} \neq 0$ only for $\boldsymbol{k} \geqq 0$.

It is easy to verify that inequalities (13) are preserved also for the functions

$$\text{(14)} \qquad f(\boldsymbol{x}) = \sum c_{\boldsymbol{k}} e^{\frac{i\pi \boldsymbol{k}\boldsymbol{x}}{l}} = \sum \delta_{\boldsymbol{k}}(f),$$

$$c_k = \frac{1}{l^n} \int_{\Delta_l} f(u) e^{-\frac{i\pi \boldsymbol{k}\boldsymbol{u}}{l}} d\boldsymbol{u}, \quad \Delta_l = \{|x_j| \leqq l;\, j = 1, \ldots, n\},$$

of arbitrary period $2l$. Here we need of course the appropriate alteration of the definition of the δ_k (the replacement of $\boldsymbol{x}$ by $\frac{\pi}{l}\boldsymbol{x}$ in the right side of (12)).

It is important to note that the constants appearing in inequalities (13) do not depend on f.

1.5.2.1. Suppose that

$$\text{(1)} \qquad f = \sum_{-\infty}^{\infty} c_k e^{ikt} = \sum_{k \geq 0} + \sum_{k<0} = f_+ + f_- \in L_p^* \quad (1 < p < \infty)$$

is a periodic function of one variable, and that the sequence $\{n_k\}$ and the functions $\delta_k(f)$ $(k = 0, 1, \ldots)$ are defined in the same way as in 1.5.2.

Further we put

$$(2)\qquad \delta_{-k}(f) = \sum_{n_{k-1}+1}^{n_k} c_{-\nu} e^{-i\nu t},$$

$$(3)\qquad \beta_k(f) = \delta_k(f) + \delta_{-k}(f), \quad k = 1, 2, \ldots,$$

$$\beta_0(f) = \delta_0(f) = c_0.$$

Then the following inequalities analogous to them follow from inequalities 1.5.2. (6):

$$(4)\qquad \|f\|_p \ll \|f_*\|_p \ll \|f\|_p, \quad 1 < p < \infty,$$

where

$$(5)\qquad f_* = \sum_0^\infty \varepsilon_k \beta_k(f) \ (\varepsilon_k = \pm 1),$$

with constants depending on α and p but not on f and on the distribution of the ε_k. Indeed,

$$\left(\sum \varepsilon_k \beta_k(f)\right)_+ = \sum \varepsilon_k \delta_k(f_+),$$

$$\left(\sum \varepsilon_k \beta_k(f)\right)_- = \sum \varepsilon_k \delta_k(f_-(-t)).$$

Moreover, $\|f_+\|_p, \|f_-\|_p \ll \|f\|_p$, so that in view of 1.5.2. (6)

$$\|f_+\|_p \ll \|\sum \varepsilon_k \delta_k(f_+)\|_p \ll \|\sum \varepsilon_k \beta_k(f)\|_p,$$

$$\|f_-\|_p = \|f_-(-t)\|_p \ll \left\|\sum \varepsilon_k \delta_k(f_-(-t))\right\|_p \ll \|\sum \varepsilon_k \beta_k(f)\|_p,$$

from which the first inequality in (4) follows.

Further

$$\|\sum \varepsilon_k \beta_k(f)\|_p \leqq \|\sum \varepsilon_k \delta_k(f_+)\|_p + \left\|\sum \varepsilon_k \delta_k(f_-(-t))\right\|_p \ll \|f_+\|_p + \|f_-\|_p \ll \|f\|_p,$$

i.e. the second inequality (4).

From (4) there follow the inequalities

$$\|f\|_p \ll \left\|\left(\sum |\beta_k(f)|^2\right)^{1/2}\right\|_p \ll \|f\|_p,$$

which are proved as in 1.5.2. (10) (replace δ by β).

Another assertion, proved in the book of Zygmund (Ch. XV, 2.15) for the one-dimensional case, may be extended by induction to the multidimensional case. It is the following.

Suppose that $f_1, f_2, \ldots \in L_{p*}$ $(1 < p < \infty)$ is a sequence of functions of $\boldsymbol{x} = (x_1, \ldots, x_n)$ of period 2π with Fourier coefficients $c_{\boldsymbol{k}}$, not equal to zero unless $\boldsymbol{k} \geqq 0$, and suppose that $S_{n,\boldsymbol{k}_n}$ denotes the Fourier sum of the f_n of order $\boldsymbol{k}_n$. Then there exists a constant A_p, not depending on f_n and N and such that

$$\int_0^{2\pi} \cdots \int_0^{2\pi} \left(\sum_{n=1}^N |S_{n,\boldsymbol{k}_n}|^2 \right)^{p/2} dx_1 \ldots dx_n \tag{6}$$

$$\leqq A_p^p \int_0^{2\pi} \cdots \int_0^{2\pi} \left(\sum_1^N |f_n|^2 \right)^{p/2} dx_1 \ldots dx_n .$$

1.5.3. The multiplicator theorem in the periodic case. For a numerical sequence $\{\lambda_l\}$ depending on a single index l, we introduce the difference $\Delta\lambda_l = \lambda_{l+1} - \lambda_l$. For a multiple sequence $\{\lambda_{\boldsymbol{k}}\}$ depending on a nonnegative integer-valued vector $\boldsymbol{k} = (k_1, \ldots, k_n)$, we will consider the difference $\Delta_j \lambda_{\boldsymbol{k}}$, taken relative to the component k_j, and the multiple differences $\Delta_{j_1} \ldots \Delta_{j_m} \lambda_{\boldsymbol{k}}$ $(m \leqq n)$.

Theorem (Marcinkiewicz). *Suppose given a multiple sequence* $\{\lambda_{\boldsymbol{k}}\}$, *subjected to the inequalities*

$$|\lambda_{\boldsymbol{k}}| \leqq M, \quad \sum_{\nu_{j_1} = \pm 2^{|k_1|-1}}^{\pm 2^{|k_1|}-1} \cdots \sum_{\nu_{j_m} = \pm 2^{|k_m|-1}}^{\pm 2^{|k_m|}-1} |\Delta_{j_1} \ldots \Delta_{j_m} \lambda_{\nu}| \leqq M \tag{1}$$

for arbitrary collections of natural numbers $j_1, \ldots, j_m$ *such that* $1 \leqq j_1 < j_2 < \ldots < j_m \leqq n$, *where* M *is a constant not depending on* $\boldsymbol{k}$ *and* $j_1, \ldots, j_m$ (*for* $k_j = 0$ *the corresponding sum is extended only to* $\nu_j = 0$); $+$ *or* $-$ *is entered according as* k_j *is positive or negative.*

Transform a function of period 2π *of the type* (see 1.5.2. (7))

$$f(\boldsymbol{x}) = \sum_{\boldsymbol{k}} c_{\boldsymbol{k}} e^{i\boldsymbol{k}\boldsymbol{x}} = \sum \delta_{\boldsymbol{k}}(f) \in L_p* \quad (1 < p < \infty) \tag{2}$$

by means of the Marcinkiewicz multipliers $\lambda_{\boldsymbol{k}}$:

$$F(\boldsymbol{x}) = \sum_{\boldsymbol{k}} \lambda_{\boldsymbol{k}} c_{\boldsymbol{k}} e^{i\boldsymbol{k}\boldsymbol{x}} = \sum \delta_{\boldsymbol{k}}(F) . \tag{3}$$

Then $F \in L_p^*$ *and there exists a constant* c_p *depending only on* p *such that*

$$\|F\|_p \leqq c_p M \, \|f\|_p . \tag{4}$$

Proof. We restrict ourselves to the case $n = 2$. Moreover, we will suppose that in (2) the series are extended only over $\boldsymbol{k} \geqq 0$, which does not affect the generality.

Putting

$$\sum_{\mu=2^{k-1}}^{s} \sum_{\nu=2^{l-1}}^{t} c_{\mu\nu} e^{i(\mu x + \nu y)} = r_{st} = r_{s,t,k,l} \tag{5}$$

and applying the Abel transformation, we get

$$(6)\qquad \delta_{kl}(F) = \sum_{2^{k-1}}^{2^k-1} \sum_{2^{l-1}}^{2^l-1} \lambda_{\mu\nu} c_{\mu\nu} e^{i(\mu x+\nu y)}$$

$$= \sum_{2^{k-1}}^{2^k-2} \sum_{2^{l-1}}^{2^l-2} r_{ij}[\lambda_{i,j} - \lambda_{i,j+1} - \lambda_{i+1,j} + \lambda_{i+1,j+1}]$$

$$+ \sum_{2^{l-1}}^{2^l-2} r_{2^k-1,j}[\lambda_{2^k-1,j} - \lambda_{2^k-1,j+1}] + \sum_{2^{k-1}}^{2^k-2} r_{i,2^l-1}[\lambda_{i,2^l-1} - \lambda_{i+1,2^l-1}]$$

$$+ r_{2^k-1,2^l-1}\lambda_{2^k-1,2^l-1} = \sum r_{ij}\gamma_{ij}.$$

We apply the Schwartz-Bunjakovskiĭ inequality and take account of (1):

$$|\delta_{kl}(F)|^2 = (\sum r_{ij}\gamma_{ij})^2 \leqq \sum |\gamma_{ij}| \sum r_{ij}^2 |\gamma_{ij}| \leqq M \sum r_{ij}^2 |\gamma_{ij}|.$$

Therefore on the basis of 1.5.2 (13) it follows ($n_k = 2^k$) that:

$$(7)\quad \|F_p\|^p \ll \Big\| \Big(\sum_{k,l} |\delta_{k,l}(F)|^2\Big)^{1/2} \Big\|_p^p \ll M^{p/2} \int_0^{2\pi}\int_0^{2\pi} \Big\{\sum_{k,l} \sum \big(r_{ij} \sqrt{|\gamma_{ij}|}\big)^2\Big\}^{p/2} dx\, dy.$$

In the curly brackets on the right side of (7) there appears the function (see also (5))

$$(8)\qquad r_{i,j,k,l} \sqrt{|\gamma_{ij}|},$$

where $\sqrt{|\gamma_{ij}|}$ is a coefficient not depending on x, y. It obviously may be considered as a segment of the Fourier series of the function

$$(9)\qquad \delta_{kl}(f) \sqrt{|\gamma_{ij}|}.$$

Accordingly, the quantity in curly brackets on the right side of (7) is a sum $\sum\limits_{k,l}\sum$ of squares of segments of the Fourier series of the functions (9).

On the basis of 1.5.2.1 (6) the integral (7) is majorized by the same integral, where the finite Fourier sums for the functions are replaced by the functions themselves:

$$\|F_p\|^p \ll M^{p/2} \int_0^{2\pi}\int_0^{2\pi} \Big\{\sum_{k,l} |\delta_{k,l}(f)|^2 \sum |\gamma_{ij}|\Big\}^{p/2} dx\, dy$$

$$\ll M^p \int_0^{2\pi}\int_0^{2\pi} \Big\{\sum_k \sum_l |\delta_{k,l}(f)|^2\Big\}^{p/2} dx\, dy \ll M^p \|f\|_p^p,$$

and we have obtained inequality (4).

It is easy to verify that inequality (4) remains valid for functions of arbitrary period $2l$ with the same constant c_p.

1.5.4. The multiplicator theorem in the nonperiodic case. We suppose given a vector of the form

$$\boldsymbol{k} = (k_1, \ldots, k_n) \quad (k_j = 0, 1; \; j = 1, \ldots, n). \tag{1}$$

The *support* of the vector $\boldsymbol{k}$ is the set

$$e_k = \{j_1, \ldots, j_m\}$$

of those indices j for which $k_j = 1$.

Theorem. *Suppose given on* $\mathbb{R} = \mathbb{R}_n$ *a function* $\lambda(\boldsymbol{x})$, *having the following properties.*

For any vector $\boldsymbol{k}$ *of the type* (1), *the derivative*

$$D_{\boldsymbol{k}}\lambda = \frac{\partial^{|\boldsymbol{k}|}\lambda}{\partial x_1^{k_1} \ldots \partial x_n^{k_n}} \tag{2}$$

exists and is continuous at any point $\boldsymbol{x} = (x_1, \ldots, x_n)$ *with* $x_i \neq 0$, *where* $i \in e_{\boldsymbol{k}}$, *and it satisfies the inequality*

$$|\boldsymbol{x}^{\boldsymbol{k}} D^{\boldsymbol{k}} \lambda| \leqq M. \tag{3}$$

Then λ *is a Marcinkiewicz multiplier. That is, there exists a constant* $\varkappa_p$ *not depending on* M *and* f *such that*

$$\|\widehat{\lambda \tilde{f}}\|_p \leqq x_p M \, \|f\|_p \quad (1 < p < \infty) \tag{4}$$

for all $f \in L_p$.

This theorem may be somewhat generalized in terms of generalized derivatives.

We note that since λ satisfies the property indicated in the theorem for $\boldsymbol{k} = 0$, then it is bounded on $\mathbb{R}$ and continuous except at points belonging to the coordinate planes. Therefore λ is a measurable function on $\mathbb{R}_n$ and it is at the same time a generalized function ($\lambda \in S'$).

Proof. We restrict ourselves to the two-dimensional case. Suppose that $f(x, y)$ is a finite infinitely differentiable function. We shall suppose that its support belongs to the square

$$\Delta_{s_0} = \{|x|, |y| < s_0\}. \tag{5}$$

Now suppose that

$$f(x, y) = \sum_{\mu, \nu} c_{\mu\nu}^s e^{i\frac{\pi}{s}(\mu x + \nu y)} \quad (s \geqq s_0), \tag{6}$$

where

$$(7)\qquad c^s_{\mu\nu} = \frac{1}{(2s)^2} \iint_{\Delta_s} f(u, v)\, e^{i\frac{\pi}{s}(\mu u + \nu v)}\, du\, dv \quad (\mu, \nu = 0, \pm 1, \ldots),$$

is its Fourier series. Put

$$(8)\qquad u_s(x, y) = \sum_{\mu,\nu} \lambda\left(\frac{\mu\pi}{s}, \frac{\nu\pi}{s}\right) c^s_{\mu\nu} e^{i\frac{\pi}{s}(\mu x + \nu y)}.$$

In view of (3) (for $\boldsymbol{k} = \boldsymbol{0}$)

$$(9)\qquad \left|\lambda\left(\frac{\mu\pi}{s}, \frac{\nu\pi}{s}\right)\right| \leqq M.$$

Now suppose that $k > 0$, $l \geqq 0$. Then

$$\sum_{\mu=2^{k-1}}^{2^k-1} \left|\lambda\left(\frac{(\mu+1)\pi}{s}, \frac{l\pi}{s}\right) - \lambda\left(\frac{\mu\pi}{s}, \frac{l\pi}{s}\right)\right| = \sum \left| \int_{\frac{\mu\pi}{s}}^{\frac{(\mu+1)\pi}{s}} \frac{\partial\lambda}{\partial x}\left(\xi, \frac{l\pi}{s}\right) d\xi \right|$$

$$(10)\qquad \leqq \frac{s}{\pi \cdot 2^{k-1}} \sum \int_{\frac{\mu\pi}{s}}^{\frac{(\mu+1)\pi}{s}} \left|\xi \frac{\partial\lambda}{\partial\xi}\left(\xi, \frac{l\pi}{s}\right)\right| d\xi \leqq \frac{1}{2^{k-1}}\, 2^k M \leqq 2M.$$

In these calculations we used the continuity of $\dfrac{\partial\lambda}{\partial x}$ relative to x for $x > 0$ and any y. The resulting inequality holds also for $k = 0$ for any l:

$$\left|\lambda\left(\frac{\pi}{s}, \frac{l\pi}{s}\right) - \lambda\left(0, \frac{l\pi}{s}\right)\right| \leqq \left|\lambda\left(\frac{\pi}{s}, \frac{l\pi}{s}\right)\right| + \left|\lambda\left(0, \frac{l\pi}{s}\right)\right| \leqq 2M.$$

Analogously, using the continuity of $\dfrac{\partial\lambda}{\partial y}$ relative to y for $y > 0$ and any x, we get

$$(11)\qquad \sum_{\nu=2^{l-1}}^{2^l-1} \left|\lambda\left(\frac{k\pi}{s}, \frac{(\nu+1)\pi}{s}\right) - \lambda\left(\frac{k\pi}{s}, \frac{\nu\pi}{s}\right)\right| \leqq 2M \quad (k, l \geqq 0).$$

Further, still for $k, l > 0$,

$$\text{(12)}\quad \sum_{2^{k-1}}^{2^k-1}\sum_{2^{l-1}}^{2^l-1}\left|\lambda\left(\frac{(\mu+1)\pi}{s}, \frac{(\nu+1)\pi}{s}\right) - \lambda\left(\frac{\mu\pi}{s}, \frac{(\nu+1)\pi}{s}\right) - \lambda\left(\frac{(\mu+1)\pi}{s}, \frac{\nu\pi}{s}\right) + \lambda\left(\frac{\mu\pi}{s}, \frac{\nu\pi}{s}\right)\right|$$

$$= \sum\sum\left|\int_{\frac{\mu\pi}{s}}^{\frac{(\mu+1)\pi}{s}}\int_{\frac{\nu\pi}{s}}^{\frac{(\nu+1)\pi}{s}} \frac{\partial^2\lambda}{\partial x\partial y}(\xi, \eta)\, d\xi\, d\eta\right|$$

$$\leqq \left(\frac{s}{\pi}\right)^2 \frac{\lambda}{2^{k+l-2}}\sum\sum\int_{\frac{\pi\pi}{s}}^{\frac{(\mu+1)\pi}{s}}\int_{\frac{\nu\pi}{s}}^{\frac{(\nu+1)\pi}{s}}\left|\xi\eta\frac{\partial^2\lambda}{\partial x\partial y}(\xi, \eta)\right| d\xi\, d\eta \leqq M.$$

Here we have used the continuity of $\dfrac{\partial^2\lambda}{\partial x\partial y}$ for $x, y > 0$. For $k > 0$ and $l = 0$ this inequality reduces to the following:

$$\sum_{2^{k-1}}^{2^k-1}\left|\lambda\left(\frac{(\mu+1)\pi}{s}, \frac{\pi}{s}\right) - \lambda\left(\frac{\mu\pi}{s}, \frac{\pi}{s}\right) - \lambda\left(\frac{(\mu+1)\pi}{s}, 0\right) + \lambda\left(\frac{\mu\pi}{s}, 0\right)\right| \leqq 4M \quad \text{(13)}$$

Here we have taken account of inequality (10), valid for any $k, l \geqq 0$ (see (10) and the paragraph following it). For $k = 0, l = 0$ the sum on the left side reduces to one term also not exceeding $4M$.

We have proved that the left sides of (9)—(12) for any $k, l \geqq 0$ do not exceed $4M$. Analogous inequalities may be proved for the three remaining quadrants: 1) $k \geqq 0, l \leqq 0$; 2) $k \leqq 0, l \geqq 0$; 3) $k, l \leqq 0$.

This proves that the Marcinkiewicz conditions are met, so that there exists a constant c_p not depending on s, M and f (see the remark at the end of subsection 1.5.3.), such that

$$\text{(14)}\quad \|u_s\|_{L_p(\Delta_s)} \leqq c_p M\,\|f\|_{L_p(\Delta_s)} = c_p M\,\|f\|_p,\ 1 < p < \infty.$$

In the case at hand the transform of the function f by means of the multiplier λ may be written in the form of an integral

$$u(x, y) = \frac{1}{2\pi}\int_{-\infty}^{\infty}\int_{-\infty}^{\infty}\lambda(\xi, \eta)\,\tilde{f}(\xi, \eta)\,e^{i(x\xi+y\eta)}\,d\xi, d\eta,$$

where

$$\tilde{f}(x, y) = \frac{1}{2\pi} \int_{\Delta_s} f(u, v)\, e^{-i(xu+yv)}\, du\, dv \quad (s \geqq s_0).$$

Obviously

(15) $$c_{kl}^{s} = \frac{\pi}{2s^2} \tilde{f}\left(\frac{k\pi}{s}, \frac{l\pi}{s}\right).$$

We shall estimate on an arbitrary fixed square $\Delta_\mu (\mu > 0)$ the difference

$$u(x, y) - u_s(x, y) = r_1 + r_2 + r_3.$$

Here

(16) $$r_1 = \frac{1}{2\pi} \int_{\Delta_N} \lambda(\xi, \eta)\, f(\xi, \eta)\, e^{i(x\xi+y\eta)}\, d\xi\, d\eta$$

$$- \sum_{|k|,|l| \leqq \alpha N} \lambda\left(\frac{k\pi}{s}, \frac{l\pi}{s}\right) c_{kl}^{s} e^{i\frac{\pi}{s}(kx+ly)}$$

$$= \frac{1}{2\pi} \int_{\Delta_N} \lambda(\xi, \eta)\, \tilde{f}(\xi, \eta)\, e^{i(x\xi+y\eta)}\, d\xi\, d\eta$$

$$- \frac{\pi}{2s^2} \sum_{|k|} \sum_{|l| \leqq \alpha N} \lambda\left(\frac{k\pi}{s}, \frac{l\pi}{s}\right) \tilde{f}\left(\frac{k\pi}{s}, \frac{l\pi}{s}\right) e^{i\frac{\pi}{s}(kx+ly)},$$

N being a natural number, and s chosen in such a way that $\alpha = \dfrac{s}{\pi}$ is a nytural number:

$$r_2 = \frac{1}{2\pi} \int_{\mathbb{R}_2 - \Delta_N} \lambda(\xi, \eta)\, \tilde{f}(\xi, \eta)\, e^{i(x\xi+y\eta)}\, d\xi,\, d\eta,$$

$$r_3 = -\frac{\pi}{2s^2} \sum{}' \lambda\left(\frac{k\pi}{s}, \frac{l\pi}{s}\right) \tilde{f}\left(\frac{k\pi}{s}, \frac{l\pi}{s}\right) e^{i\frac{\pi}{s}(kx+ly)},$$

where the sum $\sum'$ is extended over pairs (k, l) such that either $|k|$ or $|l|$ is larger than αN. The function f, being an infinitely differentiable finite function, lies in the fundamental class S. Therefore also $\tilde{f} \in S$, which means that

$$|\tilde{f}(\xi, \eta)| \leqq (1 + \xi^2)^{-1} (1 + \eta^2)^{-1}$$

and

$$|r_2| \leqslant \int\int\limits_{\mathbb{R}_2 - \Delta_N} (1+\xi^2)^{-1}(1+\eta^2)^{-1}\,d\xi\,d\eta \to 0 \quad (N \to \infty).$$

For $\alpha = \dfrac{s}{\pi} > 1$ the similar estimate holds for r_3:

$$|r_3| \leqslant \sum{}' \left[1 + \left(\frac{k\pi}{s}\right)^2\right]^{-1} \left[1 + \left(\frac{l\pi}{s}\right)^2\right]^{-1} \to 0 \quad (N \to \infty).$$

Suppose given an $\varepsilon > 0$. It is then possible to find an $N > 0$ such that for all $s > s_0$

$$|r_2|, |r_2| < \varepsilon.$$

For this N it is possible to indicate an $s_1 > s_0$ such that for all $s > s_1$ and for all $(x, y) \in \Delta_\mu$

$$|r_1| < \varepsilon.$$

We have proved that for any $\mu > 0$

$$\lim_{s\to\infty} u_s(x, y) = u(x, y) = \widehat{\lambda\tilde{f}}$$

uniformly on Δ_μ.

It follows from (14) that

$$\|u_s\|_{L_p(\Delta_\mu)} \leqq c_p \|f\|_p \quad (\mu \leqq s).$$

Passing to the limit as $s \to 0$, and then as $\mu \to \infty$, we obtain inequality (4) for finite functions $f \in S$.

Thus we have proved that λ is a Marcinkiewicz multiplier.

1.5.4.1. If the function $u(\boldsymbol{x}) = u(x_1, \ldots, x_n)$ of n variables satisfies the conditions of the theorem formulated in 1.5.4., then it also satisfies the conditions of this theorem if it is considered as a function of the k variables $x_1, \ldots, x_k$ $(k < n)$, and accordingly it is a multiplicator relative to them.

1.5.5. Examples of Marcinkiewicz multipliers (in the sense of L_p, $1 < p < \infty$).

1. $\operatorname{sign} \boldsymbol{x} = \prod\limits_1^n \operatorname{sign} x_j$.

2. $(1 + |\boldsymbol{x}|^2)^{-\lambda}$ $(\lambda > 0)$.

3. $(1 + x_j^2)^{r/2}(1 + |\boldsymbol{x}|^2)^{-r/2}$ $(r > 0;\ j = 1, \ldots, n)$.

4. $(1+|\boldsymbol{x}|^2)^{r/2}\left(1+\sum_1^n |x_j|^r\right)^{-1} \quad (r>0)$.

5. $\boldsymbol{x}^{\boldsymbol{l}}(1+|\boldsymbol{x}|^2)^{r/2} \quad (|\boldsymbol{l}| \leqq r,\ r>0,\ \boldsymbol{l} \geqq \boldsymbol{0})$.

6. $\boldsymbol{x}^{\boldsymbol{l}}(1+x_s^2)^{\frac{\varkappa r_s}{2}}\left\{\sum_1^n (1+x_j^2)^{\frac{r_j}{2}}\right\}^{-1} \left(\boldsymbol{r}>0,\ \boldsymbol{l} \geqq 0,\ \varkappa = 1-\sum_1^n \frac{l_j}{r_j} \geqq 0\right)$.

7. $(1+|\boldsymbol{x}|^2)^{r/2}\left(1+\sum_1^n |x_j|\right)^{-r}$

(r an arbitrary real number)

8. $(1+|\boldsymbol{x}|^2)^{-r/2}\,\Lambda_{\boldsymbol{r}}^{-1}(\boldsymbol{x}) \quad (r=r_1=\cdots=r_n>0)$.

9. $(1+|\boldsymbol{x}|^2)^{r/2}\,\Lambda_{\boldsymbol{r}}(\boldsymbol{x}) \quad (r=r_1=\cdots=r_n>0)$.

10. $(1+x_i^2)^{\frac{r_i}{2}}\,\Lambda_{\boldsymbol{r}}(\boldsymbol{x}) \quad (i=1,\ldots,n)$.

11. $\left\{\sum_1^n (1+x_j^2)^{\frac{r_j}{2}}\right\}^{-1} \Lambda_{\boldsymbol{r}}^{-1}(\boldsymbol{x})$.

12. $\left\{\sum_1^n (1+x_j^2)^{\frac{r_j\lambda}{2\sigma}}\right\}^{\sigma}\left\{\sum_1^n (1+x_j^2)^{\frac{r_j\delta}{2\sigma}}\right\}^{\sigma}$

$$\times\left\{\sum_1^n (1+x_j^2)^{\frac{r_j(\lambda+\delta)}{2\sigma}}\right\}^{-\sigma} \quad (r_j>0;\ \sigma,\delta,\lambda>0).$$

$$\Lambda_{\boldsymbol{r}}(\boldsymbol{x}) = \left\{\sum_{j=1}^n (1+x_j^2)^{\frac{r_j\sigma}{2}}\right\}^{-\frac{1}{\sigma}}.$$

These functions will be denoted by μ_i, $i=1,\ldots,12$. We shall need them in what follows. The proof that they are Marcinkiewicz multipliers reduces to Theorem 1.5.4 above.

Its criteria are satisfied trivially for μ_1, since μ_1 is a constant ($+1$ or -1) in each open coordinate angle.

The functions μ_i are continuous along with their partial derivatives of any order on $\mathbb{R}=\mathbb{R}_n$, with the exclusion of the cases $i=4,5,6,7$, which are continuous on $\mathbb{R}$ but whose partial derivatives are in general discontinuous on the coordinate planes.

Below we give the proof that the Marcinkiewicz criterion is satisfied for the functions μ_i. The question reduces to the verification of the fact that the functions

$$\boldsymbol{x}^{\boldsymbol{k}}\mu_i^{(\boldsymbol{k})} \quad (\boldsymbol{k} = (k_1, \ldots, k_n)), \; k_j = 0, 1)$$

are bounded on each coordinate angle of the space $\mathbb{R}$. As a consequence of the symmetry of these functions, it suffices to carry out such a verification for the positive coordinate angle. All the functions in question except for μ_6 and μ_{12}, are products $\mu_i = \lambda_i \psi_i$ of certain functions λ_i and ψ_i. By the Leibniz formula

$$\boldsymbol{x}^{\boldsymbol{k}}\mu_i^{(\boldsymbol{k})} = \boldsymbol{x}^{\boldsymbol{k}} \sum_{\alpha \leqq \boldsymbol{k}} C_{\boldsymbol{k}}^{\alpha} \lambda_i^{(\boldsymbol{k}-\alpha)} \psi_i^{(\alpha)},$$

where the sum is extended over all possible integer nonnegative vectors $\boldsymbol{\alpha} \leqq \boldsymbol{k}$. The question has been reduced to the estimation of functions of the type

$$\boldsymbol{x}^{\boldsymbol{k}} \lambda_i^{(\boldsymbol{k}-\alpha)} \psi_i^{(\alpha)} = \boldsymbol{x}^{\boldsymbol{k}} \chi_i$$

on the positive coordinate angle.

We agree to write $A \approx B$ in place of $|A| = cB$, where c is some constant. We have

$$\boldsymbol{x}^{\boldsymbol{k}} \chi_4 \approx \boldsymbol{x}^{\boldsymbol{k}} \boldsymbol{x}^{\boldsymbol{k}-\alpha} (1 + |\boldsymbol{x}|^2)^{r/2 - |\boldsymbol{k}-\alpha|} \boldsymbol{x}^{(r-1)\alpha} \left(1 + \sum_1^n x_j^r\right)^{-1-|\alpha|}$$

$$= \frac{\boldsymbol{x}^{\alpha r}}{\left(1 + \sum_1^n x_j^r\right)^{|\alpha|}} \frac{\boldsymbol{x}^{2(\boldsymbol{k}-\alpha)}}{(1 + |x|^2)^{|\boldsymbol{k}-\alpha|}} \frac{(1 + |\boldsymbol{x}|^2)^{r/2}}{1 + \sum_1^n x_j^r} \leqq 1 \cdot 1 \cdot c < \infty.$$

For $r < 1$ the function χ_4 is discontinuous when one of the coordinates x_j, where $j \in e_\alpha \subset e_k$ (e_α the support of the vector $\boldsymbol{\alpha}$), is equal to zero. But by Theorem 1.5.4. it is sufficient that the function χ_4 should be continuous for positive x_j with $j \in e_k$ and arbitrary remaining x_j, which, obviously, is the case in the situation at hand.

For the estimation of $\mu_6 = uvw$ we will have

$$(\boldsymbol{x})^{\boldsymbol{k}} (uvw)^{(\boldsymbol{k})} = \boldsymbol{x}^{\boldsymbol{k}} \sum u^{(\alpha)} v^{(\beta)} w^{(\gamma)}, \tag{1}$$

where the sum is extended over all possible vectors $\boldsymbol{\alpha}, \boldsymbol{\beta}, \boldsymbol{\gamma}$ with components equal to 1 or 0 such that $\boldsymbol{\alpha} + \boldsymbol{\beta} + \boldsymbol{\gamma} = \boldsymbol{k}$. In the estimation of any term of this sum we will suppose that $e_\alpha \subset e_l$ (otherwise one is zero), and that $\boldsymbol{\beta}$ is a vector whose s^{th} component is equal to unity and whose other components are zero. The question reduces to the estimate

$$\boldsymbol{x}^{k}\boldsymbol{x}^{l-\alpha}(1+x_s^2)^{\frac{\varkappa r_s}{2}-1} x_s \boldsymbol{x}^{\gamma} \prod_{e_\gamma} (1+x_j^2)^{\frac{r_j}{2}-1}$$

$$\times \left\{\sum_1^n (1+x_j^2)^{\frac{r_j}{2}}\right\}^{-1-|\gamma|} = \frac{\boldsymbol{x}^{l}(1+x_s^2)^{\frac{\varkappa r_s}{2}}}{\sum_1^n (1+x_j^2)^{\frac{r_j}{2}}} \frac{x_s^2}{1+x_s^2}$$

$$\times \frac{\boldsymbol{x}^{2\gamma}}{\prod_{e_\gamma}(1+x_j^2)} \frac{\prod_{e_\gamma}(1+x_j^2)^{\frac{r_j}{2}}}{\left\{\sum_1^n (1+x_j^2)^{\frac{r_j}{2}}\right\}^{|\gamma|}} \leqq 1.$$

We shall explain the estimate of the first factor in the second term of these relations. Suppose

$$(1+x_{j_0}^2)^{r_{j_0}} = \max_j (1+x_j^2)^{r_j};$$

then

$$\frac{\boldsymbol{x}^{l}(1+x_s^2)^{\frac{\varkappa r_s}{2}}}{\sum_1^n (1+x_j^2)^{\frac{r_j}{2}}} \leqq (1+x_{j_0}^2)^{\frac{r_{j_0}+\varkappa r_{j_0}}{2}} \prod_{j=1}^n (1+x_{j_0}^2)^{\frac{l_j r_{j_0}}{2r_j}} = (1+x_{j_0}^2)^0 = 1.$$

For $l_j > 1$, the product being estimated without the factor $\boldsymbol{x}^k$ is discontinuous when one of the coordinates x_j, where $j \in e_\alpha \subset e_k$, is equal to zero. But, by Theorem 1.5.4., it is sufficient that this factor should be continuous for $x_j > 0$, where $j \in e_k$, and arbitrary remaining x_j, which, in the situation at hand, is obviously the case.

For μ_7

$$\boldsymbol{x}^k \chi_7 \approx \boldsymbol{x}^k \boldsymbol{x}^{k-\alpha} (1+|\boldsymbol{x}|^2)^{r/2-|k-\alpha|} \left(1+\sum_1^n x_j\right)^{-r-|\alpha|}$$

$$= \frac{x^{2(k-\alpha)}}{(1+|\boldsymbol{x}|^2)^{|k-\alpha|}} \frac{\boldsymbol{x}^\alpha}{\left(1+\sum_{j=1}^n x_j\right)^{|\alpha|}} \left(\frac{(1+|\boldsymbol{x}|^2)^{1/2}}{1+\sum_{j=1}^n x_j}\right)^r < c < \infty.$$

Here we have applied the inequalities

$$c_1\left(1+\sum_1^n x_j\right) \leqq (1+|\boldsymbol{x}|^2)^{1/2} < 1+\sum x_j,$$

the second for $r > 0$ and the first for $r < 0$.

The function χ_7 is discontinuous on certain coordinate planes, but its limits on these from within each coordinate angle exist, so that in each such closed orthant χ_7 may be considered to be continuous.

For μ_8 we shall reason as follows. Suppose that $\boldsymbol{l}$ is a vector with components equal to 1 or 0. Applying the Leibniz formula on the differentiation of a product of functions of several variables, dropping constant coefficients and considering vectors $\boldsymbol{x}$ with nonnegative coordinates, we find, denoting by $e_{\boldsymbol{s}}$ the support of the vector $\boldsymbol{s}$, that

$$[\boldsymbol{x}^{\boldsymbol{l}} D^{\boldsymbol{l}}\{(1 + |\boldsymbol{x}|^2)^{-r/2} \Lambda_{\boldsymbol{r}}\}$$

$$\ll \sum_{\boldsymbol{s} \leq \boldsymbol{l}} \boldsymbol{x}^{\boldsymbol{l}} \boldsymbol{x}^{\boldsymbol{l}-\boldsymbol{s}} (1 + |\boldsymbol{x}|^2)^{-r/2-|\boldsymbol{l}-\boldsymbol{s}|} \leqq \left(\sum (1 + x_j^2)^{\frac{r\sigma}{2}}\right)^{1/\sigma - |\boldsymbol{s}|}$$

$$\times \prod_{j \in e_{\boldsymbol{s}}} (1 + x_j^2)^{\frac{r\sigma}{2} - 1} x_j = \sum_{\boldsymbol{s} \leq \boldsymbol{l}} \frac{x^{2(\boldsymbol{l}-\boldsymbol{s})}}{(1 + |\boldsymbol{x}|^2)^{|\boldsymbol{l}-\boldsymbol{s}|}} \frac{x^{2\boldsymbol{s}}}{\prod_{j \in e_{\boldsymbol{s}}} (1 + x_j^2)}$$

$$\times \frac{\prod_{j \in e_{\boldsymbol{s}}} (1 + x_j^2)^{\frac{r\sigma}{2}}}{\left\{\sum (1 + x_j^2)^{\frac{r\sigma}{2}}\right\}^{|\boldsymbol{s}|}} \frac{\left(\sum (1 + x_j^2)^{\frac{r\sigma}{2}}\right)^{1/\sigma}}{(1 + |\boldsymbol{x}|^2)^{r/2}} \ll 1$$

(taking into account the fact that $(\sum u_j^\sigma)^{1/\sigma} \leqq \sum u_j$, $\sigma > 0$, $u_j \geqq 0$).

For μ_9

$$\boldsymbol{x}^{\boldsymbol{l}} D^{\boldsymbol{l}}\{(1 + |\boldsymbol{x}|^2)^{-r/2} \Lambda_{\boldsymbol{r}}\}$$

$$\ll \sum \boldsymbol{x}^{\boldsymbol{l}} \boldsymbol{x}^{\boldsymbol{l}-\boldsymbol{s}} (1 + |\boldsymbol{x}|^2)^{r/2-|\boldsymbol{l}-\boldsymbol{s}|} \left(\sum (1 + x_j^2)^{\frac{r\sigma}{2}}\right)^{\frac{1}{\sigma} - |\boldsymbol{s}|}$$

$$\times \prod_{j \in e_{\boldsymbol{s}}} (1 + x_j^2)^{\frac{r\sigma}{2} - 1} \boldsymbol{x}^{\boldsymbol{s}} = \sum_{\boldsymbol{s} \leq \boldsymbol{l}} \frac{x^{2(\boldsymbol{l}-\boldsymbol{s})}}{(1 + |\boldsymbol{x}|^2)^{|\boldsymbol{l}-\boldsymbol{s}|}} \frac{x^{2\boldsymbol{s}}}{\prod_{j \in e_{\boldsymbol{s}}} (1 + x_j^2)^{|\boldsymbol{s}|}}$$

$$\times \frac{\prod_{j \in e_{\boldsymbol{s}}} (1 + x_j^2)^{\frac{r\sigma}{2}}}{\left(\sum (1 + x_j^2)^{\frac{r\sigma}{2}}\right)^{|\boldsymbol{s}|}} \frac{(1 + |\boldsymbol{x}|^2)^{r/2}}{\left(\sum (1 + x_j^2)^{\frac{r\sigma}{2}}\right)^{\frac{1}{\sigma}}} \ll 1.$$

For μ_{10}

$$\boldsymbol{x}^l D^l \left\{ (1 + x_i^2)^{\frac{r_i}{2}} \Lambda_r(\boldsymbol{x}) \right\} \ll \sum_{\boldsymbol{s} \leq \boldsymbol{l}} \boldsymbol{x}^l \left\{ \sum_{j=1}^{n} (1 + x_j^2)^{\frac{r_j \sigma}{2}} \right\}^{-\frac{1}{\sigma} - |\boldsymbol{s}|}$$

$$\times \prod_{j \in e_{\boldsymbol{s}}} (1 + x_j^2)^{\frac{r_j \sigma}{2} - 1} x_j D^{(\boldsymbol{l}-\boldsymbol{s})} (1 + x_i^2)^{\frac{r_i}{2}}$$

$$= \sum_{\boldsymbol{s} \leq \boldsymbol{l}} \frac{x^{2\boldsymbol{s}}}{\prod_{j \in e_{\boldsymbol{s}}} (1 + x_j^2)^{|\boldsymbol{s}|}} \frac{\prod_{j \in e_{\boldsymbol{s}}} (1 + x_j^2)^{\frac{r_j \sigma}{2}}}{\left\{ \sum_{j=1}^{n} (1 + x_j^2)^{\frac{r_j \sigma}{2}} \right\}^{|\boldsymbol{s}|}} \frac{x^{\boldsymbol{l}-\boldsymbol{s}} D^{(\boldsymbol{l}-\boldsymbol{s})} (1 + x_i^2)^{\frac{r_i}{2}}}{\left\{ \sum_{j=1}^{n} (1 + x_j^2)^{\frac{r_j \sigma}{2}} \right\}^{1/\sigma}}.$$

The first two fractions in the right-hand side do not exceed a constant. The third fraction also does not exceed a constant, since its numerator $\psi \equiv 0$, if there exists a $j \neq i$, $j \in e_{\boldsymbol{l}-\boldsymbol{s}}$, and $\psi = (1 + x_i^2)^{\frac{r_i}{2}}$, if $\boldsymbol{l} - \boldsymbol{s} = \boldsymbol{0}$, and lastly $\psi = 2x_i^2 (1 + x_i^2)^{\frac{r_i}{2} - 1}$, if the set $e_{\boldsymbol{l}-\boldsymbol{s}}$ consists only of one index i.

For μ_{11}

$$\boldsymbol{x}^l D^l \left\{ \Lambda_r^{-1}(\boldsymbol{x}) \left(\sum_{1}^{n} (1 + x_j^2)^{\frac{r_j}{2}} \right)^{-1} \right\}$$

$$\ll \sum_{\boldsymbol{s} \leq \boldsymbol{l}} \boldsymbol{x}^l \frac{\prod_{j \in e_{\boldsymbol{l}-\boldsymbol{s}}} (1 + x_j^2)^{\frac{r_j}{2} - 1} x_j}{\left\{ \sum_{1}^{n} (1 + x_j^2)^{\frac{r_j}{2}} \right\}^{1 + |\boldsymbol{l}-\boldsymbol{s}|}} \left\{ \sum_{1}^{n} (1 + x_j^2)^{\frac{r_j \sigma}{2}} \right\}^{1/\sigma - |\boldsymbol{s}|}$$

$$\times \prod_{j \in e_{\boldsymbol{s}}} (1 + x_j^2)^{\frac{r_j \sigma}{2} - 1} x_j$$

$$= \sum_{\boldsymbol{s} \leq \boldsymbol{l}} \frac{\prod_{j \in e_{\boldsymbol{s}}} (1 + x_j^2)^{\frac{r_j \sigma}{2}}}{\left\{ \sum_{1}^{n} (1 + x_j^2)^{\frac{r_j \sigma}{2}} \right\}^{|\boldsymbol{s}|}} \frac{x^{2(\boldsymbol{l}-\boldsymbol{s})}}{\left\{ \sum_{1}^{n} (1 + x_j^2)^{\frac{r_j}{2}} \right\}^{|\boldsymbol{l}-\boldsymbol{s}|}}$$

$$\times \frac{\left\{ \sum_{1}^{n} (1 + x_j^2)^{\frac{r_j \sigma}{2}} \right\}^{1/\sigma}}{\sum_{1}^{n} (1 + x_j^2)^{\frac{r_j}{2}}} \frac{\boldsymbol{x}^{2\boldsymbol{s}}}{\prod_{j \in e_{\boldsymbol{s}}} (1 + x_j^2)} \ll 1.$$

For μ_{12} one of the terms of the Leibniz sum (1) is estimated as follows:

$$(2)\quad u^k\left\{\sum_1^n (1+u_j^2)^{\frac{r_j\lambda}{2\sigma}}\right\}^{\sigma-|\alpha|}\left\{\sum_1^n (1+u_j^2)^{\frac{r_j\delta}{2\sigma}}\right\}^{\sigma-|\beta|}$$

$$\times\left\{\sum_1^n (1+u_j^2)^{\frac{r_j(\lambda+\delta)}{2\sigma}}\right\}^{-\sigma-|\gamma|}\prod_{j\in e_\alpha}(1+u_j^2)^{\frac{r_j\lambda}{2\sigma}-1}u_j$$

$$\times\prod_{j\in e_\beta}(1+u_j^2)^{\frac{r_j\delta}{2\sigma}-1}u_j\prod_{j\in e_\gamma}(1+u_j^2)^{\frac{r_j(\lambda+\delta)}{2\sigma}-1}u_j$$

$$=\frac{\left\{\sum_1^n (1+u_j^2)^{\frac{r_j\lambda}{2\sigma}}\right\}^{\sigma}\left\{\sum_1^n (1+u_j^2)^{\frac{r_j\delta}{2\sigma}}\right\}^{\sigma}}{\left\{\sum_1^n (1+u_j^2)^{\frac{r_j(\lambda+\delta)}{2\sigma}}\right\}^{\sigma}}\;\frac{u^{2k}}{\prod_{j\in e_k}(1+u_j^2)}$$

$$\times\frac{\prod_{j\in e_\alpha}(1+u_j^2)^{\frac{r_j\lambda}{\sigma 2}}\prod_{j\in e_\beta}(1+u_j^2)^{\frac{r_j\delta}{2\sigma}}\prod_{j\in e_\gamma}(1+u_j^2)^{\frac{r_j(\lambda+\delta)}{2\sigma}}}{\left\{\sum_1^n (1+u_j^2)^{\frac{r_j\lambda}{2\sigma}}\right\}^{|\alpha|}\left\{\sum_1^n (1+u_j^2)^{\frac{r_j\delta}{2\sigma}}\right\}^{|\beta|}\left\{\sum_1^n (1+u_j^2)^{\frac{r_j(\lambda+\delta)}{2\sigma}}\right\}^{|\gamma|}}\ll 1.$$

In the first fraction, if σ is deleted throughout, the order does not change. It does not change either if the powers λ, δ, $\lambda+\delta$ are carried outside the sign of the corresponding curly bracket.

For the proof in the case of the function μ_{12}^{-1} one explains the terms which may be written out as the right side of (2). The first fraction is changed only to the inverse quantity, which in any case will be bounded.

It is easy to see that the functions μ_{12} and μ_{12}^{-1} stop being Marcinkiewicz multipliers, if in each of its three factors the parameter σ takes on different values σ_1, σ_2, σ_3 or if in its first factor one replaces n by $m < n$. In the latter case we need to suppose in (2) that the support e_α consists of indices with numbers not exceeding m. Otherwise the corresponding term of the Leibniz sum is equal to zero. In the last term of (2), in the first factor of the numerator, n has to be replaced by m.

1.5.6. Extension of inequality 1.5.2 (13) to the nonperiodic case. Our object will be to show that for any function $f \in L_p(\mathbb{R}_n) = L_p(1 < p < \infty)$ the inequalities

$$(1)\qquad c_1\|f\|_p \leqq \|\{\sum |\delta_k(f)|^2\}^{1/2}\|_p \leqq c_2\|f\|_p,$$

are satisfied, where c_1, c_2 are constants not depending on f.

$$(2) \qquad \delta_{\boldsymbol{k}}(f) = \widehat{(1)_{\Delta_{\boldsymbol{k}}} \tilde{f}} \left((1)_e = \begin{cases} 1, & \boldsymbol{x} \in e, \\ 0, & \boldsymbol{x} \in e \end{cases} \right)$$

and for $\boldsymbol{k} \geqq \boldsymbol{0}$

$$(3) \qquad \Delta_{\boldsymbol{k}} = \Delta_{k_1, \ldots, k_n} = \{2^{k_j - 1} \leqq u_j \leqq 2^{k_j}; j = 1, \ldots, n\}$$

(if $k_j = 0$ the quantity 2^{k_j-1} is replaced by 0), and for arbitrary $\boldsymbol{k}$ the rectangle $\Delta_{\boldsymbol{k}}$ is the set of points $\{u_1 \operatorname{sign} k_1, \ldots, u_n \operatorname{sign} k_n\}$, where $\boldsymbol{u} = (u_1, \ldots, u_n) \in \Delta_{|k_1|, \ldots, |k_n|}$.

In what follows it will be proved (see 8.10.2.) that if f is regular in the sense of $L_p (1 < p < \infty)$ a generalized function (see further 1.5.10.) for which the norm appearing in the second term of (1) is finite, then $f \in L_p$.

We shall restrict ourselves to the consideration of the two-dimensional case. Suppose given an infinitely differentiable finite function $f(x, y)$ with a support belonging to

$$(4) \qquad \Delta_{s_0} = \{|x|, |y| < s_0\},$$

and the Fourier series 1.5.4. (6). In view of the fact that $f = 0$ outside Δ_{s_0} for $s > s_0$, we will have the inequalities

$$(5) \qquad \|f\|_p \ll \|(\sum |\delta_{kl}(f)|^2)^{1/2}\|_{L_p(\Delta_s)} \ll \|f\|_p,$$

where

$$\delta_{kl}(f) = \sum_{\pm(n_{|k|-1}+1)}^{\pm n_{|k|}} \sum_{\pm(n_{|l|-1}+1)}^{\pm n_{|l|}} c_{\mu\nu} e^{i\frac{\pi}{s}(\mu x + \nu y)}$$

and where we this time suppose that $n_0 = n_{-1} = 0$, $n_1 = 1$, $n_k = 2^{k-2}\beta$ $(k = 2, 3, \ldots)$, and $s > s_0$ is chosen so that $\beta = \dfrac{s}{\pi} > 2$ is an integer. The sign $+$ or $-$ is substituted depending on whether k or l is positive or negative. Condition 1.5.2. (5) is satisfied:

$$\frac{n_k + 1}{n_k} \geqq 2 \quad (k = 1, 2, \ldots),$$

so that the constants entering into the inequality (5) do not depend on $s > s_0$.

Put

$$\delta_{kl}(f) = \frac{1}{2\pi} \int_{\Delta_{k,l}} \tilde{f}(u, v)\, e^{i(xu + yv)}\, du\, dv = \widehat{(1)_{\Delta_{kl}} \tilde{f}},$$

where Δ_{kl} is the rectangle (3) (for $k_1 = k$, $k_2 = l$, $n = 2$).

Suppose

$$a = \frac{k_1\pi}{s}, \quad b = \frac{k_2\pi}{s}, \quad c = \frac{l_1\pi}{s}, \; d = \frac{l_2\pi}{s};$$

$$\Delta = \{[a, b] \times [c, d]\};\; b - a,\, d - c \geqq 1;$$

$$|\tilde{f}|, \left|\frac{\partial \tilde{f}}{\partial x}\right|, \left|\frac{\partial \tilde{f}}{\partial y}\right|, \left|\frac{\partial^2 \tilde{f}}{\partial x\, \partial y}\right| \leqq M_\Delta,\ (x, y) \in \Delta.$$

We apply the Abel transform to the sum

(6) $$\delta_\Delta = \sum_{k_1}^{k_2} \sum_{l_1}^{l_2} c_{kl} e^{i\frac{\pi}{s}(kx+ly)}$$

$$= \frac{\pi}{2s^2} \sum_{k_1}^{k_2} \sum_{l_1}^{l_2} \tilde{f}\left(\frac{k\pi}{s}, \frac{l\pi}{s}\right) e^{i\frac{\pi}{s}(kx+ly)}$$

$$= \frac{\pi}{2s^2} \sum_{l_1}^{l_2} e^{i\frac{\pi}{s}ly} \left\{ \sum_{k_1}^{k_2-1} I_k(x)\, \Delta_x \tilde{f}\left(\frac{k\pi}{s}, \frac{i\pi}{s}\right) \right.$$

$$\left. + \tilde{f}\left(\frac{k^2\pi}{s}, \frac{l\pi}{s}\right) I_{k_2}(x) \right\},$$

where

$$I_k(x) = \sum_{v=1}^{k} e^{i\frac{\pi}{s}vx} = \frac{e^{i\frac{\pi}{s}(k+1)x} - 1}{e^{i\frac{\pi}{s}x} - 1},$$

$$\Delta_x a_{kl} = a_{kl} - a_{k+1,l}.$$

A further Abel transformation leads to the equation

(7) $$\delta_\Delta = \frac{\pi}{2s^2} \left\{ \sum_{k_1}^{k_2-1} \sum_{l_1}^{l_2-1} I_k(x)\, I_l(y)\, \Delta_{xy}^2 \tilde{f}\left(\frac{k\pi}{s}, \frac{l\pi}{s}\right) \right.$$

$$+ \sum_{k_1}^{k_2-1} \Delta_x \tilde{f}\left(\frac{k\pi}{s}, \frac{l_2\pi}{s}\right) I_k(x)\, I_{l_2}(y)$$

$$+ \sum_{l_1}^{l_2-1} I_{k_2}(x)\, I_l(y)\, \Delta_y \tilde{f}\left(\frac{k_2\pi}{s}, \frac{l\pi}{s}\right)$$

$$\left. + \tilde{f}\left(\frac{k_2\pi}{s}, \frac{l_2\pi}{s}\right) I_{k_2}(x)\, I_{l_2}(y) \right\}.$$

If one takes into account the fact that

$$|I_k(x)| \leqq \frac{2}{\left|e^{i\frac{\pi}{s}x} - 1\right|} \ll \frac{s}{|x|} \quad (|x| < s),$$

then (6) and (7) imply the four inequalities

$$|\delta_\Delta| \leqq cM_\Delta\,|\Delta|\,\{1, |x|^{-1}, |y|^{-1}, |xy|^{-1}\},\ |x|, |y| < s, \tag{8}$$

where c does not depend on the various factors nor on s. The second, for example, is obtained from (6) by means of the following calculations:

$$|\delta_\Delta| \leqq \frac{1}{s^2}(l_2 - l_1)\left\{(k_2 - k_1)\frac{s}{|x|}\frac{M_\Delta}{s} + M_\Delta\frac{s}{|x|}(b - a)\right\}$$

(in the second term in curly brackets a factor $(b - a) \geqq 1$ has been added). The fourth inequality follows on using analogous calculations from (7).

Put $\Phi_1(x, y) = \min c\{1, |x|^{-1}, |y|^{-1}, |xy|^{-1}\}$. Obviously $\Phi_1 \in L_p(1 < p < \infty!)$, and it follows from (8) that

$$|\delta_\Delta| \leqq M_\Delta|\Delta|\,\Phi_1(x, y)\ \big((x, y) \in \Delta_s\big). \tag{9}$$

On the basis of (9), taking into account the fact that

$$\delta^s_{kl}(f) = \delta^s_{\Delta_{kl}},$$

$$\Delta^s_{kl} = \left\{\pm\frac{(n_{k-1} + 1)\pi}{s} \lesseqgtr x \lesseqgtr \pm\frac{n_k\pi}{s};\ \pm\frac{(n_{l-1} + 1)\pi}{s} \lesseqgtr y \lesseqgtr \pm\frac{n_l\pi}{s}\right\},$$

we get $|\delta^s_{kl}(f)| \leqq M^s_{\Delta_{kl}}|\Delta^s_{kl}|\,|\Phi_1(x, y)$.

Since the functions $\tilde{f}$, $\dfrac{\partial\tilde{f}}{\partial x}$, $\dfrac{\partial\tilde{f}}{\partial y}$ decrease to infinity faster than $(1 + |x|^\lambda + |y|^\lambda)^{-1}$, where λ is arbitrarily large then obviously

$$\sum M^s_{\Delta_{kl}}|\Delta^s_{kl}| < A < \infty,$$

where the constant A does not depend on s. Therefore

$$\{\sum |\delta^s_{kl}(f)|^2\}^{1/2} \leqq \sum |\delta^s_{kl}(f)| \leqq A\Phi_1$$

$$= \Phi(x, y) \in L_p(\mathbb{R}_n)\ \big((x, y) \in \Delta_s\big). \tag{10}$$

If we restrict ourselves for simplicity to nonnegative k, l, and choose s so that $\beta = \frac{s}{\pi}$ is an integer, we get

$$\delta^s_{kl}(f) = \frac{\pi}{2s^2} \sum_{2^{k-3}\beta+1}^{2^{k-2}\beta} \sum_{2^{l-3}\beta+1}^{2^{l-2}\beta} \tilde{f}\left(\frac{\mu\pi}{s}, \frac{\nu\pi}{s}\right) e^{i\frac{\pi}{s}(\mu x+\nu y)} \xrightarrow[s\to\infty]{}$$

$$\xrightarrow[s\to\infty]{} \frac{1}{2\pi} \int_{2^{k-3}}^{2^{k-2}} \int_{2^{l-3}}^{2^{l-2}} \tilde{f}(u, v)\, e^{i(ux+vy)}\, du\, dv = \delta_{k-2,l-2}(f)$$

uniformly relative to $(x, y) \in \Delta_N$ for $k, l \geqq 2$ (2^{-1} has to be replaced by 0), for any fixed $N > 0$. If now one of the numbers (while nonnegative) is less than 2, then the repeated sum turns into a single sum or even for $(\delta^s_{00}, \delta^s_{10}, \delta^s_{01})$ degenerates to one term. In these cases $\delta^s_{kl}(f) \to 0$ uniformly relative to Δ_N, since the function under the integral sign is continuous relative to x, y, u, v. Analogous facts are also true for numbers k, l. Therefore it is proved that for arbitrary k, l

$$\delta^s_{kl}(f) \xrightarrow[s\to\infty]{} \delta_{k-2,l-2}(f) \text{ on } \Delta_N$$

uniformly, for any $N > 0$.

From the second inequality (5) it follows that for any $N, N_1 > 0$

$$\left\| \left(\sum_{|k|,|l|<N_1} |\delta^s_{kl}(f)|^2 \right)^{1/2} \right\|_{L_p(\Delta_N)} \leqq \|f\|_p$$

and after passing to the limit as $s \to \infty$, then $N_1 \to \infty$ and then $N \to \infty$ we get

$$\|(\sum |\delta_{kl}(f)|^2)^{1/2}\|_p \leqq \|f\|_p .$$

From (10) it follows that

$$\left(\sum_{|k|,|l|<N} |\delta^s_{kl}(f)|^2 \right)^{1/2} \leqq \Phi(x, y),$$

so that after a passage to the limit, first for $s \to \infty$ and then for $N \to \infty$, we get

$$\{\sum |\delta_{kl}(f)|^2\}^{1/2} \leqq \Phi(x, y).$$

Finally,

$$| \|\{\sum |\delta_{kl}^s(f)|^2\}^{1/2}\|_{L_p(\Delta_s)}^p - \|\{\sum |\delta_{kl}(f)|^2\}^{1/2}\|_{L_p(\mathbb{R}_2)}^p|$$

$$\leqq \left| \left\| \left\{ \sum_{|k|,|l|<N} |\delta_{kl}^s(f)|^2 \right\}^{1/2} \right\|_{L_p(\Delta_N)}^p \right.$$

$$\left. - \left\| \left\{ \sum_{|k|,|l|<N} |\delta_{kl}(f)|^2 \right\}^{1/2} \right\|_{L_p(\Delta_N)}^p \right|$$

$$+ \left\| \sum_{|k|,|l|<N} |\delta_{kl}^s(f)| \right\|_{L_p(\Delta_s - \Delta_N)} + \left\| \sum_{|k|,|l|<N} |\delta_{kl}(f)| \right\|_{L_p(\mathbb{R}_2 - \Delta_N)}$$

$$+ \|\sum{}' |\delta_{kl}^s(f)|\|_{L_p(\Delta_N)} + \|\sum{}' |\delta_{kl}(f)|\|_{L_p(\Delta_N)} \equiv I_1 + \cdots + I_5,$$

where $\sum'$ is the sum over pairs of numbers k, l at least one of which is not less than N.

Here

$$I_2, I_3 \leqq \|\Phi\|_{L_p(\mathbb{R}_2 - \Delta_N)},$$

$$I_4 \leqq \sum{}' M_{\Delta_{kl}^s} |\Delta_{kl}^s| \|\Phi\|_p \leqq \varepsilon_N \|\Phi\|_p,$$

where ε_N does not depend on s and tends to zero as $N \to \infty$

$$I_5 \leqq \varepsilon_N \to 0 \ (N \to \infty).$$

Thus N may be chosen sufficiently large that $I_2, \ldots, I_5$ will be less than a given $\varepsilon > 0$, and then s_0 may by taken so that $I_1 < \varepsilon$ for all $s > s_0$.

We have proved that for any infinitely differentiable finite function f

$$\lim_{s \to \infty} \|\{\sum |\delta_{kl}^s(f)|^2\}^{1/2}\|_{L_p(\Delta_s)} = \|\{\sum |\delta_{kl}(f)|^2\}^{1/2}\|_p,$$

and then on the basis of (5), where the constants in the inequality do not depend on s, we obtain (1) (up till now for infinitely differentiable finite functions).

Now if $f \in L_p$, then we choose a sequence of infinitely differentiable finite functions $f_j (j = 1, 2, \ldots)$ such that

$$\|f - f_j\|_p \to 0 \ (j \to \infty). \tag{11}$$

This shows that for any $\varepsilon > 0$ there is a λ such that for $i, j \geqq \lambda$

$$\left\| \left\{ \sum_{|k|,|l|<N} |\overbrace{(1)_{\Delta_{kl}}(f_i - f_j)}|^2 \right\} \right\|_p \leqq \|f_i - f_j\|_p < \varepsilon.$$

After passing to the limit as $i \to \infty$ in these inequalities f_i is replaced by f. A subsequent passage to the limit as $N \to \infty$ now leads to the inequality

$$\|\{\sum |\delta_{kl}(f_j - f)|^2\}^{1/2}\|_p < \varepsilon \ (j > \lambda),$$

from which it follows that

$$(12) \qquad \|\{\sum |\delta_{kl}(f_j - f)|^2\}^{1/2}\| \to 0 \ (j \to \infty).$$

For the functions f_j inequalities (1) are satisfied. As a consequence now of (11) and (12) it is legitimate to pass to the limit as $j \to \infty$ in these inequalities, thus obtaining inequalities (1) for $f \in L_p$.

1.5.6.1. By quite analogous arguments, simpler because we have in mind the one-dimensional case, one proves that for functions $f(x) \in L_p(-\infty, \infty) = L_p(1 < p < \infty)$ the inequalities

$$(13) \qquad \|f\|_p \ll \|\{\sum |\beta_l(f)|^2\}^{1/2}\|_p \ll \|f\|_p$$

hold, where

$$\beta_l(f) = \widehat{(1)_{\Delta_l}\tilde{f}},$$

$\Delta_l = \{2^{l-1} \leqq |x| \leqq 2^l\}$, $l = 0, 1, \ldots$; 2^{l-1} for $l = 0$ is replaced by zero$\}$, and the constants in (13) do not depend on f. As the initial inequality in the periodic case we need to choose 1.5.2 (4).

1.5.7. The Fourier transform of the function sign $\boldsymbol{x}$. The function

$$\operatorname{sign} \boldsymbol{x} = \prod_{j=1}^{n} \operatorname{sign} x_j$$

is a multiplicator for $1 < p < \infty$ (see subsection 1.5.5). The functional (explanation below)

$$(1) \quad \widehat{(\operatorname{sign} \boldsymbol{x}, \varphi)} = (\operatorname{sign} \boldsymbol{x}, \hat{\varphi}) = \frac{1}{(2\pi)^{n/2}} \int \operatorname{sign} \boldsymbol{u} \, d\boldsymbol{u} \int e^{i\boldsymbol{u}\boldsymbol{t}} \varphi(\boldsymbol{t}) \, d\boldsymbol{t}$$

$$= \frac{1}{(2\pi)^{n/2}} \int_{R_+} d\boldsymbol{u} \int \varphi(\boldsymbol{t}) \prod_{j=1}^{n} (e^{it_j u_j} - e^{-it_j u_j}) \, d\boldsymbol{t}$$

$$= \left(\frac{2}{\pi}\right)^{n/2} i^n \lim_{N \to \infty} \int \varphi(\boldsymbol{t}) \, d\boldsymbol{t} \int_{\Delta_N^+} \prod_{j=1}^{n} \sin t_j u_j du_j$$

$$= \left(\frac{2}{\pi}\right)^{n/2} i^n \lim_{N \to \infty} \int \varphi(\boldsymbol{t}) \, d\boldsymbol{t} \prod_{j=1}^{n} \int_0^N \sin t_j u_j du_j =$$

$$= \left(\frac{2}{\pi}\right)^{n/2} i^n \lim_{N\to\infty} \int \varphi(t) \prod_{j=1}^{n} \frac{1-\cos Nt_j}{t_j}\, dt$$

$$= \left(\frac{2}{\pi}\right)^{n/2} i^n \lim_{N\to\infty} \int_{R_+} \Delta\varphi(t) \prod_{j=1}^{n} \frac{1-\cos Nt_j}{t_j}\, dt$$

$$= \left(\frac{2}{\pi}\right)^{n/2} i^n \int_{R_+} \frac{\Delta\varphi(t)}{t}\, dt = \left(\frac{2}{\pi}\right)^{n/2} i^n \int \frac{\varphi(t)}{t}\, dt.$$

Here $\mathbb{R}_+$ is the positive coordinate orthant,

(2) $$\Delta_N = \{0 \leqq x_j \leqq N; i = 1, \ldots, n\},$$

$$\Delta\varphi(t) = \Delta_1\Delta_2 \ldots \Delta_n\varphi(t)$$

and

(3) $$\Delta_j\varphi(t) = \varphi(t) - \varphi(t_1, \ldots, t_{j-1}, -t_j, t_{j+1}, \ldots, t_n)$$

$$(j = 1, \ldots, n).$$

In the next-to-last equation (1), in multiplying out the terms of the product, there appear the integrals

$$\int_{R_+} \frac{\Delta\varphi(t)}{t} \prod_{j=1}^{k} \cos Nt_j\, dt \to 0,$$

tending to zero as $N \to \infty$ as a consequence of the summability of $t^{-1}\Delta\varphi(t)$ on $\mathbb{R}_+$, in view of a well-known lemma in the theory of Fourier series. The integral in the last term of (1) is written in the sense of Cauchy:

(4) $$\int \frac{\varphi(t)}{t}\, dt = \lim_{\varepsilon\to 0} \int_{\mathbb{R}_+^\varepsilon} \frac{\varphi(t)}{t}\, dt,$$

where $\mathbb{R}_+^\varepsilon$ is the set of points $x \in \mathbb{R}_+$ distant from arbitrary coordinate planes by more than $\varepsilon > 0$. The functional (4) defines a generalized function, which we denote by v.p. $\frac{1}{t}$. Thus, we have proved the equation

$$\widehat{\operatorname{sign} x} = \left(\frac{2}{\pi}\right)^{n/2} i^n \text{v.p.} \frac{1}{t}.$$

For $f \in S$

$$(5)\qquad \widehat{\operatorname{sign} \boldsymbol{x}} * f = \widehat{\operatorname{sign} \boldsymbol{x}\tilde{f}} = \frac{1}{(2\pi)^n} \int \operatorname{sign} \boldsymbol{u} \int f(t)\, e^{-it\boldsymbol{u}} dt\, e^{i\boldsymbol{u}\boldsymbol{x}} d\boldsymbol{u}$$

$$= \frac{1}{(2\pi)^n} \int \operatorname{sign} \boldsymbol{u}\, d\boldsymbol{u} \int f(\boldsymbol{x} - \boldsymbol{t})\, e^{i\boldsymbol{u}\boldsymbol{t}}\, dt = \left(\frac{i}{\pi}\right)^n \int \frac{f(\boldsymbol{x} - \boldsymbol{t})}{\boldsymbol{t}}\, d\boldsymbol{t},$$

where the last equation follows from the already proved equalities between the third and last terms of (1), if one replaces $\varphi(\boldsymbol{t})$ there by $f(\boldsymbol{x} - \boldsymbol{t})$. The last integral (5) is understood in the sense of Cauchy.

The notation

$$(6)\qquad \widehat{\operatorname{sign} \boldsymbol{x}} * f = \widehat{\operatorname{sign} \boldsymbol{x}\tilde{f}} = \left(\frac{i}{\pi}\right)^n \int \frac{f(\boldsymbol{x} - \boldsymbol{t})}{\boldsymbol{t}}\, d\boldsymbol{t}$$

will also be used when $f \in L_p (1 < p < \infty)$, understanding the terms (6) as the limits in L_p to which the corresponding expressions for the finite functions f_l tend as $\|f_l - f\|_p \to 0$. In the relation of the first and second terms of (6) this was justified earlier in 1.5.1, because sign $\boldsymbol{x}$ is a multiplicator in L_p for $1 < p < \infty$. We have made now the appropriate definition for an expression written outwardly in the form of an integral. In fact it is possible to show (M. Riesz [1] for $n = 1$) that for $f \in L_p$ $(1 < p < \infty)$ this is really an integral in the sense of Cauchy, existing for almost all $\boldsymbol{x}$, but we will not dwell on this. We will not need it.

Suppose that $\boldsymbol{\mu} = (\mu_1, \ldots, \mu_n)$, $\boldsymbol{a} = (a_1, \ldots, a_n) > 0$ $(a_j > 0)$ are two fixed vectors and

$$\Delta_a = \{|x_j| < a_j;\, j = 1, \ldots, n\},$$

$$\Delta(\boldsymbol{\mu}, \boldsymbol{a}) = \{|x_j - \mu_j| < a_j;\, j = 1, \ldots, n\},\ \Delta(0, \boldsymbol{a}) = \Delta_{\boldsymbol{a}}.$$

Thus $\Delta(\boldsymbol{\mu}, \boldsymbol{a})$ is a shift of $\Delta_{\boldsymbol{a}}$ by the vector $\boldsymbol{\mu}$. We note that the characteristic function (of one variable t) of the interval (a, b) is equal to

$$(1)_{(a,b)} = \frac{1}{2} [\operatorname{sign}(t - a) - \operatorname{sign}(t - b)].$$

Hence it follows that

$$(7)\quad (1)_{\Delta(\mu,a)} = \prod_{j=1}^{n} \frac{1}{2} [\operatorname{sign}(x_j - \mu_j + a_j) - \operatorname{sign}(x_j - \mu_j - a_j)]$$

$$= \frac{(-1)^n}{2^n} \sum \operatorname{sign} \alpha \operatorname{sign}(\boldsymbol{x} - \boldsymbol{\mu} - \alpha),$$

where the sum is extended over all vectors $\alpha = (\alpha_1, \ldots, \alpha_n)$ such that $|\alpha_j| = a_j$, $j = 1, \ldots, n$.

We know that the function sign x is a multiplicator:

$$\text{(8)} \qquad \|\widehat{\operatorname{sign} x \tilde{f}}\|_p \leqq \varkappa_p \|f\|_p \quad (1 < p < \infty),$$

where $\varkappa_p$ does not depend on f (see 1.5.1), and sign $(x - a)$ is also a multiplicator with the same constant $\varkappa_p$ in the corresponding inequality (see 1.5.1.2), for any vector $a \in R$. Therefore it follows from (7) that

$$\text{(9)} \qquad \|\widehat{(1)_{\Delta(\mu,a)} \tilde{f}}\|_p \leqq \frac{1}{2^n} \sum \varkappa_p \|f\|_p = \varkappa_p \|f\|_p, \quad 1 < p < \infty,$$

because this sum is extended over 2^n terms. It is significant that the constant $\varkappa_p$ in (9) is the same as in (8). Thus it does not depend on μ and a.

It follows from (7) (see 1.5. (8)), that

$$\text{(10)} \qquad \widehat{(1)}_{\Delta(\mu,a)} = \frac{(-1)^n}{2^n} \sum \operatorname{sign} \alpha \, e^{i(\mu+a)x} \, \widehat{\operatorname{sign} x} =$$

$$= \frac{1}{2^n} e^{i\mu x} \prod_{j=1}^{n} (e^{ia_j x_j} - e^{-ia_j x_j}) \left(\frac{2}{\pi}\right)^{n/2} (-i)^n \, \text{v.p.} \frac{1}{x} =$$

$$= \left(\frac{2}{\pi}\right)^{n/2} e^{i\mu x} D_a(x),$$

where

$$\text{(11)} \qquad D_a(x) = \prod_{j=1}^{n} \sin a_j x_j \, \text{v.p.} \frac{1}{x} = \prod_{j=1}^{n} \frac{\sin a_j x_j}{x_j}.$$

In (11), by the multiplication of an ordinary function by a generalized function one has obtained an ordinary function. For example, in the one-dimensional case this is proved as follows:

$$\text{(12)} \qquad \left(\sin ax \, \text{v.p.} \frac{1}{x}, \varphi(x)\right) = \left(\text{v.p.} \frac{1}{x}, \sin ax \varphi(x)\right)$$

$$= \int_0^\infty \frac{\Delta[\sin ax \varphi(x)] \, dx}{x} = \int \frac{\sin ax}{x} \varphi(x) \, dx,$$

where $\Delta F(x) = F(x) - F(-x)$. The integral in the right side of (12) already may be understood in the sense of Lebesgue.

For functions $f \in S$ we have the equation

$$\widetilde{(1)_{\Delta(\mu,a)}\tilde{f}} = \frac{1}{\pi^n} \int e^{i\mu(x-u)} D_a(x-u) f(u)\, du,$$

and in particular for $\mu = 0$

$$F(x) = \widetilde{(1)_{\Delta_a}\tilde{f}} = \frac{1}{\pi^n} \int D_a(x-u) f(u)\, du, \tag{13}$$

where the integrals in the right sides are understood in the sense of Lebesgue. We shall consider (13) in detail. The integral (13) has a meaning also for any function $f \in L_p$ $(1 \leqq p < \infty)$, since $D_a(x) \in L_q \left(\frac{1}{p} + \frac{1}{q} = 1\right)$ and

$$\int |D_a(x-a) f(u)|\, du \leqq \|D_a\|_q \|f\|_p < \infty.$$

It is immediately clear that F is a continuous function of x (even uniformly continuous):

$$|F(x) - F(y)| \leqq \|D_a(x-u) - D(y-u)\|_q \|f\|_p \to 0 \quad (x \to y).$$

If $f_l \in S$, $\|f_l - f\|_p \to 0$ and F_l is the result of substitution of f_l in place$_y$ of f in (13), then

$$|F(x) - F_l(x)| \leqq \|D_a\|_q \|f - f_l\|_p \to 0$$

uniformly. On the other hand, $(1)_{\Delta_a}$ is a Marcinkiewicz multiplier, and therefore $\|F_k - F_l\|_p \to 0$ $(k, l \to 0)$. This shows F_l tends in the sense of L_p to the function F defined by the integral (13), and that for $f \in L_p$ equation (13) is valied, where its right side is a Lebesgue integral and the left is understood in the terms of the Marcinkiewicz multiplier (see 1.5.1.).

Indeed, $F(x)$ is an analytic function, moreover an entire function of exponential type (see further 3.6.2).

1.5.8. The functions φ_ε and ψ_ε. The function φ_ε is defined on $\mathbb{R} = \mathbb{R}_n$, depends on a small positive parameter $\varepsilon (0 < \varepsilon < \varepsilon_0)$, and has the following properties: $\varphi_\varepsilon(x)$ is infinitely differentiable and nonnegative on $\mathbb{R}$, and has its support on the cube

$$\Delta_\varepsilon = \{|x_j| < \varepsilon;\ j = 1, \dots, n\}$$

(i.e. $\varphi_\varepsilon = 0$ outside Δ_ε), and moreover satisfies the equation

$$\int_{\Delta_\varepsilon} \varphi_\varepsilon(x)\, dx = 1 \quad (0 < \varepsilon < \varepsilon_0). \tag{1}$$

It is important that $\varphi_\varepsilon \in S$ and has a compact support, i.e. is a finite function (see 1.4.1.).

If φ is an arbitrary function continuous on $\mathbb{R}$ (even locally summable on $\mathbb{R}$ and continuous at the zero point), then

(2) $$\lim_{\varepsilon \to 0} \int \varphi_\varepsilon(\boldsymbol{x})\, \varphi(\boldsymbol{x})\, d\boldsymbol{x} = \varphi(0),$$

because

$$\left| \int_{\Delta\varepsilon} \varphi_\varepsilon(\boldsymbol{x})\, \varphi(\boldsymbol{x})\, d\boldsymbol{x} - \varphi(\mathbf{0}) \right| = \int_{\Delta\varepsilon} \varphi_\varepsilon(\boldsymbol{x})\, [\varphi(\boldsymbol{x}) - \varphi(\mathbf{0})]\, d\boldsymbol{x}$$

$$\leqq \int_{\Delta\varepsilon} \varphi_\varepsilon(\boldsymbol{x}) \sup_{\Delta\varepsilon} |\varphi(\boldsymbol{x}) - \varphi(\mathbf{0})|\, d\boldsymbol{x}$$

$$= \sup_{\Delta\varepsilon} |\varphi(\boldsymbol{x}) - \varphi(\mathbf{0})| \to 0, \quad \varepsilon \to 0.$$

If $\varphi \in S$, then equation (2) may be written as follows:

(3) $$\lim_{\varepsilon \to 0} (\varphi_\varepsilon, \varphi) = (\delta, \varphi) = \varphi(\mathbf{0}),$$

where $\delta = \delta(\boldsymbol{x})$ is the Dirac delta function.

Put

$$\psi_\varepsilon(\boldsymbol{x}) = (2\pi)^{n/2}\, \tilde{\varphi}_\varepsilon(\boldsymbol{x}).$$

Since $\varphi_\varepsilon \to \delta$ $(\varepsilon \to 0)$ weakly then $\psi_\varepsilon \to (2\pi)^{n/2}\, \tilde{\delta} = 1$ weakly. Moreover $\psi_\varepsilon(\boldsymbol{x})$ as an ordinary function converges boundedly to 1 for all $\boldsymbol{x}$ as $\varepsilon \to 0$:

(4) $$\psi_\varepsilon(\boldsymbol{x}) = \int \varphi_\varepsilon(\boldsymbol{t})\, e^{-i\boldsymbol{x}\boldsymbol{t}} d\boldsymbol{t} \to 1,$$

(5) $$|\psi_\varepsilon(\boldsymbol{x})| \leqq \int \varphi_\varepsilon(\boldsymbol{t})\, d\boldsymbol{t} = 1.$$

Below it will be shown that if $f \in L_p$, $g \in L$ and $\varepsilon \to 0$, then, weakly,

(6) $$\psi_\varepsilon f \to f,$$

(7) $$\psi_\varepsilon g * f \to g * f,$$

(8) $$g * \psi_\varepsilon f \to g * f.$$

Further, if $f \in L_p$, $g \in L_q$, $\dfrac{1}{p} + \dfrac{1}{q} = 1$, then it is possible to define the convolution $g * f$ by means of the integral

$$g * f = \frac{1}{(2\pi)^{n/2}} \int g(\boldsymbol{u})\, f(\boldsymbol{x} - \boldsymbol{u})\, d\boldsymbol{u}.$$

Obviously,

$$|(g * f)(\boldsymbol{x})| \leqq \frac{1}{(2\pi)^{n/2}} \|g\|_q \|f\|_p .$$

This convolution is somewhat different from the generalization of this concept introduced in 1.5., where $g \in S'$ was a function such that $f \in L_p$ implies $g * f \in L_p$. In the case at hand, if $f \in L_p$ the function $g * f$ lies in the class $L_\infty = M$ of bounded (measurable) functions. However for such a convolution there holds a property analogous to (8):

$$g * \psi_\varepsilon f \to g * f \quad (\varepsilon \to 0) . \tag{9}$$

Proof of (6). In view of the theorem of Lebesgue

$$(\psi_\varepsilon f, \varphi) = \int \psi_\varepsilon(\boldsymbol{t})\, f(\boldsymbol{t})\, \varphi(\boldsymbol{t})\, d\boldsymbol{t} \to \int f\varphi\, d\boldsymbol{t} = (f, \varphi) .$$

Proof of (7)

$$(\psi_\varepsilon g * f, \varphi) = \frac{1}{(2\pi)^{n/2}} \iint \psi_\varepsilon(\boldsymbol{t})\, g(\boldsymbol{t})\, f(\boldsymbol{x} - \boldsymbol{t})\, \varphi(\boldsymbol{x})\, d\boldsymbol{t}\, d\boldsymbol{x}$$

$$\to \frac{1}{(2\pi)^{n/2}} \iint g(\boldsymbol{t})\, f(\boldsymbol{x} - \boldsymbol{t})\, \varphi(\boldsymbol{t})\, d\boldsymbol{t}\, d\boldsymbol{x} = (g * f, \varphi)$$

since

$$\iint |g(\boldsymbol{t})\, f(\boldsymbol{x} - \boldsymbol{t})|\, d\boldsymbol{t}\, |\varphi(\boldsymbol{x})|\, d\boldsymbol{x}$$

$$\leqq \left\| \int |g(\boldsymbol{t})\, f(\boldsymbol{x} - \boldsymbol{t})\, d\boldsymbol{t} \right\|_p \|\varphi\|_q \leqq \|g\|_L \|f\|_p \|\varphi\|_q \left(\frac{1}{p} + \frac{1}{q} = 1\right).$$

Proof of (8).

$$(g * \psi_\varepsilon f, \varphi) = \frac{1}{(2\pi)^{n/2}} \iint \psi_\varepsilon(\boldsymbol{t})\, f(\boldsymbol{t})\, g(\boldsymbol{x} - \boldsymbol{t})\, \varphi(\boldsymbol{x})\, d\boldsymbol{x}\, d\boldsymbol{t} \tag{10}$$

$$\to \frac{1}{(2\pi)^{n/2}} \iint f(\boldsymbol{t})\, g(\boldsymbol{x} - \boldsymbol{t})\, \varphi(\boldsymbol{x} - t)\, d\boldsymbol{x}\, d\boldsymbol{t},$$

since

$$\iint |f(\boldsymbol{t})\, g(\boldsymbol{x} - \boldsymbol{t})\, \varphi(\boldsymbol{x})|\, d\boldsymbol{t}\, d\boldsymbol{x} \leqq \left\| \int f(\boldsymbol{t})\, g(\boldsymbol{x} - \boldsymbol{t})\, d\boldsymbol{t} \right\|_p \|\varphi\|_q \leqq \|g\|_L \|f\|_p \|\varphi\|_q .$$

Proof of (9). The same as in the proof of (8), but we need to take into account the inequality

$$\iint |f(\boldsymbol{t})\, g(\boldsymbol{x} - \boldsymbol{t})\, \varphi(\boldsymbol{x} - \boldsymbol{t})|\, d\boldsymbol{x}\, d\boldsymbol{t} \leqq \|f\|_p \|g\|_q \|\varphi\|_L .$$

1.5.9. Operations I_r of Liouville type. Suppose that r is an arbitrary real number. The function

$$(1) \qquad (1 + |\boldsymbol{u}|^2)^{r/2} = \left(1 + \sum_{j=1}^{n} u_j^2\right)^{r/2}$$

is infinitely differentiable on $\mathbb{R}$ and has polynomial growth for any sign of r.

Put

$$(2) \qquad G_r(\boldsymbol{u}) = \widehat{(1 + |\boldsymbol{u}|^2)^{-r/2}}.$$

Since

$$(3) \qquad \widetilde{G_r(\boldsymbol{u})} = (1 + |\boldsymbol{u}|^2)^{-r/2}$$

is an infainitely differentiable function with polynomial growth, then for any generalized function $f \in S'$ one has the convolution

$$(4) \qquad F = G_r * f = \widehat{\tilde{G}_r \tilde{f}} = \widehat{(1 + |\boldsymbol{u}|^2)^{-r/2} \tilde{f}} = I_r f,$$

defining an operation I_r mapping $f \in S'$ into $F \in S'$.

Obviously,

$$(5) \qquad I_0 f = f.$$

If r and ϱ are arbitrary real numbers and $f \in S'$, then

$$(6) \qquad I_{r+\varrho} f = \widehat{(1 + |\lambda|^2)^{-r/2} (1 + |\lambda|^2)^{-\varrho/2} \tilde{f}} = \widehat{(1 + |\lambda|^2)^{-r/2} \widetilde{I_\varrho f}} = I_r I_\varrho f.$$

In particular, if $\varrho = -r$

$$(7) \qquad I_r I_{-r} f = I_0 f = f,$$

i.e. the operations I_r and I_{-r} are mutually inverse.

It is not difficult to see also that the operation I_r maps S onto S in a 1—1 way and continuously: if $\varphi_m, \varphi \in S$ and $\varphi_m \to \varphi(S)$ as $m \to \infty$, then

$$I_r \varphi_m \to I_r \varphi(S).$$

One may also introduce an operation I_r^*, defined by the formula

$$I_r^* = \widetilde{(1 + |\lambda|^2)^{-r/2} \tilde{f}},$$

which it is natural to call the adjoint to I_r. It obviously has all the properties established above for I_r, including continuity in the sense of convergence in S.

The connection between I_r and I_r^* appears in the equations

$$(I_r f, \varphi) = (f, I_r^* \varphi),$$
$$(I_r^* f, \varphi) = (f, I_r \varphi) \; (f \in S', \varphi \in S).$$

It follows directly from these that the operations I_r and I_r^* are continuous on S' (weakly continuous), i.e. from the facts that $f_m, f \in S'$, $m = 1, 2, \dots$ and

$$f_m \to f(S'),$$

it follows that

$$I_r f_m \to I_r f, \; I_r^* f_m \to I_r^* f(S').$$

Indeed, for example,

$$(I_r f_m, \varphi) = (f_m, I_r^* \varphi) \to (f, I_r^* \varphi) = (I_r f, \varphi).$$

We note that for $r = -2$ one has the important equation

$$I_{-2} f = \widehat{(1 + |\lambda|^2) \tilde{f}} = f + \widehat{\sum_{j=1}^{n} \lambda_j^2 \tilde{f}}$$

$$= f - \sum_{j=1}^{n} \widehat{(i\lambda_j)^2 \tilde{f}} = f - \sum_{j=1}^{n} \frac{\partial^2 f}{\partial x_j^2} = f - \Delta f = (1 - \Delta) f,$$

where Δ is the Laplace operator.

Accordingly, for any natural number l

(8) $$I_{-2l} f = \widehat{(1 + |\lambda|^2)^l \tilde{f}} = (1 - \Delta)^l f \; (f \in S').$$

1.5.10. Regular generalized functions. Further extension of the concept of convolution. The operation I_r may serve as a convenient means for extending the concept of convolution to a class of generalized functions which we shall call *regular*.

The function $f \in S'$ will be called *regular in the sense of* L_p, and we will write $f \in S'_p$, if for some $\varrho_0 > 0$

(1) $$I_{\varrho_0} f = F \in L_p.$$

Suppose that μ is a multiplicator in $L_p (\hat{\mu} \in L$ for $p = 1)$ (see 1.5.1, 1.5.1.1). Suppose further that f is a function regular in the sense of L_p, for which property (1) is satisfied.

For $\varrho \geqq \varrho_0$ put

(2) $$\hat{\mu} * f = I_{-\varrho} (\hat{\mu} * I_\varrho f).$$

This definition does not depend on $\varrho \geqq \varrho_0$. Indeed, suppose along with (1)

$$(3) \qquad I_{\varrho'} f = F_1 \in L_p \quad (\varrho' > \varrho).$$

Then for $\varrho' - \varrho = r$, taking account of the fact that $I_{\varrho_0} f = F \in L_p$, we get

$$I_{-\varrho'}(\hat{\mu} * I_{\varrho'} f) = I_{-\varrho} I_{-r}(\hat{\mu} * I_r I_\varrho f)$$

$$= I_{-\varrho}\widehat{(1 + |\boldsymbol{x}|^2)^{r/2} \mu (1 + |\boldsymbol{x}|^2)^{-r/2} \widetilde{I_\varrho f}} = I_{-\varrho}\widehat{\mu \widetilde{I_\varrho f}} = I_{-\varrho}(\hat{\mu} I_{\varrho_0} f)$$

(see 1.5.1 (12) for $\mu \in L$ and 1.5.1.1 (9) for $1 \leqq p < \infty$). In the third equation we have used the fact, which will be proved below (see 8.1), that

$$\widehat{(1 + |\boldsymbol{x}|^2)^{-r/2}} \in L \qquad (r > 0),$$

and the fact that the function $(1 + |\boldsymbol{x}|^2)^\lambda$ is for any real λ infinitely differentiable and of polynomial growth.

For any real r the equation $I_r \boldsymbol{x} = \boldsymbol{x} = (x_1, \ldots, x_n)$ holds, which shows that the function $\boldsymbol{x}$ does not belong to $S_p'(1 \leqq p \leqq \infty)$, although it belongs to S'. This follows from 1.5. (12) for $\boldsymbol{k} = \omega = (1, \ldots, 1)$:

$$(1 + |\boldsymbol{x}|^2)^{-r/2} \tilde{\boldsymbol{x}} = i^n (2\pi)^{n/2} (1 + |\boldsymbol{x}|^2)^{-r/2} \delta^{(\omega)}(\boldsymbol{x}) = i^n (2\pi)^{n/2} \delta^{(\omega)}(\boldsymbol{x}) = \tilde{\boldsymbol{x}}.$$

It is important to note that for a generalized function f regular in the sense of L_p the equation

$$(4) \qquad I_{-\lambda}(\hat{\mu} * I_\lambda f) = \hat{\mu} * f$$

holds for any λ, positive or negative. Indeed, for f there exists a $\varrho > 0$ such that $I_\varrho f \in L_p$. For $\lambda \geqq \varrho$ equation (4) is already proved above. If on the other hand $\lambda < \varrho$, then we put $\varrho = \lambda + \sigma$ $(\sigma > 0)$. Then the function $I_\lambda f$ is regular. Indeed $I_\sigma I_\lambda f \in L_p$. Therefore

$$I_{-\lambda}(\hat{\mu} * I_\lambda f) = I_{-\lambda} I_{-\sigma}(\hat{\mu} * I_\varrho f) = I_{-\varrho}(\hat{\mu} * I_\varrho f) = \hat{\mu} * f.$$

It follows from (4) that for functions f regular in the sense of L_p and any real r

$$(5) \qquad I_r(\hat{\mu} * f) = I_r I_{-r}(\hat{\mu} * I_r f) = \hat{\mu} * I_r f,$$

i.e. the operation I_r may be applied under the sign of the convolution to a regular function f.

It follows from (5) that if μ is a Marcinkiewicz multiplier and f is a function regular in the sense of L_p, then the convolution $\hat{\mu} * f$ is also regular. Indeed, suppose that $I_r f \in L_p$. Then (5) holds, where the right hand side belongs to L_p.

Earlier equations 1.5.1.1 (9) were proved, which we write out in convolution terms:

$$(6) \qquad \hat{\lambda} * (\hat{\mu} * f) = \hat{\mu} * (\hat{\lambda} * f) = \widehat{\lambda\mu} * f, \quad f \in L_p (1 \leqq p < \infty).$$

They are true if λ and μ are Marcinkiewicz multipliers, from which it follows that $(\lambda\mu)$ is also a Marcinkiewicz multiplier. Now suppose that f is a generalized function regular in the sense of L_p and $I_\varrho f \in L_p$ $(\varrho > 0)$. Then equation (6) will be satisfied, if one substitutes $I_\varrho f$ into it in the place of f. But for regular f it is legitimate in all the terms of (6) to carry the operation I_ϱ across the convolution sign. But then the functions under the sign of I_ϱ are equal to one another. Thus we have proved that 6) holds for generalized function regular in the sense of L_p.

Chapter 2

Trigonometric Polynomials

2.1. Theorems on Zeros. Linear Independence

The function

$$(1) \qquad T_n(z) = \frac{\alpha_0}{2} + \sum_{k=1}^{n} (\alpha_k \cos kz + \beta_k \sin kz),$$

where α_k, β_k $(k = 0, 1, \ldots, n)$ are arbitrary complex numbers, and z is a complex or real variable, is said to be a *trigonometric polynomial of the nth order*. This definition does not exclude the case $\alpha_n = \beta_n = 0$.

A trigonometric polynomial is a function of period 2π, and accordingly in its study it suffices to restrict the considerations to the variation of the independent variable $z = x + iy$ in any vertical strip $a \leqq x < a + 2\pi$ (or $a < x \leqq a + 2\pi$), $-\infty < y < \infty$ of the complex palane, of width 2π.

Using the equations

$$(2) \qquad \cos kz = \frac{e^{ikz} + e^{-ikz}}{2}, \ \sin kz = \frac{e^{ikz} - e^{-ikz}}{2i}$$

$$(k = 0, 1, 2, \ldots)$$

the trigonometric polynomial may be transformed into the more symmetric form

$$(3) \qquad T_n(z) = \sum_{k=-n}^{n} c_k e^{ikz},$$

$$c_k = \frac{\alpha_k - \beta_k i}{2}, \quad c_{-k} = \frac{\alpha_k + \beta_k i}{2} \quad (k = 1, 2, \ldots),$$

$$c_0 = \frac{\alpha_0}{2}.$$

It is clear from (3) that if the coefficients α_k, β_k of the polynomial (1) are real, then the coefficients c_k, c_{-k} for each k are pairwise complex conjugate:

$$(4) \qquad c_{-k} = \bar{c}_k, \ k = 0, 1, \ldots, n.$$

Conversely, it follows from (4) that the numbers α_k, β_k are real.

The most important property of trigonometric polynomials is expressed by the theorem below.

2.1.1. In the statement of the theorem the number a will be said to be a *zero of multiplicity m of the function f*, if $f(a) = f'(a) = \cdots = f^{(m-1)}(a) = 0$, $f^{(m)}(a) \neq 0$.

Theorem. *A trigonometric polynomial T_n of order n, for which the coefficients c_n and c_{-n} in (3) are unequal to zero, has in any strip $a \leqq x < a + 2\pi$ of the complex plane $x = x + iy$ exactly $2n$ zeros, taking account of the multiplicity.*

If one denotes these zeros by $z_1, \ldots, z_{2n}$, then the equation

$$T_n(z) = A \prod_{k=1}^{n} \sin \frac{z - z_k}{2} \tag{1}$$

holds, where $A \neq 0$ is some constant. Conversely, equation (1) defines a trigonometric polynomial of order n.

Proof. We use the representation of T_n in the form 2.1 (3). After the substitution $Z = e^{iz}$, which transforms the strip of the z-plane in question in a 1—1 way onto the entire complex Z-plane less the point $Z = 0$, we get

$$T_n(Z) = \sum_{n=-k}^{n} c_k Z^k = Z^{-n} P_{2n}(Z),$$

where

$$P_{2n}(Z) = c_{-n} + c_{-n+1} Z + \cdots + c_n Z^{2n}.$$

In view of the conditions of the theorem $c_n \neq 0$ and $c_{-n} \neq 0$, and therefore the polynomial $P_{2n}(Z)$ of degree $2n$ has in the complex Z-plane exactly $2n$ zeros, taking account of multiplicity, not equal to zero.

It therefore follows that the trigonometric polynomial T_n has in the strip in question also exactly $2n$ zeros, taking account of multiplicity. We denote the zeros of the polynomial $P_{2n}(Z)$ by $Z_k = e^{iz_k}$ $(k = 1, \ldots, 2n)$. Then

$$T_n(z) = c_n e^{-inz} \prod_{k=1}^{2n} (e^{iz} - e^{iz_k})$$

$$= c_n e^{\frac{i}{2} \sum_{k=1}^{2n} z_k} \prod_{k=1}^{2n} \left(e^{i\frac{z-z_k}{2}} - e^{i\frac{z_k - z}{2}} \right) = A \prod_{k=1}^{2n} \sin \frac{z - z_k}{2},$$

where

$$A = c_n 2^{2n} (-1)^n e^{\frac{i}{2} \sum_{k=1}^{2n} z_k}.$$

Thus, the first part of the theorem is proved. In order to verify that the function (1), where the numbers z_k, $k = 1, \ldots, 2n$ belong to some strip of the complex plane closed on one side and of width 2π, is a trigonometric polynomial of order n, it is sufficient to carry through the transformation just done in the opposite direction, starting from (1).

2.1.2. Linear independence. If the trigonometric polynomial $T_p(z)$ is equal to zero at more than $2n$ points of the vertical strip of width 2π, then on the basis of Theorem 2.1.1. all of its coefficients must be equal to zero. In particular this holds if the trigonometric polynomial of order n is identically or almost everywhere equal to zero on the real axis.

Hence it follows that the system of functions

$$1, \cos x, \sin x, \ldots, \cos nx, \sin nx \tag{1}$$

is linearly independent in C^* and L_p^* (see 1.1.1 and 1.2.1). We need to take account of the fact that the null element in L_p^* is a function almost everywhere equal to zero.

The linear independence of the system (1) follows also from the orthogonality properties of the trigonometric functions

$$(m, n = 0, 1, 2, \ldots)$$

$$\frac{1}{\pi}\int_{-\pi}^{\pi} \sin mx \sin nx \, dx = \frac{1}{\pi}\int_{-\pi}^{\pi} \cos mx \cos nx \, dx = \begin{cases} 1, & m = n \\ 0, & m \neq n, \end{cases}$$

$$\int_{-\pi}^{\pi} \sin mx \cos nx \, dx = 0.$$

2.1.3. If T_m and T_n are trigonometric polynomials of orders m and n respectively, with $m \geqq n$, then their sum and difference is a trigonometric polynomial of order not larger than m.

Their product is a trigonometrical polynomial of order not exceeding $m + n$, which follows from the equations

$$\cos mx \cos nx = \frac{1}{2}\left[\cos (m - n) x + \cos (m + n) x\right],$$

$$\sin mx \sin nx = \frac{1}{2}\left[\cos (m - n) x - \cos (m + n) x\right],$$

$$\cos mx \sin nx = \frac{1}{2}\left[\sin (m + n) x - \sin (m - n) x\right].$$

2.1.4. From the orthogonality properties of the system 2.1.2. (1) it follows that if a trigonometric polynomial is an even function, then it contains as its terms only cosines ($\beta_k = 0$), and if it is an odd function, then only the sines ($\alpha_k = 0$).

Considering the real parts of the equation

$$\cos nx + i \sin nx = (\cos x + i \sin x)^n,$$

we get

$$\cos nx = \cos^n x - C_n^2 \cos^{n-2} x (1 - \cos^2 x) + C_n^4 \cos^{n-4} x\, (1 - \cos^2 x)^2 + \cdots,$$

from which it follows that every even trigonometric polynomial of the nth order may be represented in the form $P_n(\cos x)$, where

$$P_n(z) = a_0 + a_1 z + \cdots + a_n z^n$$

is some algebraic polynomial of the nth degree.

On the other hand, it follows from the equation

$$\cos^n x = \frac{(e^{ix} + e^{-ix})^n}{2^n} = \frac{1}{2^n}\,(e^{inx} + C_n^1 e^{i(x-2)n} + \cdots + e^{-inx})$$

that

$$\text{(1)}\quad \cos^n x = \frac{1}{2^{n-1}}\left[\cos nx + C_n^1 \cos (n-2)\, x + \cdots + C_n^{\frac{n}{2}-1} \cos 2x + \frac{C_n^{\frac{n}{2}}}{2}\right]$$

for n even,

$$\text{(1')}\quad \cos^n x = \frac{1}{2^{n-1}}\left[\cos nx + C_n^1 \cos (n-2)\, x + \cdots + C_n^{\left[\frac{n}{2}\right]} \cos x\right]$$

for n odd.

Thus the function $P_n(\cos x)$, where $P_n(z)$ is an algebraic polynomial of the nth degree, is an even polynomial of the nth order.

2.2. Important Examples of Trigonometric Polynomials

From the equation

$$\sum_{k=0}^{n} e^{ikx} = \frac{e^{i(n+1)x} - 1}{e^{ix} - 1} = \frac{e^{i\left(n+\frac{1}{2}\right)x} - e^{-i\frac{x}{2}}}{e^{i\frac{x}{2}} - e^{-i\frac{x}{2}}},$$

considering in it the real and imaginary parts separately, we get

(1) $$\frac{1}{2} + \sum_{k=1}^{n} \cos kx = \frac{\sin\left(n + \frac{1}{2}\right)x}{2\sin\frac{x}{2}} = D_n(x),$$

(2) $$\sum_{k=1}^{n} \sin kx = \frac{\cos\frac{x}{2} - \cos\left(n + \frac{1}{2}\right)x}{2\sin\frac{x}{2}} = D_n^*(x).$$

In particular equation (1) shows that the polynomial $D_n(x)$ vanishes at the points

$$x_k = \frac{2\pi k}{2n+1} \quad (k = 1, \ldots, 2n)$$

of the interval $(0, 2\pi)$, so that it may also be written in the form of a product

$$D_n(x) = A \prod_{k=1}^{2n} \sin\frac{x - x_k}{2},$$

where A is a constant. Putting $x = 0$ in this equation, we obtain the relation

$$\frac{2n+1}{2} = A \prod_{k=1}^{2n} \sin\frac{x_k}{2},$$

from which we may determine A.

The trigonometric polynomial $D_n(x)$ plays a large role in the theory of Fourier series. It is known as the *Dirichlet kernel.*

We note that (explanation below)

(3) $$\|D_n\|_{L^*} = \int_0^{\pi} \left| \frac{\sin\frac{2n+1}{2}x}{\sin\frac{x}{2}} \right| dx = 2\int_0^{\pi} \frac{\left|\sin\frac{2n+1}{2}x\right|}{x}\, dx + O(1)$$

$$= 2\int_0^{\frac{(2n+1)\pi}{2}} \frac{|\sin u|}{u}\, du + O(1) = 2\int_{\pi}^{n\pi} \frac{|\sin u|}{u}\, du + O(1)$$

$$= 2\sum_{k=1}^{n-1} \int_0^{\pi} \frac{|\sin u|}{k\pi + u}\, du + O(1) = 2\sum_{k=1}^{n-1} \frac{1}{k\pi} \int_0^{\pi} \sin u\, du + O(1)$$

$$= \frac{4}{\pi} \sum_{k=1}^{n-1} \frac{1}{k} + O(1) = \frac{4}{\pi} \ln n + O(1) \quad (n = 1, 2, \ldots).$$

The quantity $\frac{1}{\pi}\|D_n\|_{L^*}$ is called the *Lebesgue constant of the nth order Fourier sum.* Here $O(1)$ denotes some bounded function of the natural number n. In the calculations which we have presented we have made use of the boundedness of the function $x^{-1} - (\sin x)^{-1}$ on $\left[0, \frac{\pi}{2}\right]$ and the fact that for $u \in [0, \pi]$

$$\sum_{k=1}^{n-1} \left(\frac{1}{k\pi} - \frac{1}{k\pi + u}\right) \leqq c \sum_{1}^{n-1} \frac{1}{k^2} < c_1 < \infty.$$

2.2.1. Separating out the real and imaginary parts of the equation

$$\sum_{k=0}^{n} e^{i\left(k+\frac{1}{2}\right)x} = \frac{e^{i\frac{x}{2}} - e^{i\left(n+\frac{3}{2}\right)x}}{1 - e^{ix}} = \frac{1 - e^{i(n+1)x}}{e^{-i\frac{x}{2}} - e^{i\frac{x}{2}}},$$

we obtain

$$(1) \qquad \sum_{k=0}^{n} \cos\left(k + \frac{1}{2}\right) x = \frac{\sin (n+1) x}{2 \sin \frac{x}{2}},$$

$$(2) \qquad \sum_{k=0}^{n} \sin\left(k + \frac{1}{2}\right) x = \frac{1 - \cos (n+1) x}{2 \sin \frac{x}{2}} = \frac{\sin^2 \frac{n+1}{2} x}{\sin \frac{x}{2}}.$$

2.2.2. Using 2.2(1) and 2.2.1(2), we get

$$(1) \qquad \frac{1}{2} + \sum_{k=1}^{n} \frac{n+1-k}{n+1} \cos kx = \frac{1}{n+1} \sum_{k=0}^{n} D_k(x)$$

$$= \frac{1}{2(n+1) \sin \frac{x}{2}} \sum_{k=0}^{n} \sin\left(k + \frac{1}{2}\right) x$$

$$= \frac{1}{(n+1)} \frac{\sin^2 \frac{n+1}{2} x}{2 \sin^2 \frac{x}{2}} = F_n(x).$$

This polynomial in the theory of Fourier series is called the *Féjer kernel* of order n.

We note that the function

(2) $$k_\nu(x) = \left(\frac{\sin\frac{\lambda x}{2}}{\sin\frac{x}{2}}\right)^{2\sigma},$$

where λ and σ are natural numbers, is a trigonometric polynomial of order $\nu = \sigma(\lambda - 1)$, since it differs from the σth power of the Féjer kernel $F_{\lambda-1}(x)$ only by a constant factor.

In what follows it will be useful to estimate the exact order of variation of the quantity

(3) $$a_\nu = \int_{-\pi}^{\pi} k_\nu(x)\,dx = 2\int_0^{\pi} k_\nu(x)\,dx,$$

when $\nu = 1, 2, \ldots,$

In the calculation below and throughout this book we suppose that $a_\lambda \sim b_\lambda (\lambda \in \mathscr{E})$, where $\mathscr{E}$ is some set of numbers λ, if there exist two positive constants c_1 and c_2 such that the inequality $c_1 a_\lambda \leqq b_\lambda \leqq c_2 a_\lambda$ holds for all $\lambda \in \mathscr{E}$.

If we take into account that

(4) $$\frac{2}{\pi}\alpha \leqq \sin\alpha \leqq \alpha \quad \left(0 \leqq \alpha \leqq \frac{\pi}{2}\right),$$

then we will have

$$2^{2\sigma+1}\int_0^{\pi}\left(\frac{\sin\frac{\lambda x}{2}}{x}\right)^{2\sigma} dx \leqq a_\nu \leqq 2\pi^{2\sigma}\int_0^{\pi}\left(\frac{\sin\frac{\lambda x}{2}}{x}\right)^{2\sigma} dx.$$

But

$$\int_0^{\pi}\left(\frac{\sin\frac{\lambda x}{2}}{x}\right)^{2\sigma} dx = \left(\frac{\lambda}{2}\right)^{2\sigma-1}\int_0^{\frac{\lambda\pi}{2}}\left(\frac{\sin t}{t}\right)^{2\sigma} dt \sim \lambda^{2\sigma-1}$$

$$(\lambda = 1, 2, \ldots).$$

Thus it is obvious that for any fixed σ

(5) $$a_\nu \sim \lambda^{2\sigma-1} \quad (\lambda = 1, 2, \ldots).$$

We introduce further the trigonometric polynomial

$$(6)\qquad d_\nu(x) = \frac{1}{a_\nu} k_\nu(x) = \frac{1}{a_\nu}\left(\frac{\sin\frac{\lambda x}{2}}{\sin\frac{x}{2}}\right)^{2\sigma} \qquad (\nu = \sigma(\lambda - 1)),$$

where $\sigma > 0$ is a fixed positive number, $\lambda = 1, 2, \ldots,$ and a_ν is the constant defined by equation (3).

2.3. The Trigonometric Interpolation Polynomial of Lagrange

If two trigonometric polynomials $T_n(x)$ and $Q_n(x)$ coincide at $2n+1$ distinct points of the half-closed interval $a \leqq x < a + 2\pi$, then their difference $T_n(x) - Q_n(x)$, being a polynomial of order n, is equal to zero at these points, and accordingly is identically equal to zero, since no non-identically zero polynomial of the nth order can have more than $2n$ zeros on the period.

Thus the trigonometrical polynomial $T_n(x)$ is completely defined by its values

$$y_0, y_n, \ldots, y_{2n},$$

corresponding to some or other $2n + 1$ distinct points

$$x_0 < x_1 < \cdots < x_{2n} < x_0 + 2\pi$$

of the period.

It is not hard to write down an effective expression for it.

Indeed, on the basis of 2.1.1 the function

$$Q^{(m)}(x) = \frac{\sin\frac{x - x_0}{2} \cdots \sin\frac{x - x_{m-1}}{2} \sin\frac{x - x_{m+1}}{2} \cdots \sin\frac{x - x_{2n}}{2}}{\sin\frac{x_m - x_0}{2} \cdots \sin\frac{x_m - x_{m-1}}{2} \sin\frac{x_m - x_{m+1}}{2} \cdots \sin\frac{x_m - x_{2m}}{2}}$$

$$(m = 0, 1, \ldots, 2\pi)$$

is a trigonometric polynomial of order n, having obviously the property

$$Q^{(m)}(x_k) = \begin{cases} 1, & k = m, \\ 0, & k \neq m \end{cases} \qquad (k, m = 0, 1, \ldots, 2m).$$

Therefore the desired trigonometric polynomial $T_n(x)$, satisfying the conditions

$$T_n(x_k) = y_k \quad (k = 0, 1, \ldots, 2n),$$

may be written out in the form

$$T_n(x) = \sum_{m=0}^{2} Q^{(m)}(x) y_m$$

$$= \sum_{m=0}^{2n} \frac{\sin \frac{x - x_0}{2} \cdots \sin \frac{x - x_{m-1}}{2} \sin \frac{x - x_{m+1}}{2} \cdots \sin \frac{x - x_{2n}}{2}}{\sin \frac{x_m - x_0}{2} \cdots \sin \frac{x_m - x_{m-1}}{2} \sin \frac{x_m - x_{m+1}}{2} \cdots \sin \frac{x_m - x_{2n}}{2}} y_m .$$

The case of equidistant interpolation nodes is particularly important. This is the case

$$x_k = \frac{2k\pi}{2n+1} \qquad (k = 0, 1, \ldots, 2n).$$

In this case one may write down a simple expression for $Q^{(m)}(x)$, if we take into account that the trigonometrical polynomial

$$(1) \qquad D_n(x) = \frac{1}{2} + \cos x + \cdots + \cos nx = \frac{\sin \frac{2n+1}{2} x}{2 \sin \frac{x}{2}}$$

has the properties

$$D_n(0) = \frac{2n+1}{2}, \quad D_n(x_k) = 0, \quad x_k = \frac{2k\pi}{2n+1}$$

$$(k = 1, 2, \ldots, 2n).$$

Hence it follows that the polynomial

$$Q^{(m)}(x) = \frac{2}{2n+1} D_n(x - x_m) \qquad (m = 0, 1, 2, \ldots)$$

satisfies the conditions

$$Q^{(m)}(x_k) = \begin{cases} 1, & k = m, \\ 0, & k \neq m. \end{cases}$$

Thus every trigonometric polynomial

$$(2) \qquad T_n(x) = \frac{a_0}{2} + \sum_{k=1}^{n} (a_k \cos kx + b_k \sin kx)$$

may be written in the form

$$(3)\qquad T_n(x) = \frac{2}{2n+1}\sum_{k=0}^{n} D_n(x - x_k)\, T_n(x_k)$$

$$= \frac{1}{2\pi+1}\sum_{k=0}^{2n} \frac{\sin\frac{2n+1}{2}(x - x_k)}{\sin\frac{x - x_k}{2}}\, T_n(x_k).$$

Substituting for $D_n(x)$ in this equation the corresponding sum, we get

$$T_n(x) = \frac{2}{2n+1}\sum_{k=0}^{2n}\sum_{i=0}^{n}{}' \cos i(x - x_k)\, T_n(x_k)$$

$$= \frac{2}{2n+1}\sum_{i=0}^{n}{}' \left[\left(\sum_{k=0}^{2n}\cos ix_k T_n(x_k)\right)\cos ix + \left(\sum_{k=0}^{2n}\sin ix_k T_n(x_k)\right)\sin ix\right].$$

Here we are supposing that $\sum_{k=0}^{n}{}' u_k = \frac{u_0}{2} + \sum_{k=1}^{n} u_k$. Comparing the coefficients on $\cos ix$ and $\sin ix$ with the corresponding coefficients of $T_n(x)$, we get

$$a_i = \frac{2}{2n+1}\sum_{k=0}^{2n}\cos ix_k T_n(x_k) \quad (i = 0, 1, \ldots, n),$$

$$b_i = \frac{2}{2n+1}\sum_{k=0}^{2n}\sin ix_k T_n(x_k) \quad (i = 1, 2, \ldots, n).$$

2.4. The Interpolation Formula of M. Riesz

This formula is due to M. Riesz [1].

If $T_n(\theta)$ is a trigonometric polynomial

$$(1)\qquad T_n(\theta) = \frac{a_0}{2} + \sum_{k=1}^{n}(a_k\cos k\theta + b_k\sin k\theta),$$

then it satisfies the identity

$$(2)\qquad T_n(\theta) = a_n\cos n\theta + \frac{\cos n\theta}{2\pi}\sum_{k=1}^{2n}(-1)^k \operatorname{ctg}\frac{\theta - \theta_k}{2}\, T_n(\theta_k),$$

where

$$\theta_k = \frac{2k-1}{2n}\pi \quad (k = 1, 2, \ldots, 2n).$$

We shall prove this.

The points θ_k are the zeros of the polynomial $\cos n\theta$, so that

$$\cos n\theta = A \prod_{k-1}^{2n} \sin \frac{\theta - \theta_k}{2}. \tag{3}$$

Hence the function

$$Q^{(m)}(\theta) = \frac{\cos n\theta}{2n} (-1)^m \cot \frac{\theta - \theta_m}{2}$$

$$= (-1)^{m+1} \frac{\cos n\theta}{2\pi} \frac{\sin \dfrac{\theta - (\pi + \theta_m)}{2}}{\sin \dfrac{\theta - \theta_m}{2}}$$

$$(m = 1, 2, \ldots, 2n)$$

is a trigonometric polynomial of order n, since it is a product of the type (3), in which the factor $\sin \frac{\theta - \theta_m}{2}$ is replaced by the factor $\sin \frac{\theta - (\pi + \theta_m)}{2}$. This polynomial obviously, is equal to zero at all the points θ_k, with the exclusion of the point θ_m, where it is equal to unity. The last point may be verified by applying L'Hôpital's rule. So,

$$Q^{(m)}(\theta_k) = \begin{cases} 1, & m = k, \\ 0, & m \neq k \end{cases} \quad (k, m = 1, 2, \ldots, 2n).$$

Hence it follows that the function

$$T_n^*(\theta) = \frac{\cos n\theta}{2n} \sum_{k=1}^{2n} (-1)^k \cot \frac{\theta - \theta_k}{2} T_n(\theta_k)$$

is a trigonometric polynomial of order n, coinciding with the original polynomial $T_n(\theta)$ at the zeros of $\cos n\theta$. In such a case, on the basis of Theorem 2.1.1 on the zeros of the trigonometric polynomial

$$T_n(\theta) = c \cos n\theta + T_n^*(\theta), \tag{4}$$

where c is a constant.

It remained for us still to prove that

$$c = a_n. \tag{5}$$

Indeed, the Fourier coefficient of the trigonometric polynomial $\cos n\theta \cot \dfrac{\theta - \theta_k}{2}$, corresponding to $\cos n\theta$, is equal to

$$\frac{1}{\pi} \int_{-\pi}^{\pi} \cos^2 n\theta \cot \frac{\theta - \theta_k}{2} \, d\theta = \frac{1}{\pi} \int_{-\pi}^{\pi} \cos^2 n(u + \theta_k) \cot \frac{u}{2} \, du$$

$$= \frac{1}{\pi} \int_{-\pi}^{\pi} \sin^2 nu \cot \frac{u}{2} \, du = 0,$$

since the integrand in the last integral is odd. In such a case the polynomial $Q^{(m)}(\theta)$, and accordingly the polynomial $T_n^*(\theta)$ as well, do not contain terms with $\cos n\theta$. Therefore (1) and (4) imply (5).

The identity (2) is proved. If we differentiate it and then put $\theta = 0$, we obtain

$$T_n'(0) = \frac{1}{4n} \sum_{k=1}^{2n} (-1)^{k+1} \frac{1}{\sin^2 \dfrac{\theta_k}{2}} T_n(\theta_k).$$

This last equation is valid for any polynomial of order n, and in particular it is valid for the polynomial $T_n(u + \theta)$, where u is a variable, and θ is fixed arbitrarily. Thus, for any θ we have

$$T_n'(\theta) = \frac{1}{4n} \sum_{k=1}^{2n} (-1)^{k+1} \frac{1}{\sin^2 \dfrac{\theta_k}{2}} T_n(\theta + \theta_k). \tag{6}$$

This is then the formula of M. Riesz.

2.5. Bernšteĭn's Inequality

If one puts $T_n(\theta) = \sin n\theta$ in Riesz' formula 2.4.(6), then for $\theta = 0$ we get

$$n = \frac{1}{4n} \sum_{k=1}^{n} \frac{1}{\sin^2 \dfrac{\theta_k}{2}}. \tag{1}$$

Therefore, from 2.4 (6), for any trigonometrical polynomial of order n there follows the inequality

(2) $$\|T_n'\|_{L_p^*} \leq n\,\|T_n\|_{L_p^*} \qquad (1 \leq p \leq \infty),$$

$$\|f\|_{L_p^*} = \left(\int_0^{2\pi} |f|^p \, d\theta\right)^{1/p},$$

called *Bernšteĭn's inequality.*

It is exact in the sense that there exists a trigonometrical polynomial for which it turns into an equation. Indeed, this is the case for the polynomial

$$T_n(\theta) = A \sin(n\theta + \alpha),$$

where A and α are arbitrary real constants.

2.6. Trigonometric Polynomials of Several Variables

A function of the form

(1) $$T_{\nu_1,\ldots,\nu_n}(z_1, \ldots, z_n) = \sum_{\substack{-\nu_l \leq k_l \leq \nu_l \\ l=1,\ldots,n}} c_{k_1,\ldots,k_n} e^{i(k_1 z_1 + \cdots + k_n z_n)},$$

where $\nu_1, \ldots, \nu_n$ are natural numbers, $z_1, \ldots, z_n$ are complex variables, and $c_{k_1,\ldots,k_n}$ are constant coefficients, generally speaking complex, depending on integers $k_1, \ldots, k_n$, is said to be a *trigonometric polynomial of orders* $\nu_1, \ldots, \nu_n$ *respectively relative to the variables* $z_1, \ldots, z_n$.

We require some notation:

$$\boldsymbol{\nu} = (\nu_1, \ldots, \nu_n), \qquad \boldsymbol{k} = (k_1, \ldots, k_n),$$

$$\boldsymbol{z} = (z_1, \ldots, z_n), \qquad \boldsymbol{kz} = \sum_1^n k_l z_l$$

Further we will write

$$T = T_{\boldsymbol{\nu}}(\boldsymbol{z}) = \sum_{\substack{-\nu_l \leq k_l \leq \nu_l \\ l=1,\ldots,n}} c_{\boldsymbol{k}} e^{i\boldsymbol{kz}}$$

and say that $T = T_{\boldsymbol{\nu}}$ is a trigonometric polynomial of $\boldsymbol{z}$ of order $\boldsymbol{\nu}$.

If the coefficients satisfy the relations

(2) $$c_{-\boldsymbol{k}} = \bar{c}_{\boldsymbol{k}},$$

i.e. if they change into their conjugate numbers on changing the signs of all the indices k_l, then for real $\boldsymbol{z} = (z_1, \ldots, z_n)$ the polynomial $T_{\boldsymbol{\nu}}$ is a real function. Indeed, if $\boldsymbol{x} = (x_1, \ldots, x_n)$ is a real point, then in view of (2)

$$\overline{T_\nu(\boldsymbol{x})} = \sum_{\substack{|k_l| \leqq \nu_l \\ l=1,\ldots,n}} \bar{c}_k e^{-ikx} = \sum_{\substack{-\nu_l \leqq -k_l \leqq \nu_l \\ l=1,\ldots,n}} c_{-k} e^{-ikx} = T_\nu(\boldsymbol{x}).$$

We will mainly have to do with polynomials satisfying the condition (2), which we naturally call *real trigonometric polynomials.*

For complex $\boldsymbol{z}$ the real polynomials $T_n(\boldsymbol{z})$ are generally speaking not real, but they become real functions if one considers them as functions of the real vector $\boldsymbol{x} = (x_1, \ldots, x_n)$.

We have defined real trigonometric polynomials T_ν as linear combinations (1) of complex functions e^{ikx} with real coefficients, satisfying the adjointness conditions (2), but they may be defined also as linear combinations with real coefficients of real functions. Such functions are all possible products of the type

$$\varphi_1(x_1)\,\varphi_2(x_2)\ldots\varphi_n(x_n), \tag{3}$$

where $\varphi_l(x_l)$ $(l = 1, \ldots, n)$ is either the function $\sin kx_l (1 \leqq k \leqq \nu_l)$, or the function $\cos kx_l (0 \leqq k \leqq \nu_l)$.

Conversely, every linear combination of functions of the type (3) with real coefficients is a sum of type (1) with coefficients satisfying the adjointness condition (2), i.e. a real trigonometric polynomial of order $\boldsymbol{\nu} = (\nu_1, \ldots, \nu_n)$.

Trigonometric polynomials T_ν are continuous periodic functions relative to each variable, and, accordingly, they enter as elements into the space $C_*^{(n)}$ and moreover into the space $L_{\underset{p}{*}}^{(n)}$ (see 1.1.1).

Various functions of type (3) satisfy the condition of orthogonality on the cube

$$\Delta^{(n)} = \{0 \leqq x_k \leqq 2\pi;\ k = 1, \ldots, n\},$$

and therefore form a linearly independent system in $C_*^{(n)}$ and in $L_{\underset{p}{*}}^{(n)} (1 \leqq p \leqq \infty)$.

As an example we note that every real trigonometric polynomial of orders μ, ν relative to x, y respectively may be written in the form

$$T_{u,\nu}(x, y) = \sum_{k=0}^{\mu} \sum_{l=0}^{\nu} (a_{kl} \cos kx \cos ly + b_{kl} \cos kx \sin ly$$
$$+ c_{kl} \sin kx \cos ly + d_{kl} \sin kx \sin ly),$$

where $a_{kl}, b_{kl}, c_{kl}, d_{kl}$ are real coefficients.

If in the polynomial $T_\nu(\boldsymbol{z})$ one fixes all the variables except one, for example z_l, then we get, obviously, a trigonometric polynomial of one variable z_l of degree ν_l. To it are applicable all the properties of trigonometric polynomials of one variable.

However trigonometric polynomials of several variables have also interesting properties which whould have been hard to predict on the basis only of the knowledge of the properties of polynomials of one variable.

2.7. Trigonometric Polynomials Relative to Certain Variables

Suppose that $\mathscr{E} = \mathbb{R}_m \times \mathscr{E}' \subset \mathbb{R}_n$ is a cylindrical set of points $\boldsymbol{x} = (\boldsymbol{u}, \boldsymbol{y})$, $\boldsymbol{u} = (x_1, \ldots, x_m) \in \mathbb{R}_m$, $\boldsymbol{y} = (x_{m+1}, \ldots, x_n) \in \mathscr{E}'$, where $\mathscr{E}'$ is a measurable $(n - m)$-dimensional set. We select from $\mathscr{E}$ the truncated cylinder

$$\mathscr{E}_* = \Delta^{(m)} \times \mathscr{E}',$$

where

$$\Delta^{(m)} = \{0 \leqq x_k \leqq 2\pi;\ k = 1, \ldots, m\}$$

is an m-dimensional cube, and we introduce the space $L_p^*(\mathscr{E})$ of functions $f = f(\boldsymbol{x})$ (real or complex), lying in $L_p(\mathscr{E}^*)$ and, for almost all $\boldsymbol{y} \in \mathscr{E}'$ (in the sense of $(n-m)$-dimensional measure), periodic of period 2π relative to each of the variables $x_1, \ldots, x_m$. Obviously $L_p^*(\mathscr{E})$ is a complete space.

We further denote by

$$T_\nu(\boldsymbol{x}) = T_\nu(\boldsymbol{u}, \boldsymbol{y}) = T_\nu(x_1, \ldots, x_m, \boldsymbol{y})$$

functions such that each of them belongs to $L_p^*(\mathscr{E})$ and for almost all $\boldsymbol{y} \in \mathscr{E}'$ a trigonometric polynomial of order $\boldsymbol{\nu} = (\nu_1, \ldots, \nu_m)$ relative to $\boldsymbol{u} = (x_1, \ldots, x_m)$. Here we should keep in mind that a function equivalent relative to $\mathscr{E}$ to the function $T_\nu(\boldsymbol{x})$ is considered as equal to $T_\nu(\boldsymbol{x})$.

The set of all such functions for a given ν will be denoted by $\mathfrak{M}_{\nu p}^*(\mathscr{E})$. It is obviously linear.

Each function $T_\nu \in L_p(\mathscr{E}*)$, and therefore by Fubini's theorem there exists a set $\mathscr{E}_1' \subset \mathscr{E}'$ of full measure such that $T_\nu(\boldsymbol{u}, \boldsymbol{y}) \in L_p(\Delta^{(m)}) \subset L(\Delta^{(m)})$ relative to $\boldsymbol{u}$ for all $\boldsymbol{y} \in \mathscr{E}_1'$ ($\Delta^{(m)}$ is bounded!). We may at the same time suppose that for all $\boldsymbol{y} \in \mathscr{E}_1'$ there is a representation

$$T_\nu(\boldsymbol{u}, \boldsymbol{y}) = \sum_{\substack{-\nu_l \leqq k_l \leqq \nu_l \\ l=1, \ldots, n}} c_{\boldsymbol{k}}(\boldsymbol{y}) e^{i\boldsymbol{k}\boldsymbol{u}}, \tag{1}$$

where the $c_k(\boldsymbol{y})$ are certain functions depending on $\boldsymbol{y}$. In view of the orthogonality properties of $e^{i\boldsymbol{k}\boldsymbol{u}}$ one has the equation

$$(2) \qquad c_k(\boldsymbol{y}) = \frac{1}{(2\pi)^m} \int\limits_{\Delta^{(m)}} T_\nu(\boldsymbol{u}, \boldsymbol{y}) e^{i\boldsymbol{k}\boldsymbol{u}} d\boldsymbol{u},$$

from which it follows in particular, by Fubini's theorem, that the $c_k(\boldsymbol{y})$ are measurable functions on $\mathscr{E}'$ because T_ν, belonging as it does to $L_p(\mathscr{E}_*)$, is in every case locally summable (even if $\mathscr{E}'$ were unbounded). Applying the general form of the Minkowski inequality, and then the Hölder inequality, we get

$$(3) \quad \|c_k(\boldsymbol{y})\|_{L_p(\mathscr{E}')} \leqq \frac{1}{(2\pi)^m} \int\limits_{\Delta^{(m)}} \|T_\nu(\boldsymbol{u}, \boldsymbol{y})\|_{L_p(\mathscr{E}')} d\boldsymbol{u}$$

$$\leqq \frac{1}{(2\pi)^m} |\Delta^{(m)}|^{1/q} \|T_\nu\|_{L_p(\mathscr{E}_*)} = c \|T_\nu\|_{L_p(\mathscr{E}_*)} \left(\frac{1}{p} + \frac{1}{q} = 1\right),$$

where $|\Delta^{(m)}|$ denotes the m-dimensional volume of $\Delta^{(m)}$ and c is a constant.

We have proved that each function $T_\nu \in \mathfrak{M}^*_{\nu p}(\mathscr{E})$ is representable in the form (1), where the $c_k(\boldsymbol{y})$ satisfy the inequalities (3). The converse is obviously also true.

Using this property of the functions of $\mathfrak{M}^*_{\nu p}(\mathscr{E})$ and the fact that the space $L_p(\mathscr{E}')$ is complete, we see easily that the following lemma is true.

2.7.1. Lemma. *The set $\mathfrak{M}^*_{\nu p}(\mathscr{E})$ is a subspace in $L^*_\nu(\mathscr{E})$.*

If $\mathscr{E} = \mathbb{R}_n$, i.e. $\mathscr{E}'$ is empty, then $\mathfrak{M}^*_{\nu p}(\mathscr{E})$ is obviously in addition a finite-dimensional subspace. If on the other hand $\mathscr{E}'$ has positive $(n-m)$-dimensional measure, then $\mathfrak{M}^*_{\nu p}(\mathscr{E})$ is not finite-dimensional.

2.7.2. For the functions

$$T_\nu = T_\nu(x_1, \boldsymbol{y}) \in \mathfrak{M}^*_{\nu p}(\mathscr{E}) = \mathfrak{M}^*_{\nu p}(\mathbb{R}_1 \times \mathscr{E}'),$$

which are trigonometric polynomials relative to x_1 of degree ν, for almost all $\boldsymbol{y} \in \mathscr{E}'$ the generalized Bernšteĭn inequality

$$(1) \qquad \left\| \frac{\partial T_\nu}{\partial x_1} \right\|_{L_p(\mathscr{E}_*)} \leqq \nu \|T_\nu\|_{L_p(\mathscr{E}_*)}$$

$$(\mathscr{E}_* = [0, 2\pi] \times \mathscr{E}'; \ x_1 \in [0, 2\pi], \ y \in \mathscr{E}')$$

is satisfied. Indeed, $T_\nu(x_1, \mathbf{y})$ is a trigonometric polynomial relative to x_1 for all $\mathbf{y} \in \mathscr{E}_1' \subset \mathscr{E}'$, where $\mathscr{E}_1'$ is a set of full measure in $\mathscr{E}'$. Therefore, on the basis of 2.5 (2), for $1 \leqq p < \infty$

$$\int_0^{2\pi} \left| \frac{\partial T_\nu(x_1, \mathbf{y})}{\partial x_1} \right|^p dx_1 \leqq \nu^p \int_0^{2\pi} |T_\nu(x_1, \mathbf{y})|^p \, dx_1 \quad (\mathbf{y} \in \mathscr{E}_1'). \tag{2}$$

Integrating both sides of this inequality relative to $\mathbf{y} \in \mathscr{E}_1$ and raising to the $(1/p)^{\text{th}}$ power, we get (1). For $p = \infty$ inequality (1) follows in an obvious way from the corresponding inequality 2.5 (2).

2.7.3. For other inequalities involving trigonometric polynomials, which we will use extensively, see 3.3 and 3.4.

Chapter 3

Entire Functions of Exponential Type, Bounded on $\mathbb{R}_n$

3.1. Preparatory Material

In this chapter we consider certain properties of entire functions of exponential type, bounded on the real space $\mathbb{R}_n = \mathbb{R}$. We see that they are very analogous to the corresponding properties of trigonometric polynomials. While trigonometric polynomials are a good means of approximation of periodic functions, entire functions of exponential type may serve as a means of approximation[1] of nonperiodic functions, given on n-dimensional space. Perhaps the reader inexperienced in these questions ought to start this chapter by reading 3.1.1, where we give in addition general information on the theory of multiple power series.

Suppose we are given a nonnegative vector $\boldsymbol{v} = (\nu_1, \ldots, \nu_n) \geqq \mathbf{0}$.

The function

$$g = g_\nu(\boldsymbol{z}) = g_{\nu_1, \ldots, \nu_n}(z_1, \ldots, z_n)$$

is called an *entire function of exponential type* $\boldsymbol{v}$, if it satisfies the following properties:

1) it is an entire function in all of its variables, i.e. it decomposes into a power series

$$g(\boldsymbol{z}) = \sum_{\boldsymbol{k} \geqq \mathbf{0}} a_{\boldsymbol{k}} \boldsymbol{z}^{\boldsymbol{k}} = \sum_{\substack{k_l \geqq 0 \\ l=1, \ldots, n}} a_{k_1, \ldots, k_n} z_1^{k_1} \ldots z_n^{k_n} \tag{1}$$

with constant coefficients $a_k (a_{k_1, \ldots, k_n})$, and converges absolutely for all complex $\boldsymbol{z} = (z_1, \ldots, z_n)$.

2) For every $\varepsilon > 0$ there exists a positive number A_ε such that for all complex $z_k = x_k + i y_k$ $(k = 1, \ldots, n)$ the inequality

$$|g(\boldsymbol{z})| \leqq A_\varepsilon \exp \sum_{j=1}^{n} (\nu_j + \varepsilon) \, |z_j| \, . \tag{2}$$

is satisfied.

---

[1] However, while a trigonometric polynomial is defined by a finite number of numerical parameters (coefficients), a function of exponential type is generally speaking defined by an infinite (countable) number of parameters (for example, the coefficients of its Taylor series). Therefore in practical calculations it would seem necessary to approximate them by simpler functions.

We shall say further that the function g_ν indicated above *belongs to the class* E_ν.

Put $\varrho = (\varrho_1, \ldots, \varrho_n)$ $(\varrho_j > 0;\ j = 1, \ldots, n)$, and suppose that

$$|z_j| \leqq \varrho_j,\ M(\varrho) = \sup_{|z_j| \leqq \varrho_j} |g(\boldsymbol{z})|.$$

Then it follows from property (2), obviously, that the inequality

$$M(\varrho) \leqq A_\varepsilon \exp \sum_{j=1}^{n} (\nu_j + \varepsilon)\, \varrho_j$$

holds, and conversely, because

$$|g(\boldsymbol{z})| \leqq M(|z_1|, \ldots, |z_n|) \leqq A_\varepsilon \exp \sum_{j=1}^{n} (\nu_j + \varepsilon) |z_j|.$$

The derivative of order $\boldsymbol{k} = (k_1, \ldots, k_n)$ of g at the point $\boldsymbol{z} = (z_1, \ldots, z_n)$ may be written in terms of the Cauchy formula

$$g^{(\boldsymbol{k})}(\boldsymbol{z}) = \frac{\boldsymbol{k}!}{(2\pi i)^n} \int_{C_1} \cdots \int_{C_n} \frac{g(\zeta_1, \ldots, \zeta_n)\, d\zeta_1 \ldots d\zeta_n}{\prod_{j=1}^{n} (\zeta_j - z_j)^{k_j+1}}, \tag{3}$$

where c_j is a circumference in the ζ_j plane with center at z_j. Accordingly, if one supposes that $\boldsymbol{z} = \boldsymbol{0}$ and c_j has the radius ϱ_j, then we obtain the Cauchy inequality

$$|a_{\boldsymbol{k}}| \leqq \frac{M(\varrho)}{\varrho^{\boldsymbol{k}}}.$$

Suppose that

$$\varrho_j = \frac{k_j}{\nu_j + \varepsilon}.$$

Then

$$|a_{\boldsymbol{k}}| \leqq A_\varepsilon \frac{e^{|\boldsymbol{k}|}(\boldsymbol{\nu} + \boldsymbol{\varepsilon})^{\boldsymbol{k}}}{\boldsymbol{k}^{\boldsymbol{k}}} \qquad \big(\boldsymbol{\varepsilon} = (\varepsilon, \ldots, \varepsilon)\big). \tag{4}$$

We have proved that it follows from (2) that for each $\varepsilon > 0$ there exists an A_ε such that (4) is satisfied. By the Stirling formula

$$\frac{e^{|\boldsymbol{k}|}}{\boldsymbol{k}^{\boldsymbol{k}}} = \frac{\left(\sqrt{2\pi}\right)^n (k_1 \ldots k_n)^{1/2}}{\boldsymbol{k}!} \prod_{j=1}^{n} (1 + \varepsilon_{k_j})^{k_j} \ (\varepsilon_{k_j} \to 0,\ k_j \to \infty),$$

so that (2) follows from (4), but generally with a different constant B_ε:

$$M(\varrho) \leqq \sum_{\boldsymbol{k}} |a_{\boldsymbol{k}}|\, \varrho^{\boldsymbol{k}} \leqq B_\varepsilon \sum_{\boldsymbol{k}} \frac{(\boldsymbol{\nu} + 2\boldsymbol{\varepsilon})^{\boldsymbol{k}}}{\boldsymbol{k}!}\, \varrho^{\boldsymbol{k}} = B_\varepsilon e^{\sum_1^n (\nu_j + 2\varepsilon)\varrho_j},$$

where B_ε is a sufficiently large number depending on ε.

It follows from what has been proved that if $g \in E_\nu$, then any of its partial derivatives satisfies $g^{(\lambda)} \in E_\nu$. The point is that it follows from (4) that the modulus of the $(\boldsymbol{k} - \boldsymbol{\lambda})^{\text{th}}$ coefficient of the power series for $g^{(\lambda)}$ satisfies the inequality

$$\left| \frac{\boldsymbol{k}!}{(\boldsymbol{k} - \boldsymbol{\lambda})!} a_{\boldsymbol{k}} \right| \leq A'_\varepsilon \frac{e^{|\boldsymbol{k}-\boldsymbol{\lambda}|}(\boldsymbol{\nu} + 2\varepsilon)^{\boldsymbol{k}-\boldsymbol{\lambda}}}{(\boldsymbol{k} - \boldsymbol{\lambda})^{\boldsymbol{k}-\boldsymbol{\lambda}}},$$

where A'_ε is sufficiently large.

It follows from what has been said that in the case of an entire function

$$f(z) = \sum_{k=0}^{\infty} a_k z^k$$

of one variable the following two conditions, each of which expresses the fact that f is of exponential type of degree ν, are equivalent:

$$\overline{\lim_{r\to\infty}} \frac{\ln M(r)}{r} \leq \nu \tag{5}$$

and

$$\overline{\lim_{n\to\infty}} \frac{n}{e} \sqrt[n]{|a_n|} = \overline{\lim_{n\to\infty}} \sqrt[n]{n!|a_n|} \leq \nu. \tag{6}$$

Denote by $\mathfrak{M}_{\nu p}(\mathbb{R}) = \mathfrak{M}_{\nu p}$ $(1 \leq p \leq \infty)$ the collection of all entire functions of exponential type ν which as functions of a real $\boldsymbol{x} \in \mathbb{R} = \mathbb{R}_n$ lie in $L_p = L_p(\mathbb{R})$. Further we put $\mathfrak{M}_{\boldsymbol{\nu}} = \mathfrak{M}_{\boldsymbol{\nu}\infty}$, so that $\mathfrak{M}_{\boldsymbol{\nu}}$ consists of all functions of type $\boldsymbol{\nu}$ which are bounded on $\mathbb{R}$.

We further note here that in what follows it will be proved (see 3.2.5 or 3.3.5) that $\mathfrak{M}_\nu \subset \mathfrak{M}_{\nu p}$ for any p, $1 \leq p \leq \infty$. Moreover, it will be clear (see 3.2.2 (10)) that for each function $g \in \mathfrak{M}_{\boldsymbol{\nu}}$ there exists a constant A not depending on $\boldsymbol{z}$ such that

$$|g(\boldsymbol{z})| \leq A e^{\sum_{j=1}^{n} \nu_j |y_k|} \qquad (z_j = x_j + iy_j). \tag{7}$$

This inequality is stronger than inequality (2). It follows directly from it that g is bounded on $\mathbb{R}_n$. Thus $\mathfrak{M}_{\boldsymbol{\nu}}$ may be defined as the class of entire functions $f(\boldsymbol{z})$ for which (7) holds.

The functions

$$e^{ikz}, \cos kz = \frac{e^{ikz} + e^{-ikz}}{2}, \sin kz = \frac{e^{ikz} - e^{-ikz}}{2i},$$

where k is a real number, lie obviously in $\mathfrak{M}_{|k|}(\mathbb{R}_1) = \mathfrak{M}_{|k|}$.

The trigonometric polynomial

$$T_\nu(\boldsymbol{z}) = T_{\nu_1,\ldots,\nu_n}(z_1, \ldots, z_n) = \sum_{\substack{|k_l| \leqq \nu_l \\ 1 \leqq l \leqq n}} c_k e^{ikz}$$

lies in $\mathfrak{M}_\nu(\mathbb{R}_n)$ but not in $\mathfrak{M}_{\nu p}(1 \leqq p < \infty)$.

The function $\sin z/z$ of one variable z lies in $\mathfrak{M}_{1p}(\mathbb{R}_1)$, $1 < p \leqq \infty$. Indeed, as a function of the real variable x it obviously belongs to L_p with the indicated restrictions on p. On the other hand, it is obviously an entire function. Further, $\sin z$ is an entire function, and it is not difficult to see that for it, for some constant c_1 the inequality

$$|\sin z| \leqq c_1 e^{|y|}.$$

holds. Therefore

$$\left|\frac{\sin z}{z}\right| \leqq c_1 e^{|y|} \quad (|z| \geqq 1).$$

On the other hand, there exists a positive constant c_2 such that

$$\left|\frac{\sin z}{z}\right| \leqq c_2 \quad (|z| \leqq 1).$$

But, since $1 \leqq e^{|y|}$, then

$$\left|\frac{\sin z}{z}\right| \leqq c_2 e^{|y|} \quad (|z| \leqq 1).$$

Thus, it follows that

$$\left|\frac{\sin z}{z}\right| \leqq c e^{|y|} \quad \text{for all } z,$$

where

$$c = \max(c_1, c_2).$$

The function e^z belongs to $E_1(\mathbb{R}_1)$, i.e. it is entire of exponential type, but it does not belong to $\mathfrak{M}_{1p}(\mathbb{R}_1)$ for any $p (1 \leqq p \leqq \infty)$. On the other hand, the function e^{iz} obviously lies in $\mathfrak{M}_{1\infty} = \mathfrak{M}_1(\mathbb{R}_1)$. The algebraic polynomial $P(z) = \sum_0^n a_k z^k$ is obviously a function of type 0, which however does not belong to $\mathfrak{M}_{0p}(\mathbb{R}_1)$ for any $p (1 \leqq p \leqq \infty)$. From what follows (see the first footnote in 3.2.2) it will be clear that if $f \in \mathfrak{M}_{0p}(\mathbb{R}_1)$, then f is a constant (equal to zero if $1 \leqq p < \infty$).

Obviously $\mathfrak{M}_{\nu p} \subset \mathfrak{M}_{\nu' p}$, if

$$\boldsymbol{\nu} = (\nu_1, \ldots, \nu_n) \leqq \boldsymbol{\nu}' = (\nu_1', \ldots, \nu_n'), \text{ i.e. } \nu_j \leqq \nu_j' \ (j = 1, \ldots, n).$$

If we suppose that g_ν denotes some function of the class E_ν, then obviously

$$g_\nu g_\mu = g_{\nu+\mu}, \quad g_\nu + g_{\nu'} = g_\mu,$$

where

$$\mu = (\mu_1, \ldots, \mu_n), \ \mu_j = \max(\nu_j, \nu_j'),$$

$$\prod_{j=1}^{n} g_{\nu j}(z_j) = g_\nu(z).$$

It is easy to see that if g is an entire function of type unity relative to all variables and $\mu_j \neq 0$ $(j = 1, \ldots, n)$, then $g(\mu_1 x_1, \ldots, \mu_n x_n)$ is an entire function of type $|\mu_1|, \ldots, |\mu_n|$. The converse assertion is also true. It is possible, using the general properties just presented, to construct from given entire functions of exponential type other such functions. For this purpose have adopted as these operations addition and multiplication, taken in finite number. An important means of construction of entire functions of exponential type is the process of integration with respect to a parameter (see 3.6.2).

3.1.1. Information on multiple power series. It suffices to carry out all the considerations on the example of double series. For series of higher multiplicity they are analogous.

By the *sum* of the series

$$S = \sum_{k=0}^{\infty} \sum_{l=0}^{\infty} u_{kl}, \tag{1}$$

where the u_{kl} are in general complex numbers, we understand the limit

$$\lim_{m,n\to\infty} \sum_{k=0}^{m} \sum_{l=0}^{n} u_{kl} = S \tag{2}$$

(if it exists), when the natural numbers m and n tend unboundedly to infinity independently of one another.

The series (1) is said to be *absolutely convergent*, if the series with terms $|u_{kl}|$ converges. Obviously an absolutely convergent series converges.

If the series (1) converges absolutely, then its terms are uniformly bounded, i.e. there exists a constant K such that

$$|u_{kl}| \leq K \ (k, l = 1, 2, \ldots).$$

However if the series (1) converges nonabsolutely, then its terms are not necessarily uniformly bounded, as is shown by the example of the series

$$\sum_{0}^{\infty} \sum_{0}^{\infty} a_{kl}, \tag{3}$$

where $a_{0l} = l!$, $a_{1l} = -l!$, $a_{kl} = 0$ for the remaining natural numbers k and l. It converges to a sum equal to zero, but not absolutely, and its terms are not uniformly bounded.

Consider the power series

$$f(\eta, \zeta) = \sum_0^\infty \sum_0^\infty c_{kl} \eta^k \zeta^l, \tag{4}$$

where the c_{kl} are complex constants and η, ζ are complex variables. Suppose that this series converges absolutely[1] at a point η_0, ζ_0, where $\eta_0 \neq 0$, $\zeta_0 \neq 0$. Then it converges also absolutely and uniformly for any η, ζ satisfying the inequalities

$$|\eta| \leqq \varrho_1, |\eta_0|, |\zeta| \leqq \varrho_2 |\zeta_0|, 0 < \varrho_1, \varrho_2 < 1. \tag{5}$$

Indeed, there exists a constant c such that

$$|c_{kl} \eta_0^k \zeta_0^l| < c \quad (k, l = 0, 1, \ldots),$$

so that for the indicated η, ζ

$$|c_{kl} \eta^k \zeta^l| = |c_{kl} \eta^k \zeta^l| \left|\frac{\eta}{\eta_0}\right|^k \left|\frac{\zeta}{\zeta_0}\right|^l \leqq c \varrho_1^k \varrho_2^l.$$

Accordingly, the terms of the series (1) do not exceed in absolute value the corresponding terms of the convergent series

$$c \sum \sum \varrho_1^k \varrho_2^l = \frac{c}{(1 - \varrho_1)(1 - \varrho_2)}.$$

It is legitimate to differentiate the series (1) termwise for the indicated η, ζ as many times as desired. Indeed, after multiple differentiation, for example relative to η, the general term of the resulting series for the indicated η, ζ will satisfy the inequalities

$$|c_{kl} k \eta^{k-1} \zeta^l| = |c_{kl} \eta_0^k \zeta_0^l| \left|\frac{k}{|\eta_0|}\right| \left|\frac{\eta}{\eta_0}\right|^{k-1} \left|\frac{\zeta}{\zeta_0}\right|^l \leqq \frac{ck}{|\eta_0|} \varrho_1^{k-1} \varrho_2^l.$$

Accordingly, the differentiated series converges uniformly in the region (5), since the series

$$\sum \sum k \varrho_1^{k-1} \varrho_2^l < \infty$$

---

[1] If we delete the word "absolutely", then this assertion is in general false. For example, if in (4) we choose as the c_{kl} the coefficients a_{kl} of the series (3), then the series (4) converges for $\eta = \zeta = 1$ and diverges for $\eta = 0$ and any $\zeta \neq 0$, since it degenerates in this case into the divergent series

$$\sum_0^\infty a_{0l} \zeta^l = \sum_0^\infty l! \zeta^l.$$

converges. It follows from what has been said that

$$c_{kl} = \frac{1}{k!l!} \frac{\partial^{k+l} f(0, 0)}{\partial \eta^k \partial \zeta^l}.$$

This shows in particular that the decomposition of the function f in question into the power series (1) is unique.

The function $f(\eta, \zeta)$, represented in the form of the absolutely[1] convergent power series (1) defined by the inequalities

$$(6) \qquad |\eta| < \varrho_1, \; |\zeta| < \varrho_2,$$

is said to be *analytic* in this region.

Suppose that $f(\eta, \zeta)$ is a function analytic in the region (6). Then for fixed $\eta (|\eta| < \varrho_1)$ and arbitrary $\zeta (|\zeta| < \varrho_2)$ the function

$$f(\eta, \zeta) = \sum_0^\infty \left(\sum_0^\infty c_{kl} \eta^k \right) \zeta^l$$

decomposes into a series converging relative to ζ. Therefore $f(\eta, \zeta)$ is an analytic function of ζ for $|\zeta| < \varrho_2$. Analogously, for fixed $\zeta (|\zeta| < \varrho_2)$, $f(\eta, \zeta)$ is an analytic function of η for $|\eta| < \varrho_1$. Hence it is possible to represent f in the form of a Cauchy integral

$$(7) \qquad f(\eta, \zeta) = \frac{1}{2\pi i} \int_{C_1} \frac{f(u, \zeta)}{u - \eta} du = \frac{1}{(2\pi i)^2} \int_{C_1} \int_{C_2} \frac{f(u, v) \, du \, dv}{(u - \eta)(v - \zeta)},$$

obtained by successive application of this representation relative to each of the variables η, ζ. Here C_1 and C_2 are circumferences in the η- and ζ-complex planes respectively with centers at the origin and radii $r_1 < \varrho_1$, $r_2 < \varrho_2$ and $|\eta| < r_1$, $|\zeta| < r_2$.

Since the series

$$\frac{1}{(u - \eta)(v - \zeta)} = \sum \sum \frac{\eta^k \zeta^l}{u^{k+1} v^{l+1}}$$

converges uniformly relative to $u \in C_1$, $v \in C_2$ and η, ζ satisfy the inequalities

$$|\eta| < r_1' < r_1, \; |\zeta| < r_2' < r_2,$$

[1] Here the word "absolutely" may be dropped, since it is possible to show that the convergence of the series (4) for all (!) η, ζ with $|\eta| < \varrho_1$, $|\zeta| < \varrho_2$ implies its absolute convergence for all the indicated η, ζ.

then the substitution of it into (7) and termwise integration lead to the original equation (4), whence

$$c_{kl} = \frac{1}{(2\pi i)^2} \int_{C_1} \int_{C_2} \frac{f(u, v)}{u^{k+1} v^{l+1}}\, du\, dv. \tag{8}$$

Now if we started from an arbitrary function $f(u, v)$ continuous on C_1, C_2 then the integral (of Cauchy type) on the right side of (7) would be equal to some function $F(\eta, \zeta)$, representable in the form of a series

$$F(u, v) = \sum_{k=0}^{\infty} \sum_{l=0}^{\infty} c_{kl} \eta^k \zeta^l, \tag{9}$$

converging absolutely and uniformly for $|\eta| < r_1'$, $|\zeta| < r_2'$ for any $r_1' < r_1$ and $r_2' < r_2$. Thus, the function $F(\eta, \zeta)$ is analytic if $|\eta| < r_1$, $|\zeta| < r_2$.

From the fact that a function f analytic in the region (6) is analytic relative to each variable, there follows the formula

$$f(\eta, \zeta) = \frac{1}{(2\pi)^2} \int_0^{2\pi} \int_0^{2\pi} f(\eta + re^{i\theta}, \zeta + \varrho e^{i\theta})\, d\theta\, d\varphi, \tag{10}$$

$$0 < r < \varrho_1 - |\eta|, \; 0 < \varrho < \varrho_2 - |\zeta|,$$

obtained from the corresponding one-dimensional formula.

We note a further property, as follows: if a sequence of functions $f_N(\eta, \zeta)$ of functions analytic in the region (6) converges uniformly as $N \to \infty$ on the set

$$|\eta| < r_1 < \varrho_1, \; |\zeta| < r_2 < \varrho_2$$

for arbitrarily chosen r_1, r_2 to a function $f(\eta, \zeta)$, then this last is analytic in the region (6). In order to verify this, we replace f by f_N in (7) and pass to the limit as $N \to \infty$. Then for the limit function $f(\eta, \zeta)$ formula (7) will be satisfied and hence f will be analytic when $|\eta| < r_1$, $|\zeta| < r_2$. Then, as a consequence of the arbitrariness of $r_1 < \varrho_1$, $r_2 < \varrho_2$, f will be analytic in the region (6).

Suppose given a sequence of power series

$$f_n(\eta, \zeta) = \sum_0^{\infty} \sum_0^{\infty} c_{kl}^{(n)} \eta^k \zeta^l \quad (n = 1, 2, \ldots),$$

absolutely converging for $\eta = \eta_0 \neq 0$, $\zeta = \zeta_0 \neq 0$ and such that

$$\sum_0^{\infty} \sum_0^{\infty} |c_{kl}^{(n)} - c_{kl}^{(m)}|\, |\eta_0|^k\, |\zeta_0|^l \to 0 \quad n, m \to \infty.$$

Then, obviously,

$$\lim_{n\to\infty} c_{kl}^{(n)} = c_{kl},$$

where c_{kl} are certain numbers. In addition the series

$$f(\eta, \zeta) = \sum_0^\infty \sum_0^\infty c_{kl}\eta^k\zeta^l$$

converges absolutely for $\eta = \eta_0$, $\zeta = \zeta_0$, and for $|\eta| \leqq \eta_0$, $|\zeta| \leqq \zeta_0$

$$|f(\eta, \zeta) - f_n(\eta, \zeta)| \leqq \sum_0^\infty \sum_0^\infty |c_{kl} - c_{kl}^{(n)}| \, |\eta_0|^k \, |\zeta_0|^l \to 0.$$

From this it is clear that the equation

$$\lim_{n\to\infty} f_n(\eta, \zeta) = f(\eta, \zeta)$$

holds uniformly in the region

$$|\eta| \leqq |\eta_0|, \; |\zeta| \leqq |\zeta_0|.$$

In this book we will operate only with entire functions

$$f(\mathbf{z}) = \sum_{\mathbf{k}\geqq 0} a_{\mathbf{k}} \mathbf{z}^{\mathbf{k}} \tag{11}$$

i.e. functions for which the series (11) converges absolutely for each complex $\mathbf{z}$.

From the properties of analytic functions noted above it follows that the power series (11) of an entire function converges uniformly on any bounded region, as well as the series obtained by successive differentiation of (11), which it is legitimate to carry out relative to arbitrary variables $z_1, \ldots, z_n$ any finite number of times. For fixed $z_{m+1}, \ldots, z_n$ the function $f(z_1, \ldots, z_m, z_{m+1}, \ldots, z_n)$ is an entire function relative to $z_1, \ldots, z_m$.

If the function $f(\mathbf{z})$ is entire, then it may be decomposed (in a unique way) into a series

$$f(\mathbf{z}) = \sum_{\mathbf{k}\geqq 0} c_{\mathbf{k}} (\mathbf{z} - \mathbf{z}_0)^{\mathbf{k}}$$

relative to the powers $(\mathbf{z} - \mathbf{z}_0)^{\mathbf{k}} = (z_1 - z_{0_1})^{k_1} \ldots (z_n - z_{0n})^{k_n}$, converging absolutely for all $\mathbf{z}$. For example, in the case $n = 2$ this assertion follows from the fact that the formal identities,

$$f(z_1, z_2) = \sum_{k=0}^{\infty} \sum_{l=0}^{\infty} a_{kl} z_1^k z_2^l =$$

$$= \sum_{k=0}^{\infty} \sum_{l=0}^{\infty} a_{kl} \sum_{s=0}^{k} C_k^s z_{10}^{k-s} (z_1 - z_{10})^s \sum_{j=0}^{l} C_l^j z_{20}^{l-j} (z_2 - z_{20})^j$$

$$= \sum_{\mu=0}^{\infty} \sum_{\nu=0}^{\infty} c_{\mu\nu} (z_1 - z_{10})^\mu (z_2 - z_{20})^\nu,$$

are essentially legitimate. The last equation is obtained after reduction of similar terms on the same products $(z_1 - z_{10})^\mu$, $(z_2 - z_{20})^\nu$. In order to justify this, it suffices to show that its left side is an absolutely convergent multiple series, i.e. that

$$\sum_{k=0}^{\infty} \sum_{l=0}^{\infty} |a_{kl}| \sum_{s=0}^{k} C_k^s x_0^{k-s} (x - x_0)^s \sum_{j=0}^{l} C_l^j y_0^{l-j} (y - y_0)^j < \infty$$

$$(x_0 = |z_{10}|,\ y_0 = |z_{20}|,\ x - x_0 = |z_1 - z_{10}|,\ y - y_0 = |z_2 - z_{20}|).$$

But this is so, because all of the terms of this series are nonnegative and their sum

$$\sum_{k=0}^{\infty} \sum_{l=0}^{\infty} |a_{kl}|\, x^k y^l < \infty$$

converges by hypothesis.

3.1.2. Fourier transforms of functions of class $\mathfrak{M}_{\nu p}$. From 3.1. we know that an entire function of one variable

$$F(z) = a_0 + \frac{a_1}{1} z + \frac{a_2}{2!} z^2 + \cdots \tag{1}$$

of type $\sigma > 0$ may be defined as an entire function, having one of the following properties:

$$\overline{\lim_{r \to \infty}} \frac{\ln M(r)}{r} \leqq \sigma \tag{2}$$

or

$$\overline{\lim_{n \to \infty}} \sqrt[n]{|a_n|} \leqq \sigma. \tag{3}$$

In view of this we may say that the function $F(z)$ defined by the series (1) has type σ if and only if the series

$$f(z) = \frac{a_0}{z} + \frac{a_1}{z^2} + \frac{a_2}{z^3} + \cdots \tag{4}$$

converges for $|z| > \sigma$.

The function $f(z)$ is said to be the *Borel transform* of the function $F(z)$. With it is connected the following integral:

$$\int_0^{\infty(\theta)} F(\zeta)\, e^{-\zeta z} d\zeta = f(z, \theta), \tag{5}$$

taken along the ray $\zeta = \varrho e^{-i\theta}$, $0 \leqq \varrho < \infty$. Indeed, it turns out as one sees from the book of N. I. Ahiezer [1], § 81, that if the entire function F is of type σ, then for it the integral (5) converges uniformly and absolutely on any set interior relative to a halfplane Δ_θ which does not contain the point $z = 0$ and whose boundary is tangent to the circumference $|z| = \sigma$ at the point $\sigma e^{i\theta}$. In addition one has the identity

$$f(z) = f(z, \theta) \quad (z \in \Delta_\theta)$$

for any (real) θ.

Suppose it is known that the function $F(z)$ is not only entire of exponential type σ, but also belongs to $L = L(-\infty, \infty)$ as a function of real x. In other words, suppose $F \in \mathfrak{M}_{\sigma 1}(R_1) = \mathfrak{M}_{\sigma 1}$. If one puts $\theta = 0, \pi$ in (5) and $z = x + iy$, then we get

$$f(x + iy) = \int_0^\infty F(\xi) e^{-\xi(x+iy)} d\xi \quad (x \geqq 0), \tag{6}$$

$$f(x + iy) = -\int_{-\infty}^0 F(\xi) e^{-\xi(x+iy)} d\xi \quad (x \leqq 0). \tag{7}$$

We note that on the basis of the general considerations stated above, we could have said only that the integrals (6), (7) converge for $x > \sigma$ and $x < \sigma$ respectively. However in the case at hand one is considering a function $F \in L$. For it is directly clear that the integrals (6) and (7) converge in the wider regions $x \geqq 0$ and $x \leqq 0$ respectively. The integrals obtained from (6) and (7) by formal differentiation relative to $z = x + iy$ again obviously converge absolutely for $x \geqq 0$ and $x \leqq 0$ respectively. This shows that the integrals (6) and (7) define analytic functions for $x > 0$ and $x < 0$ respectively. They accordingly coincide on the indicated regions with the function $f(z)$—the Borel transform of the function F.

For $\varepsilon > 0$ it follows from (6) and (7) that

$$f(\varepsilon + iy) - f(-\varepsilon + iy) = \int_{-\infty}^{\infty} F(\xi) e^{-i\xi y} e^{-\varepsilon|\xi|} d\xi,$$

from which, passing to the limit as $\varepsilon \to 0$, we get

$$0 = \int_{-\infty}^{\infty} F(\xi) e^{-i\xi y} d\xi \quad (|y| > \sigma),$$

i.e. the Fourier transform of a function $F \in \mathfrak{M}_{\sigma 1}$ is a continuous function, identically equal to zero outside the segment $[-\sigma, \sigma]$.

If $\boldsymbol{\sigma} = (\sigma_1, \ldots, \sigma_n)$ is a positive vector and $F \in \mathfrak{M}_{\sigma 1}(\mathbb{R}_n) = \mathfrak{M}_{\sigma 1}$, then $\tilde{F}(\boldsymbol{x})$ is continuous on $\mathbb{R}_n$. Since $F(u_1, \ldots, u_n) = F(\boldsymbol{u})$ is entire of type σ_1 in u_1, lying in $L(\mathbb{R}_1) = L(-\infty, \infty)$ for almost all $\boldsymbol{u}' = (u_2, \ldots, u_n)$ of the corresponding $(n-1)$-dimensional space, then for such $\boldsymbol{u}'$

$$\int F(u_1, u') e^{-ix_1 u_1} du_1 = 0, \quad |x_1| > \sigma_1,$$

But then $\tilde{F}(\boldsymbol{x}) = 0$, if $|x_1| > \sigma_1$. This consideration may be carried out for all $x_j (j = 1, \ldots, n)$. Thus we have proved the following assertion.

3.1.3. Theorem. *If $F \in \mathfrak{M}_{\sigma 1}$, then $\tilde{F}(\boldsymbol{x})$ is a continuous function equal to zero outside*

$$\Delta_\sigma = \{|x_j| \leqq \sigma_j,\ j = 1, \ldots, n\}.$$

3.1.4. Below we formulate a theorem of Paley-Wiener[1] without proof.

Theorem. If $F \in \mathfrak{M}_{02}$, $\boldsymbol{\sigma} = (\sigma_1, \ldots, \sigma_n)$, *then the function*

(1) $$\tilde{F}(\boldsymbol{x}) = \frac{1}{(2\pi)^{n/2}} \int F(\boldsymbol{u}) e^{-i\boldsymbol{x}\boldsymbol{u}} d\boldsymbol{u},$$

where the integral is understood in the sense of convergence in the mean:

(2) $$\left\| \tilde{F}(\boldsymbol{x}) - \frac{1}{(2\pi)^{n/2}} \int_{\Delta_n} f(\boldsymbol{u}) e^{-i\boldsymbol{x}\boldsymbol{u}} d\boldsymbol{u} \right\|_{L_2(\Delta_n)} \to 0 \quad (N \to \infty),$$

$$\Delta_N = \{|x_j| < N;\ j = 1, \ldots, 2\},$$

not only belongs to $L_2(\mathbb{R}_n)$, as follows from (2), but, moreover, satisfies $\tilde{F}(\boldsymbol{x}) = 0$ almost everywhere outside Δ_σ. Conversely, if φ is any function

---

[1] Paley and Wiener [1] for $n = 1$. For the proof in this case see, for example, the book of N. I. Ahiezer [1]. Plancherel and Polya [1] for $n > 1$.

of $L_2(\Delta_\sigma)$, then the function

$$F(\boldsymbol{x}) = \frac{1}{(2\pi)^{n/2}} \int\limits_{\Delta_\sigma} \varphi(\boldsymbol{u})\, e^{i\boldsymbol{x}\boldsymbol{u}} d\boldsymbol{u} \tag{3}$$

lies in $\mathfrak{M}_{\sigma 2}$, and the function $\tilde{F}$ defined by formula (1) *satisfies $\tilde{F} = \varphi$ almost everywhere.*

It is easily verified that if one supposes that $F(\boldsymbol{x}) \in \mathfrak{M}_{\sigma 2}$ represents a generalized function ($F \in S'$), then the function $\tilde{F}(\boldsymbol{x})$ represents the transform $\tilde{F}$ (in the sense of S) and $F = \hat{\tilde{F}}$ (see 1.5.).

Thus the Fourier transform of a function $F \in \mathfrak{M}_{\sigma 2}$ may be considered as a generalized and an ordinary function, and moreover lying ln $L_2(\Delta_\sigma)$.

From what follows it will be clear that if $1 \leqq p \leqq 2$ and $F \in \mathfrak{M}_{\sigma p}$ then $F \in \mathfrak{M}_{\sigma 2}$. Therefore $\tilde{F}$ has its support on Δ_σ and $\tilde{F} \in L_2(\Delta_\sigma)$. But if $2 < p \leqq \infty$, then the Fourier transform of the function $F \in \mathfrak{M}_{\sigma p}$ may turn out to be an essentially generalized function. For example, $1 \in \mathfrak{M}_{\sigma\infty} = \mathfrak{M}_\sigma$, but $\tilde{1} = (2\pi)^{n/2}\, \delta(\boldsymbol{x})$ (see 1.5.). Therefore for $p > 2$ the assertion that $\tilde{F}$ has support in Δ_σ may be formulated only in the language of generalized functions.

We shall say that the generalized function $\Phi \in S'$ *has support on* Δ_σ if for any basic function $\varphi \in S$ such that $\varphi \equiv 0$ on $\Delta_{\sigma+\varepsilon}$, where $\boldsymbol{\sigma} + \boldsymbol{\varepsilon} = \{\sigma_1 + \varepsilon, \ldots, \sigma_n + \varepsilon\}$, the equation $(\Phi, \varphi) = 0$ holds.

3.1.5. We shall prove the following theorem, due to L. Schwartz.

Theorem. *If $g \in \mathfrak{M}_{\sigma,p}(1 \leqq p \leqq \infty)$, then $\tilde{g}$ has support on Δ_σ.*

Proof. We introduce the functions φ_ε (see 1.5.8.) and $\psi_\varepsilon = (2\pi)^{n/2}\tilde{\varphi}_\varepsilon$. Since $\varphi_\varepsilon \in S$, then $\psi_\varepsilon \in S \subset L_q \left(\frac{1}{p} + \frac{1}{q} = 1\right)$ so that $\psi_\varepsilon g \in L$. Moreover, the function ψ_ε is an entire exponential function of type ε, so that $\psi_\varepsilon g$ is an entire exponential function of type $\boldsymbol{\sigma} + \boldsymbol{\varepsilon} = (\sigma_1 + \varepsilon, \ldots, \sigma_n + \varepsilon)$. Accordingly $\psi_\varepsilon g \in \mathfrak{M}_{\sigma+\varepsilon,1}$. This means that if $\varphi \in S$ and $\varphi = 0$ on $\Delta_{\sigma+\varepsilon}$, then (see 3.1.3.)

$$\left(\widetilde{\psi_\varepsilon g}, \varphi\right) = 0.$$

After passing to the limit as $\varepsilon \to 0$ (see 1.5.8 (6)), we get

$$(\tilde{g}, \varphi) = 0,$$

as we were required to prove.

As to inverting this theorem, for our goals it will be sufficient to know that the Fourier transform of an ordinary function equal to zero outside Δ_σ and lying in $L_2(\Delta_\sigma)$ is, on the basis of the above-cited theorem of Paley and Wiener, a function of class $\mathfrak{M}_{\sigma,2}$.

3.2. Interpolation Formula

Suppose (see Civin [1]) that $\omega_\nu(\boldsymbol{t})$ is a continuous function of period $2\nu > 0$ relative to each of the variables and that $\boldsymbol{a} = (a_1, \ldots, a_n)$ is a vector such that the Fourier series

$$(1)\qquad e^{i\boldsymbol{a}\boldsymbol{x}}\omega_\nu(\boldsymbol{x}) = \sum_{\boldsymbol{k}} c_{\boldsymbol{k}}^\nu e^{i\frac{k\pi}{\nu}\boldsymbol{x}} \quad (|x_j| < \nu),$$

$$(2)\qquad c_{\boldsymbol{k}}^\nu = \frac{1}{(2\nu)^n} \int_{\Delta_\nu} \omega_\nu(\boldsymbol{u}) e^{i\left(a - \frac{k\pi}{\nu}\right)u} \, d\boldsymbol{u},$$

converges absolutely, i.e.

$$(3)\qquad \sum_{\boldsymbol{k}} |c_{\boldsymbol{k}}^\nu| < \infty.$$

We shall show that if moreover $f, \tilde{f} \in L$ (thus, f and $\tilde{f}$ are continuous and bounded on $\mathbb{R}$), then

$$(4)\qquad \widehat{\omega_\nu(\boldsymbol{t})\,\tilde{f}(\boldsymbol{t})} = \lim_{N\to\infty} \sum_{|k_j|<N} \widehat{c_{\boldsymbol{k}}^\nu e^{i\left(\frac{k\pi}{\nu} - a\right)t} \tilde{f}(\boldsymbol{t})}$$

$$= \lim_{N\to\infty} \sum_{|k_j|<N} c_{\boldsymbol{k}}^\nu f\left(\boldsymbol{x} + \frac{k\pi}{\nu} - \boldsymbol{a}\right) = \sum_{\boldsymbol{k}} c_{\boldsymbol{k}}^\nu f\left(\boldsymbol{x} + \frac{k\pi}{\nu} - \boldsymbol{a}\right),$$

where the series on the right converges uniformly relative to $\boldsymbol{x} \in \mathbb{R}$. The first equation in (4) follows from the fact that the partial sums

$$\lambda_N(\boldsymbol{x}) = \sum_{|k_j|<N} c_{\boldsymbol{k}}^\nu e^{i\left(\frac{k\pi}{\nu} - a\right)t}$$

converge uniformly as $N \to \infty$ to $\omega_\nu(\boldsymbol{x})$. Indeed, taking account of the fact that $\tilde{f} \in L$, we get

$$|\widehat{\omega_\nu \tilde{f}} - \widehat{\lambda_N \tilde{f}}| = |\widehat{(\omega_\nu - \lambda_N)\tilde{f}}|$$

$$\leqq \frac{1}{(2\pi^{n/2})} \int |\omega_\nu - \lambda_N| \; |\tilde{f}| \, dt \to 0 \quad (N \to \infty).$$

The second equation (4) follows from formula 1.5. (19). The third is obvious.

3.2.1. Theorem. *Suppose that $\Omega(\boldsymbol{x}) = \Omega(x_1, \ldots, x_n)$ is an infinitely differentiable function of polynomial growth, even or odd. If $\Omega(\boldsymbol{x})$ is even, we will suppose that for any $\nu > 0$ the function*

(1) $$\omega_\nu(\boldsymbol{x}) = \Omega(\boldsymbol{x}) \ \ (|x_j| < \nu, j = 1, \ldots, n)$$

periodic of period 2ν relative to each of its variables, decomposes on Δ_ν into an absolutely convergent Fourier series (3.2 (1) if $\boldsymbol{a} = \boldsymbol{0}$). If on the other hand $\Omega(\boldsymbol{x})$ is an odd function, then we will suppose that the series 3.2 (1), for $\boldsymbol{a} = \boldsymbol{a}_\nu = \left(\frac{\pi}{2\nu}, \ldots, \frac{\pi}{2\nu}\right)$, converges absolutely.

We shall suppose further that

(2) $$|c_{\boldsymbol{k}}^\nu| \leqq c_{\boldsymbol{k}} \ (\nu \leqq \nu_0)$$

and

(3) $$\sum_{\boldsymbol{k}} c_{\boldsymbol{k}} < \infty .$$

Suppose further that $g(\boldsymbol{x}) \in \mathfrak{M}_{\nu p}(\mathbb{R}_n) = \mathfrak{M}_{\nu p}$.
Then we have the equation

(4) $$\widehat{\Omega(\boldsymbol{t})\tilde{g}} = \sum_{\boldsymbol{k}} c_{\boldsymbol{k}}^\nu g\left(\boldsymbol{x} - \boldsymbol{a}_\nu + \frac{\boldsymbol{k}\pi}{\nu}\right)$$

($\boldsymbol{a}_\nu = \boldsymbol{0}$ *for even Ω, $\boldsymbol{a}_\nu = \left(\frac{\pi}{2\nu}, \ldots, \frac{\pi}{z\nu}\right)$ for odd Ω, where the series converges in the sense of L_p.*

It is not hard to see that if $\Omega(\boldsymbol{x})$ is an odd function not identically equal to zero, then the periodic function $\omega_\nu(\boldsymbol{x})$ corresponding to it is in general discontinuous, and without multiplication by e^{iat} its Fourier series surely cannot converge absolutely.

Proof. Suppose given $\varepsilon_1 > 0$ and suppose that $0 < \varepsilon < \varepsilon_1$. We introduce the functions φ_ε and ψ_ε. In 3.1.5 it was shown that $\psi_\varepsilon g \in \mathfrak{M}_{\nu+\varepsilon,1}$, so that $\widetilde{\psi_\varepsilon g}$ is a continuous function with support in $\Delta_{\nu+\varepsilon}$. Thus $\widetilde{\psi_\varepsilon g} \in L$ and it is legitimate to apply formula 3.2 (5) to $\psi_\varepsilon g$.

We have

(5) $$\Lambda_\varepsilon(\boldsymbol{x}) = \widehat{\Omega(\boldsymbol{t})\widetilde{\psi_\varepsilon g}} = \widehat{\omega_{\nu+\varepsilon_1}\widetilde{\psi_\varepsilon g}}$$
$$= \sum_{\boldsymbol{k}} c_{\boldsymbol{k}}^{\nu+\varepsilon_1}\psi_\varepsilon\left(\boldsymbol{x} - \boldsymbol{a}_{\nu+\varepsilon_1} + \frac{\boldsymbol{k}\pi}{\nu+\varepsilon_1}\right) g\left(\boldsymbol{x} - \boldsymbol{a}_{\nu+\varepsilon_1} + \frac{\boldsymbol{k}\pi}{\nu+\varepsilon_1}\right)$$

($\boldsymbol{a}_{\nu+\varepsilon_1} = 0$ for an even function Ω and $\boldsymbol{a}_{\nu+\varepsilon_1} = \left(\frac{\pi}{z(\nu+\varepsilon_1)}, \ldots\right)$ for an odd function Ω).

In these relations there always figure ordinary functions; the first equation holds because $\Omega = \omega_{\nu+\varepsilon}$ on $\Delta_{\nu+\varepsilon}$ and thus on the support of the function $\tilde{\psi}_\varepsilon g$; the second equation is true in view of 3.2 (4).

Since $\widetilde{\psi_\varepsilon g} \in L$, then $\psi_\varepsilon g$ is a function bounded on R and the series (5) converges uniformly on $\mathbb{R}$ to its sum, which we have denoted by $\Lambda_\varepsilon(\boldsymbol{x})$. On the other hand,

$$|\psi_\varepsilon(\boldsymbol{x})| = \left| \int_{\Delta_\varepsilon} \varphi_\varepsilon(\boldsymbol{t}) e^{-i\boldsymbol{x}\boldsymbol{t}} d\boldsymbol{t} \right| \leqq \left| \int_{\Delta_\varepsilon} \varphi_\varepsilon d\boldsymbol{t} \right| = 1$$

and by hypothesis $g \in L_p$. Therefore the series (5) converges to $\Lambda_\varepsilon(\boldsymbol{x})$ also in the sense of L_p. We shall indeed understand the convergence of the series (5) in this sense.

We note that

$$\psi_\varepsilon(\boldsymbol{x}) = \int_{\Delta_\varepsilon} \varphi_\varepsilon(\boldsymbol{t}) e^{-i\boldsymbol{x}\boldsymbol{t}} d\boldsymbol{t} \to 1 \quad (\varepsilon \to 0),$$

so that

$$\lim_{\varepsilon \to 0} \Lambda_\varepsilon(\boldsymbol{x}) = \sum_{\boldsymbol{k}} c_{\boldsymbol{k}}^{\nu+\varepsilon_1} g\left(\boldsymbol{x} - \boldsymbol{a}_{\nu+\varepsilon_1} + \frac{\boldsymbol{k}\pi}{\nu + \varepsilon_1}\right) \tag{6}$$

in the sense of L_p. Indeed,

$$\left\| \Lambda_\varepsilon(\boldsymbol{x}) - \sum_{\boldsymbol{k}} c_{\boldsymbol{k}}^{\nu+\varepsilon_1} g\left(\boldsymbol{x} - \boldsymbol{a}_{\nu+\varepsilon_1} + \frac{\boldsymbol{k}\pi}{\nu + \varepsilon_1}\right) \right\|_{L_p}$$

$$\leqq \sum_{|k_j|<N} |c_{\boldsymbol{k}}^{\nu+\varepsilon_1}| \left\| \left[\psi_\varepsilon\left(\boldsymbol{x} - \boldsymbol{a}_{\nu+\varepsilon_1} + \frac{\boldsymbol{k}\pi}{\nu + \varepsilon_1}\right) - 1 \right] \right.$$

$$\left. \times g\left(\boldsymbol{x} - \boldsymbol{a}_{\nu+\varepsilon_1} + \frac{\boldsymbol{k}\pi}{\nu + \varepsilon_1}\right) \right\|_{L_p} + 2 \sum{}' |c_{\boldsymbol{k}}^{\nu+\varepsilon_1}| \, \|g\|_{L_p},$$

where the primes in the second sum indicate that the sum is extended over all $\boldsymbol{k}$, which did not enter into the first sum. N may always be chosen so large that the second sum is less than any $\eta > 0$ given in advance. Thereupon one may choose ε_0 so small that the first sum will be less than η for positive $\varepsilon < \varepsilon_0$, which is possible by a theorem of Lebesgue[1]. Both sides of equation (6) in fact do not depend on ε_1—this is clear from (5). From (6), after a passage to the limit as $\varepsilon_1 \to 0$, it follows finally that

$$\lim_{\varepsilon \to \infty} \Lambda_\varepsilon(\boldsymbol{x}) = \sum_{\boldsymbol{k}} c_{\boldsymbol{k}}^{\nu} g\left(\boldsymbol{x} - \boldsymbol{a}_\nu + \frac{\boldsymbol{k}\pi}{\nu}\right), \tag{7}$$

---

[1] $|\psi_\varepsilon(x)| \leqq 1$, $\psi_\varepsilon(x) \to 1$, $\varepsilon \to 0$.

where the series on the right converges in the sense of L_p, and the limit on the left, as we mentioned above, also is understood in the sense of L_p. Indeed, the norm in the sense of L_p of the difference of the right sides of (6) and (7) does not exceed

$$\sum_{|k_j|<N} |c_{\boldsymbol{k}}^{\nu+\varepsilon_1} - c_{\boldsymbol{k}}^{\nu}| \, \|g\|_{L_p} + \sum_{|k_j|<N} |c_{\boldsymbol{k}}^{\nu}| \left\| g\left(\boldsymbol{x} - \boldsymbol{a}_{\nu+\varepsilon_1} + \frac{\boldsymbol{k}\pi}{\nu+\varepsilon_1}\right) \right.$$

$$\left. - g\left(\boldsymbol{x} - \boldsymbol{a}_{\nu} + \frac{\boldsymbol{k}\pi}{\nu}\right) \right\|_{L_p} + \|g\|_{L_p} \cdot 2 \sum{}' |c_k| ,$$

where we have taken into account inequalities (2) and (3), in which we need to suppose that $\nu_0 = \nu + \varepsilon^0$ ($\varepsilon_1 < \varepsilon^0$). Again N may be chosen so large that the third sum is less than η. For this N the first and second sums will, for sufficiently small ε_1, be also smaller than η, because $\boldsymbol{a}_{\nu+\varepsilon_1} \to \boldsymbol{a}_\nu$ and $c_{\boldsymbol{k}}^{\nu+\varepsilon_1} \to c_{\boldsymbol{k}}^{\nu}$ ($\varepsilon_1 \to 0$) (here we take into account the fact that Ω is infinitely differentiable, and hence summable, on $\Delta_{\nu+\varepsilon}$).

Thus (7) is proved.

On the other hand (see (5)),

$$(\Lambda_\varepsilon, \varphi) = (\widehat{\Omega\widetilde{\psi_\varepsilon g}}, \varphi) = \left(\psi_\varepsilon g, \widetilde{\Omega\hat{\varphi}}\right) \underset{\varepsilon\to 0}{\longrightarrow} \left(g, \widetilde{\Omega\hat{\varphi}}\right) = \left(\widehat{\Omega\tilde{g}}, \varphi\right),$$

and we have proved the interpolation formula (4).

3.2.2. Interpolation formula for an arbitrary entire function of exponential type. This formula will be obtained as a special case of the general formula 3.2.1 (4). Suppose that $g_\nu(x) = g(x) \in \mathfrak{M}_{\nu p}(\mathbb{R}_1) = \mathfrak{M}_{\nu p}$, i.e. it is an entire function of one variable of degree ν, bounded on the real axis. For its derivative, the following formula holds (1.5. (10)):

$$g'(x) = \widehat{it\tilde{g}}. \tag{1}$$

The function it is infinitely differentiable, odd and has polynomial growth. We consider the function

$$e^{iat} it = \sum_{k=-\infty}^{\infty} c_k^\nu e^{\frac{ik\pi}{\nu}t} \quad \left\{ |t| < \nu, \, a = \frac{\pi}{2\nu} \right\},$$

$$c_k^\nu = \frac{i}{2\nu} \int_{-\nu}^{\nu} u e^{i\left(\frac{\pi}{2\nu} - \frac{k\pi}{\nu}\right)u} \, du$$

$$= -\frac{1}{\nu} \int_0^\nu u \sin\left(k - \frac{1}{2}\right)\frac{\pi}{\nu} u \, du = \frac{\nu(-1)^{k-1}}{\pi^2 \left(k - \frac{1}{2}\right)^2},$$

periodic of period 2ν. It is clear that

$$|c_k^\nu| \leqq |c_k^{\nu_0}| \quad (0 < \nu \leqq \nu_0)$$

and

$$\sum_k |c_k^{\nu_0}| < \infty.$$

Therefore the function *it* satisfies all the requirements which were imposed on $\Omega(t)$ in 3.2.1. In view of 3.2.1(4) we have the interpolation formula

$$g'(x) = \widehat{it\bar{g}} = \frac{\nu}{\pi^2} \sum_{-\infty}^{\infty} \frac{(-1)^{k-1}}{\left(k - \frac{1}{2}\right)^2} g\left(x + \frac{\pi}{\nu}\left(k - \frac{1}{2}\right)\right), \tag{2}$$

where the series converges in the sense of L_p. It may be considered as an analogue of the formula of M. Riesz for trigonometric polynomials.

Later on we shall show that $\mathfrak{M}_{\nu p} \subset \mathfrak{M}_{\nu\infty} = \mathfrak{M}_\nu$. Thus in fact the series (2) converges not only in the sense of L_p but uniformly.

If in (2) we put $g(x) = \sin x \in \mathfrak{M}_{1\infty}$, and then substitute $x = 0$, we get

$$1 = \frac{1}{\pi^2} \sum_{-\infty}^{\infty} \frac{1}{\left(k - \frac{1}{2}\right)^2}. \tag{3}$$

Therefore for any function $g \in \mathfrak{M}_{\nu p}$ $(1 \leqq p \leqq \infty)$ we have the inequality

$$\|g'\|_{L_p} \leqq \nu \|g\|_{L_p}, \tag{4}$$

which bears the name *Bernšteĭn's inequality*[1]. It was proved by S. N. Bernšteĭn ([1], 269—270) for $p = \infty$.

If $z = x + iy$ is an arbitrary complex number, then $g(z) = \sum_0^\infty \frac{(iy)^s}{s!} g^{(s)}(x)$, so that

$$\|g(x + iy)\|_{L_p} \leqq e^{\nu|y|} \|g\|_{L_p}, \tag{5}$$

where the norm is taken relative to $x \in \mathbb{R}_1$.

Since the function $g(u + iy)$ is for any y an entire function of exponential type σ relative to u, then it follows from (5) that if $g(z) \in \mathfrak{M}_{\nu p}$, then also $g(u + iy) \in \mathfrak{M}_{\nu p}$. But then equation (2) is true, if one replaces x

---

[1] Inequality (4) is true for $\nu = 0$ as well. In fact from (4) and the fact that $g_0 \in \mathfrak{M}_{0p} \subset \mathfrak{M}_{\nu p}$, it follows that $\|g'\|_p = 0$, and g is a constant, obviously equal to zero for finite p.

in it by any complex z:

$$g'(z) = \frac{\nu}{\pi^2} \sum_{-\infty}^{\infty} \frac{(-1)^{k-1}}{\left(k - \frac{1}{2}\right)^2} g\left(z + \frac{\pi}{\nu}\left(k - \frac{1}{2}\right)\right). \tag{6}$$

The series (6) converges relative to $x (z = x + iy)$ in the sense of $L_p(\mathbb{R}_1)$. We have already warned the reader that we shall prove later on (see 3.3.5.) that $\mathfrak{M}_{\nu p} \subset \mathfrak{M}_{\nu\infty} = \mathfrak{M}_\nu$, from which it follows directly that the series (6) converges uniformly relative to x $(z = x + iy)$. In view of (5) it easily follows as well that it converges uniformly on any strip $\{y_1 < y < y_2\}$, where y_1 and y_2 are arbitrary real numbers.

Suppose that $g(\boldsymbol{x}) = g(x_1, \boldsymbol{x}')$ is a function defined on a measurable set $\mathscr{E} = \mathbb{R}_1 \times \mathscr{E}'$ $(\boldsymbol{x}_1 \in \mathbb{R}_1, x' \in \mathscr{E}')$, belonging to $L_p(\mathscr{E})$, entire exponential of type ν relative to x_1 for almost all (in the sense of $(n - 1)$-dimensional measure) $\boldsymbol{x}' \in \mathscr{E}'$. In view of Fubini's theorem one may say that for the indicated $\boldsymbol{x}'$ the function $g(x_1, \boldsymbol{x}') \in \mathfrak{M}_{\nu p}(\mathbb{R}_1)$ relative to x_1, and therefore, as a consequence of (4),

$$\left\|\frac{\partial g}{\partial x_1}\right\|^p_{L_p(\mathbb{R}_1)} \leqq \nu^p \|g\|^p_{L_p(\mathbb{R}_1)} \quad (1 \leqq p < \infty).$$

After integration of both sides of this inequality with respect to $x' \in \mathscr{E}'$ and raising to the $(1/p)^{\text{th}}$ power we get

$$\left\|\frac{\partial g}{\partial x_1}\right\|_{L_p(\mathscr{E})} \leqq \nu \|g\|_{L_p(\mathscr{E})} \quad (1 \leqq p \leqq \infty). \tag{7}$$

We intend here also the obvious case $p = \infty$.

If the function $g(\boldsymbol{x}) = g(x_1, \ldots, x_n) \in \mathfrak{M}_{\nu p}(\mathbb{R}_n)$, then, taking account of the fact that any of its partial derivatives $g^{(\lambda)}$ is an entire function of exponential type $\boldsymbol{\nu}$ (see 3.1.), we easily obtain on the basis of (7) $\mathscr{E} = \mathbb{R}_n$) the inequality

$$\|g^{(\lambda)}\|_{L_p(\mathbb{R}_n)} \leqq \boldsymbol{\nu}^\lambda \|g\|_{L_p(\mathbb{R}_n)}. \tag{8}$$

It may be generalized further, supposing that $g(\boldsymbol{x}) = g(u, \boldsymbol{x}') = g(x_1, \ldots, x_m, \boldsymbol{x}) \in L_p(\mathscr{E})$, $\mathscr{E} = \mathbb{R}_m \times \mathscr{E}' \subset \mathbb{R}_n$ is a measurable set, $m < n$, and g is for almost all $x' \in \mathscr{E}'$, relative to $x_1, \ldots, x_m$, an entire function of exponential type $\boldsymbol{\nu} = (\nu_1, \ldots, \nu_m)$. Then, if $\boldsymbol{\lambda} = (\lambda_1, \ldots, \lambda_m, 0, \ldots, 0)$, we get

$$\|g^{(\lambda)}\|_{L_p(\mathscr{E})} \leqq \boldsymbol{\nu}^\lambda \|g\|_{L_p(\mathscr{E})}. \tag{9}$$

If we take into account the fact that for almost all $\boldsymbol{x}'$

$$g(\boldsymbol{u} + i\boldsymbol{y}, \boldsymbol{x}') = g(\boldsymbol{z}) = g(\boldsymbol{u} + i\boldsymbol{y}) = \sum_{\lambda \geqq 0} \frac{g^{(\lambda)}(\boldsymbol{u})}{\lambda!} (i\boldsymbol{y})^{\lambda},$$

then we easily find from (9) that

$$(10) \qquad \|g(\boldsymbol{u} + i\boldsymbol{y}, \boldsymbol{x}')\|_{L_p(\mathscr{E})} \leqq \|g(x)\|_{L_p(\mathscr{E})} e^{\sum_{j=1}^{m} \nu_j |y_j|}.$$

From (2) we may obtain as a special case the theorem of M. Riesz (see 2.4.). To prove it we need only take into account that a trigonometric polynomial of order ν is an entire function of exponential type $\nu (T_\nu \in \mathfrak{M}_\nu)$, bounded on the real axis, so that formula (2) is applicable to it. We need to take into account further the fact that T_ν is a function of period 2π.

3.2.3. Inequality 3.2.2. (4) may be extended to more general norms[1]. Suppose that E is the Banach space of functions $f(x, \boldsymbol{w})$, defined and measurable on $\mathscr{E} = \mathbb{R}_1 \times \mathscr{E}_1$ having the following properties:

1) the addition of two functions of E and the multiplication of a function by a number are defined in the usual way. Two functions f_1 and f_2, equal almost everywhere on $\mathscr{E}$, are considered equal as elements of E, and one writes $f_1 = f_2$;

2) if $f = f(x, \boldsymbol{w}) \in E$, then $f_{x_0} = f(x + x_0, \boldsymbol{w}) \in E$ for each real value of x_0 and $\|f(x, \boldsymbol{w})\| = \|f(x + x_0, \boldsymbol{w})\|$;

3) if it is true that $f_n \in E (n = 1, 2, \ldots)$, $f \in E$, $\|f_n - f\| \to 0$ and $f_n(x, \boldsymbol{w}) \to \psi(x, \boldsymbol{w})$ $(n \to \infty)$ for $x \in \mathbb{R}_1$ and almost all $\boldsymbol{w} \in \mathscr{E}_1$, then it follows that $\psi = f$.

If the function $g_\nu(x, \boldsymbol{w}) \in E$ and if for almost all $\boldsymbol{w}$ it is a bounded entire function of type ν relative to x, then it will satisfy the equation

$$(1) \qquad g'(x, \boldsymbol{w}) = \frac{\nu}{\pi^2} \sum_{-\infty}^{\infty} \frac{(-1)^{k-1}}{\left(k - \frac{1}{2}\right)^2} g\left(x + \frac{\pi}{\nu}\left(k - \frac{1}{2}\right), \boldsymbol{w}\right)$$

for almost all $\boldsymbol{w}$ in the sense of ordinary convergence. On the other hand, as a consequence of property 2) the sum of the norms of the terms of the series (1) does not exceed $\nu \|g_\nu\|$, and thus the right side of the series (1) converges relative to the norm in question to some function $\psi \in E$. But, in view of property 3), the function ψ must necessarily be equal to $\frac{\partial g_\nu}{\partial x}$. This is based on the inequality

$$(2) \qquad \left\| \frac{\partial g_\nu}{\partial x} \right\| \leqq \nu \|g_\nu\|.$$

---

[1] See at the end of the book the Remark to 3.2.3.

3.2.4. Basing ourselves on 2.7.2, we may obtain a generalized inequality analogous to 3.2.3 (2), for trigonometric polynomials as well. To this end it suffices to recall that the space E consists of functions $f(x, \boldsymbol{w})$ of period 2π relative to x with norms subjected only to properties 1) and 2).

3.2.5. Theorem[1]. *Suppose that* $1 \leq p < \infty$ *and that the entire function*

$$g = g_\nu(\boldsymbol{z}) = g_\nu(x + iy)$$

of exponential type $\boldsymbol{\nu} = (\nu_1, \ldots, \nu_n) > \boldsymbol{0}$ *belongs to the class* $L_p(\mathbb{R}_n)$. *Then*

$$\lim_{|\boldsymbol{x}|\to\infty} g_\nu(\boldsymbol{x}) = 0. \tag{1}$$

Hence it follows in particular that $g_\nu(\boldsymbol{x})$ is bounded on $\mathbb{R}_n$.

Proof. It suffices to prove the theorem in the case $\nu_1 = \cdots = \nu_n = 1$, to which it is possible to reduce our function by replacing it by $g_\nu\left(\frac{z_1}{\nu_1}, \ldots, \frac{z_n}{\nu_n}\right)$. We confine ourselves to the two-dimensional case. The case $n > 2$ is treated analogously.

So, suppose given an entire function $g(z_1, z_2) = g$ of type $(1,1)$, belonging to $L_p(\mathbb{R}_2)$, where $1 \leqq p < \infty$. As always, we will suppose x_1, x_2 real.

We have the equation (see 3.1.1.)

$$g(x_1, x_2) = \frac{1}{(2\pi)^2} \int_0^{2\pi}\int_0^{2\pi} g(x_1 + \varrho_1 e^{i\theta_1}, x_2 + \varrho_2 e^{i\theta_2})\, d\theta_1\, d\theta_2,$$

where $\varrho_1, \varrho_2 > 0$. We multiply both of its sides by $\varrho_1\varrho_2$ and integrate over the rectangle $0 \leqq \varrho_1, \varrho_2 \leqq \delta$. Then we get

$$g(x_1, x_2)\,\frac{\delta^4}{4} =$$

$$= \frac{1}{(2\pi)^2} \int_\sigma\int_\sigma g(x_1 + \xi_1 + i\eta_1, x_2 + \xi_2 + i\eta_2)\, d\xi_1\, d\eta_1\, d\xi_2\, d\eta_2,$$

where (ξ_1, η_1) and (ξ_2, η_2) serve as Cartesian coordinates, and σ is a disk of radius δ with center at the origin of coordinates.

---

[1] Plancherel and Polya [1].

Hence

$$|g(x_1, x_2)| \leq \frac{1}{\delta^4 \pi^2} \int_\sigma \int_\sigma |g(x_1 + \xi_1 + i\eta_1, x_2 + \xi_2 + i\eta_2)| \, d\xi_1 \, d\eta_1 \, d\xi_2 \, d\eta_2$$

(2) $$\leq \frac{2}{\delta^{4\left(1-\frac{1}{q}\right)} \pi^2} \left(\int_{-\delta}^{\delta} \int_{-\delta}^{\delta} d\eta_1 \, d\eta_2 \int_{x_1-\delta}^{x_1+\delta} \int_{x_2-\delta}^{x_2+\delta} |g(\boldsymbol{\xi} + i\boldsymbol{\eta})^p \, d\xi_1 d\xi_2 \right)^{1/p},$$

$$\boldsymbol{\xi} + i\boldsymbol{\eta} = (\xi_1 + i\eta_1, \xi_2 + i\eta_2).$$

We shall prove that the integral

$$I(g) = \int_{-\delta}^{\delta} \int_{-\delta}^{\delta} d\xi_1, d\xi_2 \int_{-\infty}^{\infty} \int_{-\infty}^{\infty} |g(\boldsymbol{\xi} + i\boldsymbol{\eta})|^p \, d\eta_1 \, d\eta_2$$

is finite, from which it will follow that the right side of (2) tends to zero when $|\boldsymbol{x}| \to \infty$, and the theorem will be proved.

Indeed (see 3.2.2 (10)),

$$I(g) = \int_{-\delta}^{\delta} \int_{-\delta}^{\delta} \|g(\xi + i\eta)\|^p_{L_p(\mathbb{R}_2)} \, d\eta_1 \, d\eta_2$$

$$\leq \|g(\xi)\|^p_{L_p(\mathbb{R}_2)} \int_{-\delta}^{\delta} \int_{-\delta}^{\delta} e^{(|\eta_1|+|\eta_2|)p} \, d\eta_1 \, d\eta_2 < \infty.$$

For $p = \infty$ this theorem is false, which is shown by the example of the function $\sin z \in \mathfrak{M}_{1,\infty}(\mathbb{R}_1)$.

3.2.6. Entire functions of spherical exponential type. Concerning the entire function

$$g(\boldsymbol{z}) = g(z_1, \ldots, z_n)$$

we will say that it is of *spherical exponential type* $\sigma \geq 0$, if for any $\varepsilon > 0$ it is possible to find a constant $A_\varepsilon > 0$ such that

(1) $$|g(\boldsymbol{z})| < A_\varepsilon \exp\left\{ (\sigma + \varepsilon) \sqrt{\sum_1^n |z_j|^2} \right\}$$

for all $\boldsymbol{z}$. The collection of all functions of a given type $\sigma \geq 0$ will be denoted by SE_σ. Since

$$\frac{1}{\sqrt{n}} \sum_{j=1}^{n} |z_j| \leq \sqrt{\sum_{j=1}^{n} |z_j|^2} \leq \sum_{j=1}^{n} |z_j|,$$

then

$$E_{\sigma/\sqrt{n}} \subset SE_\sigma \subset E_\sigma .$$

The set of functions $g \in SE_\sigma$, which as functions of the real vector $\boldsymbol{x} \in \mathbb{R}_n$ lie in $L_p(\mathbb{R}_n) = L_p$, will be denoted by $S\mathfrak{M}_{\sigma p}$.

Suppose that $\boldsymbol{\omega} = (\omega_1, \ldots, \omega_n)$ is an arbitrary real unit vector. We denote by

$$D_\omega f(\boldsymbol{x}) = f'_\omega(\boldsymbol{x}) = \sum_{j=1}^{n} \frac{\partial f}{\partial x_j}(x)\, \omega_j$$

the derivative of f at the point $\boldsymbol{x}$ in the direction $\boldsymbol{\omega}$, and by

$$f_\omega^{(l)}(\boldsymbol{x}) = D_\omega f_\omega^{(l-1)}(\boldsymbol{x}) = \sum_{|k|=l} f^{(k)}(\boldsymbol{x})\boldsymbol{\omega}^{\boldsymbol{k}} \quad (l = 1, 2, \ldots)$$

the derivative of order l of f at $\boldsymbol{x}$ in the direction ω. We introduce the transformation

$$\boldsymbol{x} = (x_1, \ldots, x_n) \leftrightarrows (\xi_1, \ldots, \xi_n) = \boldsymbol{\xi},$$

where $\xi_1, \ldots, \xi_n$ are the corrdinates of $\boldsymbol{x}$ in the new (real) rectangular system of coordinates, which is chosen such the increase of ξ_1 for fixed $\xi_2, \ldots, \xi_n$ will lead to a motion of the point $\boldsymbol{x}$ in the direction $\boldsymbol{\omega}$. The coordinate transformation

(2) $$x_k = \sum_{s=1}^{n} \alpha_{ks}\xi_s \quad (k = 1, \ldots, n)$$

is defined by a real orthogonal matrix. This matrix defines also a transformation

$$z_k = \sum_{s=2}^{n} \alpha_{ks} w_s$$

of the complex systems $\boldsymbol{w} = (w_1, \ldots, w_n)$ into the systems $\boldsymbol{z} = (z_1, \ldots, z_n)$. Here, evidently, one has the equation

$$\sum_{s=1}^{n} |z_s|^2 = \sum_{s=1}^{n} |w_s|^2 .$$

Put

$$g(\boldsymbol{z}) = g(z_1, \ldots, z_n) = g_*(w_1, \ldots, w_n) = g_*(\boldsymbol{w}),$$

and suppose that $g \in S\mathfrak{M}_{\sigma p}$. Then also $g_* \in S\mathfrak{M}_{\sigma p}$, because g_* is obviously an entire function and

$$|g_*(\boldsymbol{w})| = |g(\boldsymbol{z})| < A_\varepsilon e^{(\sigma+\varepsilon)\sqrt{\sum_{j=1}^{n} |z_j|^2}} = A_\varepsilon e^{(\sigma+\varepsilon)\sqrt{\sum_{j=1}^{n} |w_j|^2}} .$$

From this inequality it is clear that g_* is a function of type σ relative to w_1 and that inequality 3.2.2. (4) is applicable to it.

Further

$$g_\omega^{(l)}(\boldsymbol{x}) = \frac{\partial^l}{\partial \xi_1^l} g_*(\boldsymbol{\xi}),$$

so that

(3) $$\|g_\omega^{(l)}(\boldsymbol{x})\|_{L_p} = \left\| \frac{\partial^l g_*(\boldsymbol{\xi})}{\partial \xi_1^l} \right\|_{L_p} \leqq \sigma^l \|g_*(\boldsymbol{\xi})\|_{L_p} = \sigma^l \|g(\boldsymbol{x})\|_{L_p}.$$

Putting $\boldsymbol{z} = (x_1 + iy_1, \ldots, x_n + iy_n)$, we get

$$g(\boldsymbol{z}) = \sum_{l=0}^{\infty} \frac{1}{l!} \sum_{|\boldsymbol{k}|=l} g^{(\boldsymbol{k})}(\boldsymbol{x})\,(i\boldsymbol{y})^{\boldsymbol{k}} = \sum_{l=0}^{\infty} \frac{1}{l!} g_\omega^{(l)}(x)\,(i|\boldsymbol{y}|)^l,$$

$$\left(|\boldsymbol{k}| = \sum_{j=1}^{n} k_j, \quad |\boldsymbol{y}|^2 = \sum_{j=1}^{n} |y_j|^2, \quad \omega = \frac{\boldsymbol{y}}{|\boldsymbol{y}|}\right),$$

where $g_\omega^{(l)}$ is thus the derivative of order l in the direction $\boldsymbol{\omega} = \boldsymbol{y}/|\boldsymbol{y}|$, i.e. of $\boldsymbol{y}$.

But then

(4) $$\|g(\boldsymbol{x} + i\boldsymbol{y})\|_{L_p} \leqq \|g(\boldsymbol{x})\|_{L_p} \sum_{l=0}^{\infty} \frac{(\sigma|\boldsymbol{y}|)^l}{l!}$$

$$= \|g(\boldsymbol{x})\|_{L_p} \exp\left(\sigma \sqrt{\sum_1^n y_j^2}\right).$$

The Fourirr transform $\tilde{g}$ of the function $g \in \mathfrak{M}_{\sigma p}$ has a support belonging to the sphere $v_\sigma \subset \mathbb{R}_n$ of radius σ with center at the origin (L. Schwarz [1]).

Indeed, if $g \in L_1$, then, taking account of the fact that under the orthogonal coordinate transformation (2)

$$\boldsymbol{x} \leftrightarrows \boldsymbol{\xi}, \ \boldsymbol{u} \leftrightarrows \boldsymbol{v},$$

we find that $\boldsymbol{xu} = \boldsymbol{\xi v}$, $d\boldsymbol{u} = d\boldsymbol{v}$. Recalling then that $\boldsymbol{v}' = (v_2, \ldots, v_n)$, we get

$$\tilde{g}(\boldsymbol{x}) = \frac{1}{(2\pi)^{n/2}} \int g(\boldsymbol{u}) e^{-i\boldsymbol{xu}} d\boldsymbol{u} = \frac{1}{(2\pi)^{n/2}} \int g_*(\boldsymbol{v}) e^{-i\boldsymbol{\xi v}}\, d\boldsymbol{v}$$

$$= \frac{1}{(2\pi)^{n/2}} \int e^{-i\boldsymbol{\xi}'\boldsymbol{v}'}\, d\boldsymbol{v}' \int g_*(v_1, \boldsymbol{v}') e^{-iv_1\xi_1} dv_1 = \tilde{g}_*(\boldsymbol{\xi}).$$

But $g_*(\xi_1, \boldsymbol{\xi}')$ is of type σ relative to ξ_1, and for almost all $\boldsymbol{\xi}'$ lies in $L(\mathbb{R}_1)$. Hence (see 3.1.3) $\tilde{g}_*(\boldsymbol{\xi})$ is equal to zero outside the strip $|\xi_1| \leqq \sigma$ for any choice of coordinates $(\xi_2, \ldots, \xi_n)$. But then $\tilde{g}(\boldsymbol{x}) = 0$ outside the sphere

v_σ. Our assertion, if $g \in S\mathfrak{M}_{\sigma 1}$, is proved. If $g \in S\mathfrak{M}_{\sigma p}$, then we introduce the functions $\varphi_\varepsilon(\boldsymbol{x})$ and $\psi_\varepsilon(\boldsymbol{x})$ and argue as in 3.1.5.

If f is a generalized function and $e \subset \mathbb{R}$ is an open set, then we will write

$$(5) \qquad (f)_e = 0,$$

if

$$(f, \varphi) = 0$$

for all $\varphi \in S$ having a support lying in e.

Suppose that $0 < \lambda < \sigma$ and that v_σ is the same sphere as before with the center at the origin and radius σ. We shall show that *if* $f \in L_p(1 \leqq p \leqq \infty)$ *and*

$$(\tilde{f})_{v_\sigma} = 0,$$

then for any entire function g of spherical degree λ lying in L, one has the equation

$$(6) \qquad g * f = \frac{1}{(2\pi)^{n/2}} \int g(\boldsymbol{x} - \boldsymbol{u})\, f(\boldsymbol{u})\, d\boldsymbol{u} = 0.$$

We introduce the functions φ_ε and $\psi_\varepsilon = (2\pi)^{n/2}\tilde{\varphi}_\varepsilon$ defined in 1.5.8. For $\varphi \in S$ and an $\varepsilon > 0$ such that $\lambda + \varepsilon < \sigma$

$$(7) \qquad (\psi_\varepsilon g * f, \varphi) = (\widetilde{\psi_\varepsilon g}\tilde{f}, \varphi) = (\tilde{f}, \widetilde{\psi_\varepsilon g}\varphi) = 0,$$

because g along with its derivatives is a bounded infinitely differentiable function, $\psi_\varepsilon \in S$, $\psi_\varepsilon g \in S$, $\widetilde{\psi_\varepsilon g} \in S$, so that $\widetilde{\psi_\varepsilon g}\varphi \in S$, and has support lying strictly in v_σ. After a passage to the limit in equation as $\varepsilon \to 0$ (see 1.5.8 (7)), we obtain (6).

3.3. Inequalities of Different Metrics for Entire Functions of Exponential Type

In this section we shall be interested in the classes $\mathfrak{M}_{\nu p}(\mathbb{R}_n)$ of entire functions.

The central place is taken here by the inequality of different metrics, with which the norm of the function $g_\nu(\boldsymbol{x})$ in the metric $L_{p'} = L_{p'}(\mathbb{R}_n)$ is estimated in terms of its norm in the metric of $L_p(1 \leqq p \leqq p' \leqq \infty)$ and the product of powers of some of the $\nu_1, \ldots, \nu_n$. This inequality will play an essential role in what follows in the study of differentiable functions of more general classes.

Obviously $\mathfrak{M}_{\nu p}(\mathbb{R}_n)$ is a linear set. It is infinite-dimensional. For example, the functions $\frac{1}{x^2}\sin^2\frac{x}{2k}$ $(k = 1, 2, \ldots)$ belong to $\mathfrak{M}_{1p}(\mathbb{R}_1)$, $1 \leqq p \leqq \infty$, and form a linearly independent system. Therefore already from general considerations of functional analysis it is possible to conclude that the unit sphere in $\mathfrak{M}_{\nu p}(\mathbb{R}_n)$ is not compact in the metric $L_p(\mathbb{R}_n) = L_p$. However we shall see that it is compact in a weakened sense (see 3.3.6).

3.3.1. Theorem. *Suppose that* $1 \leqq p \leqq \infty$, $h > 0$, $x_k = kh$ $(k = 0, \pm 1, \pm 2, \ldots)$, *is an entire function of one variable of type* ν *and*

$$((g_\nu))_{L_p} = \sup_u \left(h \sum_{-\infty}^{\infty} |g_\nu(x_k - u)|^p\right)^{1/p} < \infty$$

or $\|g_\nu\|_{L_p} < \infty$. *Then the following inequality holds:*

$$\|g_\nu\|_{L_p} \leqq ((g_\nu))_{L_p} \leqq (1 + h\nu)\, \|g_\nu\|_{L_p}. \tag{1}$$

Proof. For $p = \infty$ the theorem is trivial. Suppose that $1 \leqq p < \infty$ and $\|g_\nu\|_{L_\nu} < \infty$. Then

$$\int_{-\infty}^{\infty} |g_\nu|^p\, dx = \sum_{-\infty}^{\infty} \int_{x_k}^{x_{k+1}} |g_\nu|^p\, dx = h \sum_{-\infty}^{\infty} |g_\nu(\xi_k)|^p,$$

where the numbers ξ_k satisfy the inequalities $x_k < \xi_k < x_{k+1}$. Using the generalized Bernšteĭn inequality, the Hölder inequality, and also the inequality $|\, \|x\| - \|y\|\, | \leqq \|x - y\|$,

$$\left|\left(h \sum_{-\infty}^{\infty} |g_\nu(\xi_k)|^p\right)^{1/p} - \left(h \sum_{-\infty}^{\infty} |g_\nu(x_k)|^p\right)^{1/p}\right|$$

$$\leqq \left(h \sum_{-\infty}^{\infty} |g_\nu(\xi_k) - g_\nu(x_k)|^p\right)^{1/p} \leqq \left[h \sum_{-\infty}^{\infty} \left|\int_{x_k}^{x_{k+1}} g_\nu'(t)\, dt\right|^p\right]^{1/p}$$

$$\leqq \left(h \sum_{-\infty}^{\infty} \int_{x_k}^{x_{k+1}} |g_\nu'|^p\, dt h^{p/q}\right)^{1/p} = h\, \|g_\nu'\|_{L_p(\mathbb{R}_1)} \leqq h\nu \|g_\nu\|_{L_p(\mathbb{R}_1)}.$$

Therefore

$$\left(h \sum_{-\infty}^{\infty} |g_\nu(x_k)|^p\right)^{1/p} \tag{2}$$

$$= \left[\left(h \sum_{-\infty}^{\infty} |g_\nu(x_j)|^p\right)^{1/p} - \left(h \sum_{-\infty}^{\infty} |g_\nu(\xi_k)|^p\right)^{1/p}\right]$$

$$+ \left(h \sum_{-\infty}^{\infty} |g_\nu(\xi_k)|^p\right)^{1/p} \leqq (1 + h\nu)\, \|g_\nu\|_{L_p(\mathbb{R}_1)}.$$

If one recalls that for each fixed u the function $g_\nu(x-u)$, considered as a function of x, is an entire function of type ν, then from inequality (2), replacing $g_\nu(x)$ in it by $g_\nu(x-u)$, we obtain the second inequality (1).

On the other hand, if $((g_\nu))_{L_p} < \infty$, then

$$(3)\qquad \int_{-\infty}^{\infty} |g_\nu|^p\,dx = \sum_{-\infty}^{\infty} \int_{x_k}^{x_{k+1}} |g_\nu|^p dx = \sum_{-\infty}^{\infty} \int_0^h |g_\nu(x_{k+1}-u)|^p\,du =$$

$$= \int_0^h \sum_{-\infty}^{\infty} |g_\nu(x_{k+1}-u)|^p\,du \leqq h \sup_u \sum_{-\infty}^{\infty} |g_\nu(x_k-u)|^p,$$

where the change of the order of summation and integration is legitimate in view of the fact that we are operating with nonnegative functions. With this the first inequality (1) is proved.

3.3.2. Theorem[1]. *Suppose that* $1 \leqq p \leqq \infty$, $h_l > 0$, $x_k^{(l)} = kh_l$ $(l=1,\ldots, n, k=0, \pm 1, \pm 2, \ldots)$, $g = g_\nu$ *is an entire function of the type* $\boldsymbol{\nu} = (\nu, \ldots, \nu_n)$,

$$(1)\qquad ((g))_p^{(n)} = \sup_{u_l} \left(\prod_{m=1}^{n} h_m \sum_{l_1=-\infty}^{\infty} \cdots \right.$$

$$\left. \cdots \sum_{l_n=-\infty}^{\infty} |g(x_{l_1}^{(1)} - u_1, \ldots, x_{l_n}^{(n)} - u_n)|^p \right)^{1/p} < \infty$$

or

$$\|g\|_{L_p(\mathbb{R}_n)} = \|g\|_p < \infty.$$

Then

$$(2)\qquad \|g_\nu\|_{L_p(\mathbb{R}_n)} \leqq ((g_\nu))_p^{(n)} \leqq \prod_1^n (1 + h_i\nu_i)\, \|g_\nu\|_{L_p(\mathbb{R}_n)}.$$

Proof. For $p = \infty$ inequalities (2) are trivial. Suppose that $1 \leqq p < \infty$. Then

$$\int |g|^p\,dx$$

$$= \sum_{l_1=-\infty}^{\infty} \cdots \sum_{l_n=-\infty}^{\infty} \int_0^{h_1} \cdots \int_0^{h_n} |g(x_{l_1}^{(1)} - u_1, \ldots, x_{l_n}^{(n)} - u_n|^p\,du$$

$$= \int_0^{h_1} \cdots \int_0^{h_n} \sum \cdots \sum |g(x_{l_1}^{(1)} - u_1, \ldots, x_{l_n}^{(n)} - u_n)|\,du \leqq ((g))_p^{(n)},$$

and we have proved the first inequality (2) under the hypothesis that the second term in (2) is finite. Now suppose that the third term in (2) is finite. In order to prove the second inequality (2), we note that

[1] S. M. Nikol'skiĭ [3].

$g(\boldsymbol{z}-\boldsymbol{u}) = g(z_1 - u_1, \ldots, z_n - u_n)$ for any fixed u_l is an entire function of type $\boldsymbol{v}$ relative to $\boldsymbol{z}$, for which $\|g(\boldsymbol{x}-\boldsymbol{u})\|_p = \|g(\boldsymbol{x})\|_p$. Therefore it is sufficient to prove the inequality

$$\left(\prod_{l=1}^{n} h_l \sum_{l_1=-\infty}^{\infty} \cdots \sum_{l_n=-\infty}^{\infty} |g(x_{l_1}^{(l)}, \ldots, x_{l_n}^{(n)})|^p\right)^{1/p} \leqq \prod_{l=1}^{n} (1 + h_l \nu_l) \| g\|_p .$$

But it was already proved in the preceding theorem for $n = 1$. We suppose that its truth has been established for $m = n - 1$. Then, in view of the fact that for any fixed x_1 the function g is entire of type $\nu_2, \ldots, \nu_n$ relative to $x_2, \ldots, x_n$ respectively, we will have

$$\prod_{l=1}^{n} (1 + h_l \nu_l)^p \int \cdots \int |g(x_1, x_2, \ldots, x_n)|^p \, dx_2 \cdots dx_n$$
$$\geqq \prod_{l=2}^{n} h_l \sum_{l_2=-\infty}^{\infty} \cdots \sum_{l_n=-\infty}^{\infty} |g(x_l, x_{l_2}^{(2)}, \ldots, x_{l_n}^{(n)})|^p ,$$

so that after integration with respect to x_1 and raising to the power p^{-1} we get

$$\prod_{l=2}^{n} (1 + h_l \nu_l) \|g\|_p \geqq \left(\prod_{l=2}^{n} h_l\right)^{1/p}$$
$$\times \left(\sum_{l_2=-\infty}^{\infty} \cdots \sum_{l_n=-\infty}^{\infty} \left(\int |g(x_1, x_{l_2}^{(2)}, \ldots, x_{l_n}^{(n)})|^p\right) dx_1\right)^{1/p}$$
$$\geqq \frac{1}{1 + h_1 \nu_1} \left(\prod_{l=1}^{n} h_l \sum_{l_1=-\infty}^{\infty} \cdots \sum_{l_n=-\infty}^{\infty} |g(x_{l_1}^{(1)}, \ldots, x_{l_n}^{(n)})|^p\right)^{1/p} .$$

The last inequality holds in view of the first inequality 3.3.1 (1), since g is an entire function of type ν_1 relative to x_1.

3.3.3. Lemma[1]. *For any* $a_k \geqq 0$, $k = 1, 2, \ldots$,

$$\left(\sum_1^{\infty} a_k^{p'}\right)^{1/p'} \leqq \left(\sum_1^{\infty} a_k^{p}\right)^{1/p} \quad (1 \leqq p \leqq p' \leqq \infty). \tag{1}$$

Proof. It suffices to suppose that

$$\sum_1^{\infty} a_k^p = 1 .$$

Then

$$a_k \leqq 1, \quad \sum_1^{\infty} a_k^{p'} \leqq \sum_1^{\infty} a_k^p = 1 ,$$

[1] See Hardy, Littlewood and Polya [1].

Hence inequality (1) follows for $1 \leqq p < p' < \infty$. In order to obtain (1) for $p' = \infty$, it suffices to pass to the limit as $p' \to \infty$.

3.3.4. Theorem. *Under the hypotheses of Theorem 3.3.2. the following inequality holds:*

$$(1) \qquad \big((g)\big)_{p'}^{(n)} \leqq \left(\prod_{l=1}^{n} h_l\right)^{\frac{1}{p'}-\frac{1}{p}} \big((g)\big)_p^{(n)} \quad (1 \leqq p \leqq p' \leqq \infty).$$

This follows directly from the definition of $\big((g)\big)_p^{(n)}$ and the preceding lemma.

3.3.5. Theorem[1]. *If $1 \leqq p \leqq p' \leqq \infty$, then for an entire function $g = g_\nu \in L_p(\mathbb{R}_n)$ of exponential type $\boldsymbol{\nu} = (\nu_1, \ldots, \nu_n)$, the following inequality holds (different metrics):*

$$(1) \qquad \|g_\nu\|_{L_{p'}(\mathbb{R}_n)} \leqq 2^n \left(\prod_{1}^{n} \nu_k\right)^{\frac{1}{p}-\frac{1}{p'}} \|g_\nu\|_{L_p(\mathbb{R}_n)}.$$

For fixed n and arbitrary ν_k this inequality is exact in the sense of order of magnitude, i.e., if one replaces the $\frac{1}{p} - \frac{1}{p'}$ in the right side of inequality (1) by $\frac{1}{p} - \frac{1}{p'} - \varepsilon$, $\varepsilon > 0$, then it is no longer true for arbitrary g_ν.

Proof. On the basis of 3.3.2(2) and 3.3.4, putting $\omega = \frac{1}{p} - \frac{1}{p'}$, we get

$$(2) \quad \|g\|_{L_{p'}(\mathbb{R}_n)} \leqq \big((g)\big)_{p'}^{(n)} \leqq \left(\prod_{l=1}^{n} h_l\right)^{-\omega} \big((g)\big)_p^{(n)} \leqq \prod_{1}^{n} \frac{1 + h_l\nu_l}{h_l^{\omega}} \|g\|_{L_p(\mathbb{R}_n)}$$

$$= \prod_{1}^{n} \frac{1 + \alpha_l}{\alpha_l^{\omega}} \left(\prod_{1}^{n} \nu_l\right)^{\omega} \|g\|_{L_p(\mathbb{R}_n)}, \qquad (\alpha_l = h_l\nu_l).$$

The function

$$\psi(\alpha) = \frac{1 + \alpha}{\alpha^{\omega}}$$

on the semiaxis $0 < \alpha < \infty$ takes on its minimum, equal to

$$(3) \qquad \lambda_\omega = \frac{1}{\omega^{\omega}(1 - \omega)^{1-\omega}} \leqq 2.$$

Thus we may write

$$\|g\|_{L_{p'}(\mathbb{R}_n)} \leqq (\lambda_\omega)^n \left(\prod_{1}^{n} \nu_l\right)^{\omega} \|g\|_{L_p(\mathbb{R}_n)},$$

so that, in view of (3), we obtain (1).

[1] S. M. Nikol'skiĭ [3]; see Remarks 3.3—3.4.3 at the end of the book.

In order to prove the second assertion of the theorem, we consider the function

(4) $$F_\nu = \prod_1^n \frac{\sin^2 \frac{\nu_k z_k}{2}}{z_k^2},$$

which obviously belongs to $L_p(\mathbb{R}_n)$ for any p satisfying the inequalities $1 \leqq p \leqq \infty$ and which is an entire function of type $\boldsymbol{\nu} = (\nu_1, \ldots, \nu_n)$. Its norm is equal to

$$\|F_\nu\|_{L_p(\mathbb{R}_n)} = \left(2^n \prod_1^n \int_0^\infty \left|\frac{\sin^2 \frac{\nu_k t}{2}}{t^2}\right|^p dt\right)^{1/p} = c_p \left(\prod_1^n \nu_k\right)^{2-\frac{1}{p}}$$

$$(1 \leqq p \leqq \infty),$$

where c_p is a positive constant not depending on ν_l. Accordingly,

$$\|F_\nu\|_{L_{p'}(\mathbb{R}_n)} = \frac{c_{p'}}{c_p} \left(\prod_1^n \nu_k\right)^{\frac{1}{p} - \frac{1}{p'}} \|F_\nu\|_{L_p(\mathbb{R}_n)},$$

as we were required to prove.

3.3.6. Compactness Theorem[1]. *From any sequence of functions $g_{(k)} \in \mathfrak{M}_{\nu p}(\mathbb{R}_n)$ $(1 \leqq p \leqq \infty,\ k = 1, 2, \ldots)$, bounded in the metric of $L_p(\mathbb{R}_n)$, it is possible to select a subsequence $g_{(k_l)}$ $(l = 1, 2, \ldots)$ and to define a function $g \in \mathfrak{M}_{\nu p}(\mathbb{R}_n)$, such that the equation*

$$\lim_{l \to \infty} g_{(k_l)}(\boldsymbol{z}) = g(\boldsymbol{z})$$

holds uniformly on any bounded set.

Proof. By hypothesis there exists a constant A_1 such that

(1) $$\|g_{(k)}\|_{L_p(\mathbb{R}_n)} \leqq A_1, \quad k = 1, 2, \ldots$$

Hence, in view of 3.3.5. (1)

(2) $$|g_{(k)}(x)| \leqq 2^n \left(\prod_1^n \nu_k\right)^{1/p} \|g_{(k)}\|_{L_p(\mathbb{R}_n)} \leqq A,$$

where the constant A also does not depend on k.

[1] $p = \infty$, $n = 1$, S. N. Bernšteĭn [1], pages 269–270.

Decompose $g_{(k)}(\boldsymbol{z})$ into Taylor series:

$$g_{(k)}(\boldsymbol{z}) = \sum_{\alpha \geqq 0} \frac{c_\alpha^{(k)} \boldsymbol{z}^\alpha}{\boldsymbol{\alpha}!},$$

where the $\boldsymbol{\alpha} = (\alpha_1, \ldots, \alpha_n)$ are systems of nonnegative integers and

$$c_\alpha^{(k)} = \frac{\partial^{\alpha_1 + \cdots + \alpha_n} g_{(k)}(\mathbf{0})}{\partial x_1^{\alpha_1} \cdots \partial x_n^{\alpha_n}}.$$

Because of (2) and the Bernšteĭn inequality (3.2.2 (8))

$$|c_\alpha^{(k)}| \leqq A \boldsymbol{v}^\alpha, \quad k = 1, 2, \ldots. \tag{3}$$

Thus the coefficients $c_\alpha^{(k)}$, $k = 1, 2, \ldots$, are uniformly bounded for each fixed system $\boldsymbol{\alpha}$. It is then possible using the diagonal process to obtain a subsequence of natural numbers $k_1, k_2, \ldots$, such that

$$\lim_{s \to \infty} c_\alpha^{(k_s)} = c_\alpha. \tag{4}$$

Put

$$g(\boldsymbol{z}) = \sum_{\alpha \geqq 0} \frac{c_\alpha}{\boldsymbol{\alpha}!} \boldsymbol{z}^\alpha. \tag{5}$$

Then

$$|g(\boldsymbol{z})| \leqq A \sum_{\alpha \geqq 0} \frac{\boldsymbol{v}^\alpha |z_1|^{\alpha_1} \cdots |z_n|^{\alpha_n}}{\boldsymbol{\alpha}!} = A e^{\sum_1^n v_j |z_j|},$$

because the numbers c_α satisfy the inequality

$$|c_\alpha| \leqq A \boldsymbol{v}^\alpha. \tag{6}$$

Accordingly $g(\boldsymbol{z})$ is an entire function of exponential type $\boldsymbol{v}$.

Further, recalling that $|\alpha|^2 = \sum_{j=1}^n \alpha_j^2$, we get

$$|g(\boldsymbol{z}) - g_{(k_s)}(\boldsymbol{z})| \leqq \sum_{|\alpha| < N} \frac{|c_\alpha - c_\alpha^{(k_s)}|}{\alpha!} |z_1|^{\alpha_1} \cdots |z_n|^{\alpha_n}$$

$$+ \sum_{|\alpha| \geqq N} \frac{|c_\alpha| + |c_\alpha^{(k_s)}|}{\alpha!} |z_1|^{\alpha_1} \cdots |z_n|^{\alpha_n} = \sigma_1 + \sigma_2.$$

But in view of (3) and (6) for $\sqrt{\sum_1^n |z_j|^2} \leqq K$

$$\sigma_2 \leqq 2A \sum_{|\alpha| \geqq N} \frac{|v_1 z_1|^{\alpha_1} \cdots |v_n z_n|^{\alpha_n}}{\alpha!} < \varepsilon$$

for sufficiently large N. If this sufficiently large N is fixed, then in view of (4) it is possible to indicate an s_0 such that $|\sigma_1| < \varepsilon$ for all $s > s_0$ and $|\boldsymbol{z}| \leqq K$.

Thus

$$\lim_{s\to\infty} g_{k_s}(\boldsymbol{z}) = g(\boldsymbol{z})$$

uniformly for all $\boldsymbol{z}$ satisfying the inequality $|\boldsymbol{z}| \leqq K$, where K is any positive number.

Finally, if $V_\varrho \subset \mathbb{R}_n$ is a sphere of radius ϱ with center at the origin, then in consequence of (7) and (1)

$$\|g\|_{L_p(V_\varrho)} = \lim_{s\to\infty} \|g_{(k_s)}\|_{L_p(V_\varrho)} \leqq A_1,$$

so that after passing to the limit as $\varrho \to \infty$ we get

$$\|g\|_{L_p(\mathbb{R}_n)} \leqq A_1$$

and $g \in \mathfrak{M}_{\nu p}(\mathbb{R}_n)$.

3.3.7. Example of the application of Theorem 3.3.5. Suppose given numbers $1 \leqq p_1, p_2, \ldots, p_n \leqq \infty$, and consider the space $L_{(p_1,\ldots,p_n)}(\mathbb{R}_n) = L_p(\mathbb{R}_n)$ of functions $f(\boldsymbol{x}) = f(x_1, \ldots, x_n)$, measurable on $\mathbb{R}_n$ and having the finite norm

$$(1) \qquad \|f\|_{L_p(\mathbb{R}_n)} = \left\{\int \left[\ldots \left(\int \left(|f|^{p_n} dx_n\right)^{\frac{p_{n-1}}{p_n}} dx_{n-1}\right)^{\frac{p_{n-2}}{p_{n-1}}} \ldots\right]^{\frac{p_1}{p_2}} dx_1\right\}^{\frac{1}{p}},$$

where all the integrals are taken from $-\infty$ to $+\infty$.

Here, if $p_{n'} = p_{n'+1} = \cdots = p_n = \infty$, we need to suppose that

$$(2) \qquad \left\{\int \left[\ldots \left(\left(\int |f|^{p_n} dx_n\right)^{\frac{p_{n-1}}{p_n}} dx_{n-1}\right)^{\frac{p_{n-2}}{p_{n-1}}}\right]^{\frac{p_{n'}}{p_{n'+1}}} dx_{n'}\right\}^{\frac{1}{p'}}$$

$$= \sup_{x_{n'},\ldots,x_n} |f(x_1, \ldots, x_{n'-1}, x_{n'}, \ldots, x_n)|$$

for arbitrary fixed $x_1, \ldots, x_{n'+1}$.

Suppose first that $1 \leqq p \leqq p_1 \leqq p_2 \leqq \cdots \leqq p_n$ and that $g_\nu = g_{\nu_1,\ldots,\nu_n} = g$ is an entire function of exponential type $\boldsymbol{\nu}$, bounded on $\mathbb{R}_n$. Considering it as a function of the variable x_n alone, we may write

$$\left(\int |g|^{p_n} dx_n\right)^{\frac{1}{p_n}} \leqq 2\nu_n^{\frac{1}{p_{n-1}} - \frac{1}{p_n}} \left(\int |g|^{p_{n-1}} dx_n\right)^{\frac{1}{p_{n-1}}},$$

where the integrals are always agreed to be taken with the infinite limit $(-\infty, \infty)$.

Hence

$$\left(\int\left(\int |g|^{p_n}\,dx_n\right)^{\frac{p_{n-1}}{p_n}} dx_{n-1}\right)^{\frac{1}{p_{n-1}}}$$

$$\leqq 2\nu_n^{\frac{1}{p_{n-1}}-\frac{1}{p_n}}\left(\int\int |g|^{p_{n-1}}\,dx_{n-1}dx_n\right)^{\frac{1}{p_{n-1}}}$$

$$\leqq 2^{1+2}\nu_n^{\frac{1}{p_{n-1}}-\frac{1}{p_n}}(\nu_{n-1}\nu_n)^{\frac{1}{p_{n-2}}-\frac{1}{p_{n-1}}}\left(\int\int |g|^{p_{n-2}}dx_{n-1}dx_n\right)^{\frac{1}{p_{n-2}}}.$$

The first inequality here follows from Theorem 3.3.5 for $n = 1$ and $p = p_{n-1}$, $p' = p_n$, and the second from Theorem 3.3.5 for $n = 2$ and $p = p_{n-2}$, $p' = p_{n-1}$. Carrying out this process to the end, we obtain the inequality[1]

$$(3)\qquad \|g_\nu\|_{L(p_1,\ldots,p_n)(\mathbb{R}_n)} \leqq 2^{\frac{n(n+1)}{2}}\nu_n^{\frac{1}{p_{n-1}}-\frac{1}{p_n}}(\nu_{n-1}\nu_n)^{\frac{1}{p_{n-2}}-\frac{1}{p_{n-1}}}$$

$$\cdots \times (\nu_1\cdots\nu_n)^{\frac{1}{p}-\frac{1}{p_1}}\|g_\nu\|_{L_p(\mathbb{R}_n)} = 2^{\frac{n(n+1)}{2}}\prod_1^n \nu_k^{\frac{1}{p}-\frac{1}{p_k}}\|g_\nu\|_{L_p(\mathbb{R}_n)}.$$

In order to obtain this inequality, inequality 3.3.5 (1) was essentially applied n times in the corresponding special cases.

In order to prove inequality (3) in the general case $1 \leqq p \leqq p_1, \ldots, p_n \leqq \infty$, it is sufficient to note that

$$(4)\qquad \|f\|_{L(p_1,\ldots,p_n)(\mathbb{R}_n)} \leqq \|f\|_{L(q_1,\ldots,q_n)(\mathbb{R}_n)},$$

where $q_1, \ldots, q_n$ is a permutation of the numbers $p_1, \ldots, p_n$ in non-decreasing order. Inequality (4) follows from the generalized Minkowski inequality (see 1.3.2). For example, for $n = 2$, $p_2 \leqq p_1$ we have

$$\left[\int\left(\int |f(x_1, x_2)|^{p_2}\,dx_2\right)^{\frac{p_1}{p_2}} dx_1\right]^{\frac{1}{p_1}} \leqq \left[\int\left(\int |f(x_1, x_2)|^{p_1}\,dx_1\right)^{\frac{p_2}{p_1}} dx_1\right]^{\frac{1}{p_2}}.$$

Inequality (3) is exact in the sense of order of magnitude as one verifies on the functions F_ν (see 3.3.5. (4)).

---

[1] S. M. Nikol'skiĭ [5, 13, 14].

3.4. Inequalities of Different Dimensions for Entire Functions of Exponential Type

These inequalities will also have an essential significance for what follows. With their aid the norm of an entire function of exponential type, calculated for a subspace $\mathbb{R}_m \subset \mathbb{R}_n$ $(m < n)$, is estimated in terms of its norm calculated for the entire space $\mathbb{R}_n$. We shall see in the sequel that inequalities of different dimensions serve as the basis for the study of stable boundary properties of differentiable functions.

3.4.1. Suppose that $\mathscr{E} = \mathbb{R}_m \times \mathscr{E}' \subset \mathbb{R}_n$ is a cylindrical measurable set of points $\boldsymbol{x} = (\boldsymbol{u}, \boldsymbol{y})$,

$$\boldsymbol{u} = (x_1, \ldots, x_m) \in \mathbb{R}_m,$$

and

$$\boldsymbol{y} = (x_{m+1}, \ldots, x_n) \in \mathscr{E}' \subset \mathbb{R}_{n-m}$$

$$\boldsymbol{\nu} = (\nu_1, \ldots, \nu_m).$$

By definition the function $g(\boldsymbol{x}) \in \mathfrak{M}_{\nu p}(\mathscr{E})$, if it belongs to $L_p(\mathscr{E})$ and, for almost all $\boldsymbol{y} \in \mathscr{E}'$ is a function of exponential type $\boldsymbol{\nu}$ relative to $\boldsymbol{u}$.

For the functions

$$g = g_\nu \in \mathfrak{M}_{\nu p}(\mathscr{E}) = \mathfrak{M}_{\nu p}(\mathbb{R}_m \times \mathscr{E}')$$

one has the inequality

$$(1) \qquad \big\| \|g_\nu(\boldsymbol{u}, \boldsymbol{y})\|_{L_{p'}(\mathbb{R}_m)} \big\|_{L_p(\mathscr{E}')} \leqq 2^m \left(\prod_1^m \nu_k\right)^{\frac{1}{p} - \frac{1}{p'}} \|g_\nu\|_{L_p(\mathscr{E})},$$

$$1 \leqq p \leqq p' \leqq \infty,$$

where in the left side the inside norm is calculated relative to the variable $\boldsymbol{u} \in \mathbb{R}_m$, and the outside one relative to the variable $\boldsymbol{y} \in \mathscr{E}'$. Indeed, on the basis of the inequality (3.3.5 (1)) of different metrics, which is applied for almost all $\boldsymbol{y} \in \mathscr{E}'$,

$$\left(2^m \left(\prod_1^m \nu_k\right)^{\frac{1}{p} - \frac{1}{p'}} \|g(\boldsymbol{u}, \boldsymbol{y})\|_{L_p(\mathscr{E})}\right)^p = \int_{\mathscr{E}'} \left(2^m \left(\prod_1^m \nu_k\right)^{\frac{1}{p} - \frac{1}{p'}} \|g(\boldsymbol{u}, \boldsymbol{y})\|_{L_p(\mathbb{R}_m)}\right)^p dy$$

$$\geqq \int_{\mathscr{E}'} \|g(\boldsymbol{u}, \boldsymbol{y})\|^p_{L_{p'}(\mathbb{R}_m)} d\boldsymbol{y},$$

so that, raising both sides of the resulting inequality to the power $1/p$, we obtain (1).

We put $p' = \infty$ in formula (1), and take into account the fact that for some set $\mathscr{E}_1' \subset \mathscr{E}'$ of full measure the following property holds: for each $\boldsymbol{y} \in \mathscr{E}_1'$ the function $g(\boldsymbol{u}, \boldsymbol{y})$ is of type $\boldsymbol{v}$ relative to $\boldsymbol{u}$ and of finite norm:

$$(2)\quad \|g(\boldsymbol{u}, \boldsymbol{y})\|_{L_\infty(\mathbb{R}_m)} = \sup_{\boldsymbol{u} \in \mathbb{R}_m} \operatorname{vrai} |g(\boldsymbol{u}, \boldsymbol{y})| = \lim_{\varrho \to \infty} \max_{\boldsymbol{u} \in V_\varrho} |g(\boldsymbol{u}, \boldsymbol{y})| \geqq |g(\boldsymbol{u}, \boldsymbol{y})|$$

$$(\boldsymbol{u} \in \mathbb{R}_m),$$

where we denote by V_ϱ a ball with center at the origin and of radius ϱ, belonging to $\mathbb{R}_m$.

Thus inequality (2) is valid for all $\boldsymbol{y} \in \mathscr{E}_1'$ and $\boldsymbol{u} \in \mathbb{R}_m$, so that

$$\|g(\boldsymbol{u}, \boldsymbol{y})\|_{L_p(\mathscr{E}')} \leqq \| \, \|g(\boldsymbol{u}, \boldsymbol{y})\|_{L_\infty(\mathbb{R}_m)}\|_{L_p(\mathscr{E}')},$$

and we find, taking account of (1), the following inequality:

$$(3)\qquad \|g(\boldsymbol{u}, \boldsymbol{y})\|_{L_p(\mathscr{E}')} \leqq 2^m \left(\prod_1^m \nu_k\right)^{1/p} \|g_{\boldsymbol{v}}\|_{L_p(\mathscr{E})}.$$

3.4.2. Theorem[1]. *If $1 \leqq p \leqq \infty$ and $1 \leqq m < n$, then for each entire function $g_{\boldsymbol{v}}(z) = g_{\nu_1, \ldots, \nu_n}(z_1, \ldots, z_n) \in L_p(\mathbb{R}_n)$ of exponential type $\boldsymbol{v}$ the following inequality (of different dimensions) holds:*

$$(1)\qquad \left(\int_{-\infty}^{\infty} \cdots \int_{-\infty}^{\infty} |g(u_1, \ldots, u_m, x_{m+1}, \ldots, x_n)|^p \, du_1 \ldots du_m\right)^{1/p}$$

$$\leqq 2^{n-m} \left(\prod_{m+1}^n \nu_k\right)^{1/p} \left(\int_{-\infty}^{\infty} \cdots \int_{-\infty}^{\infty} |g|^p \, du_1 \ldots du_n\right)^{1/p}.$$

For fixed n and m and arbitrary $\boldsymbol{v} = (\nu_1, \ldots, \nu_n)$ this inequality is exact in the sense of order.

Proof. The space $\mathbb{R}_n$ may be considered as a topological product

$$\mathbb{R}_n = \mathbb{R}_{n-m} \times \mathbb{R}_m,$$

where $(x_1, \ldots, x_m) \in \mathbb{R}_m$, $(x_{m+1}, \ldots, x_n) \subset \mathbb{R}_{n-m}$. If now we suppose in inequality 3.4.1 (3) that $\mathscr{E} = \mathbb{R}_n$, $\mathscr{E}' = \mathbb{R}_m$ and replace $\mathbb{R}_m$ by $\mathbb{R}_{n-m}$, then we obtain the desired inequality.

The exactness of inequality (1) in the sense of order relative to $\boldsymbol{v}$ may be verified on the functions $F_{\boldsymbol{v}}$ (see 3.3.5 (4)), which were already used in 3.3.5 for a similar end.

[1] S. M. Nikol'skiĭ [3].

Remark. Putting $p_1 = \ldots = p_n = p$ and $p_{m+1} = \ldots = p_n = \infty$ in 3.3.7 (3), we obtain the inequality

$$\left(\int_{-\infty}^{\infty} \cdots \int_{-\infty}^{\infty} \sup_{x_{m+1}, \ldots, x_n} |g(u_1, \ldots, u_m, x_{m+1}, \ldots, x_n)|^p \, du_1 \ldots du_m\right)^{1/p}$$

$$\leqq 2^{\frac{n(n+1)}{2}} \left(\prod_{m+1}^{n} \nu_k\right)^{1/p} \left(\int_{-\infty}^{\infty} \cdots \int_{-\infty}^{\infty} |g(\boldsymbol{u})|^p \, du_1 \ldots du_n\right)^{1/p},$$

which sharpens inequality (1) in the sense that on the left we have

$$\sup_{x_{m+1}, \ldots, x_n} |g(u_1, \ldots, u_m, x_{m+1}, \ldots, x_n)|^p.$$

in place of $|g(u_1, \ldots, u_m, x_{m+1}, \ldots, x_n)|^p$

3.4.3. Inequalities of different metrics and dimensions for trigonometric polynomials. These are analogous to the corresponding inequalities for functions of exponential type.

Suppose that $T = T_{\boldsymbol{\nu}}(\boldsymbol{x}) \in \mathfrak{M}_{\boldsymbol{\nu}}^*(\mathbb{R}_n)$, i.e. T is a trigonometric polynomial of order $\boldsymbol{\nu}$ of n variables, and

$$(1) \qquad ((T))_p^{*(n)} = \max_{u_l} \left(\prod_{l=1}^{n} h_l \sum_{l_1=1}^{N_1} \cdots \sum_{l_n=1}^{N_n} |T(x_{l_1}^{(1)} - u_1, \ldots, x_{l_n}^{(n)} - u_n)|^p\right)^{1/p}$$

$$\left(h_l = \frac{2\pi}{N_l}, \; N_l = 1, 2, \ldots; \; l = 1, \ldots, n, \; 1 \leqq p \leqq \infty\right).$$

Then the following inequalities hold[1]:

$$(2) \qquad \|T_\nu\|_{L_p^*}^{(n)} \leqq ((T_\nu))_p^{*(n)} \leqq \prod_{1}^{n} (1 + h_i \nu_i) \|T_\nu\|_{L_p^*}^{(n)},$$

$$(3) \qquad \|T_\nu\|_{L_{p'}^*}^{(n)} \leqq 3^n \left(\prod_{1}^{n} \nu_i\right)^{\frac{1}{p} - \frac{1}{p'}} \|T_\nu\|_{L_p^*}^{(n)},$$

$$(4) \qquad \left(\int_{0}^{2\pi} \cdots \int_{0}^{2\pi} |T(u_1, \ldots, u_m, x_{m+1}, \ldots, x_n)|^p \, du_1 \ldots du_m\right)^{1/p}$$

$$\leqq 3^{n-m} \left(\prod_{m+1}^{n} \nu_i\right)^{1/p} \left(\int_{0}^{2\pi} \cdots \int_{0}^{2\pi} |T|^p \, du_1 \ldots du_m\right)^{1/p}.$$

---

[1] See Remarks 3.3—3.4.3 at the end of the book.

They are analogous to the corresponding inequalities proved above for entire functions of exponential type and are proved analogously. Now, in the proof all the sums ($\sum$) are extended, as in (1), over a finite number of terms ($N_1, \ldots, N_m$), the integrals are taken over periods and the Bernšteĭn inequality for trigonometric polynomials is applied. However if one brings in arguments in analogy with what was done for functions of exponential type, one obtains a (rounded-off) constant 3 instead of 2, explained by the fact that in the periodic case it is necessary to seek the minimum of $\psi(\alpha)$ among discrete values of α. But, of course, in both cases the constants given are excessive.

For trigonometric polynomials it is possible to obtain analogies to the other inequalities presented in 3.3 as well.

The exactness of the indicated inequalities in the sense of order is verified this time on Féjer kernels (see 2.2.2).

3.5. Subspaces of Functions of Given Exponential Type

Theorem. *The space* $\mathfrak{M}_{\nu p}(\mathscr{E}) = \mathfrak{M}_{\nu p}(\mathbb{R}_m \times \mathscr{E}')$ (see 3.4.1) *is a subspace of the space* $L_p(\mathscr{E})$, *i.e. a linear set closed in* $L_p(\mathscr{E})$.

Proof. The linearity of $\mathfrak{M}_{\nu p}(\mathscr{E})$ is obvious.

Suppose for the sequence $g_k = g_{\nu k} \in \mathfrak{M}_{\nu p}(\mathscr{E})$ ($k = 1, 2, \ldots$) that the condition

$$\lim_{k,l\to\infty} \|g_k - g_l\|_{L_p(\mathscr{E})} = 0. \tag{1}$$

is satisfied. Then there exists a function $f \in L_p(\mathscr{E})$ such that

$$\lim_{k\to\infty} \|f - g_k\|_{L_p(\mathscr{E})} = 0. \tag{2}$$

Obviously, it is possible to indicate a set $\mathscr{E}_1' \subset \mathscr{E}$ of full measure, one and the same for all $k = 1, 2, \ldots$, such that $g_k(\boldsymbol{u}, \boldsymbol{y})$ will be an entire function of exponential type $\boldsymbol{\nu}$ relative to $\boldsymbol{u}$ for all $\boldsymbol{y} \in \mathscr{E}_1'$. At the same time we may suppose that $\mathscr{E}_1'$ has the additional property

$$\lim_{s\to\infty} \|f(\boldsymbol{u}, \boldsymbol{y}) - g_{k_s}(\boldsymbol{u}, \boldsymbol{y})\|_{L_p(R_m)} = 0 \quad \text{for all} \quad \boldsymbol{y} \in \mathscr{E}_1', \tag{3}$$

where k_s is some subsequence of natural numbers, one and the same for all $\boldsymbol{y} \in \mathscr{E}_1'$ (this follows from (2) on the basis of Lemma 1.3.8). Further, from (3), as a consequence of inequality 3.2.2 (10) ($p = \infty$) and the inequality of different metrics, it follows that ($\boldsymbol{y} \in \mathscr{E}_1'$)

$$(4)\quad |g_{k_s}(\boldsymbol{u} + i\boldsymbol{v}, \boldsymbol{y}) - g_{k_{s'}}(\boldsymbol{u} + i\boldsymbol{v}, \boldsymbol{y})|$$

$$\leqq \sup_u |g_{k_s}(\boldsymbol{u}, \boldsymbol{y}) - g_{k_{s'}}(\boldsymbol{u}, \boldsymbol{y})| \, e^{\sum_{j=1}^{m} \nu_j |v_j|}$$

$$\leqq 2^m \prod_{j=1}^{m} \nu_j^{1/p} \|g_{k_s}(\boldsymbol{u}, \boldsymbol{y}) - g_{k_{s'}}(\boldsymbol{u}, \boldsymbol{y})\|_{L_p(\mathbb{R}_m)} e^{\sum_{j=1}^{m} \nu_j |v_j|} \to 0$$

$$s, s' \to \infty.$$

This shows that $g_{k_s}(\boldsymbol{z}, \boldsymbol{y})$, for each fixed $\boldsymbol{y} \in \mathscr{E}_1'$ tends uniformly as $s \to \infty$ on any bounded set of complex $\boldsymbol{z}$ to some function $g(\boldsymbol{z}, \boldsymbol{y})$, which is obviously entire relative to $\boldsymbol{z}$. Suppose that

$$\Delta_N = \{|x_j| \leqq N; j = 1, \ldots, n\}.$$

From what has been said it follows that $g_{k_s}(\boldsymbol{x}) \to g(\boldsymbol{x})$ $(s \to \infty)$ almost everywhere on $\mathscr{E}\Delta_N$. It then follows from (2) that $g(\boldsymbol{x}) = f(\boldsymbol{x})$ almost everywhere on $\mathscr{E}\Delta_N$, and therefore, since N is arbitrary, on $\mathscr{E}$ as well.

Finally, we consider the following inequality, analogous to (4):

$$(5)\qquad |g_{k_s}(\boldsymbol{z}, \boldsymbol{y})| \leqq 2^m \prod_{j=1}^{m} \nu_j^{1/p} \|g_{k_s}(\boldsymbol{u}, \boldsymbol{y})\|_{L_p(\mathbb{R}_m)} (y \in \mathscr{E}_1'),$$

Passing to the limit in (5) as $s \to \infty$, we obtain the same inequality, but now for g. This shows that for any $\boldsymbol{y} \in \varepsilon_1'$ the function g is of exponential type $\boldsymbol{\nu}$ relative to $\boldsymbol{u}$.

We have proved that the function f figuring in (2) may be perturbed on a set of n-dimensional measure zero in such a way that it is an entire function of exponential type $\boldsymbol{\nu}$ relative to $\boldsymbol{u}$ for almost all $\boldsymbol{y}$. Since $f \in L_p(\mathbb{R}_n)$, then $f \in \mathfrak{M}_{\nu p}(\mathscr{E})$. The theorem is proved.

3.6. Convolutions with Entire Functions of Exponential Type

3.6.1. Lemma. *Suppose that*

$$(1)\qquad g(t) = \sum_0^\infty c_{2k} t^{2k}$$

is an even entire function of one variable of exponential type ν. Then the function

$$(2)\qquad g_*(\boldsymbol{x}) = g(|\boldsymbol{x}|) = \sum_0^\infty c_{2k} |\boldsymbol{x}|^{2k}$$

is an entire spherical function of type ν.

Proof. The series (1) converges absolutely for any t. The polynomial

$$|\boldsymbol{x}|^{2k} = (x_1^2 + \cdots + x_n^2)$$

has positive coefficients. Therefore the series (2), after expanding the parentheses in each of its terms, is absolutely convergent for any $\boldsymbol{x} = (x_1, \ldots, x_n)$ as a power series in $x_1, \ldots, x_n$. Accordingly $g_*(\boldsymbol{z})$ is an entire function. It is exponential spherical of type ν, because

$$|g_*(\boldsymbol{z})| = \left|g\left(\sqrt{\sum_1^n |z_j|^2}\right)\right| \leq A_\varepsilon e^{(\nu+\varepsilon)\sqrt{\sum_1^n |z_j|^2}}.$$

3.6.2. Theorem. *Suppose that g is an entire function of exponential type ν_j relative to z_j $(j = 1, \ldots, n)$ (or of spherical type ν), lying in $L_q(\mathbb{R}_n)$, $1 \leq q \leq \infty$, and $f \in L_p(\mathbb{R}_n)$ $\left(\frac{1}{p} + \frac{1}{q} = 1\right)$.* Then the function

$$\omega(\boldsymbol{x}) = \int g(\boldsymbol{x} - \boldsymbol{u})\, f(\boldsymbol{u})\, d\boldsymbol{u}$$

lies in $\mathfrak{M}_{\nu\infty}$ (respectively in $S\mathfrak{M}_{\nu\infty}$). That is, it is an entire exponential function of type ν_j relative to z_j (respectively of spherical degree ν), bounded on $\mathbb{R} = \mathbb{R}_n$.

If $g \in L$, $f \in L_p$ $(1 \leq p \leq \infty)$, then $\omega \in L_p$ (1.3.3.), so that $\omega \in \mathfrak{M}_{\nu p}$ $\times (S\mathfrak{M}_{\nu p})$.

Proof. The boundedness of ω on R follows from the inequality

$$(1) \qquad |\omega(\boldsymbol{x})| \leq \|g(\boldsymbol{x} - \boldsymbol{u})\|_q \, \|f(\boldsymbol{u})\|_p = \|g\|_q \, \|f\|_p .$$

In view of the fact that g is an entire function, there is an expansion into Taylor series

$$(2) \qquad g(\boldsymbol{z} - \boldsymbol{u}) = \sum_{k \geq 0} \frac{g^{(\boldsymbol{k})}(-\boldsymbol{u})}{\boldsymbol{k}!} \boldsymbol{z}^{\boldsymbol{k}},$$

converging absolutely for any $\boldsymbol{u} \in \mathbb{R}$ and any complex $\boldsymbol{z} = (z_1, \ldots, z_n)$.

We have

$$\sum_{|\boldsymbol{k}| < N} \left|\frac{g^{(\boldsymbol{k})}(-\boldsymbol{u})\, f(\boldsymbol{u})\, z^{\boldsymbol{k}}}{\boldsymbol{k}!}\right| \leq \sum_{\boldsymbol{k} \geq 0} \left|\frac{g^{(\boldsymbol{k})}(-\boldsymbol{u})\, f(\boldsymbol{u})\, \boldsymbol{z}^{\boldsymbol{k}}}{\boldsymbol{k}!}\right| = \Phi(\boldsymbol{u}),$$

and then $\left(\frac{1}{p} + \frac{1}{q} = 1\right)$

$$\int \Phi(u)\, du \leqq \sum_k \frac{|z^k|}{k!} \|g^{(k)}\|_q \, \|f\|_p$$

$$\leqq \|g\|_q \, \|f\|_p \sum_k \frac{|z^k|\, \nu^k}{k!} = \|g\|_q \, \|f\|_p e^{\sum\limits_{j=1}^{n} \nu_j |z_j|}.$$

This inequality shows that equation (2), after multiplication by $f(u)$, may be legitimately (on the basis of the theorem of Lebesgue) integrated term by term:

(3) $$\omega(z) = \int g(z - u)\, f(u)\, du = \sum_k \frac{c_k}{k!} z^k,$$

$$c_k = \int g^{(k)}(-u)\, f(u)\, du,$$

and in addition the following inequality holds:

$$|\omega(z) \leqq \|g\|_q \, \|f\|_p e^{\sum\limits_{i=1}^{n} \nu_j |z_j|}.$$

If g is not only of type ν relative to each variable, but is also of spherical type ν, then it is necessary to show in addition that ω is also of spherical type ν.

Indeed, for real x, u, y

$$g(x + iy - u) = \sum_{k \geqq 0} \frac{g^{(k)}(x - u)}{k!} (iy)^k$$

$$= \sum_{l=0}^{\infty} i^l \sum_{|k|=l} \frac{g^{(k)}(x - u)}{k!} y^k = \sum_{l=0}^{\infty} i^l \frac{g_y^{(l)}(x - u)}{l!} |y|^l,$$

where $g_y^{(l)}$ is the derivative of g of order l in the direction y. Therefore, reasoning as in (3), we will have

$$\omega(x + iy) = \sum_{l=0}^{\infty} \frac{\int g_y^{(l)}(x - u)\, f(u)\, du}{l!} (i|y|)^l.$$

Taking account of the inequality

$$\left| \int g_y^{(l)}(x - u)\, f(u)\, du \right| \leqq \|g_y^{(l)}\|_q \, \|f\|_p \leqq \nu^l \, \|g\|_q \, \|f\|_p,$$

which is derived on the basis of 3.2.6 (3), we get

$$|\omega(\boldsymbol{x} + i\boldsymbol{y})| \leqq \|g\|_q \|f\|_p \sum^{\infty} \frac{(\nu|\boldsymbol{y}|)^l}{l!} = \|g\|_q \|f\|_p e^{\nu|\boldsymbol{y}|}.$$

We have proved that $\omega \in \mathfrak{M}_{\nu\infty}$.

3.6.3. Theorem. *Suppose that* $q > 0$, $\frac{1}{p} + \frac{1}{q} = 1$. *Put*

$$(1) \qquad \omega(\boldsymbol{z}) = \int k(|\boldsymbol{z} - \boldsymbol{u}|) f(\boldsymbol{u}) \, d\boldsymbol{u}, \quad \int = \int_{\mathbb{R}_m},$$

where

$$(2) \qquad k(t) = \left(\frac{\sin \frac{t}{\lambda}}{t} \right)^{\lambda},$$

λ being a natural even number satisfying the inequalities

$$(3) \qquad (\lambda - \mu) q - m > 0$$

and

$$(4) \qquad \int |f(\boldsymbol{u}) (1 + |\boldsymbol{u}|)^{-\mu}|^p \, d\boldsymbol{u} < \infty \quad (\mu > 0).$$

Then $\omega(\boldsymbol{z})$ is an entire function of exponential spherical type 1.

Proof. Put

$$A = \left(\sum_{j=1}^{m} |z_j|^2 \right)^{1/2}.$$

Applying the Hölder inequality, we find

$$(5) \qquad |\omega(\boldsymbol{z})| \leqq I(\boldsymbol{z}) \| f(\boldsymbol{u}) (1 + |\boldsymbol{u}|)^{-\mu} \|_{L_p(\mathbb{R}_m)},$$

where

$$(6) \qquad I(\boldsymbol{z}) = \left(\int |k(|\boldsymbol{z} - \boldsymbol{u}|)|^q (1 + |\boldsymbol{u}|)^{q\mu} \, d\boldsymbol{u} \right)^{1/q}$$

$$\leqq \left(\int_{|\boldsymbol{u}|<1} \right)^{1/q} + \left(\int_{|\boldsymbol{u}|<3A} \right)^{1/q} + \left(\int_{|\boldsymbol{u}|>1, 3A} \right)^{1/q} = I_1 + I_2 + I_3.$$

We note that

$$(7) \qquad |\boldsymbol{z} - \boldsymbol{u}|^2 = |\sum z_j^2 - 2 \sum z_j u_j + \sum u_j^2|$$

$$\geqq |\boldsymbol{u}|^2 - 2A|\boldsymbol{u}| - A^2 = (|\boldsymbol{u}| - A)^2 - 2A^2 = \psi(|\boldsymbol{u}|).$$

The functions $\left(\sin \frac{t}{\lambda}\right)^\lambda$ and $k(t)$ of the single variable t are entire of type 1, bounded on the real axis, so that $\left(\sin \frac{|\boldsymbol{z}|}{\lambda}\right)^\lambda$ and $k(|\boldsymbol{z}|)$ are entire spherical functions of type 1, bounded on $\mathbb{R}_m$ (see (3.6.1.)). In such a case, taking account of the fact that the $\boldsymbol{u} \in \mathbb{R}_m$ are real points and $\boldsymbol{z} = (x_1 + iy_1, \ldots, x_m + iy_m)$, we will have (see 3.2.6. (4)):

$$k(|\boldsymbol{z} - \boldsymbol{u}|) \ll \exp |\boldsymbol{y}| \leqq \exp A, \tag{8}$$

$$\left(\sin \frac{|\boldsymbol{z} - \boldsymbol{u}|}{\lambda}\right)^\lambda \ll \exp |\boldsymbol{y}| \leqq \exp A. \tag{9}$$

Therefore in view of (8)

$$I_1 \ll \exp A \left(\int_{|\boldsymbol{u}|<1} (1 + |\boldsymbol{u}|)^{q\mu}\, d\boldsymbol{u}\right)^{1/q} \ll \exp A, \tag{10}$$

$$I_2 \ll \exp A \left(\int_{|\boldsymbol{u}|<3A} (1 + |\boldsymbol{u}|)^{q\mu}\, d\boldsymbol{u}\right)^{1/q} \ll \exp\{(1 + \varepsilon) A\}, \tag{11}$$

where $\varepsilon > 0$ is an arbitrarily small number number, the constant in the second inequality (11) depends on ε, and in view of (7) and (9) (explanation below)

$$I_3 \ll \exp A \left(\int_{\varrho>1,3A} \frac{\varrho^{\mu q+m-1}}{\psi(\varrho)^{\frac{\lambda q}{2}}}\, d\varrho\right)^{1/q} \ll \exp A \quad (\varrho = |u|). \tag{12}$$

The function

$$\frac{\varrho^2}{\psi(\varrho)} = \varphi(\varrho) \quad (\varrho \geqq 3A)$$

is bounded, because it is positive and its derivative is negative. Accordingly, putting

$$\nu = \frac{1}{2}(\lambda q - \mu q - m + 1)$$

and $\varrho = tA$, we find that the integral entering into (12), does not, up to a constant factor, exceed

$$\int_{\varrho>3A,1} \frac{d\varrho}{\psi(\varrho)^\nu} = \frac{1}{A^{2\nu-1}} \int_{t>3,\frac{1}{A}} \frac{dt}{\psi(t)^\nu}$$

$$\ll \frac{1}{A^{2\nu-1}} \int_{t>3,\frac{1}{A}} \frac{dt}{t^{2\nu}} \ll \begin{cases} \int_3^\infty t^{-2\nu}\, dt \ll 1 & (A \geqq 1), \\ \frac{1}{A^{2\nu-1}} A^{2\nu-1} = 1 & (A < 1), \end{cases}$$

i.e. the second inequality (12) is valid. From the estimates just obtained it follows that for any $\varepsilon > 0$ there exists a constant c_ε, not depending on f, such that

$$(13) \qquad |\omega(\boldsymbol{z})| \leqq c_\varepsilon \|f(\boldsymbol{u}) (1 + |\boldsymbol{u}|)^{-\mu}\|_{L_p(\mathbb{R}_m)} \exp\{(1 + \varepsilon)A\}.$$

It remains to prove that $\omega(\boldsymbol{z})$ is an entire function. Suppose that $f_N = f$ for $|\boldsymbol{u}| < N$ and $f_N = 0$ for $|\boldsymbol{u}| > N$ and

$$\omega_N(\boldsymbol{z}) = \int k(|\boldsymbol{z} - \boldsymbol{u}|) f_N(\boldsymbol{u}) \, d\boldsymbol{u}.$$

Suppose given an arbitrary number $\Lambda > 0$, and suppose that ε_Λ is the set of points $\boldsymbol{z} = (z_1, \ldots, z_m)$, for which $|z_j| < \Lambda$. For such points

$$A = \sqrt{\sum_1^m |z_j|^2} \leqq m\Lambda$$

and in view of (13)

$$|\omega(\boldsymbol{z}) - \omega_N(\boldsymbol{z})| < c_\varepsilon m\Lambda \, \|(f - f_N) (1 + |\boldsymbol{u}|)^{-\mu}\|_{L_p(\mathbb{R}_m)} \to 0,$$

$$N \to \infty$$

i.e. $\omega_N(\boldsymbol{z}) \to \omega(\boldsymbol{z})$, $N \to \infty$ uniformly on any $\mathscr{E}_\Lambda$. Accordingly, $\omega(\boldsymbol{z})$ is entire (3.1.1). The assertion is proved (see further 4.2.2).

Chapter 4

The Function Classes W, H, B

4.1. The Generalized Derivative

Suppose given in the space $\mathbb{R} = \mathbb{R}_n$ an open set g. We denote by g_1 its orthogonal projection on the hyperplane $x_1 = 0$. Suppose given on g a real (complex) measurable function $f(\boldsymbol{x}) = f(x_1, \boldsymbol{y})$. For fixed $\boldsymbol{y}$ it is a function of x_1, defined on the corresponding open one-dimensional set. If f is absolutely continuous on any closed finite segment belonging to this set, then we will say that it is for the indicated $\boldsymbol{y}$ *locally absolutely continuous* relative to x_1.

By definition the function f has a generalized derivative $\frac{\partial f_1}{\partial x_1}$ *relative to x_1, if f is measurable on g and if there exists a function f_1 equivalent to it which is locally absolutely continuous for almost all admissible $\boldsymbol{y}$ (i.e. $\boldsymbol{y} \in g_1$). The function f_1 will have almost everywhere on g (in the sense of n-dimensional measure) an ordinary partial derivative* $\frac{\partial f_1}{\partial x_1}$. *Any function equivalent to it in the sense of n-dimensional measure will be called the generalized derivative of f on g relative to x_1, and it will be denoted by* $\frac{\partial f_1}{\partial x_1}$.

If $\varphi(t)$ is a function of one variable t and Ω is an open set of points t, the fact that φ has on Ω a generalized derivative $\varphi'(t)$ may be expressed as follows: there exists a function φ_1, equivalent to φ relative to Ω, locally absolutely continuous on Ω. Then φ_1 has, as is well known, an ordinary derivative $\varphi_1'(t)$ almost everywhere on Ω. Any function equivalent to $\varphi_1'(t)$ is also, by definition, a generalized derivative of $\varphi(t)$ on Ω.

In order to avoid misunderstandings, we shall explain in somewhat more detail why the ordinary partial derivative $\frac{\partial f_1}{\partial x_1}$ exists almost everywhere on g.

The projection g_1 of the open set g onto the subspace of points $\boldsymbol{y} = (0, x_2, \ldots, x_n)$ is obviously an open set relative to that $(n-1)$-dimensional subspace. To every fixed point $\boldsymbol{y} \in g_1$ there corresponds a one-dimensional open (in the one-dimensional sense) set $e_{\boldsymbol{y}}$ of points of the form $(x_1, \boldsymbol{y}) \in g$. The set g may be considered as the set-theoretic union

$$g = \bigcup_{\boldsymbol{y} \in g_1} e_{\boldsymbol{y}}.$$

By hypothesis the function $f_1(x_1, \boldsymbol{y})$ is for almost all $\boldsymbol{y} \in g_1$ absolutely continuous relative to x_1 on each closed segment of variation of x_1 lying in $e_{\boldsymbol{y}}$. Hence it follows that for almost all points $\boldsymbol{y} \in g_1$ the function $f_1(x_1, \boldsymbol{y})$ has for almost all $x_1 \in e_\alpha$ an ordinary partial derivative $\frac{\partial f_1}{\partial x_1} = f_{1x_1}$. We denote by g' the set of all points $\boldsymbol{x} = (x_1, \boldsymbol{y}) \in g$, for which there does not exist a partial derivative f_{1x_1}. The set g' is measurable, since it is the complement to the set of those points $\boldsymbol{x} \in g$ for which the limit of the ratio

$$\lim_{h \to 0} \frac{f_1(x_1 + h, \boldsymbol{y}) - f_1(x_1, \boldsymbol{y})}{h} = f_{1x_1}(x_1, \boldsymbol{y}),$$

exists, this ratio for each h being a measurable function (f is measurable on g by hypothesis!). We need to keep in mind the fact that the set of points of convergence of a sequence of measurable functions on a measurable set $\mathscr{E}$ is measurable[1].

On the other hand,

$$g' = \bigcup_{\boldsymbol{y} \in g_1} e'_{\boldsymbol{y}},$$

while for almost all $\boldsymbol{y} \in g_1$, in the sense of $(n - 1)$-dimensional measure, $\mu e'_{\boldsymbol{y}} = 0$. Hence, by Fubini's theorem, g' has n-dimensional measure $\mu g' = 0$. Thus the function f_1 has almost everywhere on g an ordinary partial derivative $\frac{\partial f_1}{\partial x_1}$, which we have called the generalized partial derivative of f relative to x_1.

The function $\frac{\partial f}{\partial x_1}$ (measurable on the open[2] set g) may in its turn have a generalized derivative relative to x_1. I.e. it may happen that

---

[1] Suppose that $\{F_k\}$ is a sequence of measurable functions given on a measurable set $\mathscr{E}$, $e_{nm} = \left\{x: |F_k(x) - F_l(x)| < \frac{1}{m}\right\}$ for any $k, l \geqq n$; $n, m = 1, 2, \ldots$. Then $A = \bigcap_m \bigcup_n e_{nm}$ is the set of points of convergence of $\{F_k\}$, and obviously measurable.

[2] In the definition of $\frac{\partial f}{\partial x_1}$, we may choose in place of a set $g \subset \mathbb{R}_n$ open in $\mathbb{R}_n$ a measurable set $\mathscr{E} \subset \mathbb{R}_n$ which is open relative to the variable x_1. More precisely, one may choose a measurable set $\mathscr{E} \subset \mathbb{R}_n$ whose projection $\mathscr{E}_1$ on the subspace of points $\boldsymbol{y} = (0, x_2, \ldots, x_n)$ is measurable in the $(n - 1)$-dimensional sense, such that

$$\mathscr{E} = \bigcup_{\boldsymbol{y} \in \mathscr{E}_1} e_{\boldsymbol{y}},$$

there exists a function φ defined on the set g and equivalent to it in the of n-dimensional measure, absolutely continuous relative to x_1 on any closed segment of variation of x_1, for almost all $\boldsymbol{y} \in g_1$. The ordinary derivative of φ relative to x_1, existing almost everywhere, or a function equivalent to it, will be denoted by $\frac{\partial^2 f}{\partial x_1^2}$. Similarly we define $\frac{\partial^k f}{\partial x_1^k} = f_{x_1}^{(k)}$ $(k = 0, 1, 2, \ldots, f_{x_1}^{(0)} = f)$. It is not hard to see that if the generalized derivative $\frac{\partial^k f}{\partial x^k}$, exists on g, it is always possible to reduce the function f to a function φ defined on g and corresponding to it such that the derivatives $\frac{\partial \varphi}{\partial x_1}, \frac{\partial^2 \varphi}{\partial x_1^2}, \ldots, \frac{\partial^k \varphi}{\partial x_1^k}$ exist in the ordinary sense almost everywhere on g, and such that furthermore the derivatives $\frac{\partial^i \varphi}{\partial x_1^i}$ $(i = 0, 1, \ldots, k - 1)$ are absolutely continuous relative to x_1 on any closed segment of variation of x_1 for all $\boldsymbol{y}$ belonging to one and the same set $g_1' \subset g_1$, differing from g_1 by a set of measure zero.

Analogously one defines the generalized derivatives $\frac{\partial^k f}{\partial x_i^k}$ $(i = 2, \ldots, n)$ on g. Mixed derivatives of the second and higher orders are defined by induction. For example, the derivative $\frac{\partial^2 f}{\partial x_1 \partial x_2}$ is defined by the equation

$$\frac{\partial^2 f}{\partial x_1 \partial x_2} = \frac{\partial}{\partial x_1} \frac{\partial f}{\partial x_2}.$$

Obviously the fact that the function $f(x)$ of one variable has on (a, b) a generalized derivative of order k reduces to the statement that it has, after being altered on a set of measure zero, ordinary derivatives up to order $k - 1$ inclusive which are absolutely continuous on any closed segment $[c, d] \subset (a, b)$. This implies further the existence of the derivative $f^k(x)$ of order k almost everywhere on the interval (a, b).

Throughout this book we shall be dealing with generalized derivatives. Therefore we shall frequently call them derivatives without adjoining the words "generalized".

Although the definition of generalized derivative given above is very general, it is in itself sufficiently effective in the applications in integration by parts. Suppose that the function f has on g a generalized

---

where $e_{\boldsymbol{y}}$ is an open one-dimensional set of points of the form $(x_1, \boldsymbol{y})$ with the number x_1 varying. In particular, a set of the form $\mathscr{E} = \mathbb{R}_1 \times \mathscr{E}_1 \subset \mathbb{R}_n$, with $\mathscr{E}_1$ a set measurable in the $(n - 1)$-dimensional-sense, is of this type.

derivative $\frac{\partial f}{\partial x_1}$. Here we will suppose that f has already been altered on a set of n-dimensional measure zero, as required by the definition. Suppose moreover that $\varphi(x)$ is a function continuous on g along with its derivative $\frac{\partial \varphi}{\partial x_1}$. Then for almost all $\boldsymbol{y} = (x_2, \ldots, x_n)$ for any segment $[a, b] \times \boldsymbol{y}$ belonging to g it is legitimate to integrate by parts:

$$(1) \qquad \int_a^b f(x_1, \boldsymbol{y}) \frac{\partial \varphi}{\partial x_1}(x_1, \boldsymbol{y})\, dx_1$$

$$= f(b, \boldsymbol{y})\, \varphi(b, \boldsymbol{y}) - f(a, \boldsymbol{y})\, \varphi(a, \boldsymbol{y}) - \int_a^b \frac{\partial f}{\partial x_1}(x_1, \boldsymbol{y})\, \varphi(x_1, \boldsymbol{y})\, dx.$$

It is frequently necessary to integrate this expression with respect to $\boldsymbol{y}$. But for this the measurability of $f(\boldsymbol{x}) = f(x_1, \boldsymbol{y})$ on g is not sufficient; there must be additional conditions on f. Such effective conditions may be summability or local summability of f and $\frac{\partial f}{\partial x_1}$ or of $\frac{\partial f}{\partial x_1}$ alone on g.

The concept of generalized derivative is found in the investigations of Beppo Levi [1], who considered generalized derivatives with integrable square on g. Later on many mathematicians arrived at this concept, frequently independently from their predecessors.

S. L. Sobolev [1, 2] came to the definition of generalized derivative from the point of view of the concept of generalized function introduced by him. The definition of Sobolev consists in the following. *Suppose that f and λ are functions locally summable on an open set g. If for any infinitely differentiable function φ with compact support in g the equation*

$$\int \lambda \varphi\, d\boldsymbol{x} = (-1)^{|s|} \int f \varphi^{(s)} d\boldsymbol{x},$$

is satisfied, then λ is the generalized derivative $f^{(s)}$ of f.

If the function f is locally summable on g along with its derivative $\frac{\partial f}{\partial x_1}$, understood in the sense of the first definition, then for a C^∞ function φ in g with compact support we will have (see (1))

$$\int_g f \frac{\partial \varphi}{\partial x_1}\, d\boldsymbol{x} = \int_{g_1} d\boldsymbol{y} \int f(x, \boldsymbol{y}) \frac{\partial \varphi}{\partial x_1}(x_1, \boldsymbol{y})\, dx_1$$

$$= -\int_{g_1} d\boldsymbol{y} \int \frac{\partial f}{\partial x_1}(x_1, \boldsymbol{y})\, \varphi(x_1, \boldsymbol{y})\, dx_1 = -\int \frac{\partial f}{\partial x_1} \varphi\, d\boldsymbol{x},$$

and we will have proved that the first definition of the derivative $\frac{\partial f}{\partial x_1}$ implies the second. The converse is also true. It is convenient to present the proof of this assertion later (see 4.5.2), and until then we will start from the first definition. We note also that both definitions of the non-mixed generalized derivative $\frac{\partial^\varrho f}{\partial x_1^\varrho}$ also coincide. But this is already not the case for the mixed derivative. The first definition implies the second, but not conversely, as is shown by the example of the function $f(x_1, x_2) = \varphi(x_1) + \psi(x_2)$, where $\varphi(x_1)$ and $\psi(x_2)$ are continuous nowhere differentiable functions. In the sense of S. L. Sobolev $\frac{\partial^2 f}{\partial x_1 \partial x_2} = 0$, and in the sense of the first definition the derivative $\frac{\partial^2 f}{\partial x_1 \partial x_2}$ does not exist.

Both definitions of $f^{(s)}$ coincide in every case when $f^{(s)} = f^{(s_1, \ldots, s_n)}$, and also $f^{(s_1, \ldots, s_{n-1}, 0)}$, $f^{(s_1, \ldots, s_{n-2}, 0, 0)}$, $\ldots$, $f^{(s_1, 0, \ldots, 0)}$ and f are locally summable. This holds for functions of the classes W, H, B, L, which we shall study in this book (see for example 4.4.6.).

We shall present a characteristic problem leading in a natural way to the concept of generalized derivative.

Suppose given on g a sequence of continuously differentiable functions $f_k(\boldsymbol{x})$ $(k = 1, 2, \ldots)$, having the following property: for any bounded region $\Omega \subset \bar{\Omega} \subset g$

(2) $$\|f_k - f_l\|_{L_p(\Omega)} \to 0 \quad (k, l \to \infty),$$

(3) $$\left\| \frac{\partial f_k}{\partial x_1} - \frac{\partial f_l}{\partial x_1} \right\|_{L_p(\Omega)} \to 0 \quad (k, l \to \infty).$$

It is required to characterize the local properties of a function f to which the f_k tend in the mean (locally):

(4) $$\lim_{k \to \infty} \|f_k - f\|_{L_p(\Omega)} = 0.$$

These properties (see 4.4.5) consist in that the function f has a generalized derivative $\frac{\partial f}{\partial x_1}$ on g and that f and $\frac{\partial f}{\partial x_1}$ are locally summable to the pth power on g.

We present still another problem closely connected with our objectives, leading to the concept of generalized derivative.

We denote by Ω_h the set of points $\boldsymbol{x} \in \Omega$, distant from the boundary Γ of an open set Ω by a distance larger than $h > 0$. Put

$$(5)\quad M_\alpha[f] = \sup_h \frac{\|f(x_1 + h, \boldsymbol{y}) - f(x_1, \boldsymbol{y})\|_{L_p(\Omega_h)}}{|h|^\alpha} < \infty \quad (0 < \alpha \leq 1).$$

Suppose further that there is given on Ω a sequence of continuously differentiable functions $f_k(\boldsymbol{x})$, having the property (5) and such that

$$(6)\qquad M_\alpha\left[\frac{\partial f_k}{\partial x_1} - \frac{\partial f_1}{\partial x_e}\right] \to 0 \quad (k, l \to \infty).$$

It is required to characterize the properties of the function f for which (4) holds. These properties (see 4.7.) consist in that f has on Ω a generalized derivative $\frac{\partial f}{\partial x_1}$ and that the quantity $M_\alpha\left[\frac{\partial f}{\partial x_1}\right]$ is finite.

4.2. Finite Differences and Moduli of Continuity

Suppose that $g \subset \mathbb{R}_n$ is an open set and $\boldsymbol{h} = (h_1, \ldots, h_n) \in \mathbb{R}_n$ any vector. We denote by g_h the set of points $\boldsymbol{x} \in g$ such that if $\boldsymbol{x}$ belongs to g then the point $\boldsymbol{x} + t\boldsymbol{h}$ does as well for all $0 \leqq t \leqq 1$, i.e. the entire segment joining $\boldsymbol{x}$ to $\boldsymbol{x} + \boldsymbol{h}$ lies in g.

We will also require the notation g_δ, where $\delta > 0$. This is the set of points $\boldsymbol{x} \in g$ distant from the boundary of g by a distance greater than δ. The sets g_h and g_δ may be empty. Obviously $g_{|\boldsymbol{h}|} \subset g_{\boldsymbol{h}}$.

Suppose that f is a function defined on g. If $\boldsymbol{x} \in g_{\boldsymbol{h}}$, then the (*first*) *difference*

$$\Delta_{\boldsymbol{h}} f = \Delta_{\boldsymbol{h}} f(x) = f(\boldsymbol{x} + \boldsymbol{h}) - f(\boldsymbol{x})$$

of the function f at the point $\boldsymbol{x}$ with (*vector*) *step $\boldsymbol{h}$* has a meaning.

By induction one introduces the concept of the kth *difference of the function f at the point $\boldsymbol{x}$ with step $\boldsymbol{h}$*:

$$\Delta_{\boldsymbol{h}}^k f = \Delta_{\boldsymbol{h}}^k f(\boldsymbol{x}) = \Delta_{\boldsymbol{h}} \Delta_{\boldsymbol{h}}^{k-1} f(\boldsymbol{x}) \quad (\Delta_{\boldsymbol{h}}^0 f = f,\ \Delta_{\boldsymbol{h}}^1 = \Delta_{\boldsymbol{h}},\ k = 1, 2, \ldots).$$

This is at least defined on the set $g_{k\boldsymbol{h}}$.

Obviously,

$$(1)\qquad \Delta_{\boldsymbol{h}}^k f(\boldsymbol{x}) = \sum_{l=0}^{k} (-1)^{l+k} C_k^l f(\boldsymbol{x} + l\boldsymbol{h}) \quad (k = 0, 1, \ldots).$$

If s is a natural number, then obviously

$$\Delta_{s\boldsymbol{h}} f(\boldsymbol{x}) = \sum_{l=0}^{s-1} \Delta_{\boldsymbol{h}} f(\boldsymbol{x} + l\boldsymbol{h})$$

and (by induction)

(2) $$\Delta^k_{sh} f(\boldsymbol{x}) = \sum_{l_1=0}^{s-1} \cdots \sum_{l_n=0}^{s-1} \Delta^k_{h} f(\boldsymbol{x} + l_1 \boldsymbol{h} + \cdots + l_n \boldsymbol{h}).$$

Let $\boldsymbol{h}$ be a unit vector, $|\boldsymbol{h}| = 1$. The *modulus of continuity of order k of the function f in the metric of $L_p(g)$ along the direction of* $\boldsymbol{h}$ is the quantity

(3) $$\omega^k(\delta) = \omega^k_{\boldsymbol{h}}(f, \delta) = \sup_{|t| \leq \delta} \|\Delta^k_{t\boldsymbol{h}} f(\boldsymbol{x})\|_{L_p(g_{kt\boldsymbol{h}})},$$

$$\omega(\delta) = \omega_{\boldsymbol{h}}(f, \delta) = \omega^1_{\boldsymbol{h}}(f, \delta).$$

(If $\mathscr{E}$ is the empty set, then we suppose $\| \cdot \|_{L_p(\mathscr{E})} = 0$.) In order that the quantity (3) have a meaning it is necessary that the norms inside the sup sign be finite, which will be the case for example if $f \in L_p(g)$. Below we shall dwell on certain characteristic properties of the modulus $\omega^k(\delta)$.

It is well known (see 1.3.12) that if the function $f \in L_p(g)$ and $1 \leq p < \infty$, then

(4) $$\lim_{t \to 0} \omega(t) = 0.$$

For $p = \infty$ this property does not in general hold. However it is satisfied in a trivial way if f is uniformly continuous on g.

The following inequalities hold:

(5) $$0 \leq \omega(\delta_2) - \omega(\delta_1) \leq \omega(\delta_2 - \delta_1) \quad (0 < \delta_1 < \delta_2).$$

The first of them is obvious. The second may be proved in the following way. If $\delta_1, \delta_2 \geq 0$, then every t with $|t| \leq \delta_1 + \delta_2$ may be represented in the form $t = t_1 + t_2$, where t_1 and t_2 are of the same sign as t and $|t_1| \leq \delta_1$, $|t_2| \leq \delta_2$.

Therefore

$$\omega(\delta_1 + \delta_2) = \sup_{\substack{|t'| \leq \delta_1 \\ |t''| \leq \delta_2}} \|f(\boldsymbol{x} + (t' + t'')\boldsymbol{h}) - f(\boldsymbol{x})\|_{L_p(g_{t\boldsymbol{h}})}$$

$$\leq \sup_{\substack{|t'| \leq \delta_1 \\ |t''| \leq \delta_2}} \|f(\boldsymbol{x} + (t' + t'')\boldsymbol{h}) - f(\boldsymbol{x} + t''\boldsymbol{h})\|_{L_p(g_{t\boldsymbol{h}})}$$

$$+ \sup_{|t''| \leq \delta_2} \|f(\boldsymbol{x} + t''\boldsymbol{h}) - f(\boldsymbol{x})\|_{L_p(g_{t\boldsymbol{h}})}$$

$$\leq \sup_{|t'| \leq \delta_1} \|f(\boldsymbol{x} + t'\boldsymbol{h}) - f(\boldsymbol{x})\|_{L_p(g_{t'\boldsymbol{h}})} + \omega(\delta_2) = \omega(\delta_1) + \omega(\delta_2).$$

Replacing δ_2 and $\delta_1 + \delta_2$ in this inequality by $\delta_2 - \delta_1$ and δ_2 respectively, we obtain (5).

It follows from (4) and (5) that the function $\omega(t)$ is continuous for any $t \geqq 0$, whenever $1 \leqq p < \infty$ or $p = \infty$ and f is uniformly continuous on g.

The following further property follows from the second inequality in (5):

$$\omega(l\delta) \leqq l\omega(\delta) \quad (\delta > 0;\ l = 1, 2, \ldots).$$

It may be obtained also, and then in a more general form, from equation (2), with $s = l$:

$$\omega^k(l\delta) \leqq l^k \omega^k(\delta) \quad (k, l = 1, 2, \ldots). \tag{6}$$

Obviously

$$\omega^k(\delta) \leqq \omega^k(\delta') \quad (0 < \delta < \delta'). \tag{7}$$

Inequality (6) generalizes to arbitrary, not necessarily integer, $l > 0$. To show this we select a natural number m such that $m \leqq l < m + 1$. Then

$$\omega^k(l\delta) \leqq \omega^k[(m+1)\,\delta] \tag{8}$$

$$\leqq (m+1)^k\, \omega^k(\delta) \leqq (l+1)^k \omega^k(\delta) \quad (l > 0,\ \ k = 1, 2, \ldots).$$

We note further that for any $\boldsymbol{L}$ with $|\boldsymbol{h}| = 1$

$$\sup_{|t| \leqq \delta} \|\Delta_{t\boldsymbol{h}}^{s+k} f(\boldsymbol{x})\|_{L_p(g_{t(s+k)\boldsymbol{h}})} \leqq \sup_{|t| \leqq \delta} 2^s \|\Delta_{t\boldsymbol{h}}^k f(\boldsymbol{x})\|_{L_p(g_{tk\boldsymbol{h}})}, \tag{9}$$

so that

$$\omega^{s+k}(\delta) \leqq 2^s \omega^k(\delta). \tag{10}$$

Suppose that $1 \leqq m \leqq n$, $\boldsymbol{x} = (\boldsymbol{u}, \boldsymbol{y})$, $\boldsymbol{u} = (x_1, \ldots, x_m) \in \mathbb{R}_m$, $\boldsymbol{y} = (m+1, \ldots, x_n)$. We will also denote by $\mathbb{R}_m$ the set of points $(x_1, \ldots, x_m, 0, \ldots, 0)$ of the space $\mathbb{R}_n$, so that $\mathbb{R}_m \subset \mathbb{R}_n$.

We introduce the quantity

$$\Omega_{\mathbb{R}_m}^k(f, \delta)_{L_p(g)} = \sup_{\substack{\boldsymbol{h} \in \mathbb{R}_m \\ |\boldsymbol{h}| = 1}} \omega_{\boldsymbol{h}}^k(f, \delta)_{L_p(g)}, \tag{11}$$

which we will call the *modulus of continuity of order k of the function f in the direction of the subspace* $\mathbb{R}_m \subset \mathbb{R}_n$. If g is a bounded set and d is its diameter, then it is easy to see that for $\delta > d$ the function $\Omega_{\mathbb{R}_m}^k(f, \delta)$ is a constant.

Now suppose that the function f has arbitrary derivatives relative to $\boldsymbol{u} \in \mathbb{R}_m$, of order ϱ. Then it makes sense on g to speak of the derivative in the direction of any unit vector $\boldsymbol{h} \in \mathbb{R}_m$:

(12)
$$f_{\boldsymbol{h}}^{(\varrho)} = \sum_{|\boldsymbol{s}|=\varrho} f^{(\boldsymbol{s})} \boldsymbol{h}^{\boldsymbol{s}}$$

$$(\boldsymbol{h} = (h_1, \ldots, h_m, 0, \ldots, 0), |\boldsymbol{h}| = 1,$$

$$\boldsymbol{h}^{\boldsymbol{s}} = h_1^{s_1} \ldots h_m^{s_m} = h_1^{s_1} \ldots h_m^{s_m} 0^0 \ldots 0^0,$$

where we are employing the usual convention $0^0 = 1$. Put

$$\Omega_{\mathbb{R}_m}(f^{(\varrho)}, \delta) = \sup_{\boldsymbol{h} \in \mathbb{R}_m} \omega_{\boldsymbol{h}}^k(f_{\boldsymbol{h}}^{(\varrho)}, \delta).$$

We will call this quantity the *modulus of continuity of the derivatives (all of them) of order* ϱ *of the function* f.

Since in view of (8)

(13)
$$\omega_{\boldsymbol{h}}^k(f_{\boldsymbol{h}}^{(\varrho)}, l\delta) \leqq (1 + l)^k \, \omega_{\boldsymbol{h}}^k(f_{\boldsymbol{h}}^{(\varrho)}, \delta),$$

then the upper limits of these quantities relative to $\boldsymbol{h} \in \mathbb{R}_m$ stand in the same relation:

(14)
$$\Omega_{\mathbb{R}_m}^k(f^{(\varrho)}, l\delta) \leqq (1 + l)^k \Omega_{\mathbb{R}_m}^k(f^{(\varrho)}, \delta).$$

Inequalities (8) and (14) show that the finiteness of the modulus of continuity for small δ guarantees its finiteness for large δ as well.

In view of the fact that $f_{\boldsymbol{h}}^{(\varrho)}$ is a finite linear combination of the derivatives $f^{(\boldsymbol{s})}$, $|\boldsymbol{s}| = \varrho$ (relative to the coordinate directions) with bounded coefficients $\boldsymbol{h}^{s}$, with $|\boldsymbol{h}^s| \leqq 1$ and not depending on $\boldsymbol{x}$, we have

(15)
$$\Omega_{\mathbb{R}_m}^k(f_{\boldsymbol{h}}^{(\varrho)}, \delta) = \sup_{\substack{|\boldsymbol{h}|=1 \\ \boldsymbol{h} \in \mathbb{R}_m}} \sup_{|t| \leqq \delta} \|\Delta_{t\boldsymbol{h}}^k f_{\boldsymbol{h}}^{(\varrho)}(\boldsymbol{x})\|_{L_p(g_{t\boldsymbol{h}})}$$
$$\leqq \sup_{\boldsymbol{h}} \sup_{t} \sum_{|\boldsymbol{s}|=\varrho} \|\Delta_{t\boldsymbol{h}}^k f^{(\boldsymbol{s})}(\boldsymbol{x})\|_{L_p(g_{t\boldsymbol{h}})}$$
$$= \sup_{\boldsymbol{h}} \sum_{|\boldsymbol{s}|=\varrho} \omega_{\boldsymbol{h}}^k(f^{(\boldsymbol{s})}, \delta) = \sum_{|\boldsymbol{s}|=\varrho} \Omega_{\mathbb{R}_m}^k(f^{(\boldsymbol{s})}, \delta),$$

where the sums $\sum$ are extended over all the coordinate derivatives $f^{(s)}$ of order ϱ with $\boldsymbol{s} = (s_1, \ldots, s_m, 0, \ldots, 0)$.

4.2.1. If $\mathscr{E} = \mathbb{R}_1 \times \mathscr{E}' \subset \mathbb{R}_n (\boldsymbol{x} = (x_1, \boldsymbol{y}), x_1 \in \mathbb{R}_1, \boldsymbol{y} \in \mathscr{E}')$, is a cylindrical measurable set, and f is a function defined on $\mathscr{E}$ with period 2π relative to x_1, then in this case the norm of the function f in $L_p^*(\mathscr{E})$ is understood in the sense

$$\|f\|_{L_p^*(\mathscr{E})} = \Big(\int_{\mathscr{E}_*} |f|^p \, dx\Big)^{1/p},$$

where $\mathscr{E}_* = \{(0, 2\pi) \times \mathscr{E}'\}$. Therefore in this case

$$\omega^k_{\substack{*\\x_1}}(t) = \sup_{|h| \leq t} \|\Delta^k_{x_1 h} f(x)\|_{L_p(\mathscr{E}_*)},$$

where h is a numerical increment in x_1.

The properties of the modulus of continuity $\omega^k_*(t)$ are analogous to the properties of $\omega^k(t)$.

4.2.2. Growth of a function with a bounded difference. Suppose that $\mathscr{E} = \mathbb{R}_m \times \mathscr{E}'$ is a cylindrical set of points $\boldsymbol{x} = (\boldsymbol{u}, \boldsymbol{y})$, $\boldsymbol{u} = (x_1, \ldots, x_m)$, $\boldsymbol{y} = (x_{m+1}, \ldots, x_n)$, $\boldsymbol{u} \in \mathbb{R}_m$. $\boldsymbol{y} \in \mathscr{E}'$. For conciseness we will write (in this subsection)

$$\|\cdot\|_\Delta = \|\cdot\|_{L_p(\Delta \times \mathscr{E}')},$$

$$\|\cdot\| = \|\cdot\|_{\mathbb{R}_m} = \|\cdot\|_{L_p(\mathscr{E})}.$$

Suppose we are given a natural number k and a positive number $\delta > 0$.

Suppose given on $\mathscr{E}$ a function $f(\boldsymbol{x})$, satisfying the conditions

(1) $$\|f(\boldsymbol{x})\|_{\{|\boldsymbol{u}| < \delta(k+m)\}} < A,$$

(2) $$\|\Delta^k_{\boldsymbol{h}} f(\boldsymbol{x})\| < B$$

for any $\boldsymbol{h} \in \mathbb{R}_m$, $|\boldsymbol{h}| = \delta$.

Put

(3) $$\sigma_N = \{N < |\boldsymbol{u}| < N + 1, \boldsymbol{y} \in \mathscr{E}'\}.$$

We shall prove the existance of a constant $c = c_{\delta k}$, for which the inequality

(4) $$\|f\|_{\sigma_N} < cN^{\frac{m-1}{p}}(A + (A + B)N^k), \quad N = 1, 2, \ldots$$

is satisfied.

We note that for $\varepsilon > 0$ it follows from (4) that

$$\frac{\|f\|^p_{\sigma_N}}{N^{m+(k+\varepsilon)p}} < \frac{c_1}{N^{1+\varepsilon_1}},$$

where c_1 does not depend on $N = 1, 2, \ldots$, and $\varepsilon_1 > 0$ depends on ε. Hence, taking (1) into account, we obtain for $\varepsilon > 0$ the inequality

(5) $$\left\| \frac{f(\boldsymbol{x})}{\left(1 + |\boldsymbol{u}|^{\frac{m}{p}+k+\varepsilon}\right)} \right\|_{L_p(\mathscr{E})} < \infty,$$

in which it is not possible to take ε equal to zero, as is shown by the example of the function of one variable $x^k (k = 1, 2, \ldots)$.

In the proof we will suppose $\delta = 1$ for the sake of simplicity.

Suppose given an arbitrary unit vector $\boldsymbol{u}' \in \mathbb{R}_m$. We define in $\mathbb{R}_m$ an $(m-1)$-dimensional cube orthogonal to $\boldsymbol{u}'$ with center at the origin, and having edges of unit length. With this cube as base and the vector $\boldsymbol{u}'$ as altitude we construct a unit cube $\omega = \omega_{\boldsymbol{u}'} \in \mathbb{R}_m$.

Further we suppose given a natural number N, and suppose that $\omega_N = \omega_{N\boldsymbol{u}'}$ denotes the unit cube consisting of points of the form $N\boldsymbol{u}' + \boldsymbol{u}$, where $\boldsymbol{u}$ runs through ω.

We note that for a function $\psi(\boldsymbol{x})$ locally summable to the pth power, if p is finite, or locally bounded, if $p = \infty$, one has

$$\psi(N\boldsymbol{u}' + \boldsymbol{u}, y) = \psi(\boldsymbol{u}, \boldsymbol{y}) + \sum_{j=0}^{N-1} \Delta\psi(j\boldsymbol{u}' + \boldsymbol{u}, \boldsymbol{y}),$$

$$\Delta\psi(j\boldsymbol{u}' + \boldsymbol{u}, \boldsymbol{y}) = \psi\big((j+1)\,\boldsymbol{u}' + \boldsymbol{u}, \boldsymbol{y}\big) - \psi(j\boldsymbol{u}' + \boldsymbol{u}, \boldsymbol{y}),$$

so that

$$(6) \qquad \|\psi\|_{\omega_N} \leqq \|\psi\|_\omega + \sum_{j=0}^{N-1} \|\Delta\psi\|_{\omega_j}.$$

We shall prove the inequality

$$(7) \qquad \|\Delta_{\boldsymbol{u}'}^{k-s} f\|_{\omega_{N\boldsymbol{u}'}} < c(A + (A+B)N^s)$$

$$(c = c_{k,s};\ N = 0, 1, \ldots;\ s = 0, 1, \ldots, k).$$

For $s = 0$ this follows directly from (2) ($\delta = 1$). Suppose that (7) is true for s; we shall prove its truth for $s+1$. We will suppose that

$$\psi(\boldsymbol{x}) = \Delta_{\boldsymbol{u}'}^{k-s-1} f(\boldsymbol{x}).$$

Then

$$\|\psi\|_\omega \leqq \left\| \sum_{l=0}^{k-s-1} (-1)^{l+k-s-1} C_{k-s-1}^l f(\boldsymbol{u} + l\boldsymbol{x}', \boldsymbol{y}) \right\|_\omega$$

$$\leqq A \sum_{l=0}^{k-s-1} C_{k-s-1}^l = 2^{k-s-1} A,$$

$$\|\Delta_{\boldsymbol{u}'}\psi\|_{\omega_j} \leqq \|\Delta_{\boldsymbol{u}'}^{k-s} f\|_{\omega_j} < c(A + (A+B)j^s).$$

Therefore on the basis of (6)

$$\|\Delta_{\boldsymbol{u}'}^{k-s-1} f\|_{\omega_N} \leqq 2^{k-s-1} A + c \sum_{j=0}^{N-1} (A + (A+B)j^s)$$

$$\leqq c_1(A + (A+B)N^{s+1}).$$

We have proved (7). Putting $s = k$ in (7), we get

(8) $$\|f\|_{\omega_{N\boldsymbol{u}'}} \leqq c(A + (A + B)N^k).$$

It follows from (8) and (3) that there exists a constant c_2 such that

$$\|f\|^p_{\sigma_N} \leqq c_2 N^{m-1}(A + (A + B)N^k)^p \quad (N = 1, 2, \ldots)$$

or (4). The point is that the region σ_N may be covered by cubes of the form $\omega_{Ns\boldsymbol{u}'}$, where $s = N - 1, N, N + 1$, the number of which is of order N^{m-1}.

4.3. The Classes W, H, B

We begin with the definition of the concept of imbedding, widely employed in this book.

If E and E' are two normed spaces with $E \subset E'$, and if in addition there exists a constant c not depending on $\boldsymbol{x}$ such that

(1) $$\|\boldsymbol{x}\|_{E'} \leqq c\|\boldsymbol{x}\|_E,$$

where $\|\cdot\|_{E'}$, $\|\cdot\|_E$ are the norms in the sense of E' and E respectively, then we will say that the *imbedding* $E \to E'$ *holds*. If $E \to E'$ and $E' \to E$, then we will write $E \rightleftarrows E'$.

If the elements of one and the same linear set are normed in the sense of different metrics E and E_1 and $E \rightleftarrows E_1$, the we will write $E \rightleftarrows E_1$ and even $\|\boldsymbol{x}\|_E = \|\boldsymbol{x}\|_{E_1}$, adjoining the statement that this equation holds up to an equivalence in the cases when confusion could arise.

Suppose that $\mathbb{R}_n$ is considered as the direct product $\mathbb{R}_n = \mathbb{R}_m \times \mathbb{R}_{n-m}$ of the coordinate subspaces $\mathbb{R}_m$ and $\mathbb{R}_{n-m}$, $1 \leqq m \leqq n$. Then any point $\boldsymbol{x} \in \mathbb{R}_n$ may be written in the form $\boldsymbol{x} = (\boldsymbol{u}, \boldsymbol{y})$, where $\boldsymbol{u} \in \mathbb{R}_m$, $\boldsymbol{y} \in \mathbb{R}_{n-m}$. In particular, $\boldsymbol{x} = \boldsymbol{u}$ for $m = n$. Suppose further that $g \subset \mathbb{R}_n$ is an open set and $1 \leqq p \leqq \infty$. In this section we define the classes

$$W^l_{\boldsymbol{u}p} = W^l_{\boldsymbol{u}p}(g) \ \left(l = 0, 1, \ldots;\ W^0_{\boldsymbol{u}p}(g) = L_p(g)\right),$$

$$H^r_{\boldsymbol{u}p} = H^r_{\boldsymbol{u}p}(g) \ (r > 0),$$

$$B^r_{\boldsymbol{u}p\theta} = B^r_{\boldsymbol{u}p\theta}(g) \ (r > 0, 1 \leqq \theta < \infty;\ B^r_{\boldsymbol{u}pp} = B^r_{\boldsymbol{u}p}).$$

For $m = n$ in this notation the letter $\boldsymbol{u}$ will be dropped and then we obtain: $W^l_p, H^r_p, B^r_{p\theta}, B^r_p(\theta = p)$. In another important case, when $\mathbb{R}_m = \mathbb{R}_{x_j}$ $(j = 1, \ldots, m)$, we will write $W^l_{x_jp}, H^r_{x_jp}, B^r_{x_jp\theta}, B^r_{x_jp}$.

We will call these classes *isotropic in the direction of* $\mathbb{R}_m$, because their differential properties relative to the various directions of $\mathbb{R}_m$ are one and the same. If $m = n$ we simply call them isotropic. For an integer vector $\boldsymbol{l} = (l, \ldots, l_m) \geqq \mathbf{0}$ $(l_j \geqq 0)$ and a vector $\boldsymbol{p} = (p_1, \ldots, p_m)$, where $1 \leqq p_j \leqq \infty$, we will define further the class (*anisotropic* for distinct l_j and p_j)

$$W_{\boldsymbol{p}}^{\boldsymbol{l}}(g) = \bigcap_{j=1}^{n} W_{x_j p_j}^{l_j}(g) \quad \left(W_{\boldsymbol{p}}^{\boldsymbol{l}} = W_{p}^{\boldsymbol{l}} \quad \text{for} \quad \boldsymbol{p} = (p, \ldots, p)\right)$$

as the intersection of the classes $W_{x_j p}^{l_j}(g)$. Analogously one defines the classes

$$H_{\boldsymbol{p}}^{\boldsymbol{r}}(g) = \bigcap_{j=1}^{n} H_{x_j p_j}^{r_j}(g), \; B_{p\theta}^{\boldsymbol{r}}(g) = \bigcap_{j=1}^{n} B_{x_j p_j \theta}^{r_j}(g),$$

where $\boldsymbol{r} = (r_1, \ldots, r_n) > \mathbf{0}$ $\left(H_{\boldsymbol{p}}^{\boldsymbol{r}} = H_{p}^{\boldsymbol{r}}, B_{\boldsymbol{p}\theta}^{\boldsymbol{r}} = B_{p\theta}^{\boldsymbol{r}} \text{ for } \boldsymbol{p} = (p, \ldots, p)\right)$.

The (n-dimensional) classes $W_p^l(g)$ $(l = 0, 1, \ldots)$ are said to be *Sobolev classes*, after S. L. Sobolev[1], who studied their basic properties and first obtained for them fundamental imbedding theorems relative to regions g which are star-shaped relative to some ball and to finite sums of such regions. These classes consist of functions which are integrable to the pth power on g along with their partial derivatives (generalized) of order l.

The (n-dimensional) classes $H_p^r(g)$, $H_p^{\boldsymbol{r}}(g)$ are defined for any $r > 0$ or $r_j > 0$. They consist of functions belonging to $L_p(g)$ and having on g partial derivatives of definite orders, satisfying in the L_p metric a Hölder condition (Lipschitz condition if $p = \infty$) or, for integer r, r_j, a generalization of such a condition (Zygmund condition), in which the first difference is replaced by a difference of a higher order.

H-classes in their complete form were defined in the papers of S. M. Nikol'skiĭ[2], who found imbedding theorems for them. It turned out that these theorems form a closed system, and in particular that the imbedding theorems for different dimensions are inversible.

The classes $B_{p\theta}^r$, $B_{p\theta}^{\boldsymbol{r}}$ in their complete form were defined by O. V. Besov[3], who found for them a closed system of imbedding theorems. The imbedding theorems for these classes also are inversible.

In what follows, for the classes H and B there will be given equivalent definitions in the terms of best approximation by functions of exponential type. As applied to the classes B they will be wider, including the case $\theta = \infty$. We will see that it is natural to suppose that

$$B_{\boldsymbol{u}p\infty}^r = H_{\boldsymbol{u}p}^r.$$

---

[1] S. L. Sobolev [3, 4]. For the anisotropic Sobolev classes W_p^l see S. M. Nikol'skiĭ [10].

[2] S. M. Nikol'skiĭ [3, 5, 10].

[3] O. V. Besov [2, 3]. For imbedding theorems for the classes $B_{p\theta}^r$ see V. P. Il'in and V. A. Solonnikov [1, 2].

In what follows it will be proved (see 9.3.) that for sufficiently general regions ($\theta = p$)[1]

(2) $$B^l_{up} \to W^l_{up} \quad (1 \leqq p \leqq 2),$$

(3) $$W^l_{up} \to B^l_{up} \; (2 \leqq p \leqq \infty, B^l_{u\infty} = H^l_{u\infty}) \quad (l = 1, 2, \ldots).$$

Thus in particular

(4) $$B^l_{u2} = W^l_{u2} \quad (l = 1, 2, \ldots).$$

Equation (4) indicates a certain connection between the classes B and W, appearing for $p = 2$. But there is another connection as well, appearing already for arbitrary p. It follows from the properties of the traces of the functions of the classes in question (see 9.1.).

Historically the presence of these connections gave a reason for calling[2] the classes which are denoted here by B^r_p, $B^{\mathbf{r}}_p$ ($\theta = p$), in the case of fractional (not integer) r, $\mathbf{r}$, W^r_p and $W^{\mathbf{r}}_p$ respectively. Here it was considered, apparently, that indeed these classes are the natural extensions of the Sobolev classes W^l_p, $W^{\mathbf{l}}_p$ with integer l, $\mathbf{l}$. The problem, of course is not here in the notations, but nowadays, when all the basic questions on the mutual relationships of the classes in question have been finally clarified, it is clear that the natural (or perhaps true) extensions of the Sobolev classes to the n-dimensional case are something else, the so-called Liouville classes, constructed on the basis of a direct generalization of the concept of fractional derivative in the sense of Liouville (or of Weyl in the periodic case). We speak here of the n-dimensional case because in the one-dimensional case it was always considered in that way—the problem of traces did not arise there.

Thus, we shall start with the following notations. Having at hand the Sobolev classes W^l_p, defined for integer $l = 0, 1, \ldots$, they are "immersed" in the fractional Liouville classes, denoted by L^l_p (l a real number). Thus, $W^l_p = L^l_p$ ($l = 0, 1, \ldots$). We shall see that the classes L^r_p are connected by the fact that the functions belonging to them have a unique integral representation (in terms of the convolution of the Bessel-MacDonald kernel with the functions $f \in L_p$; see 9.1). We shall see as well that the classes L^r_p form a closed system relative to the imbedding theorems for different metrics. The closedness is made clear by the fact that the imbedding theorems for different metrics for the classes L^r_p are expressed completely in terms of these classes, and, in addition, these theorems have the transitivity property (see also 7.1).

---

[1] O. V. Besov [3, 5].

[2] L. N. Slobodeckiĭ [1].

However the classes L_p^r, for $p \neq 2$, do not form a closed system relative to the imbedding theorems of different dimensions, and here there are no distinctions between integer and non-integer r.

Precise imbedding theorems for different dimensions for the classes L_p^r for $p \neq 2$ already cannot be expressed in the terms of these classes. In order to express them, there appears the necessity of enlisting the classes B_p^r. However the case $p = 2$ constitutes an exception. This was studied in the papers of Aronszajn [1] and L. N. Slobodeckiĭ[1]. The imbedding theorems for different metrics for the classes B_r^2 (in the notation of Slobodeckiĭ W_2^r), where $p = 2$ does not change, are closed in themselves. In themselves the classes B_p^r form a closed system relative to the imbedding theorems for different metrics and dimensions (and to some others), and have unique representations in terms of the MacDonald kernel (see 8.9.1). At the same time these classes play a useful role in the problem of traces of functions of the classes L_p^r (or W_p^l for r a natural number l), which is solved by the imbedding theorems for different dimensions. In this there is included the connection between the classes L and B. Another connection, which was mentioned above, consists in that $L_2^l = B_2^l$ $(l = 0, 1, \ldots)$. Similar interrelationships hold also for the corresponding anisotropic classes.

After what has been said above it might appear appropriate either to suppose that W_p^r for fractional r denotes a Liouville class, and not employ the sign L_p^r at all, or else to stay with only the notation L_p^r for all r, rejecting the special notation W_p^l for the Sobolev classes. But I have not done this in this book, because I feared to be like a person who thinks it appropriate to rename a street and renames it, without asking the opinion of those inhabiting the street.

We will see in 6.1 that for any $\varepsilon > 0$ one has the imbeddings

$$(5) \qquad H_{up}^{r+\varepsilon} \to B_{up\theta}^r \to H_{up}^r \quad (r > 0,\ 1 \leqq \theta \leqq \infty),$$

$$(6) \qquad H_{up}^{l+\varepsilon} \to W_{up}^l \to H_{up}^l \quad (l = 1, 2, \ldots).$$

The indicated classes are normed linear spaces. This will be directly clear from their definitions. As will be clear in what follows, they are complete and therefore Banach spaces (see 4.7).

We shall see that the norm in the sense of W, H, B is formed from two numbers

$$(7) \qquad \|f\|_W = \|f\|_{L_p} + \|f\|_w,\ \|f\|_H = \|f\|_{L_p} + \|f\|_h, \ldots,$$

where the second term, which we shall call the seminorm, characterizes the purely differential properties of f. The seminorm may be considered

---

[1] L. N. Slobodeckiĭ [1, 2]. See further V. M. Babič and L. N. Slobodeckiĭ [1].

as a norm in the corresponding space w, h, b, where one does not distinguish functions differing from one another by polynomials of certain degrees (relative to $x_1, \ldots, x_m$).

In what follows (see 8.9.2, 9.2) the classes in question will be defined in the case $g = \mathbb{R}_n$, and for zero and negative values of r, but they consist in general of generalized functions (regular in the sense of L_p).

4.3.1. The classes W. Suppose that $g \in \mathbb{R}_n$ is an open set, l a nonnegative whole number, $1 \leq p \leq \infty$ and $\boldsymbol{x} = (\boldsymbol{u}, \boldsymbol{y})$, $\boldsymbol{u} = (x_1, \ldots, x_m) \in \mathbb{R}_m$, $\boldsymbol{y} = (x_{m+1}, \ldots, x_n)$; We denote by $\mathbb{R}_m$ also a subspace of points of the type $(\boldsymbol{u}, \boldsymbol{0})$.

By definition $f \in W^l_{\boldsymbol{u}p}(g)$ $(W^l_{\boldsymbol{u}p}(g) = W^l_p(g)$[1] *for* $m = n$, $W^0_{\boldsymbol{u}p}(g) = L_p(g))$, *if the norm*

$$\|f\|_{W^l_{\boldsymbol{u}p}(g)} = \|f\|_{L_p(g)} + \|f\|_{w^l_{\boldsymbol{u}p}(g)} \quad (l = 1, 2, \ldots), \tag{1}$$

$$\|f\|_{W^0_{\boldsymbol{u}p}(g)} = \|f\|_{L_p(g)},$$

$$\|f\|_{w^l_{\boldsymbol{u}p}(g)} = \sum_{|\boldsymbol{s}|=l} \|f^{(\boldsymbol{s})}\|_{L_p(g)}$$

$$\left(\boldsymbol{s} = (s_1, \ldots, s_m, 0, \ldots, 0),\ |\boldsymbol{s}| = \sum_1^m s_j\right), \tag{2}$$

is finite, where the sum is thus extended over all derivatives (generalized), mixed and unmixed, of order l relative to $\boldsymbol{u}$. Thus we suppose that f has generalized derivatives relative to $\boldsymbol{u}$ of orders less than l. But we do not assume *a priori* that they lie in $L_p(g)$. But we shall see that in any case they are locally summable on g, and that moreover, the derivatives of order l included, they do not depend on the order in which the differentiation is carried out (see 4.5.1).

It is possible to consider the space $w^l_{\boldsymbol{u}p}(g)$ (for $m = n$ the space $w^l_p(g)$) of functions f for which the seminorm (2) is finite, i.e. to suppose that $w^l_{\boldsymbol{u}p}(g)$ consists of measurable functions f, possibly not belonging to $L_p(g)$, but such that the generalized derivative on g of order l has meaning for them and belongs to $L_p(g)$. Obviously $w^l_{\boldsymbol{u}p}(g)$ is a linear set. It will be a normed space, if one supposes that two functions $f_1, f_2 \in w^l_{\boldsymbol{u}p}(g)$, differing by a polynomial of degree $l - 1$, define the same element of the space $w^l_p(g)$. In other words, the null element in $w^l_p(g)$ is any polynomial

$$P_{l-1}(\boldsymbol{x}) = \sum_{|\boldsymbol{k}| \leq l-1} a_{\boldsymbol{k}} \boldsymbol{x}^{\boldsymbol{k}}, \quad \boldsymbol{k} = (k_1, \ldots, k_m; 0, \ldots, 0),$$

of degree $l - 1$ with coefficients $a_k = a_k(\boldsymbol{y})$, depending on $\boldsymbol{y} = (x_{m+1}, \ldots, x_n)$.

---

[1] S. L. Sobolev [3, 4].

The norm (1) is equivalent to the following norm:

$$\|f\|_{W^l_{\boldsymbol{u}p}(g)} = \left(\int_g \left(|f|^p + \sum_{|\boldsymbol{s}|=l} |f^{(\boldsymbol{s})}|^p\right) dx\right)^{1/p}, \tag{3}$$

$$\boldsymbol{s} = (s_1, \ldots, s_m, 0, \ldots, 0).$$

The advantage of this last consists in that for $p = 2$ it is Hilbertian. The scalar product, generated with $p = 2$ by this norm, has the form

$$(f, \varphi) = \int_g \left(f\varphi + \sum_{|\boldsymbol{s}|=l} f^{(\boldsymbol{s})}\varphi^{(\boldsymbol{s})}\right) d\boldsymbol{x}. \tag{4}$$

One may speak also of the classes $W^l_{x_j p}(g)$ of functions f for which the norm

$$\|f\|_{W^l_{x_j p}} = \|f\|_{L_p(g)} + \left\|\frac{\partial^l f}{\partial x_j^l}\right\|_{L_p(g)} \quad (j = 1, \ldots, n), \tag{5}$$

is finite, and also of the classes[1]

$$W^{\boldsymbol{r}}_{\boldsymbol{u}\boldsymbol{p}}(g) = \bigcap_{j=1}^{n} W^{r_j}_{x_j p_j}(g) \quad (W^{\boldsymbol{r}}_{up} = W^{\boldsymbol{r}}_{up} \quad \text{if} \quad p = p_1 = \cdots = p_n), \tag{6}$$

$$\boldsymbol{r} = (r_1, \ldots, r_n) > \boldsymbol{0}, \boldsymbol{p} = (p_1, \ldots, p_n), \quad 1 \leqq p_j \leqq \infty,$$

with the norm

$$\|f\|_{W^{\boldsymbol{r}}_{\boldsymbol{u}\boldsymbol{p}}(g)} = \sum_1^m \left(\|f\|_{L_{p_j}(g)} + \left\|\frac{\partial^{r_j} f}{\partial x_j^{r_j}}\right\|_{L_{p_j}(g)}\right). \tag{7}$$

We introduce also another class $'W^l_{\boldsymbol{u}p}$: the function $f \in {}'W^l_{\boldsymbol{u}p}(g)$, if the norm

$$\|f\|_{'W^l_{\boldsymbol{u}p}(g)} = \|f\|_{L_p(g)} + \sup_{\boldsymbol{u} \in \mathbb{R}_m} \|f^l_{\boldsymbol{u}}\|_{L_p(g)}, \tag{8}$$

makes sense for it, where

$$f^\varrho_{\boldsymbol{u}} = \sum_{|\boldsymbol{s}|=\varrho} f^{(\boldsymbol{s})}\boldsymbol{u}^{\boldsymbol{s}} \left(\boldsymbol{u}^{\boldsymbol{s}} = (u_1^{s_1} \ldots u_m^{s_m}), |\boldsymbol{u}| = 1\right), \tag{9}$$

is the derivative of f of order ϱ in the direction of $\boldsymbol{u}$.

In what follows (9.2.) it will be proved that

$$W^{l, \ldots, l}_p(\mathbb{R}_n) \to W^l_p(\mathbb{R}_n).$$

---

[1] S. M. Nikol'skiĭ [10].

If the region g is such that the extension theorem holds for it (see the remark to 4.3.6 at the end of the book)

$$W_p^{l,\ldots,l}(g) \to W_p^{l,\ldots,l}(\mathbb{R}_n),$$

then

$$'W_p^l(g) \to W_p^{l,\ldots,l}(g) \to W_p^{l,\ldots,l}(\mathbb{R}_n) \to W_p^l(\mathbb{R}_n) \to W_p^l(g),$$

where the first imbedding is explained by the fact that the derivative $f_{x_j}^l$ is at the same time the derivative in the direction of x_j. The inverse imbedding

$$W_p^l(g) \to {}'W_p^l(g),$$

is obviously also valid, so that in the presence of the extension theorem

$$W_p^l(g) \rightleftarrows {}'W_p^l(g).$$

As will be clear in the sequel, for many sufficiently "nice" sets g, it automatically follows from the fact that $f \in W_{\boldsymbol{u}p}^{\varrho}(g)$ that all the partial derivatives of f relative to $\boldsymbol{u}$ of order $\varrho - 1$ inclusive belong to $L_p(g)$. However for an arbitrary open set g this is in general not so.

4.3.2. *Example.* A function $f(x)$ of one variable x, given on a set $g = \sum_1^\infty \sigma_k$, which is the set-theoretical sum of intervals $\sigma_k = (a_k < x < b_k)$ of length $\delta_k = k^{-2}$.

Put

$$f(x) = \frac{(x - a_k)}{\delta_k^\alpha} \quad (k = 1, 2, \ldots).$$

Then, if $\frac{1}{2p} \leqq \alpha < 1$, then

$$\|f\|_{L_p(g)} = \left(\sum_1^\infty \int_{\sigma_k} \left(\frac{x - a_k}{\delta_k^\alpha}\right)^p dx\right)^{1/p}$$

$$= \frac{1}{(p+1)^{1/p}} \left(\sum_1^\infty \delta_k^{p(1-\alpha)+1}\right)^{1/p} = \frac{1}{(p+1)^{1/p}} \left(\sum_1^\infty \frac{1}{k^{2[p(1-\alpha)+1]}}\right)^{1/p} < \infty,$$

$$\|f^{(l)}\|_{L_p(g)} = 0 \quad \text{for} \quad l \geqq 2,$$

$$\|f'\|_{L_p(g)} = \left(\sum_1^\infty \int_{\sigma_k} \left|\frac{1}{\delta_k^\alpha}\right|^p dx\right)^{1/p} = \sum_1^\infty \frac{1}{k^{2(1-\alpha p)}} = +\infty.$$

Here the condition $\delta_k = k^{-2}$ shows that the set g can be bounded.

Thus $f \in W_p^{(l)}(g)$ $(l \geqq 2)$, but the norm of its first derivative in the metric of $L_p(g)$ is equal to $+\infty$.

4.3.3. The classes H. We continue with the notations introduced at the beginning of 4.3.1. Suppose that $1 \leqq p \leqq \infty$, $r > 0$ and that the numbers k, ϱ are nonnegative integers, satisfying the inequalities $k > r - \varrho > 0$. We shall call such pairs (k, ϱ) *admissible.*

By definition *the function* $f \in H_{up}^r(g)$[1], *if it belongs to* $L_p(g)$ *and if the derivatives* $f^{(s)}$ *of order* $\boldsymbol{s} = (s_1, \ldots, s_m, 0, \ldots, 0)$ *with* $|\boldsymbol{s}| = \sum_1^m s_j = \varrho$ *make sense for it, and if the inequalities*

$$(1) \qquad \|\Delta_{\boldsymbol{h}}^k f^{(\boldsymbol{s})}(\boldsymbol{x})\|_{L_p(g_{k\boldsymbol{h}})} \leqq M|\boldsymbol{h}|^{r-\varrho},$$

are satisfied for them, where M *does not depend on* $\boldsymbol{h} \in \mathbb{R}_m$, *or alternatively the inequalities*

$$(1') \qquad \Omega^k(f^{(\boldsymbol{s})}, \delta) = \sup_{\boldsymbol{h} \in \mathbb{R}_m} \omega_{\boldsymbol{h}}^k(f^{(\boldsymbol{s})}, \delta) \leqq M\delta^{r-\varrho},$$

which are equivalent to them.

Here we put

$$(2) \qquad \|f\|_{H_{up}^r(g)} = \|f\|_{L_p(g)} + \|f\|_{h_{up}^r(g)},$$

where the seminorm

$$(3) \qquad \|f\|_{h_{up}^r(g)} = M_f = \inf M$$

is the lower bound of all M for which for all $\boldsymbol{h} \in \mathbb{R}_m$ and any of the indicated $\boldsymbol{s}$ the inequality (1) is satisfied.

This definition in fact still depends on the admissible pair (k, ϱ). But it will be shown in 5.5.3 that the norms (2) (but not in general the norms (3)), for a measurable set $g = \mathbb{R}_m \times g'$ and distinct admissible pairs are pairwise equivalent. For other sets g the equivalence will depend on the possibility of extension of functions beyond the limits of g to R_n with the preservation of the corresponding norms. Here, see the remarks at the end of the book, referring to 4.3.6.

Suppose that $r = \bar{r} + \alpha$, where $\bar{r}$ is an integer and $0 \leqq \alpha \leqq 1$. If $\alpha < 1$, then, choosing as an admissible pair the numbers $\varrho = \bar{r}$, $k = 1$, we obtain the following special form of the inequality (1):

$$(4) \qquad \|\Delta_{\boldsymbol{h}} f^{(\bar{\boldsymbol{r}})}(\boldsymbol{x})\|_{L_p(g_{\boldsymbol{h}})} \leqq M|\boldsymbol{h}|^\alpha \quad (0 < \alpha < 1).$$

---

[1] For $p = \infty$ we mean here that the function f is equivalent to some function, denoted again by f, for which (1) is satisfied.

Now if $\alpha = 1$ this pair will not do. But it is possible to choose as an admissible pair $\varrho = \bar{r}$, $k = 2$, and then inequality (1) will have the form

$$\|\Delta_{\boldsymbol{h}}^2 f^{(\bar{r})}(\boldsymbol{x})\|_{L_p(g_{2\boldsymbol{h}})} \leqq M|\boldsymbol{h}|. \tag{5}$$

Usually one uses definitions[1] (4) and (5) or simply the single definition

$$\|\Delta_{\boldsymbol{h}}^2 f^{(\bar{r})}(\boldsymbol{x})\|_{L_p(g_{2\boldsymbol{h}})} \leqq M|\boldsymbol{h}|^\alpha \quad 0 < \alpha \leqq 1, \tag{6}$$

suitable for all the α in question.

It is possible to alter these definitions, choosing as the seminorm M_f the lower bound of those M for which (1) is satisfied for all $\boldsymbol{h} \in \mathbb{R}_m$ satisfying the inequality $|\boldsymbol{h}| \leqq \eta$, where η is a given positive number. The norm altered in this way is also, as we shall see, equivalent to the norms defined above in every case for regions of the type $g = \mathbb{R}_m \times g'$.

Finally, a further definition is possible. The function $f \in H_{\boldsymbol{u}p}^r(g)$, if the derivative $f_{\boldsymbol{h}}^\varrho$ of order ϱ relative to any direction $\boldsymbol{h} \in \mathbb{R}_m$, makes sense, and

$$\|\Delta_{\boldsymbol{h}}^k f_{\boldsymbol{h}}^\varrho\|_{L_p(g_{k|\boldsymbol{h}|})} \leqq M|\boldsymbol{h}|^{r-\varrho}, \tag{7}$$

where (k, ϱ) is an admissible pair and M does not depend on $\boldsymbol{h} \in \mathbb{R}_m$. This inequality is equivalent to the following:

$$\Omega^k(f^\varrho; \delta) = \sup_{|\boldsymbol{h}|=1} \sup_{|t| \leqq \delta} \|\Delta_{t\boldsymbol{h}}^k f_{\boldsymbol{h}}^\varrho\|_{L_p(g_{k\boldsymbol{h}})} \leqq M\delta^{r-\varrho}. \tag{8}$$

The norm of f is defined in analogy with (2).

If $\mathbb{R}_m$ ($m = 1$) is the coordinate axis x_j, then the corresponding class $H_{\boldsymbol{u}p}^r(g)$ will be denoted by $H_{x_jp}^r(g)$ ($j = 1, \ldots, m$), and the norm by

$$\|f\|_{H_{x_jp}^r(g)} = \|f\|_{L_p(g)} + M_{x_jf}, \tag{9}$$

$$M_{x_jf} = \|f\|_{h_{x_jp}^r(g)}. \tag{10}$$

Finally, if $\boldsymbol{r} = (r_1, \ldots, r_m)$, $\boldsymbol{p} = (p_1, \ldots, p_m)$ $(r_j > 0,\ 1 \leqq p_j \leqq \infty;$ $j = 1, \ldots, m \leqq n)$, then we put[2]

$$H_{\boldsymbol{u}\boldsymbol{p}}^{\boldsymbol{r}}(g) = \bigcap_{j=1}^m H_{x_jp_j}^{r_j}(g) \quad (H_{\boldsymbol{u}\boldsymbol{p}}^{\boldsymbol{r}} = H_{\boldsymbol{u}p}^{\boldsymbol{r}} \text{ if } p = p_1 = \cdots = p_m) \tag{11}$$

---

[1] S. M. Nikol'skiĭ [5].

[2] S. M. Nikol'skiĭ [10].

with the norm

(12) $$\|f\|_{H^{r}_{\boldsymbol{u}\boldsymbol{p}}(g)} = \max_{1\leq j\leq m} \|f\|_{L_{p_j}(g)} + \|f\|_{h^{r}_{\boldsymbol{p}}(g)},$$

(13) $$\|f\|_{h^{r}_{\boldsymbol{u}\boldsymbol{p}}(g)} = \max_{1\leq j\leq m} \|f\|_{h^{r_j}_{x_j p_j}(g)}.$$

In (13) one may replace the $\max_j$ by $\sum_j$, obtaining an equivalent norm.

4.3.4. The classes B. We keep to the notations introduced at the beginning of 4.3.1 and introduce a further parameter θ, where $1 \leqq \theta < \infty$. Suppose that $r > 0$ and that the numbers k, σ, forming an admissible pair, are nonnegative integers satisfying the inequalities $k > r - \varrho > 0$.

By definition *the function f belongs to the class $B^{r}_{\boldsymbol{u}p\theta}(g)$*[1] *(for $m = n$ simply $B^{r}_{p\theta}(g)$) if $f \in L_p(g)$, there exist generalized partial derivatives relative to $\boldsymbol{u} \in \mathbb{R}_m$ of f of orders $\boldsymbol{s} = (s_1, \ldots, s_m, 0, \ldots, 0)$ ($|\boldsymbol{s}| \leqq \varrho$), and one of the following seminorms is finite;*

(1) $$^{1}\|f\|_{b^{r}_{\boldsymbol{u}p\theta}(g)} = \sum_{|\boldsymbol{s}|=\varrho} \left(\int_0^\infty t^{-1-\theta(r-\varrho)} \Omega^{k}_{\mathbb{R}_m}(f^{(\boldsymbol{s})}, t)^{\theta}_{L_p(g)}\, dt \right)^{1/\theta},$$

(2) $$^{2}\|f\|_{b^{r}_{\boldsymbol{u}p\theta}(g)} = \left(\int_0^\infty t^{-1-\theta(r-\varrho)} \Omega^{k}_{\mathbb{R}_m}(f^{\varrho}, t)^{\theta}_{L_p(g)}\, dt \right)^{1/\theta},$$

(3) $$^{3}\|f\|_{b^{r}_{\boldsymbol{u}p\theta}(g)} = \sum_{|\boldsymbol{s}|=\varrho} \left(\int_{\mathbb{R}_m} |\boldsymbol{u}|^{-m-\theta(r-\varrho)} \, \|\Delta^{k}_{\boldsymbol{u}} f^{(\boldsymbol{s})}(\boldsymbol{x})\|^{\theta}_{L_p(g_{k\boldsymbol{u}})}\, d\boldsymbol{u} \right)^{1/\theta},$$

(4) $$^{4}\|f\|_{b^{r}_{\boldsymbol{u}p\theta}(g)} = \left(\int_{\mathbb{R}_m} |\boldsymbol{u}|^{-m-\theta(r-\varrho)} \, \|\Delta^{k}_{\boldsymbol{u}} f^{\varrho}_{\boldsymbol{u}}(\boldsymbol{x})\|^{\theta}_{L_p(g_{k\boldsymbol{u}})}\, d\boldsymbol{u} \right)^{1/\theta}$$

(see 4.2. (12), (13)). Here we put

(5) $$^{j}\|f\|_{B^{r}_{\boldsymbol{u}p\theta}(g)} = \|f\|_{L_p(g)} + {}^{j}\|f\|_{b^{r}_{\boldsymbol{u}p\theta}(g)} \quad (j = 1, 2, 3, 4).$$

All four of the seminorms (1)—(4) just presented depend further on the admissible pairs k, ϱ. Moreover, they may be altered by taking the integrals appearing in them relative to bounded regions (relative to $0 \leqq t \leqq \eta$ or $|\boldsymbol{u}| \leqq \eta$), respectively), and nevertheless the norms defined in terms of them turn out to be equivalent for regions of the type $g = \mathbb{R}_m \times g' \subset \mathbb{R}_n$ (see further 5.6), and accordingly for regions g from which the functions may be extended to $\mathbb{R}_n$ with preservation of the indicated norms.

---

[1] See note to page 159.

Frequently these norms are given in the following situation[1]. For a given $r > 0$ one defines an integer $\bar{r}$ such that $r = \bar{r} + \alpha$ and $0 < \alpha \leqq 1$. If $\alpha < 1$, then it is sufficient to choose the admissible pair $\varrho = \bar{r}$, $k = 1$. If $\alpha = 1$, then $\varrho = \bar{r}$, $k = 2$ or, in order to join these two cases, we may choose $\varrho = \bar{r}$, $k = 2$.

If g is a bounded set and d its diameter, then for $t > d$ each of the functions Ω in (1) and (2) is equal to some constant and the remainders $\int_d^\infty$ of the integrals on the right sides of (1) and (2) are finite (since θ, $r - \varrho > 0$). Therefore the finiteness of the seminorms (1), (2) depends exclusively on the properties of the indicated moduli for small t.

The classes $B^r_{x_jp}$ $(j = 1, \ldots, m)$ correspond to the case when $\mathbb{R}_m$ is replaced by the coordinate axis x_j.

Put[2]

$$\|f\|_{B^{\boldsymbol{r}}_{\boldsymbol{up}\theta}(g)} = \sum_{j=1}^{m} \|f\|_{B^{r_j}_{x_jp_j\theta}(g)} \quad (B^{\boldsymbol{r}}_{\boldsymbol{up}\theta} = B^{\boldsymbol{r}}_{\boldsymbol{p}\theta} \text{ for } m = n).$$

We note the simplest inequalities among the seminorms (1)—(4) (for one and the same pair k, ϱ):

$$(6) \qquad {}^3\|f\|_b \ll {}^1\|f\|_b, \; {}^4\|f\|_b \ll {}^2\|f\|_b, \; {}^2\|f\|_b \ll {}^1\|f\|_b.$$

The last inequality follows from the inequality 4.2. (15). The first two are obtained directly, of one introduces polar coordinates $\boldsymbol{u} = (t, \sigma)$, $t = |\boldsymbol{u}|$, $d\boldsymbol{u} = t^{m-1}\,dt\,d\sigma$ and takes account of the inequalities

$$(7) \qquad \|\Delta^k_{\boldsymbol{u}} f^{(s)}(\boldsymbol{x})\|_{L_p(g_{k\boldsymbol{u}})} \leqq \Omega^k_{\mathbb{R}_m}(f^{(s)}, t)_{L_p(g)},$$

$$(8) \qquad \|\Delta^k_{\boldsymbol{u}} f^{\varrho}(\boldsymbol{x})\|_{L_p(g_{k\boldsymbol{u}})} \leqq \Omega^k_{\mathbb{R}_m}(f^{\varrho}, t)_{L_p(g)}.$$

4.3.5. Periodic classes. The periodic classes $\overset{l}{W}{}^*_{x_jp}(\mathscr{E})$, $\overset{r}{H}{}^*_{x_jp}(\mathscr{E})$, $\overset{r}{B}{}^*_{x_jp\theta}(\mathscr{E})$ are defined on the set $\mathscr{E} = \mathbb{R}_j \times \mathscr{E}^j \subset \mathbb{R}_n$, where $\mathbb{R}_j$ is the real axis x_j $(j = 1, \ldots, n)$. These are the classes of functions $f(x_j, \boldsymbol{y}^j)$, $\boldsymbol{y}^j = (x_1, \ldots, x_{j-1}, x_{j+1}, \ldots, x_n)$ of period 2π relative to x_j. They are defined in exactly the same way as are the corresponding classes $W^l_{x_jp}(\mathscr{E}), \ldots$ of nonperiodic functions, except that we have throughout to replace the norm $\|\cdot\|_{L_p(\mathscr{E})}$ by the norm $\|\cdot\|_{L_p(\mathscr{E}_*)}$, where $\mathscr{E}_* = [0, 2\pi] \times \mathscr{E}^j$. Analogously one defines the periodic classes $\overset{\boldsymbol{r}}{W}{}^*_{\boldsymbol{up}}(\mathscr{E})$, $\overset{\boldsymbol{r}}{H}{}^*_{\boldsymbol{up}}(\mathscr{E})$, $\overset{\boldsymbol{r}}{B}{}^*_{\boldsymbol{up}}(\mathscr{E})$, where $\mathscr{E} = \mathbb{R}_m \times \mathscr{E}' \subset \mathbb{R}_n$; here we drop the symbol $\boldsymbol{u}$ when $m = n$.

---

[1] O. V. Besov [3, 5]. In these papers the norms (1) and (5) were considered.

[1] O. V. Besov [3, 5] in the case $p_1 = \cdots = p_n$, V. P. Il'in and V. A. Solonnikov [1, 2] in the general case.

4.3.6. Extension of functions with preservation of the class. We make a further important remark. Suppose that $\Lambda(g)$ denotes one of the classes $W(g)$, $H(g)$, $B(g)$, with some or other parameters $r, p, \ldots$. If the region $g \subset \mathbb{R}_n$ is such that to every function $f \in \Lambda(g)$ one may assign a function $\tilde{f}$ defined on $\mathbb{R}_n$, such that $\tilde{f} = f$ on g and

$$\|\tilde{f}\|_{\Lambda(\mathbb{R}_n)} \leqq c\|f\|_{\Lambda(g)},$$

where c does not depend on f, then we will say that *the function f of the class $\Lambda(g)$ may be extended from g with preservation of the class* (*or of the norm*). We will also say in this case that one has the imbedding

$$\Lambda(g) \to \Lambda(\mathbb{R}_n).$$

Our classes are constructed in such a way that if a function $f \in \Lambda(\mathbb{R}_n)$, then its restriction to g is a function $f \in \Lambda(g)$ and

$$\|f\|_{\Lambda(g)} \leqq \|f\|_{\Lambda(\mathbb{R}_n)}.$$

In connection with this we say that one has the imbedding

$$\Lambda(\mathbb{R}_n) \to \Lambda(g).$$

Now we suppose that for some region g there are given two classes $\Lambda(g)$ and $\Lambda'(g)$ and

$$\Lambda(g) \to \Lambda(\mathbb{R}_n) \to \Lambda'(\mathbb{R}_n). \tag{1}$$

Then

$$\Lambda(g) \to \Lambda'(g). \tag{2}$$

In this book we give our principal attention to the study of the classes indicated above in the case when $g = \mathbb{R}_n$ or $g = \mathbb{R}_m \times g'$, where $1 \leqq m < n$ and g' is a measurable $(n-m)$-dimensional set. In the remarks at the end of the book relating to 4.3.6 the reader will find the formulations of several general theorems on extension with preservation of the class. The presence of the imbeddings (1) implies automatically the imbedding (2).

4.4. Representation of an Intermediate Derivate in Terms of a Derivative of Higher Order and the Function. Corollaries

In this section we introduce certain modifications of Taylor's formula, on the basis of which we shall obtain some inequalities.

4.4.1. Consider on a finite interval (a, b) a function $f(x)$, having on any segment interior to (a, b) absolutely continuous derivatives to order

$(\varrho - 2)$ inclusive, and accordingly having almost everywhere on $[a, b]$ a derivative of order $\varrho - 1$. For this function we have for almost all x_0 the Taylor's formula

$$(1) \qquad f(x) = \sum_{j=0}^{\varrho-1} f^{(j)}(x_0) \frac{(x - x_0)^j}{j!} + \mathrm{R}(x, x_0) \quad (a < x, x_0 < b)$$

which is purely formal[1], since under the indicated conditions it is not possible to say anything about the behavior of the remainder term $R(x, x_0)$.

We shall denote the interval $\{a < x, x_0 < b\}$ by Δ. We divide the segment $[a, b]$ into 2ϱ equal partial segments

$$\Delta_0, \ldots, \Delta_{2\varrho-1}.$$

Suppose that g denotes the ϱ-dimensional cube of points $(x_1, \ldots, x_\varrho)$, whose coordinates belong respectively to the partial segments Δ_{2k}:

$$x_k \in \Delta_{2k} \ (k = 1, \ldots, \varrho).$$

Transposing $R(x, x_0)$ in (1) to the left side and substituting the numbers $x_1, \ldots, x_\varrho$ in the place of x, we obtain a linear system of ϱ equations:

$$(2) \qquad \sum_{j=0}^{\varrho-1} \frac{(x_k - x_0)^j}{j!} f^{(j)}(x_0) = f(x_k) - \mathrm{R}(x_k, x_0) \quad (k = 1, \ldots, \varrho)$$

with ϱ unknowns $f^{(j)}(x_0)$ and determinant

$$(3) \quad W = W(x_1 - x_0, \ldots, x_\varrho - x_0)$$

$$= \begin{vmatrix} 1 & (x_1 - x_0) & \cdots & \dfrac{(x_1 - x_0)^{\varrho-1}}{(\varrho-1)!} \\ \cdot & \cdot & \cdot & \cdot \\ 1 & (x_\varrho - x_0) & \cdots & \dfrac{(x_\varrho - x_0)^{\varrho-1}}{(\varrho-1)!} \end{vmatrix}$$

$$= \sum_{k=1}^{\varrho} \alpha_{jk}(x_1 - x_0, \ldots, x_\varrho - x_0) \frac{(x_k - x_0)^j}{j!} = \sum_{k=1}^{\varrho} \alpha_{jk} \frac{(x_k - x_0)^j}{j!},$$

where α_{jk} is the cofactor of the determinant W corresponding to its element $(x_k - x_0)^j (j!)^{-1}$.

---

[1] By this I mean that it is not possible to say anything in general about $R(x, x_0)$ except that it is equal almost everywhere to the difference $f(x) - \sum_{j=0}^{\varrho-1} f^{(j)}(x_0) \dfrac{(x - x_0)^j}{j!}$.

It follows from (2) and (3) that

$$(4)\qquad f^{(j)}(x_0) = \frac{1}{W}\sum_{k=1}^{\varrho} \alpha_{jk}[f(x_k) - \mathrm{I\!R}(x_k, x_0)]\ (j = 0, 1, \ldots, \varrho - 1).$$

The function W can differ only by a constant factor from the Vandermonde determinant, equal to the product of all possible factors of the form $(x_k - x_l)$, where $k < l$ and k and l run over the set $1, \ldots, \varrho$. Since the x_k, x_l lie at distances from one another larger than a positive constant, then the function $1/W$ is bounded. The functions α_{jk} are also bounded, so that from (4) one obtains the inequality

$$(5)\qquad |f^{(j)}(x_0)| \leqq c_1\left(\sum_{k=1}^{\varrho} |f(x_k)| + |\mathrm{I\!R}(x_k, x_0)|\right)$$

$$(x_0 \in [a, b],\ (x_1, \ldots, x_n) \in g).$$

Since the left side of (5) does not depend on x_k $(k = 1, \ldots, \varrho)$, then obviously

$$(6)\qquad |f^{(j)}(x_0)| \leqq c_2 \sum_{k=1}^{\varrho} (\|f(x_k)\|_{L_p(g)} + \|\mathrm{I\!R}(x_k, x_0)\|_{L_p(g)})$$

$$\leqq c_3(\|f\|_{L_p(a,b)} + \|\mathrm{I\!R}(x, x_0)\|_{L_{p,x}(a,b)})\ (j = 0, 1, \ldots, \varrho - 1),$$

where the sign $L_{p,x}$ denotes the fact that the norm is calculated relative to the variable x.

Finally, from (6) it follows that

$$(7)\qquad \|f^{(j)}\|_{L_p(a,b)} \leqq c(\|f\|_{L_p(a,b)} + \|\mathrm{I\!R}\|_{L_p(\Delta)})\ (j = 0, 1, \ldots, \varrho - 1).$$

For $p = \infty$ this is obvious, and for p finite it is obtained if one raises the left and right sides of (6) to the pth power, applies to the right side the inequality

$$a + b \leqq 2^{1-\frac{1}{p}}\,(a^p + b^p)^{1/p}\ (a, b > 0, 1 \leqq p \leqq \infty),$$

integrates both sides of the inequality relative to x_0, and, finally, raises them to the power $1/p$.

4.4.2. We note that if $\|\mathrm{I\!R}\|_{L(\Delta)} < \infty$, then, putting the expressions 4.4.1 (4) for the derivatives $f^{(j)}(x_0)$ into equation 4.4.1 (1), integrating both of its sides over the cube g of points $(x_1, \ldots, x_\varrho)$, and dividing by the value of its volume $\varkappa$, we obtain the formula[1]

$$(1)\qquad f(x) = P(x) + F(x),$$

[1] S. M. Nikol'skiĭ [11].

where

(2) $$P(x) = \sum_{j=0}^{\varrho-1} \sum_{k=1}^{\varrho} \frac{1}{\varkappa} \int_g \frac{\alpha_{jk}(x_1 - x_0, \ldots, x_\varrho - x_0)}{W(x_1, \ldots, x_\varrho)} \times [f(x_k) - \mathbb{R}(x_k, x_0)] \frac{(x - x_0)^j}{j!} \, dg$$

is a polynomial of degree $\varrho - 1$ and

(3) $$F(x) = \frac{1}{\varkappa} \int_g \mathbb{R}(x, x_0) \, dg.$$

Formula (1) shows that the function f may be represented in the form of the sum of some polynomial $P(x)$ of degree $\varrho - 1$ and a remainder $F(x)$. Here P and F are explicitly expressed only in terms of the function f itself and its Taylor remainder term $\mathbb{R}$.

Thus, in the right side, there do not appear explicitly anywhere the intermediate derivatives $f^{(1)}, \ldots, f^{(\varrho-1)}$, which makes it possible to estimate the norms of these derivatives in terms of the norms of f and $f^{(\varrho)}$.

4.4.3. We consider some important special cases of formulas 4.4.1 (6) and (7).

If the function $f \in W_p^\varrho(a, b)$, then it is equivalent to a fully defined continuous function, which we again denote by f. For this the Taylor's formula 4.4.1 (1) with remainder term

(1) $$\mathbb{R}(x, x_0) = \frac{1}{(\varrho - 1)!} \int_{x_0}^{x} (x - u)^{\varrho - 1} f^{(\varrho)}(u) \, du$$

is valid, where $f^{(\varrho)} \in L_p(a, b)$.

From (1) follows the inequality

(2) $$|\mathbb{R}(x, x_0)| \leqq c_1 \|f^{(\varrho)}\|_{L_p(a,b)}, \quad a \leqq x, x_0 \leqq b.$$

Moreover

(3) $$\|\mathbb{R}\|_{L_p(\Delta)} \leqq c_2 \|f^{(\varrho)}\|_{L_p(a,b)},$$

where the constant c_2 depends on $b - a$, p, and ϱ. In such a case one obtains from 4.4.1. (6) and (7) respectively the inequalities

(4) $$|f^{(j)}(x_0)| \leqq c_3(\|f\|_{L_p(a,b)} + \|f^{(\varrho)}\|_{L_p(a,b)}) = c_3 \|f\|_{W_p^{(\varrho)}(a,b)},$$

(5) $$\|f^{(j)}\|_{L_p(a,b)} \leqq c_3 \|f\|_{W_p^{(\varrho)}(a,b)} \quad (j = 0, 1, \ldots, \varrho).$$

Both of the inequalities just obtained extend immediately to the case of the class of functions $W^{\varrho}_{xp}(\mathscr{E})$, where $\mathscr{E} = [a, b] \times \mathscr{E}_1$ ($x \in [a, b]$, $\mathbf{y} \in \mathscr{E}_1$, $\mathscr{E} \subset \mathbb{R}_n$) is a measurable set:

(6) $$\|f^{(j)}_{x_0}(x_0, \mathbf{y})\|_{L_p(\mathscr{E}_1)} \leqq c_4 \|f\|_{W^{\varrho}_{xp}(\mathscr{E})},$$

(7) $$\|f^{(j)}_x\|_{L_p(\mathscr{E})} = \left(\int_{\mathscr{E}_1} \|f^{(j)}_x(x, \mathbf{y})\|^p_{L_p(a,b)}\, d\mathbf{y}\right)^{1/p}$$

$$\leqq c_3 \left(\int_{\mathscr{E}_1} \|f\|^p_{W^{\varrho}_{xp}(a,b)}\, d\mathbf{y}\right)^{1/p} \leqq c_4 \|f\|_{W^{\varrho}_{xp}(\mathscr{E})}$$

$$(j = 0, 1, \ldots, \varrho).$$

If $f \in H^r_{xp}(\mathscr{E})$ and $\bar{r} = \varrho - 1$, then f may be written out by formula 4.4.1 (1), where

$$\mathbb{R}(x, x_0) = \frac{1}{(\varrho - 2)!} \int_{x_0}^{x} (u - x_0)^{\varrho-2} \left[f^{(\varrho-1)}_x(u, \mathbf{y}) - f^{(\varrho-1)}_x(x_0, \mathbf{y})\right] du.$$

Hence

$$\int_a^b |\mathbb{R}|^p\, dx_0 < c \left(\left| \int_a^x \int_{x_0}^x |f^{(\varrho-1)}_x(u, \mathbf{y}) - f^{(\varrho-1)}_x(x_0, \mathbf{y})|^p\, du\, dx_0 \right| \right.$$

$$\left. + \left| \int_x^b \int_x^{x_0} |f^{(\varrho-1)}_x(u, \mathbf{y}) - f^{(\varrho-1)}_x(x_0, \mathbf{y})|^p\, du\, dx_0 \right| \right)$$

$$= c \left(\left| \int_a^x \int_0^{x-x_0} |f^{(\varrho-1)}_x(x_0 + h, \mathbf{y}) - f^{(\varrho-1)}_x(x_0, \mathbf{y})|^p\, dh\, dx_0 \right| \right.$$

$$\left. + \left| \int_x^b \int_0^{x_0-x} |f^{(\varrho-1)}_x(x_0 - h, \mathbf{y}) - f^{(\varrho-1)}_x(x_0, \mathbf{y})|^p\, dh\, dx_0 \right| \right)$$

$$= c \left| \int_0^{x-a} \int_a^{x-h} |f^{(\varrho-1)}_x(x_0 + h, \mathbf{y}) - f^{(\varrho-1)}_x(x_0, \mathbf{y})|^p\, dh\, dx_0 \right|$$

$$+ c \left| \int_0^{b-x} \int_{x+h}^{b} |f^{(\varrho-1)}_x(x_0 - h, \mathbf{y}) - f^{(\varrho-1)}_x(x_0, \mathbf{y})|^p\, dx_0\, dh \right|$$

and

$$\left(\int_{\mathscr{E}_1}\int_a^b |\mathbb{R}|^p\,dx_0\,dy\right)^{1/p}$$

$$\ll\left[\int_0^{x-a}\left|\int_{\mathscr{E}_1}\int_a^{x-h}|f_x^{(\varrho-1)}(x_0+h,y)-f_x^{(\varrho-1)}(x_0,\mathbf{y})|^p\,dx_0\,d\mathbf{y}\right|dh\right]^{1/p}$$

$$+\left[\int_0^{b-x}\left|\int_{\mathscr{E}_1}\int_{x+h}^{b}|f_x^{(\varrho-1)}(x_0-h,\mathbf{y})-f_x^{(\varrho-1)}(x_0,\mathbf{y})|^p\,dx_0\,d\mathbf{y}\right|dh\right]^{1/p}.$$

Hence, taking into account the fact that for $h > 0$

$$\left(\int_{\mathscr{E}_1}\int_a^{b-h}|f_x^{(\varrho-1)}(x_0+h,\mathbf{y})-f_x^{(\varrho-1)}(x_0,\mathbf{y})|^p\,dx_0\,d\mathbf{y}\right)^{1/p}\leqq Mh^{\alpha},$$

where

$$M=\|f\|_{h_p^r(\mathscr{E})},$$

we get

$$\|\mathbb{R}\|_{L_p(\mathscr{E})}=\left(\int_a^b\int_{\mathscr{E}_1}\int_a^b|\mathbb{R}|^p\,dx_0\,d\mathbf{y}\,dx\right)^{1/p}$$

$$\leqq\left(\int_a^b\int_0^{x-a}M^ph^{\alpha p}\,dh\,dx\right)^{1/p}+\left(\int_a^b\int_0^{b-x}M^ph^{\alpha p}\,dh\,dx\right)^{1/p}<c_2M.$$

Therefore it follows from 4.5.1 (7) that

$$\|f_x^{(j)}\|_{L_p(\mathscr{E})}\leqq c_3(\|f\|_{L_p(\mathscr{E})}+\|\mathbb{R}_{L_p(\mathscr{E})})\tag{8}$$

$$\leqq c_4(\|f\|_{L_p(\mathscr{E})}+M)\leqq c_4\|f\|_{H_p^r(\mathscr{E})}\quad(j=1,\ldots,\bar{r}).$$

Inequalities (4), (5), as well as (8), are true also for $a=-\infty$, $b=+\infty$. In the case (4) this is obvious. In the cases (5) and (8) this follows from 6.1 (2) and (8). In the case (5) ($1<p<\infty$) it already follows from 9.2.2. The corresponding inequality for the interval (a,∞) reduces to the preceding one on application of the extension theorem 4.3.6.

4.4.4. We note that in the definition of the functions of the classes $W_p^{\varrho}(\mathscr{E})$ and $H_p^r(\mathscr{E})$ one assumed the existence on $\mathscr{E}$ of generalized partial derivatives $f_x^{(j)}$ of orders $j=1,\ldots,\varrho-1$ ($\bar{r}-1$), but did not assume that they have finite norm in the sense of $L_p(\mathscr{E})$.

Inequalities 4.4.3 (7) and (8) show that the finiteness of the norms of the indicated derivatives follows from the definition of the corresponding classes. But then the derivatives $f_x^{(j)}$ ($j = 0, 1, \ldots, \varrho - 1$) are for almost all $\boldsymbol{y} \in \mathscr{E}_1$ absolutely continuous relative to the variable x on the closed segment $[a, b]$. Thus, for almost all $\boldsymbol{y} \in \mathscr{E}_1$ one has a decomposition of f relative to the Taylor formula

$$(1) \qquad f(a, \boldsymbol{y}) = \sum_0^{\varrho-1} \frac{f_x^{(k)}(a, \boldsymbol{y})}{k!} (x - a)^k + \frac{1}{(\varrho - 1)!} \int_a^x f_x^{(\varrho)}(u, \boldsymbol{y}) (x - u)^{\varrho-1} \, du$$

in the neighborhood of the endpoint a of the segment $[a, b]$ and a corresponding decomposition in the neighborhood of the other endpoint b. We note the inequality

$$(2) \qquad \|\Delta_{x_j,h}^{\varrho} f\|_{L_p(g_{x_j\varrho|h|})} \leqq |h|^\varrho \left\| \frac{\partial^\varrho f}{\partial x_j^\varrho} \right\|_{L_p(g)},$$

which may be treated as follows: if the right side of (2) has meaning, then the left side does as well, and the inequality (2) then holds.

In 4.8. we will obtain the reverse of inequality (2) for $\varrho = 1$.

Proof of (2). Suppose first that $\varrho = 1$. Then in view of the equation

$$\Delta_{x_1,h} f(x) = \int_0^h f'_{x_1}(x_1 + t, \boldsymbol{y}) \, dt, \quad \boldsymbol{x} = (x_1, \boldsymbol{y}) \in g_{|h|},$$

which holds for almost all admissible $\boldsymbol{y} = (x_2, \ldots, x_n)$, and for all x_1, h admissible for each such $\boldsymbol{y}$, we get (see 1.3.2.)

$$\|\Delta_{x_1 h} f\|_{L_p(g_{|h|})} \leqq \left| \int_0^h \|f'_{x_1}(x_1 + t, \boldsymbol{y})\|_{L_p(g_{|h|})} \, dt \right|$$

$$\leqq \left| \int_0^h \|f'_{x_1}\|_{L_p(g)} \, dt \right| = |h| \, \|f'_{x_1}\|_{L_p(g)}.$$

Therefore for any ϱ

$$\|\Delta_{x_1,h}^{\varrho} f\|_{L_p(g_{x_1\varrho|h|})} = \|\Delta_{x_1,h} \Delta_{x_1,h}^{\varrho-1} f\|_{L_p(g_{x_1\varrho|h|})} \leqq |h| \|\Delta_{x_1,h}^{\varrho-1} f'_{x_1}\|_{L_p(g_{x_1(\varrho-1)|h|})}$$

$$\leqq |h|^2 \|\Delta_{x_1,h}^{\varrho-2} f''_{x_1}\|_{L_p(g_{x_1(\varrho-2)|h|})} \leqq \cdots \leqq |h|^\varrho \|f_{x_1}^{(\varrho)}\|_{L_p(g)}.$$

Corollary 1. *For the function* $g_\nu(\boldsymbol{x}) = g_\nu(x_1, \ldots, x_n) \in \mathfrak{M}_{x,\nu p}(\mathscr{E})$, $\mathscr{E} = \mathbb{R}_1 \times \mathscr{E}_1$, i.e. *lying in* $L_p(\mathscr{E})$ *and entire of degree* ν *relative to* x_1, (see 3.4.1.) *one has the inequality*[1]

$$(3) \qquad \|\Delta^{\varrho}_{x_1,h} g_\nu\|_{L_p(\mathscr{E})} \leqq |h|^\varrho \left\| \frac{\partial^\varrho g_\nu}{\partial x_1^\varrho} \right\|_{L_p(\mathscr{E})} \leqq (\nu h)^\varrho \|g_\nu\|_{L_p(\mathscr{E})}$$

(see 3.2.2. (7)). *Here we need to take into account the fact that* $\mathscr{E}_{x_1,\delta} = \mathscr{E}$, *since* $\mathscr{E}$ *is a set cylindrical in the direction of* x_1.

Corollary 2. *If* r *is a positive integer, then*

$$(4) \qquad W^{(r)}_{x_j p}(g) \to H^{(r)}_{x_j p}(g).$$

This follows from the fact that

$$\frac{1}{h} \|\Delta^2_{x_i h} f^{(r-1)}_{x_i}\|_{L_p(g_{x_i 2|h|})} \leqq \|\Delta_{x_i h} f^{(r)}_{x_i}\|_{L_p(g_{x_i |h|})} \leqq 2 \|f^{(r)}_{x_i}\|_{L_p(g)}.$$

4.4.5. Lemma. *Suppose given a sequence of functions* f_l $(l = 1, 2, \ldots)$, *lying in* $W^\varrho_{x_1,p}(g)$, *where* $g \subset \mathbb{R}_n$ *is an open set.*

If for two functions $f, \varphi \in L_p(g)$

$$(1) \qquad \|f - f_l\|_{L_p(g)} \to 0, \ l \to \infty,$$

$$(2) \qquad \left\| \varphi - \frac{\partial^\varrho f_l}{\partial x_1^\varrho} \right\|_{L_p(g)} \to 0, \ l \to \infty,$$

then (in the generalized sense)

$$(3) \qquad \varphi = \frac{\partial^\varrho f}{\partial x_1^\varrho} \ \text{on } g.$$

Proof. First suppose that $g = [a, b]$. Since $f_l \in W^\varrho_{xp}[a, b]$ $(l = 1, 2, \ldots)$ it follows that for it or for some function equivalent to it, which we again denote by f_l, for any $x, x_0 \in [a, b]$ the following decomposition of f_l by Taylor's formula holds:

$$(4) \qquad f_l(x) = \sum_0^{\varrho-1} \frac{f_l^{(k)}(x_0)}{k!} (x - x_0)^k + \frac{1}{(\varrho - 1)!} \int_{x_0}^{x} (x - t)^{\varrho-1} f_l^{(\varrho)}(t)\, dt.$$

In view of 4.4.3 (4) and the hypotheses of the lemma,

$$|f_k^{(j)}(x_0) - f_l^{(j)}(x_0)| \leq c[\|f_k - f_l\|_{L_p(a,b)} + \|f_k^{(\varrho)} - f_l^{(\varrho)}\|_{L_p(a,b)}] \to 0$$

as $k, l \to \infty$, i. e. we have the uniform convergence

$$\lim_{l\to\infty} f_l^{(j)}(x_0) = \lambda_j(x_0) \quad (a \leqq x_0 \leqq b;\, j = 0, 1, \ldots, \varrho - 1)$$

[1] Inequality (3) is also true for trigonometric polynomials g_ν of order ν relative to x_1, if one replaces $L_p(\mathscr{E})$ by $L_{p*}(\mathscr{E})$.

on the segment $[a, b]$. But then, after passing to the limit in (4) as $l \to \infty$ we get

$$f(x) = \sum_{0}^{\varrho-1} \frac{\lambda_k(x_0)}{k!} (x - x_0)^k + \frac{1}{(\varrho - 1)!} \int_{x_0}^{x} (x - t)^{\varrho-1} \varphi(t) dt.$$

i.e.

$$\lambda_j(t) = f^{(j)}(t), \quad j = 0, 1, \ldots, \varrho - 1,$$

$$\varphi(t) = f^{(\varrho)}(t) \quad [a \leqq t \leqq b],$$

and the lemma is proved.

In the general case the lemma will obviously be proved if the validity of equation (3) is proved on any rectangular parallelepiped $\Delta \subset g$.

We will suppose that $\Delta = [a, b] \times \Delta_1$, where $x_1 \in [a, b]$. In view of the conditions imposed on the functions f_l and the fact that they form a countable set, we may regard them as having been altered on a set of measure zero, such that there exists a set $\Delta_1' \subset \Delta_1$ of full measure such that for all $\boldsymbol{y} \in \Delta_1'$ all the functions f_l are locally absolutely continuous relative to x. It follows from (1) and (2) that for almost all $\boldsymbol{y} \in \Delta_1'$, for some subsequence of indices l_ϱ depending on $\boldsymbol{y}$, one has (see 1.3.8)

$$\|f - f_{l_s}\|_{L_p(a,b)} \to 0,$$

$$\left\| \varphi - \frac{\partial^\varrho f_{l_s}}{\partial x_1^\varrho} \right\|_{L_p(a,b)} \to 0.$$

But then for the indicated $\boldsymbol{y}$

$$\varphi(x_1, \boldsymbol{y}) = \frac{\partial^\varrho f}{\partial x_1^\varrho} (x_1, \boldsymbol{y})$$

for almost all $x_1 \in [a, b]$. And this reduces to the assertion of the lemma.

4.4.6. Theorem. *Suppose that $g \subset \mathbb{R}_n$ is an open set and that g_1 is another bounded set such that $g_1 \subset \bar{g}_1 \subset g$. Then, if $f \in W_p^l(g)$,*

$$\|f^{(\boldsymbol{s})}\|_{L_p(g_1)} \leqq c_{g_1} \|f\|_{W_p^l(g)} \quad (|\boldsymbol{s}| \leqq l), \tag{1}$$

where c_{g_1} is a constant depending on p, l, and g_1, but not on f.

This theorem easily follows by induction from the inequality 4.4.3 (7). Taking account of the fact that g_1 may be covered by a finite number of cubes $\Delta \subset g$ with edges parallel to the coordinate axes, it is sufficient to prove the theorem for one of them.

4.4.7. Lemma. *Suppose given a sequence of functions*

$$f_k = f_k(x_1, \ldots, x_n) = f_k(\boldsymbol{x}) \quad (k = 1, 2, \ldots),$$

integrable to the pth *power on* g $(1 \leqq p \leqq \infty)$ *along with their derivatives figuring below in* (1) *up to order* ϱ *inclusive. Suppose moreover given functions*

$$f, f_{\alpha_1}, f_{\alpha_1\alpha_2}, \ldots, f_{\alpha_1 \ldots \alpha_s}$$

$(\alpha_1 + \cdots + \alpha_s \leqq \varrho$, α_j *positive integers*, $1 \leqq s \leqq n)$, *such that*

$$\lim_{k\to\infty} \|f - f_k\|_{L_p(g)} = 0, \tag{1}$$

$$\lim_{k\to\infty} \left\| f_{\alpha_1} - \frac{\partial^{\alpha_1} f_k}{\partial x_1^{\alpha_1}} \right\|_{L_p(g)} = 0,$$

. . . .

$$\lim_{k\to\infty} \left\| f_{\alpha_1 \ldots \alpha_s} - \frac{\partial^{\alpha_1 + \cdots + \alpha_s} f_k}{\partial x_1^{\alpha_1} \ldots \partial x_s^{\alpha_s}} \right\|_{L_p(g)} = 0.$$

Then (in the generalized sense)

$$f_{\alpha_1} = \frac{\partial^{\alpha_1} f}{\partial x_1^{\alpha_1}}, \; f_{\alpha_1\alpha_2} = \frac{\partial^{\alpha_1+\alpha_2} f}{\partial x_1^{\alpha_1} \partial x_2^{\alpha_2}}, \; \ldots, \; f_{\alpha_1 \ldots \alpha_s} = \frac{\partial^{\alpha_1 + \cdots + \alpha_s} f}{\partial x_1^{\alpha_1} \ldots \partial x_s^{\alpha_s}}. \tag{2}$$

For the case $s = 1$ this is Lemma 4.4.5. The passage to the general case is accomplished without difficulty by induction.

4.4.8. If the functions f_k considered in 4.4.7. and their partial derivatives of the appropriate orders are continuous on g, then these partial derivatives do not depend on the order of differentiation, so that also the generalized derivatives 4.4.7 (2) do not depend on the order of differentiation almost everywhere on g.

4.4.9. Theorem. *Suppose that the functions* $f_1, f_2, \ldots$ *are continuous with their partial derivatives to order* ϱ *inclusive, and along with the function* f *satisfy the hypotheses of Lemma* 4.4.7. *Suppose further that equations* (1) *are satisfied for any* $\alpha = (\alpha_1, \ldots, \alpha_n)$ *with* $|\alpha| \leqq \varrho$. *Suppose also that the domain* g *of the variables* $x_1, \ldots, x_n$ *is mapped in a* $1-1$ *way onto a domain* $\tilde{g}$ *of variables* $(t_1, \ldots, t_n)$ *by means of the functions*

$$x_j = \varphi_j(t_1, \ldots, t_n). \tag{1}$$

We suppose those functions to be continuous and to have continuous bounded partial derivatives on $\tilde{g}$ of orders not exceeding ϱ, and such that the Jacobian

$$D(t) = \frac{D(x_1, \ldots, x_n)}{D(t_1, \ldots, t_n)} > k > 0.$$

Then the function

$$F(t_1, \ldots, t_n) = f(\varphi_1, \ldots, \varphi_n)$$

is integrable to the pth power on $\tilde{g}$ along with its partial derivatives of order up to ϱ inclusive. These partial derivatives are computed by the classical formulas as if the function f had continuous partial derivatives.

Proof. Indeed, it follows from the hypotheses of the lemma that $f^{(\boldsymbol{s})} \in L_p(g)$ for all s with $|s| \leqq \varrho$, and one has

$$\int_g |f^{(\boldsymbol{s})}(\boldsymbol{x})|^p \, d\boldsymbol{x} = \lim_{k\to\infty} \int_g |f_k^{(\boldsymbol{s})}(\boldsymbol{x})|^p \, d\boldsymbol{x} = \lim_{k\to\infty} \int_{\tilde{g}} |f_k^{(\boldsymbol{s})}(\varphi_1, \ldots, \varphi_n)^p \, D(t) \, dt$$

$$= \int_{\tilde{g}} |f^{(\boldsymbol{s})}(\varphi_1, \ldots, \varphi_n)|^p \, D(t) dt,$$

Since $D(t)$ is bounded below by a positive constant, then $f^{(\boldsymbol{s})}(\varphi_1, \ldots, \varphi_n) \in L_p(\tilde{g})$. Put

$$F_k(t) = f_k(\varphi_1, \ldots, \varphi_n), \; F(t) = f(\varphi_1, \ldots, \varphi_n).$$

By the classical formula the derivative of F_k of order $\boldsymbol{l} = (l_1, \ldots, l_n)$ has the form

$$F_k^{(\boldsymbol{l})}(t) = \sum_{|\boldsymbol{s}| \leqq |\boldsymbol{l}|} \alpha_{\boldsymbol{s}} f_k^{(\boldsymbol{s})}(\varphi_1, \ldots, \varphi_n), \tag{2}$$

where the α_s are functions defined by the transformations (1), continuous and bounded on $\tilde{g}$. Since $f_k^{(\boldsymbol{s})} \to f^{(\boldsymbol{s})}$ $(k \to \infty)$ in the sense of $L_p(g)$, then on the basis of what has been said $f_k^{(\boldsymbol{s})}(\varphi_1, \ldots, \varphi_n) \to f^{(\boldsymbol{s})}(\varphi_1, \ldots, \varphi_n)$ in the sense of $L_p(\tilde{g})$. From (2) it follows after a passage to the limit as $k \to \infty$ that

$$F^{(\boldsymbol{l})}(t) = \sum_{|\boldsymbol{s}| \leqq |\boldsymbol{l}|} \alpha_{\boldsymbol{s}} f^{(\boldsymbol{s})} \, (\varphi_1, \ldots, \varphi_n) \tag{2'}$$

for almost all $t \in \tilde{g}$. The derivatives entering into (2) are, generally speaking, generalized derivatives. For $p = \infty$ nothing new comes out of the hypotheses of the lemma, because then $f_k^{(\boldsymbol{s})}(|\boldsymbol{s}| \leqq \varrho)$ converges uniformly to $f^{(\boldsymbol{s})}$.

4.5. More on Sobolev Averages[1]

Suppose that $g \subset \mathbb{R} = \mathbb{R}_n$ is an open set, $1 \leqq p \leqq \infty$, the function $f \in L_p(g)$ and

$$(1) \qquad f_\varepsilon(\boldsymbol{x}) = \frac{1}{\varepsilon^n} \int \varphi\left(\frac{\boldsymbol{x} - \boldsymbol{u}}{\varepsilon}\right) f(\boldsymbol{u}) d\boldsymbol{u} \quad (f = 0 \text{ on } \mathbb{R} - g)$$

is its ε-average (see 1.4.).

Obviously $f_\varepsilon(\boldsymbol{x})$ is infinitely differentiable on $\mathbb{R}$ and

$$(2) \qquad f_\varepsilon^{(\boldsymbol{s})}(\boldsymbol{x}) = \frac{1}{\varepsilon^{n+|\boldsymbol{s}|}} \int \varphi^{(\boldsymbol{s})}\left(\frac{\boldsymbol{x} - \boldsymbol{u}}{\varepsilon}\right) f(\boldsymbol{u}) d\boldsymbol{u}$$

for any integer vector $\boldsymbol{s} = (s_1, \ldots, s_n) \geqq \boldsymbol{0}$.

4.5.1. We denote, as usual, by g_ε the set of points $\boldsymbol{x} \in g$ distant from the boundary of g by a distance more than $\varepsilon > 0$.

Suppose that $f \in L_p(g)$ and $\dfrac{\partial f}{\partial x_1} \in L_p(g)$. If $\boldsymbol{x} \in g_\varepsilon$, then in the equation

$$\frac{\partial}{\partial x_1} f_\varepsilon(\boldsymbol{x}) = \frac{1}{\varepsilon^{n+1}} \int \varphi'_{x_1}\left(\frac{\boldsymbol{x} - \boldsymbol{u}}{\varepsilon}\right) f(\boldsymbol{u}) d\boldsymbol{u}$$

the function f in the integrand is absolutely continuous relative to u_1 for almost all $(u_2, \ldots, u_n)$. The integral is extended over a solid sphere of radius ε with center at $\boldsymbol{x}$. Because of the absolute continuity this integral may be integrated by parts relative to u_1. If $\boldsymbol{x} \notin g_\varepsilon$ this is in general not so, because f can be essentially discontinuous in the indicated sphere.

Taking account of the fact that

$$\frac{\partial}{\partial u_1} \varphi\left(\frac{\boldsymbol{x} - \boldsymbol{u}}{\varepsilon}\right) = -\frac{1}{\varepsilon} \varphi'_{x_1}\left(\frac{\boldsymbol{x} - \boldsymbol{u}}{\varepsilon}\right)$$

and that $\varphi = 0$ outside the solid sphere, we get

$$\frac{\partial}{\partial x_1} f_\varepsilon(\boldsymbol{x}) = -\frac{1}{\varepsilon^n} \int \frac{\partial}{\partial u_1} \varphi\left(\frac{\boldsymbol{x} - \boldsymbol{u}}{\varepsilon}\right) f(\boldsymbol{u}) d\boldsymbol{u}$$

$$= \frac{1}{\varepsilon^n} \int \varphi\left(\frac{\boldsymbol{x} - \boldsymbol{u}}{\varepsilon}\right) \frac{\partial f}{\partial x_1} (\boldsymbol{u}) d\boldsymbol{u} = \left(\frac{\partial f}{\partial x_1}\right)_\varepsilon (\boldsymbol{x}).$$

[1] S. L. Sobolev [4].

More generally, if we consider the functions f, $\frac{\partial f}{\partial x_{j_1}}$, $\frac{\partial^2 f}{\partial x_{j_1} \partial x_{j_2}}$, ..., then, arguing by induction, we get

$$(1) \qquad D^{s}(f_\varepsilon) = (D^{s} f)_\varepsilon \ (x \in g_\varepsilon), \quad D^{s} = \frac{\partial^{|s|}}{\partial x_1^{s_1} \ldots \partial x_n^{s_n}}.$$

In the definition of the class $W_p^l(g)$ we supposed that each function f lying in it lies in $L_p(g)$ along with its partial derivatives $f^{(l)}$ of order l. As to the subordinate derivatives $L_p(g)$, they are naturally supposed to exist, in the generalized sense, on g, but not necessarily summable to the pth power on g.

In 4.4.6 it was shown that if $f \in W_p^l(g)$ and $\sigma \subset g$ is any n-dimensional sphere, then $f^{(s)} \in L_p(\sigma)$ $(|s| \leq l)$. But then for sufficiently small $\varepsilon > 0$ equation (1) holds on σ. In view of 1.4 (5) it therefore follows for $1 \leq p < \infty$ (or for $p = \infty$ under the hypothesis that $D^{s} f$ is uniformly continuous on $\mathbb{R}_n$), that

$$(2) \qquad \|D^{s}(f_\varepsilon) - D^{s} f\|_{L_p(\sigma)} = \|(D^{s} f)_\varepsilon - D^{s} f\|_{L_p(\sigma)} \to 0 \ (\varepsilon \to 0, |s| \leq l).$$

Recalling that f_ε is an infinitely differentiable function, so that the result of the operation $D^{s} f_\varepsilon$ does not depend on the order of differentiation (relative to $s_1, \ldots, s_n$), and also that $\sigma \subset g$ is an arbitrary sphere, we arrive at the following deduction.

If $f \in W_p^l(g)$, then for the indicated vectors s *the derivatives $f^{(s)}$ almost everywhere do not depend on the order of differentiation.*

4.5.2. Theorem. *Suppose that f and λ are locally summable functions on g. If the function λ is the Sobolev derivative of f relative to x_1 on g, then it is also the derivative*

$$(1) \qquad \lambda = \frac{\partial f}{\partial x_1}$$

in the sense required by us (see the beginning of § 4.1).

Proof. Suppose that

$$(2) \qquad f_\varepsilon(x) = \int \varphi_\varepsilon(x - u) f(u) du,$$

is the ε-average of f. Then

$$(3) \qquad f'_\varepsilon(x) = \int \frac{\partial}{\partial x_1} \varphi_\varepsilon(x - u) f(u) du$$

$$= -\int \frac{\partial}{\partial u_1} \varphi_\varepsilon(x - u) f(u) du = \int \varphi_\varepsilon(x - u) \lambda(u) du \quad (x \in g_\varepsilon).$$

This last equation holds in view of the fact that λ is the derivative of f relative to x_1 in the sense of Sobolev, and from the fact that $\varphi_\varepsilon(\boldsymbol{x} - \boldsymbol{u})$, for fixed $\boldsymbol{x} \in g_\varepsilon$, is a finite function relative to $\boldsymbol{u}$.

Since f and λ are locally summable on g, then it follows from (2) and (3) (see 4.5.1. (2)) that

$$\|f_\varepsilon - f\|_{L(\sigma)} \to 0, \ \varepsilon \to 0,$$

$$\|f'_\varepsilon - \lambda\|_{L(\sigma)} \to 0, \ \varepsilon \to 0,$$

on any closed solid sphere $\sigma \subset g$. But then (1) follows from Lemma 4.4.5.

4.6. Estimate of the Increment Relative to a Direction

We consider a linear transformation

$$x_i = \sum_{k=1}^{n} a_{ik} t_k \quad (i = 1, \ldots, n) \tag{1}$$

with a determinant not equal to zero, mapping the points $\boldsymbol{x} = (x_1, \ldots, x_n) \in g$ in a $1 - 1$ way into the points $\boldsymbol{t} = (t_1, \ldots, t_n) \in \tilde{g}$. It satisfies the requirements of Theorem 4.4.9. If $f \in W_p^l(g)$, then we already know that for each sphere $\sigma \subset \bar{\sigma} \subset g$ 4.5.1 (2) is satisfied. But then, as a consequence of Theorem 4.4.9. the function $\bar{f}(\boldsymbol{t}) = f(x_1(\boldsymbol{t}), \ldots, x_n(\boldsymbol{t}))$ transformed under (1) has on each solid sphere $\sigma \subset g$, and accordingly on g as well, all derivatives $\bar{f}^{(\boldsymbol{s})}(\boldsymbol{t})$ with respect to $\boldsymbol{t}$, where $|\boldsymbol{s}| \leq l$, the derivatives furthermore being computed relative to the classical rules. Clearly $\bar{f} \in W_p^l(\tilde{g})$.

In order to define the derivative of the function $f \in W_p^l(g)$ in the direction of the vector $\boldsymbol{h} \in R_n$, we introduce an orthogonal transformation (1) such that variation of t_1 in the positive direction for fixed $t_2, \ldots, t_n$ induces variation of $\boldsymbol{x}$ in the direction of $\boldsymbol{h}$. We will suppose that the derivative of f in the direction of $\boldsymbol{h}$ is defined by the equation

$$\frac{\partial f}{\partial \boldsymbol{h}} = \frac{\partial \bar{f}}{\partial t_1} = \sum_{j=1}^{n} \frac{\partial f}{\partial x_j} \cos(\boldsymbol{h}, x_j), \tag{2}$$

$$\frac{\partial^s f}{\partial \boldsymbol{h}^s} = \frac{\partial^s \bar{f}}{\partial t_1^s} = \sum_{|\boldsymbol{s}|=s} f^{(\boldsymbol{s})} \boldsymbol{h}^{\boldsymbol{s}} \tag{3}$$

$$(\boldsymbol{s} = (s_1, \ldots, s_n);\ |\boldsymbol{s}| \leq l,\ \boldsymbol{h}^{\boldsymbol{s}} = h_1^{s_1} \ldots h_n^{s_n},\ |\boldsymbol{h}| = 1).$$

Obviously this definition does not depend on the choice of the orthogonal transformation (1), subjected to the indicated requirements.

We have the equation

$$(4)\qquad \Delta_h f(x) = f(x+h) - f(x) = |h| \int_0^1 f'_h(x+th)\,dt,$$

so that

$$(5)\qquad \|\Delta_h f(x)\|_{L_p(g_h)} \leqq |h| \int_0^1 \|f'_h(x+th)\|_{L_p(g_h)}\,dt = |h|\,\|f'_h\|_{L_p(g)},$$

where h is an arbitrary vector. It is easy to obtain also a more general inequality, which contains in particular the analogue to 4.4.4 (2):

$$(6)\qquad \|\Delta_h^\varrho f(x)\|_{L_p(g_{\varrho h})} \leqq |h|^\varrho \|f_h^{(\varrho)}\|_{L_p(g)}.$$

4.7. Completeness of the Spaces W, H, B

Theorem. *For any open set $g \subset \mathbb{R}_n$, the spaces*

$$W^l_{up}(g),\ W^{\boldsymbol{l}}_{\boldsymbol{up}}(g),\ H^r_{up}(g),\ H^{\boldsymbol{r}}_{\boldsymbol{up}}(g),\ B^r_{up\theta}(g),\ B^{\boldsymbol{r}}_{\boldsymbol{up}\theta}(g)$$

are complete.

There exist a number of variants of the definitions of the indicated classes. For an arbitrary open set g they are in general not equivalent. We shall prove the completeness for the variants 4.3.1 (1) for the spaces W^r_{up}, 4.3.3 (5) for H^r_{up}, 4.3.4 (2) or 4.3.4 (4) for $B^r_{up\theta}$. For other variants the proof is not essentially different.

Proof. We will suppose that $g_1 \subset \bar{g}_1 \subset g$ is a bounded open set. Suppose given a sequence of functions $f_k \in W^l_{\boldsymbol{up}}(g)$ $(k = 1, 2, \ldots)$, satisfying the Cauchy condition in the metric of $W^l_{\boldsymbol{up}}(g)$. Then (see 4.4.6 and 4.4.7)

$$(1)\qquad \|f_k^{(s)} - f_j^{(s)}\|_{L_p(g_1)} \leqq c_{g_1}\|f_k - f_l\|_{W^l_{\boldsymbol{up}}(g)} \to 0$$

$$k, j \to \infty,\ \boldsymbol{s} = (s_1, \ldots, s_m, 0, \ldots, 0),\ |\boldsymbol{s}| \leqq l,$$

and there exists a function f for which

$$(2)\qquad \|f_k^{(s)} - f^{(s)}\|_{L_p(g_1)} \to 0 \quad (k \to \infty).$$

It belongs to $W^l_{\boldsymbol{up}}(g_1)$, because $f_k \in W^l_{\boldsymbol{up}}(g_1)$. Given an $\varepsilon > 0$, it is possible to indicate a positive integer N such that for $k, j > N$

$$(3)\qquad \|f_k - f_j\|_{L_p(g_1)} + \sum_{|\boldsymbol{s}|=l} \|f_k^{(s)} - f_j^{(s)}\|_{L_p(g_1)} \leqq \|f_k - f_j\|_{W^l_{\boldsymbol{up}}(g)} < \varepsilon$$

for any g_1. The passage to the limit as $j \to \infty$ in the first term of (3) leads to the same expression, where we need to replace f_i by f, i.e. we have

$$\|f_k - f\|_{W^l_{up}(g_1)} \leqq \varepsilon \ (k > N)$$

for any g_1 and accordingly for g as well. In addition f belongs, along with the f_k, to $W^l_{up}(g)$. Thus the completeness of $W^l_{up}(g)$, and in particular of $W^{l_j}_{x_jp}(g)$, is proved. But then $W^{\boldsymbol{l}}_{up}(g)$ is obviously complete as well.

Now we shall prove the completeness of $B^r_{up\theta}(g)$ $(1 \leqq \theta \leqq \infty$, $B^r_{up\infty} = H^r_{up})$. It is possible to prove (see the remark to 4.3.6 at the end of the book) that the functions of the classes $B^{r,\dots,r}(g) = B^{r,\dots,r}_{up\theta}(g)$ may be extended from $g_1 \subset \bar{g}_1 \subset g$ to $\mathbb{R}$ with preservation of the norm (relative to g), i.e. it is possible for each $f \in B^{r,\dots,r}(g)$ to define an extension $\bar{f}$ ($\bar{f} = f$ in g_1) such that

$$\|\bar{f}\|_{B^{r,\dots,r}(\mathbb{R})} \leqq c\|f\|_{B^{r,\dots,r}(g)}. \tag{4}$$

But further on it will be proved (5.6.2) that

$$B^{r,\dots,r}(\mathbb{R}) = B^r(\mathbb{R}). \tag{5}$$

Therefore

$$0 \leftarrow \|f_k - f_j\|_{B^r(g)} \gg \|f_k - f_j\|_{B^{r,\dots,r}(g_1)} \tag{6}$$

$$\gg \|\bar{f}_k - \bar{f}_j\|_{B^{r,\dots,r}(\mathbb{R})} \gg \|\bar{f}_k - \bar{f}_j\|_{B^r(\mathbb{R})} \gg \|f_k^{(\boldsymbol{s})} - f_j^{(\boldsymbol{s})}\|_{L_p(g_1)}$$

$(\boldsymbol{s} = (s_1, \dots, s_m, 0, \dots, 0), |\boldsymbol{s}| \leqq \bar{r}, r = \bar{r} + \alpha, \bar{r}$ an integer, $0 < \alpha \leqq 1)$

The first inequality is trivial ($B^r \to B^{r,\dots,r}$). The second is valid in view of the theorem mentioned above on extension. Here $\bar{f}_j, \bar{f}_k$ are the functions extending, according to that theorem, the functions f_j, f_k respectively from the set g_1. The third inequality follows directly from (5). The fourth will be proved later (6.2 (8)). We note that in the case of $H^r_{x_jp}(g)$ $(m = 1)$ the inequality between the first and last terms of (6) follows directly from (4.4.3 (8)) without bringing in the extension theorem.

It is obvious from (6) that for $\varrho = \bar{r}$

$$\|f^\varrho_{k\boldsymbol{u}} - f^\varrho_{j\boldsymbol{u}}\|_{L_p(g_1)} \to 0 \quad (k, j \to \infty, \boldsymbol{u} \in \mathbb{R}_m), \tag{7}$$

where $f^\varrho_{k\boldsymbol{u}}$ is the derivative of f_k of order ϱ relative to the direction $\boldsymbol{u}$, for any $g_1 \subset \bar{g}_1 \subset g$.

Now, in considering the class $H^r_{\boldsymbol{u}p}(g)$ and for simplicity supposing that also $0 < \alpha < 1$, and choosing $\varepsilon > 0$, we find ($g_1 \subset \bar{g}_1 \subset g_{\boldsymbol{u}}$) that

(8)
$$\|f_k - f_j\|_{L_p(g)}$$
$$+ \frac{\left\|\left(f^\varrho_{k\boldsymbol{u}}(\boldsymbol{x}+\boldsymbol{u}) - f^\varrho_{j\boldsymbol{u}}(\boldsymbol{x}+\boldsymbol{u})\right) - \left(f^\varrho_{k\boldsymbol{u}}(\boldsymbol{x}) - f^\varrho_{j\boldsymbol{u}}(\boldsymbol{x})\right)\right\|_{L_p(g_1)}}{|\boldsymbol{u}|^\alpha}$$
$$\leqq \|f_k - f_j\|_{H^r_{\boldsymbol{u}p}(g)} \leqq \varepsilon \quad (k, j > N,\ \boldsymbol{u} \in \mathbb{R}_m),$$

where N is sufficiently large. If we fix k and let $j \to \infty$, then in the limit the first term of (8), in view of (7), turns into the same expression, where we need to replace f_j by f. In it we replace g_1 by $g_{\boldsymbol{u}}$, which is obviously legitimate. Taking the upper bound of the resulting expression relative to $\boldsymbol{u}$, we find that

$$\|f_k - f\|_{H^r_{\boldsymbol{u}p}(g)} \leqq \varepsilon \quad (k > N),$$

Since from the fact that $f_k \in H^r_{\boldsymbol{u}p}(g)$, it follows that $f \in H^r_{\boldsymbol{u}p}(g)$, then the completeness of $H^r_{\boldsymbol{u}p}(g)$ for $\alpha < 1$ is proved. For $\alpha = 1$ we need to replace the first difference of $f^\varrho_{k\boldsymbol{u}} - f^\varrho_{j\boldsymbol{u}}$ in (8) by the second difference, reasoning analogously.

Now for the classes $B^r_{\boldsymbol{u}p\theta}(g)$ ($1 \leqq \theta < \infty$), for any $\varepsilon > 0$ ($g_1 \subset \bar{g}_1 \subset g_{k\boldsymbol{u}}$)

(9)
$$\|f_\mu - f_\nu\|_{L_p(g)} + \left(\int_{\lambda < |\boldsymbol{u}| < \varkappa} |\boldsymbol{u}|^{-m-\theta\alpha} \|\Delta^k_{\boldsymbol{u}}(f^\varrho_\mu - f^\varrho_\nu\|^\theta_{L_p(g_1)} d\boldsymbol{u}\right)^{1/\theta} \leqq \varepsilon$$

($\mu, \nu > N$, $\boldsymbol{u} \in \mathbb{R}_m$, $\varrho = \bar{r}$, $k \geqq 2$, for $\alpha < 1$, $k \geqq 1$), where $0 < \lambda < \varkappa$ are arbitrary numbers.

The passage to the limit as $\nu \to \infty$ leads to the same inequality, where we need to replace f^ϱ_ν by f^ϱ. This follows from the fact that now we can apply Lebesgue's theorem on the limit under the integral sign. The point is that the norm under the integral sign in (9), which depends on $\boldsymbol{u}$, tends boundedly to the same norm, where f^ϱ replaces f^ϱ_ν (see (6), (7)). In the resulting inequality, true for any of the indicated λ, $\varkappa$, one may obviously put $\lambda = 0$, $\varkappa = \infty$ and replace g_1 by $g_{k\boldsymbol{u}}$, which implies that

$$\|f_\mu - f\|_{B^r_{\boldsymbol{u}p\theta}} \leqq \varepsilon, \quad \mu > N.$$

Moreover, from the fact that $f_\mu \in B^r_{\boldsymbol{u}p\theta}(g)$ it follows that $f \in B^r_{\boldsymbol{u}p\theta}(g)$. The completeness of $B^r_{\boldsymbol{u}p\theta}(g)$ is proved.

4.8. Estimates of the Derivative by the Difference Quotient

Theorem (Reversing inequality 4.4.4 (2)). *Suppose that the function $f(\boldsymbol{x}) = f(x_1, \boldsymbol{y})$ is given on an open set g, locally summable on it and satisfies the inequality*

$$\int_{g_h} \left| \frac{\Delta_{x_1,h} f(\boldsymbol{x})}{h} \right|^p d\boldsymbol{x} \leqq M \quad (1 < p < \infty), \tag{1}$$

where M does not depend on h.

Then on g there exists a derivative $\frac{\partial f}{\partial x_1}$, having the property

$$\int_g \left| \frac{\partial f}{\partial x_1} \right|^p d\boldsymbol{x} \leqq M. \tag{2}$$

Proof. Suppose given two strictly imbedded, one into the other, open cubes $\Delta \subset \Delta_1 \subset \bar{\Delta}_1 \subset g$, with boundaries parallel to the coordinate axes. We have

$$\frac{\Delta_{x_1,h} f_\varepsilon(\boldsymbol{x})}{h} = \left(\frac{\Delta_{x_1,h} f(x)}{h} \right)_\varepsilon$$

($(\cdot)_\varepsilon$ is the ε-average). Therefore from (1) and 1.4 (7) it follows for sufficiently small h and ε that

$$\int_\Delta \left| \frac{\Delta_{x_1,h} f_\varepsilon(\boldsymbol{x})}{h} \right|^p d\boldsymbol{x} \leqq \int_{\Delta_1} \left| \frac{\Delta_{x_1,h} f(\boldsymbol{x})}{h} \right|^p d\boldsymbol{x} \leqq M.$$

Passing to the limit as $h \to \infty$, we get

$$\int_\Delta \left| \frac{\partial f_\varepsilon}{\partial x_1} \right|^p d\boldsymbol{x} \leqq M. \tag{3}$$

We have

$$f_\varepsilon(x_1', \boldsymbol{y}) - f_\varepsilon(x_1, \boldsymbol{y}) = \int_{x_1}^{x_1'} \frac{\partial f_\varepsilon}{\partial x_1}(t, \boldsymbol{y})\, dt, \tag{4}$$

where we suppose that $\boldsymbol{y} = (\xi_2, \ldots, \xi_n)$ runs through the rectangular parallelepiped

$$\Delta_* = \{x_j \leqq \xi_j \leqq x_j + h_j;\ j = 2, \ldots, n\}$$

and

$$[x_1, x_1'] \times \Delta_* \subset \Delta.$$

Integrating (4) over $\boldsymbol{y} \in \Delta_*$, we get

(5) $$\int_{\Delta_*} [f_\varepsilon(x_1', \boldsymbol{y}) - f_\varepsilon(x_1, \boldsymbol{y})] d\boldsymbol{y} = \int_{\Delta} d\boldsymbol{y} \int_{x_1}^{x_1'} \frac{\partial f_\varepsilon}{\partial x_1} (t, \boldsymbol{y}) dt.$$

It follows from (3) that there exists a sequence of numbers $\varepsilon_k' \to 0$ and a function $\psi \in L_p(\Delta)$ such that $\frac{\partial f_{\varepsilon_k'}}{\partial x_1} \to \psi$ weakly in the sense of $L_p(\Delta)$ (see 1.3.11). On the other hand, from the fact that $\|f_{\varepsilon_k'} - f\|_{L_p(\Delta)} \to 0$, it is possible to select from the sequence $\{\varepsilon_k'\}$ a subsequence $\{\varepsilon_k\}$ such that

$$\int_{\Delta_*} f_{\varepsilon_k}(x_1, \boldsymbol{y}) d\boldsymbol{y} \to \int_{\Delta_*} f(x_1, \boldsymbol{y}) d\boldsymbol{y}, \quad \varepsilon_k \to 0,$$

for all x_1 from some set $\mathscr{E} \subset [a, b] = \mathrm{pr}_{x_1} \Delta$ (projection of Δ onto the x_1 axis) of (linear) measure $b - a$. In such a case, if one puts $\varepsilon = \varepsilon_k$ in (5), we find in the limit as $\varepsilon_k \to 0$, for any $x_1, x_1' \in \mathscr{E}$ that

$$\int_{\Delta_*} [f(x_1', \boldsymbol{y}) - f(x_1, \boldsymbol{y})] d\boldsymbol{y} = \int_{\Delta_*} d\boldsymbol{y} \int_{x_1}^{x_1'} \psi(t, \boldsymbol{y}) dt.$$

If we divide both sides of this equation by $h_1, \ldots, h_n$ and pass to the limit as $h_1 \to 0$, and then $h_2 \to 0$, and so forth, then we obtain for almost all $\boldsymbol{y} = (x_2, \ldots, x_n)$ and

$$x_1 \in \mathscr{E}, x_1' \in \mathscr{E} \left((x_1, \boldsymbol{y}), (x_1', \boldsymbol{y}) \in \Delta\right):$$

(6) $$f(x_1', \boldsymbol{y}) - f(x_1, \boldsymbol{y}) = \int_{x_1}^{x_1'} \psi(t, \boldsymbol{y}) dt.$$

Indeed this equation is true for almost all admissible $\boldsymbol{y}$ and all $x_1, x_1' \in [a, b]$, inasmuch as its right hand side is continuous relative to x_1, x_1'. It denotes the existence on Δ of the (generalized) partial derivative $\frac{\partial f}{\partial x_1} = \psi \in L_p(\Delta)$. In view of the arbitrariness of Δ it implex the existence of $\frac{\partial f}{\partial x_1} \in L_p(\Omega)$ for any open $\Omega \subset \bar{\Omega} \subset g$.

Since now we already know that the integrand in (1) tends almost everywhere on Ω to $\left|\frac{\partial f}{\partial x_1}\right|^p$, then, by Fatou's theorem (1.3.10),

$$\int_{\Omega} \left|\frac{\partial f}{\partial x_1}\right|^p dx \leq \sup_h \int_{\Omega} \left|\frac{\Delta_{x_1,h} f}{h}\right|^p d\boldsymbol{x} \leq M,$$

In view of the arbitrariness of $\Omega \subset \bar{\Omega} \subset g$ we get (2).

4.8.1. Theorem 4.8 for $p = \infty$, $n = 1$ is a well-known theorem from the theory of functions of a real variable: *if a function f satisfies on the interval (a, b) a Lipschitz constant with constant M, then it has almost everywhere on (a, b) a derivative satisfying the inequality $|f'(x)| \leqq M$* (see P. S. Aleksandrov and A. N. Kolmogorov [1]).

4.8.2. Theorem 4.8, for $p = 1$, $n = 1$, goes over into the following: *if for a function f which is locally summable on (a, b) the inequality*

$$\int_a^{b-h} |f(x+h) - f(x)|dx \leqq Mh \quad (0 < h < b - a),$$

is satisfied, then it is equivalent to some function, which we again denote by f, of bounded variation on (a, b) and in fact satisfying

$$\operatorname*{Var}_{(a,b)} f \leqq M.$$

Indeed, reasoning as at the beginning of the proof of Theorem 4.8, we get

$$\int_\Delta |f'_\varepsilon| dx \leqq M,$$

where Δ is any interval such that $\Delta \subset \Delta_1 \subset \bar{\Delta}_1 \subset (a, b)$. Therefore

$$\operatorname*{Var}_{(a,b)} f_\varepsilon = \int_a^b |f'_\varepsilon| dx \leqq M. \tag{1}$$

Since $f \in L(\Delta_1)$, then $\int_{\Delta_1} |f_\varepsilon - f| dx \to 0$, and, in view of the arbitrariness of $\Delta_1 \subset \bar{\Delta}_1 \subset (a, b)$, there exists a sequence $\varepsilon_k \to 0$ such that

$$f_{\varepsilon_k}(x) \to f(x) \tag{2}$$

almost everywhere on (a, b). But by Helly's theorem (see I. P. Natanson [1]), from condition (1) and the fact that (2) is satisfied at even one point of the interval (a, b), it follows that there exists a subsequence $\{\varepsilon'_k\}$ of the sequence $\{\varepsilon_k\}$ such that $f'_{\varepsilon k}$ tends everywhere on (a, b) to some function ψ, bounded on (a, b) and

$$\operatorname*{Var}_{(a,b)} \psi \leqq M.$$

But then ψ and f are equivalent on (a, b).

Chapter 5

Direct and Inverse Theorems of the Theory of Approximation. Equivalent Norms

5.1. Introduction

Throughout this chapter we suppose that $\boldsymbol{u} = (x_1, \ldots, x_m) \in \mathbb{R}_m$, $\boldsymbol{y} = (x_{m+1}, \ldots, x_n)$, and we consider a measurable cylindrical set $\mathscr{E} = \mathbb{R}_m \times \mathscr{E}'$ of points $\boldsymbol{x} = (\boldsymbol{u}, \boldsymbol{y}) = (x_1, \ldots, x_n)$, where $\boldsymbol{u} \in \mathbb{R}_m$, $\boldsymbol{y} \in \mathscr{E}'$. We denote by $\mathbb{R}_m$ also the subspace of $\mathbb{R}_n$ consisting of the points $(\boldsymbol{u}, 0) = (x_1, \ldots, x_m, 0, \ldots, 0)$. For $m = n$ $\mathscr{E} = \mathbb{R}_n$. The case $m = 0$ is of little interest.

This chapter is devoted to the study of approximations of functions of the classes H, W, and B (see Chapter 4), given on the indicated cylindrical set $\mathscr{E}$. Functions of the classes H_p and W_p will be approximated by entire functions of exponential type (relative to $\boldsymbol{u}$) in the metric of L_p, and periodic functions of the classes H_p^* and W_p^* by trigonometric polynomials (relative to $\boldsymbol{u}$) in the metric of L_p^*.

For the classes H and W we shall prove direct theorems of the theory of approximation (of the type of Jackson), showing that the numbers r or systems of numbers $(r_1, \ldots, r_n)$ defining the class define also the order of approximation of the functions belonging to the class.

We shall prove also inverse theorems of the theory of approximation, of Bernšteĭn type, showing that the order of approximation of a given function f by means of functions of finite degree or trigonometric polynomials frequently determines completely that class H (but not W) to which the function f belongs. In a number of cases interesting for analysis we will obtain, in the language of orders of approximations, necessary and sufficient conditions for a function f to belong to this or that H-class. An important means of expressing these theorems is the concept of best approximation, whose idea goes back to P. L. Čebyšev.

From this point of view we shall consider also the classes $B_{p\theta}^r$. The functions belonging to them are also completely characterized by the behavior of their best approximations by entire functions of exponential type or (in the periodic case) trigonometric polynomials. Indeed, for a function to belong to a given class B, it is necessary and sufficient that a certain series, made up of its best approximations, converge. We will see that the definition of the classes $B^r_{p\theta}$ on the language of best approximations naturally extends to the case $\theta = \infty$ and leads to the equivalence $B^{(r)}_{p\infty} = H^{(r)}_p$.

In this chapter, on the basis of the method of approximations, we will obtain many different equivalent definitions of the norm in the classes H and B. The fact of equivalency itself reduces to certain inequalities, in particular inequalities among the partial derivatives of one and the same function.

We will consider functions $g_{\boldsymbol{\nu}}(\boldsymbol{x}) = g_\nu(\boldsymbol{u}, \boldsymbol{y})$, $\boldsymbol{\nu} = (\nu_1, \ldots, \nu_m)$, defined on $\mathscr{E} = \mathbb{R}_m \times \mathscr{E}'$ which are for almost all $\boldsymbol{y} \in \mathscr{E}'$ entire functions of exponential type $\boldsymbol{\nu}$ relative to the variables $\boldsymbol{u} = (x_1, \ldots, x_m)$. The collection of all such functions $g_\nu \in L_p(\mathscr{E})$ for a given ν defines a subspace $\mathfrak{M}_{\nu p}(\mathscr{E}) \subset L_p(\mathscr{E})$ (see 3.5).

Suppose given a function $f \in L_p(g)$ $(1 \leqq p \leqq \infty)$. The *best approximation of f in terms of the functions $g_\nu \in \mathfrak{M}_{\nu p}(\mathscr{E})$, where $\nu = (\nu_1, \ldots, \nu_m)$ is a given system of numbers, is the quantity*

$$(1) \qquad E_\nu(f) = E_\nu(f)_{L_p(\mathscr{E})} = \inf_{g_\nu \in \mathfrak{M}_{\nu p}(\mathscr{E})} \|f - g_\nu\|_{L_p(\mathscr{E})} = \inf_{g_\nu} \|f - g_\nu\|_{L_p(\mathscr{E})}.$$

For $m = n$ the lower bound of (1) is attained for some (best) function. Indeed, it follows from (1) that there exists a sequence of functions $g_{\nu s} \in \mathfrak{M}_{\nu p}$ for which the following inequality is satisfied:

$$\|f - g_{\nu s}\|_{L_p(\mathbb{R}_n)} \leqq E_\nu(f)_{L_p(\mathbb{R}_n)} + \varepsilon_s = d + \varepsilon_s \ (\varepsilon_s \to 0).$$

In view of 3.3.6, we may select from this sequence a subsequence, which we shall again denote by $\{g_{\nu s}\}$, which converges uniformly to some function $g_\nu \in \mathfrak{M}_{\nu p}\mathbb{R}_m)$ on any bounded region $g \subset \mathbb{R}_n$. But then

$$\|f - g_\nu\|_{L_p(g)} = \lim_{s\to\infty} \|f - g_{\nu s}\|_{L_p(g)} \leqq \lim_{s\to\infty} \|f - g_{\nu s}\|_{L_p(\mathbb{R}_n)} = d.$$

Accordingly

$$\|f - g_\nu\|_{L_p(\mathbb{R}_n)} = d.$$

Since $\mathfrak{M}_{\nu p}(\mathscr{E})$ is a subspace of the space $L_p(\mathscr{E})$, then under the condition $1 < p < \infty$ the lower bound of (1) is achieved for a unique (best) function $g_\nu \in \mathfrak{M}_{\nu p}(\mathscr{E})$. Sometimes it is convenient to consider functions which we shall denote by $g_{u\nu}(x)$ $(\nu > 0)$. These are functions defined on $\mathscr{E}$, which for almost all $\boldsymbol{y} = (x_{m+1}, \ldots, x_n)$ are entire of exponential type relative to $\boldsymbol{u} = (x_1, \ldots, x_m)$ and of spherical degree ν.

The best approximation of a function of a function $f \in L_p(\mathscr{E})$ by means of the functions $g_{u\nu}$ (for a given $\nu > 0$) *is the quantity*

$$(2) \qquad E_{u\nu}(f) = E_{u\nu}(f)_{L_p(\mathscr{E})} = \inf_{g_{u\nu}} \|f - g_{u\nu}\|_{L_p(\mathscr{E})},$$

where the lower bound is extended over all $g_{u\nu} \in L_p(\mathscr{E})$ for the given ν.

A special case of these concepts is the quantity

(3) $$E_{x_j\nu}(f)_{L_p(\mathscr{E})} = \inf_{g_{x_j\nu}} \|f - g_{x_j\nu}\|_{L_p(\mathscr{E})},$$

where $\mathscr{E} = \mathbb{R}_j \times \mathscr{E}^j$ $(j = 1, \ldots, n)$, $\mathbb{R}_j$ is the x_j coordinate axis and $g_{x_j\nu}$ is a function of $L_p(\mathscr{E})$ of exponential type ν relative to x_j.

5.2. Approximation Theorem

5.2.1. Direct theorem on approximation by entire functions of exponential type. Suppose that $g(\xi)$ is a nonnegative even function of one variable, of exponential type 1, satisfying the condition

(1) $$\int_{\mathbb{R}_m} g(|\boldsymbol{u}|)\, d\boldsymbol{u} = 1.$$

Let $\mathscr{E} = \mathbb{R}_m \times \mathscr{E}'$.

For an arbitrary function $\varphi(\boldsymbol{x})$, defined on $\mathscr{E}$, vector $\boldsymbol{h} \in \mathbb{R}_m$, and natural number l, one has the equation

(2) $$(-1)^{l+1}\Delta_{\boldsymbol{h}}^l \varphi(\boldsymbol{x}) = \sum_{j=0}^{l} (-1)^{j-1} C_l^j \varphi(\boldsymbol{x} + j\boldsymbol{h}) = \sum_{j=1}^{l} d_j \varphi(\boldsymbol{x} + j\boldsymbol{h}) - \varphi(\boldsymbol{x}),$$

where

(3) $$\sum_{j=1}^{l} d_j = 1.$$

Suppose given a function $f \in L_p(\mathscr{E})$. Then for almost all $\boldsymbol{y} \in \mathscr{E}'$ the function $f(\boldsymbol{u}, \boldsymbol{y})$ of $\boldsymbol{u}$ belongs to $L_p(\mathbb{R}_m)$, and the function

(4) $$g_\nu(\boldsymbol{x}) = g_\nu(\boldsymbol{u}, \boldsymbol{y}) = \int_{\mathbb{R}_m} g(|\boldsymbol{t}|) \{(-1)^{l-1} \Delta_{\boldsymbol{t}/\nu}^l(\boldsymbol{x}) + f(\boldsymbol{x})\}\, d\boldsymbol{t}$$

$$= \int_{\mathbb{R}_m} g(|\boldsymbol{t}|) \sum_{j=1}^{l} d_j f\left(\boldsymbol{u} + j\frac{\boldsymbol{t}}{\nu}, \boldsymbol{y}\right) d\boldsymbol{t} = \int_{\mathbb{R}_m} K_\nu(\boldsymbol{t} - \boldsymbol{u})\, f(\boldsymbol{t}, \boldsymbol{y})\, d\boldsymbol{t},$$

where

(5) $$K_\nu(\boldsymbol{u}) = \sum_{j=1}^{l}{}' d_j \left(\frac{\nu}{j}\right)^m g\left(\frac{|\boldsymbol{u}|\nu}{j}\right),$$

has a meaning.

In view of (1)

(6) $$g_\nu(\boldsymbol{x}) - f(\boldsymbol{x}) = (-1)^{l-1} \int_{\mathbb{R}_m} g(|\boldsymbol{t}|) \Delta_{\boldsymbol{t}/\nu}^l f(\boldsymbol{x})\, d\boldsymbol{t}.$$

Now we suppose that the function f has on $\mathscr{E}$ derivatives relative to $\boldsymbol{u}$ of order ϱ, lying in $L_p(\mathscr{E})$ and $k = l - \varrho$. Then it follows from (6) (explanation below) that

$$
\begin{aligned}
(7)\quad E_{\boldsymbol{u}\nu}(f)_{L_p(\mathscr{E})} &\leqq \|f - g_\nu\|_{L_p(\mathscr{E})} = \left\| \int_{\mathbb{R}_m} g(|\boldsymbol{t}|)\Delta^l_{\boldsymbol{t}/\nu} f(\boldsymbol{x})\, d\boldsymbol{t} \right\|_{L_p(\mathscr{E})} \\
&\leqq \int_{\mathbb{R}_m} g(|\boldsymbol{t}|)\|\Delta^l_{\boldsymbol{t}/\nu} f(\boldsymbol{x})\|_{L_p(\mathscr{E})} d\boldsymbol{t} \\
&\leqq \int_{\mathbb{R}_m} g(|\boldsymbol{t}|)\left(\frac{|\boldsymbol{t}|}{\nu}\right)^\varrho \|\Delta^k_{\boldsymbol{t}/\nu} f^\varrho_{\boldsymbol{t}}\|_{L_p(\mathscr{E})}\, d\boldsymbol{t} \\
&\leqq \frac{1}{\nu^\varrho}\int_{\mathbb{R}_m} g(|\boldsymbol{t}|)\, |\boldsymbol{t}|^\varrho\, \Omega^k_{\mathbb{R}_m}\left(f^\varrho, \frac{|\boldsymbol{t}|}{\nu}\right) d\boldsymbol{t} \\
&\leqq \frac{1}{\nu^\varrho}\int_{\mathbb{R}_m} g(|\boldsymbol{t}|)\, |\boldsymbol{t}|^\varrho\, (1 + |\boldsymbol{t}|)^k\, d\boldsymbol{t}\, \Omega^k_{\mathbb{R}_m}\left(f^\varrho, \frac{1}{\nu}\right)_{L_p(\mathscr{E})} = \\
&= \frac{c}{\nu^\varrho}\, \Omega^k_{\mathbb{R}_m}\left(f^\varrho, \frac{1}{\nu}\right)_{L_p(\mathscr{E})} \qquad (\nu > 0)\,,
\end{aligned}
$$

if the right side is finite.

In particular it follows from (7) that if $f \in H^r_{\boldsymbol{u}p}(\mathscr{E})$, then

$$(8)\qquad E_{\boldsymbol{u}\nu}(f) \leqq \frac{c_1}{\nu^r}\,.$$

We have applied the generalized Minkowski inequality and inequality 4.6 (6). $f^\varrho_{\boldsymbol{t}}$ is the derivative of f of order ϱ in the direction of $\boldsymbol{t}$, $\Omega^k_{\mathbb{R}_m}(f^\varrho, \delta)$ the modulus of continuity of f relative to all derivatives of order ϱ. Property 4.2. (14) is applicable to that modulus of continuity. Finally we recall that the function g is chosen so that the integral

$$\int_{-\infty}^{\infty} g(t)t^{\varrho+k+m-1}dt$$

is finite. As g we may choose a function of the form

$$(9)\qquad \mu\left(\frac{\sin\dfrac{t}{\lambda}}{t}\right)^{\lambda},$$

where $\lambda \geqq \varrho + k + m + 2$ is a even number and μ is a constant, for which (1) holds.

Since $g(\xi)$ is an entire function of one variable of exponential type 1, then in view of (5) the function $g_\nu(\boldsymbol{x})$ is in its turn an entire function of spherical type ν relative to $\boldsymbol{u} = (x_1, \ldots, x_m)$ (see 3.6.2), lying in $L_p(\mathscr{E})$.

5.2.1.1. Suppose that we know about the function f only that its modulus of continuity

(1) $$\Omega^k_{\mathbb{R}_m}(f^\varrho, \delta) < \infty$$

is finite for some $\delta > 0$. Then, arguing as above (from right to left), we can obtain for $\frac{1}{\nu} \leqq \delta$ the whole chain of relations 5.2.1 (7), excluding for the time being the first inequality. The difference $f - g_\nu$ will represent the formal writing-out of the function inside the integral in the third term of 5.2.1 (7). However if it is known that the function f is locally integrable to the pth power on $\mathscr{E}$ (or even somewhat less: see below), then one may conclude that f is integrable to the pth power on $\mathscr{E}$ with a certain weight, and g_ν, with a choice of an appropriate kernel g, is an entire function of spherical type ν, integrable with the same weight. Indeed, from (1), for any $\boldsymbol{h} \in \mathrm{R}_m$ with $|\boldsymbol{h}| \leqq \delta$ it follows that

$$\|\Delta^l_{\boldsymbol{h}} f(\boldsymbol{x})\|_{L_p(\mathscr{E})} \leqq \delta^\varrho \|\Delta^k_{\boldsymbol{h}} f^\varrho_{\boldsymbol{h}}(\boldsymbol{x})\|_{L_p(\mathscr{E})} \leqq \delta^\varrho \Omega^k_{\mathbb{R}_m}(f^\varrho, \delta)$$

and in view of 4.2.2 (5), with k replaced by l,

(2) $$\|f(\boldsymbol{x})\,(1 + |\boldsymbol{u}|^{-\mu})\|_{L_p(\mathscr{E})} < \infty,$$

where

(3) $$\mu = \frac{m}{p} + l + \varepsilon \quad (\varepsilon > 0).$$

But then for almost all $\boldsymbol{y} \in \mathscr{E}'$.

$$\|f(\boldsymbol{u}, \boldsymbol{y})\,(1 + |\boldsymbol{u}|)^{-\mu}\|_{L_p(\mathscr{E})} < \infty.$$

And in view of 3.6.2 it is possible to choose a kernel $g(\boldsymbol{t})$ of the type 5.2.1 (9), choosing λ sufficiently large, in such a way that the function

(4) $$g_\nu(\boldsymbol{x}) = \int_{\mathbb{R}_m} K_\nu(\boldsymbol{t} - \boldsymbol{u})\, f(\boldsymbol{t}, \boldsymbol{y})\, d\boldsymbol{t}$$

becomes *a fortiori* of spherical type for all $\boldsymbol{y}$ (see 5.2.1 (4), (5)).

Now the first term of formula 5.2.1 (7) acquires a meaning as well. This is the term $E_{\boldsymbol{u}\nu}(f)_{L_p(\mathscr{E})}$. It may be considered as the best approximation in the metric of $L_p(\mathscr{E})$ of the function f in question in terms of entire functions of spherical type ν (generally not belonging to $L_p(\mathscr{E})$). We have shown that if for a function f locally summable to the pth power the modulus (1) has a meaning, then it makes sense to approximate it in the metric of $L_p(\mathscr{E})$ by functions of spherical type ν relative to $\boldsymbol{u}$.

Indeed (see 4.2.2 (4)), in place of local summability of $|f|^p$ (for $p = \infty$ local boundedness and measurability of f) it is sufficient to suppose the existence of $\|f\|_{L_p(v \times \mathscr{E}')}$ where $v = |\boldsymbol{u}| < \delta(l + m)$.

5.2.2. Other estimates of approximation. Below we shall carry out other estimates, based on formula 5.2.1(6). If f has a generalized derivative relative to $\boldsymbol{u} = (u_1, \ldots, u_m)$ to order ϱ inclusive, then it follows from 5.2.1 (6) that in every case, formally, for any integer nonnegative vector $\boldsymbol{s} = (s_1, \ldots, s_m, 0, \ldots, 0)$ with $|\boldsymbol{s}| \leqq \varrho$, one has the equation

$$g_\nu^{(\boldsymbol{s})} - f^{(\boldsymbol{s})} = (-1)^{l-1} \int_{\mathbb{R}_m} g(|\boldsymbol{t}|) \Delta^l_{\boldsymbol{t}/\nu} f^{(\boldsymbol{s})}(\boldsymbol{x})\, dt \tag{1}$$

and the inequality

$$\|g_\nu^{(\boldsymbol{s})} - f^{(\boldsymbol{s})}\|_{L_p(\mathscr{E})} \leqq \int_{\mathbb{R}_m} g(|\boldsymbol{t}|)\, \|\Delta^l_{\boldsymbol{t}/\nu} f^{(\boldsymbol{s})}(\boldsymbol{x})\|_{L_p(\mathscr{E})}\, dt. \tag{2}$$

If for any $\boldsymbol{s}$ with $|\boldsymbol{s}| \leqq \varrho$ the integral in the right side of (2) is finite, then the already nonformal equation (1) and inequality (2) hold.

Further we shall make use of the inequality

$$\|\Delta^l_{\boldsymbol{h}} \varphi(\boldsymbol{x})\|_{L_p(\mathscr{E})} \leqq c \|\Delta^{l'}_{\boldsymbol{h}} \varphi(\boldsymbol{x})\|_{L_p(\mathscr{E})} \qquad (0 < l' < l), \tag{3}$$

where $c = 2^{l-l'}$, $\boldsymbol{h} \in \mathbb{R}_m$.

Then (explanation below) for $k = l - \varrho$, $|\boldsymbol{s}| \leqq \varrho$

$$\begin{aligned} \|g_\nu^{(\boldsymbol{s})} - f^{(\boldsymbol{s})}\|_{L_p(\mathscr{E})} &\leqq \int_{\mathbb{R}_m} g(|\boldsymbol{t}|) \left(\frac{|\boldsymbol{t}|}{\nu}\right)^{\varrho - |\boldsymbol{s}|} \left\| \Delta^{k+|\boldsymbol{s}|}_{\boldsymbol{t}/\nu} \frac{\partial^{\varrho-|\boldsymbol{s}|}}{\partial t^{\varrho-|\boldsymbol{s}|}} f^{(\boldsymbol{s})}(\boldsymbol{x}) \right\|_{L_p(\mathscr{E})} dt \\ &\ll \frac{1}{\nu^{\varrho-|\boldsymbol{s}|}} \int_{\mathbb{R}_m} g(|\boldsymbol{t}|)\, |\boldsymbol{t}|^{\varrho-|\boldsymbol{s}|} \sum \left\| \Delta^{k+|\boldsymbol{s}|}_{\boldsymbol{t}/\nu} f^{(\varrho)} \right\|_{L_p(\mathscr{E})} \\ &\ll \frac{1}{\nu^{\varrho-|\boldsymbol{s}|}} \int_{\mathbb{R}_m} g(|\boldsymbol{t}|)\, |\boldsymbol{t}|^{\varrho-|\boldsymbol{s}|} \sum \Omega^{k+|\boldsymbol{s}|}_{\mathbb{R}_m}\left(f^{(\varrho)}, \frac{|\boldsymbol{t}|}{\nu}\right) dt \\ &\ll \frac{1}{\nu^{\varrho-|\boldsymbol{s}|}} \int_{\mathbb{R}_m} g(|\boldsymbol{t}|)\, |\boldsymbol{t}|^{\varrho-|\boldsymbol{s}|} (1 + |\boldsymbol{t}|^{k+|\boldsymbol{s}|})\, dt \sum \Omega^k_{\mathbb{R}_m}\left(f^{(\varrho)}, \frac{1}{\nu}\right) \\ &\ll \frac{1}{\nu^{\varrho-|\boldsymbol{s}|}} \int_0^\infty g(t)\, (1 + t)^{\varrho+k+m-1}\, dt \sum \Omega^k_{\mathbb{R}_m}\left(f^{(\varrho)}, \frac{1}{\nu}\right) \\ &\leqq \frac{c}{\nu^{\varrho-|\boldsymbol{s}|}} \sum_{|\boldsymbol{s}|=\varrho} \Omega^k_{\mathbb{R}_m}\left(f^{(\boldsymbol{s})}, \frac{1}{\nu}\right) \qquad (\nu > 0). \end{aligned} \tag{4}$$

The first inequality is obtained on the basis of 4.4 (6). Here $\frac{\partial^{\varrho-|s|}}{\partial t^{\varrho-|s|}}$ denotes the derivative of order $\varrho - |s|$ in the direction of t. The second is obtained in view of the fact that that derivative is a linear combination of ordinary derivatives (along the coordinate directions) of the same order with bounded coefficients, not depending on x, the sum being extended over all derivatives $f^{(\varrho)}$ of order ϱ. The third follows from the definition 4.2 (13). The fourth, in view of 4.2 (14) for $\varrho = 0$, f being replaced by $f^{(\varrho)}$. In the last inequality it is necessary to suppose g chosen so that 5.2.1 (1) is satisfied.

We note further that any derivative $g_\nu^{(\varrho)}$ relative to $u \in \mathbb{R}_m$ may be written (see 5.2.1 (4)) in the form

$$g_\nu^{(\varrho)}(x) = \int_{\mathbb{R}_m} g(|t|) \sum_{j=1}^{l} d_j f_u^{(\varrho)}\left(u + \frac{jt}{\nu}, y\right) dt,$$

so that

$$\|\Delta_h^k g_\nu^{(\varrho)}(x)\|_{L_p(\mathscr{E})} = \int_{\mathbb{R}_m} g(|t|) \sum_{j=1}^{l} |d_j| \left\| \Delta_h^k f^{(\varrho)}\left(x + \frac{jt}{\nu}\right)\right\|_{L_p(\mathscr{E})} dt$$

$$\leqq c \int_{\mathbb{R}_m} g(|t|)\, dt\, \omega_{R_m}^{(k)}(|h|), f^{(\varrho)})_{L_p(\mathscr{E})} \leqq c_1 \omega_{\mathbb{R}_m}^{(k)}(|h|, f^{(\varrho)})_{L_p(\mathscr{E})}$$

or

$$\Omega_{\mathbb{R}_m}^k(g_\nu^{(\varrho)}, \delta) \leqq c\Omega_{\mathbb{R}_m}^k(f^{(\varrho)}, \delta), \tag{5}$$

where c does not depend on ν and f.

This shows that the differential properties of f are transmitted to g_ν uniformly relative to ν.

5.2.3. We turn again to the kernel 5.2.1 (5), which will interest us this time for $m = 1$. Thus we shall suppose that $\mathscr{E} = \mathbb{R}_1 \times \mathscr{E}' \subset \mathbb{R}_n$.

Put

$$K_{\nu,l}(u) = \sum_{j=1}^{l} d_j \frac{\nu}{j} g\left(\frac{u\nu}{j}\right) \quad (l = \varrho + k, -\infty < u < \infty). \tag{1}$$

In view of 5.2.1 (7)

$$\left\|f - \int K_{\nu,l}(t - x_1) f(t, y) dt\right\|_{L_p(\mathscr{E})} \leqq \frac{b_l}{\nu^\varrho} \omega_{x_1}^k \left(f_{x_1}^{(\varrho)}, \frac{1}{\nu}\right)_{L_p(\mathscr{E})} \tag{2}$$

of course under the assumption that the right side of (2) has meaning. Of course, we suppose as in 5.2.1 that the positive even entire function

$g(t)$ of exponential type 1 in its one variable is chosen so that condition 5.2.1 (1) is satisfied for $m = 1$, which guarantees the estimate (2). We emphasize that it follows in any case from condition 5.2.1 (1) that

$$\int_{-\infty}^{\infty} K_{\nu,l}(t)dt = 1, \tag{3}$$

$$\int_{-\infty}^{\infty} |K_{\nu,l}(t)|\, dt \leqq c_l < \infty \quad (\nu > 0), \tag{4}$$

where c_j does not depend on $\nu > 0$.

Now suppose that $\mathscr{E} = \mathbb{R}_m \times \mathscr{E}' \subset \mathbb{R}_n$, $g(\boldsymbol{x})$ is a function which for almost all $\boldsymbol{y} \in \mathscr{E}'$ is entire of exponential type $\boldsymbol{\nu} = (\nu_1, \ldots, \nu_m)$ relative to $(x_1, \ldots, x_m)$. As always we will denote it by

$$g_\nu = g_{\nu_1, \ldots, \nu_m}(x_1, \ldots, x_m).$$

This definition has a meaning when the ν_i, $i = 1, \ldots, m$, are positive finite numbers. We extend this definition to the case when some of the ν_i, not all, are equal to ∞. Indeed, if $\nu_i = \infty$ for some $i \leqq m$, we will suppose that the function g is not necessarily entire relative to the variable x_i. For example, $g_{\nu_1, \nu_2, \infty, \ldots, \infty}$, where ν_1, ν_2 are finite, denotes that this function, for almost all, in the sense of $(n-2)$-dimensional measure, $(x_3, \ldots, x_n)$, is relative to x_1, x_2 an entire function of exponential types ν_1, ν_2 respectively.

5.2.4. Theorem. *Suppose that ϱ_j, $k_j (j = 1, \ldots, m)$ are natural numbers, $l_j = \varrho_j + k_j$, $f(\boldsymbol{x}) = f(\boldsymbol{u}, \boldsymbol{y}) \in L_p(\mathscr{E})$, and that there is given a system of functions*

$$g_{\nu_1, \infty, \ldots, \infty}(\boldsymbol{x}) = \int_{-\infty}^{\infty} K_{\nu_1, l_1}(u)\, f(x_1 + u, x_2, \ldots, x_m; \boldsymbol{y})\, du, \tag{1}$$

$$g_{\nu_1, \nu_2, \infty, \ldots, \infty}(\boldsymbol{x}) = \int_{-\infty}^{\infty} \int_{-\infty}^{\infty} K_{\nu_1, l_1}(u_1) K_{\nu_2, l_2}(u_2)$$

$$\times f(x_1 + u_1, x_2 + u_2, x_3 \ldots, x_m; \boldsymbol{y}) du_1\, du_2,$$

. . . .

$$g_{\nu_1, \ldots, \nu_m}(\boldsymbol{x}) = \int_{-\infty}^{\infty} \cdots \int_{-\infty}^{\infty} K_{\nu_1, l_1}(u_1) \cdots$$

$$\cdots K_{\nu_m, l_m}(u_m)\, f(x_1 + u_1, \ldots, x_m + u_m; \boldsymbol{y})\, du_1, \ldots, du_m,$$

where $\nu_j > 0$.

Then each of the functions $g_{\nu_1,\ldots,\nu_l,\infty,\ldots,\infty}$ *(obviously) belongs to* $L_p(\mathscr{E})$ *and is entire of exponential type* $\nu_1, \ldots, \nu_l$ *relative to* $x_1, \ldots, x_l$ *respectively* $(1 \leqq l \leqq m)$. *Moreover, the following inequalities hold:*

$$\|f - g_{\nu_1,\infty,\ldots,\infty}\|_{L_{p1}(\mathscr{E})} \leqq \frac{c\omega_{x_1}^{k_1}\left(f_{x_1}^{(\varrho_1)}, \frac{1}{\nu_1}\right)_{L_{p1}(\mathscr{E})}}{\nu_1^{\varrho_1}},$$

(2) $$\|g_{\nu_1,\infty,\ldots,\infty} - g_{\nu_1,\nu_2,\infty,\ldots,\infty}\|_{L_{p2}(\mathscr{E})} \leqq \frac{c\omega_{x_2}^{k_2}\left(f_{x_2}^{\varrho_2}, \frac{1}{\nu_2}\right)_{L_{p2}(\mathscr{E})}}{\nu_2^{\varrho_2}},$$

. . . .

$$\|g_{\nu_1,\ldots,\nu_{m-1},\infty} - g_{\nu_1,\ldots,\nu_m}\|_{L_{pm}(\mathscr{E})} \leqq \frac{c\omega_{x_m}^{k_m}\left(f_{x_m}^{(\varrho_m)}, \frac{1}{\nu_m}\right)_{L_{pm}(\mathscr{E})}}{\nu_m^{\varrho_m}},$$

(3) $$\|g_\nu\|_{L_{p_j}(\mathscr{E})} \leqq c\|f\|_{L_{p_j}(\mathscr{E})},$$

(4) $$\omega_{x_j}^{k_j}\left(g_{\nu x_j}^{(\varrho_j)}, \delta\right)_{L_{p_j}(\mathscr{E})} \leqq c\omega_{x_j}^{k_j}\left(f_{x_j}^{(\varrho_j)}, \delta\right)_{L_{p_j}(\mathscr{E})} \quad (j = 1 \ldots, m).$$

From (1), in particular, it obviously follows for $p = p_1 = \ldots = p_m$ that

(5) $$\|f - g_\nu\|_{L_p(\mathscr{E})} \leqq c \sum_{j=1}^{m} \frac{\omega_{x_j}^{k_j}\left(f_{x_j}^{(\varrho_j)}, \frac{1}{\nu_j}\right)_{L_p(\mathscr{E})}}{\nu_j^{\varrho_j}}.$$

Proof. We shall carry out the proof of the theorem in the case $m = 3$; for $m > 3$ it is analogous.

The first inequality (2) is obtained on the basis of 5.2.3.(2):

$$\|f - g_{\nu_1,\infty,\ldots,\infty}\|_{L_p(\mathscr{E})} \leqq \frac{b_{l_1}}{\nu_1^{\varrho_1}} \omega_{x_1}^{(k)}\left(f_{x_2}^{(\varrho_1)}, \frac{1}{\nu_1}\right)_{L_p(\mathscr{E})}.$$

The second inequality (2) is obtained using the following computations:

(6) $$\|g_{\nu_1,\infty,\infty} - g_{\nu_1,\nu_2,\infty}\|_{L_{p2}(\mathscr{E})}$$

$$= \left\| \int_{-\infty}^{\infty} K_{\nu_1,l_1}(u_1)\, h_1(x_1 + u_1, x_2, x_3; \mathbf{y}) du_1 \right\|_{L_{p2}(\mathscr{E})},$$

where

$$h_1(x_1, x_2, x_3; \mathbf{y})$$

$$= f(x_1, x_2, x_3; \mathbf{y}) - \int_{-\infty}^{\infty} K_{\nu_2,l_2}(u_2)\, f(x_1, x_2 + u_2, x_3; \mathbf{y})\, du_2 .$$

In view of 5.2.3 (2)

$$\|h_1\|_{L_{p_2}(\mathscr{E})} \leqq \frac{b_{l_2} \omega_{x_2}^{k_2}\left(f_{x_2}^{(\varrho_2)}, \frac{1}{\nu_2}\right)}{\nu_2^{\varrho_2}} .$$

Then, applying the generalized Minkowski inequality to (6), we find (see further 5.2.3 (4))

$$\|g_{\nu_1,\infty,\infty} - g_{\nu_1,\nu_2,\infty}\|_{L_{p_2}(\mathscr{E})}$$

$$\leqq \int_{-\infty}^{\infty} |K_{\nu_1,l_1}(u_1)|\, \|h_1(x_1 + u_1, x_2, x_3; \mathbf{y})\|_{L_{p_2}(\mathscr{E})}\, du_1$$

$$= \|h_1\|_{L_{p_2}(\mathscr{E})} \int_{-\infty}^{\infty} |K_{\nu_1,l_1}(u_1)|\, du_1 \leqq \frac{c_{l_1} b_{l_2} \omega_{x_2}^{k_2}\left(f_{x_2}^{(\varrho_2)}, \frac{1}{\nu_2}\right)}{\nu_2^{\varrho_2}} .$$

Finally, the third inequality (2) is obtained by means of the following considerations:

$$(g_{\nu_1,\nu_2,\infty} - g_{\nu_1,\nu_2,\nu_3})\,(x_1, x_2, x_3; \mathbf{y})$$

$$= \int_{-\infty}^{\infty}\int_{-\infty}^{\infty} K_{\nu_1,l_1}(u_1)\, K_{\nu_2 l_2}(u_2)\, h_2(x_1 + u_1, x_2 + x, x_3; \mathbf{y})\, du_1\, du_2 ,$$

where

$$h_2(x_1, x_2, x_3; \mathbf{y}) = f(x_1, x_2, x_2; \mathbf{y}) - \int_{-\infty}^{\infty} K_{\nu_3,l_3}(u)\, f(x_1, x_2, x_3 + u; \mathbf{y})\, du .$$

Therefore, applying the generalized Minkowski inequality and relations 5.2.3 (2) and (4), we get

$$\|g_{\nu_1,\nu_2,\infty} - g_{\nu_1,\nu_2,\nu_3}\|_{L_{p_3}(\mathscr{E})} \leqq \frac{c_{l_1} c_{l_2} b_{l_3} \omega_{x_3}^{k_3}\left(f_{x_3}^{(\varrho_3)}, \frac{1}{\nu_3}\right)}{\nu_3^{\varrho_3}} .$$

Inequalities (2) are proved. Inequality (3) (for $m = 3$) is obtained immediately, if one applies the generalized Minkowski inequality to the integral in the right side of the last equation (1):

$$\|g_{\nu_1,\nu_2,\nu_3}\|_{L_{p_j}(\mathscr{E})} \leqq \int_{-\infty}^{\infty}\int_{-\infty}^{\infty}\int_{-\infty}^{\infty} |K_{\nu_1,l_1}(u_1)\, K_{\nu_2,l_2}(u_2)\, K_{\nu_3,l_3}(u_3)|\, du_1 du_2 du_3 \|f\|_{L_{p_i}(\mathscr{E})}$$

$$\leqq c_{l_1} c_{l_2} c_{l_3} \|f\|_{L_{p_j}(\mathscr{E})} \quad (i = 1, 2, 3).$$

Finally, if one differentiates the last equation (1) ϱ_1 times relative to x_1 and applies the k_1^{st} difference relative to x_1, then in view of the Minkowski inequality we find that

$$\left\| \Delta_{x,h}^k \frac{\partial^{\varrho_1} g_\nu}{\partial x_1^{\varrho_1}} \right\|_{L_{p_1}(\mathscr{E})} \leqq \int_{-\infty}^{\infty}\int_{-\infty}^{\infty}\int_{-\infty}^{\infty} K_{\nu_1,l_1}(u_1)\, K_{\nu_2,l_2}(u_2)\, K_{\nu_3,l_3}(u_3)$$

$$\times \|\Delta_{x_1,h}^{k_1} f_{x_1}^{(\varrho_1)}(x_1 + u_1, x_2 + u_2, x_3 + u_3, \mathbf{y})\|_{L_p(\mathscr{E})}\, du_1 du_2 du_3$$

$$\leqq \int_{-\infty}^{\infty}\int_{-\infty}^{\infty}\int_{-\infty}^{\infty} |K_{\nu_1,l_1} K_{\nu_2,l_2} K_{\nu_3,l_3}|\, du_1 du_2 du_3 \omega_{x_1}^{(k_1)}(|h|, f_{x_1}^{(\varrho_1)}),$$

from which (4) follows. For $i = 2, 3$ the proof is analogous.

5.3. Periodic Classes

The theorems of § 5.2 remain valid, with some alterations in the proofs, if in their formulations the functions being considered are taken to be of period 2π, and the approximating entire functions g_ν are replaced by trigonometric polynomials T_ν. As always, in this case we need to replace the norms $\|\cdot\|_{L_p(\mathscr{E})}$ ($\mathscr{E} = \mathbb{R}_m \times \mathscr{E}_1 \subset \mathbb{R}_n$) by the norms $\|\cdot\|_{L_p(\mathscr{E}_*)}$ ($\mathscr{E}^* = \Delta^{(m)} \times \mathscr{E}_1$), where $\Delta^{(m)} = \{0 \leqq x_j \leqq 2\pi;\, j = 1, \ldots, m\}$. The first simplest direct approximation theorems were obtained in the periodic case. Indeed, Jackson showed that a periodic function of one variable of period 2π, having a continuous derivative of order r, may be approximated by trigonometric polynomials $T_n(x)$ $(n = 1, 2, \ldots)$ in such a way that the deviation in the uniform metric of C satisfies the inequality

$$|f(x) - T_n(x)| \leqq c_r \frac{\omega^*\left(f^{(r)}, \dfrac{1}{n}\right)}{n^r} \quad (n = 1, 2, \ldots),$$

where $\omega_*(f^{(r)}, \delta)$ is the modulus of continuity of $f^{(r)}$. The method of approximation of periodic functions by trigonometric polynomials which will be considered below is a modernized Jackson method. In the simplest cases (see 5.3.1 (6), (8), $l = 1$, $\sigma = 2$, $n = 1$) it coincides with the method of Jackson. On the other hand, it is an analogue of the method of 5.2.1 (4) considered above of approximation by entire functions of exponential type.

5.3.1. The first two equations 5.2.1 (4) for $m = 1$, $-\infty < x < \infty$ may be further written in the form

$$(1) \qquad g_\nu(x, y) = \int_{-\infty}^{\infty} \nu g(\nu t) \{(-1)^{l+1} \Delta_{x,t}^l f(x, y) + f(x, y)\} \, dt$$

$$= \int_{-\infty}^{\infty} \nu g(\nu t) \sum_{k=1}^{l} d_k f(x + kt, \nu) \, dt,$$

where

$$(2) \qquad d_k = (-1)^{k-1} C_l^k \quad (k = 1, \ldots, l)$$

and

$$q_\nu(t) = \nu g(\nu t)$$

is a nonnegative entire function of exponential type ν, satisfying (see 5.2.1 (1), (2)) the following condition:

$$(3) \qquad \int_{-\infty}^{\infty} q_\nu(t) \, dt = 1.$$

By analogy we introduce trigonometric polynomials $\tau_\nu(t)$ $(\nu = 0, 1, 2, \ldots)$ of order not higher than ν, having the following properties:

$$(4) \qquad \int_{-\infty}^{\infty} \tau_\nu(t) \, dt = 1,$$

$$(5) \qquad \int_{-\infty}^{\infty} |\tau_\nu(t)| \, dt \leqq c, \quad (\nu = 1, 2, \ldots),$$

where c is a constant not depending on ν.

Obviously

$$\tau_0(t) = \frac{2}{\pi}.$$

For $\nu > 0$ the polynomials τ_ν are defined nonuniquely.

One may obtain such polynomials, for example (see 2.2.2 (2)), by means of the formula

$$(6) \qquad d_\nu(t) = \frac{1}{a_\nu} \left(\frac{\sin \frac{\lambda t}{2}}{\sin \frac{t}{2}} \right)^{2\sigma},$$

where σ is a positive integer not depending on λ, the natural number λ satisfying the inequalities

(7) $$2(\lambda - 1)\,\sigma \leqq \nu < 2\lambda\sigma .$$

Here the constants a_ν are chosen so that equation (4) is satisfied. In the example (6) the polynomials $\tau_\nu(t)$ are nonnegative, so that property (5) follows automatically from property (4).

In analogy with (1) we define the function

(8) $$T_\nu(x, y) = \int_0^{2\pi} \tau_\nu(t)\,\{(-1)^{l+1}\,\Delta^l_{x,t} f(x, \mathbf{y}) + f(x, \mathbf{y})\}\,dt$$

$$= \int_0^{2\pi} \tau_\nu(t) \sum_{k=1}^{l} d_k f(x + kt, \mathbf{y})\,dt,$$

where this time $f(\mathbf{x}) = f(x, y)$ is defined on $\mathscr{E} = \mathbb{R}_1 \times \mathscr{E}' \subset \mathbb{R}_n$, periodic of period 2π relative to x, integrable to the pth power on $\mathscr{E}_* = [0, 2\pi] \times \mathscr{E}'$.

We note that

(8′) $$T_0(x, \mathbf{y}) = \frac{2}{\pi} \int_0^{2\pi} \sum_{k=1}^{l} d_k f(x + kt, \mathbf{y})\,dt$$

$$= \int_0^{2\pi} f(u, \mathbf{y})\,du \sum_{k=1}^{l} (-1)^{k+1} C_l^k = \int_0^{2\pi} f(u, \mathbf{y})\,du = T_0(\mathbf{y}).$$

Thus, for fixed $\mathbf{y}$ the function $T_0(x, \mathbf{y})$ is a constant representing the mean of the value of $f(x, \mathbf{y})$ over the period 2π.

In view of the periodicity of f one may write further:

(9) $$T_\nu(x, \mathbf{y}) = \sum_{k=1}^{l} \frac{d_k}{k} \int_0^{2k\pi} \tau_\nu\left(\frac{u}{k}\right) f(x + u, \mathbf{y})\,du$$

$$= \sum_{k=1}^{l} \frac{d_k}{k} \sum_{s=0}^{k-1} \int_{2s\pi}^{2(s+1)\pi} \tau_\nu\left(\frac{u}{k}\right) f(x + u, \mathbf{y})\,dc$$

$$= \sum_{k=1}^{l} \frac{d_k}{k} \sum_{s=0}^{k-1} \int_0^{2\pi} \tau_\nu\left(\frac{t + 2s\pi}{k}\right) f(x + t, \mathbf{y})\,dt = \int_0^{2\pi} K_\nu(t)\,f(x + t, \mathbf{y})\,dt,$$

where

$$K_\nu(t) = \sum_{k=1}^{l} \frac{d_k}{k} \sum_{s=0}^{k-1} \tau_\nu\left(\frac{t+2s\pi}{k}\right). \tag{10}$$

We shall show that the function $K_\nu(t)$ is a trigonometric polynomial of order not higher than ν, from which it follows that $T_\nu(x, \mathbf{y})$, for almost all $\mathbf{y}$, is also a trigonometric polynomial of order not higher than ν relative to x.

Indeed, the trigonometric polynomial τ_ν may be written out in the form of a certain linear combination

$$\tau_\nu(t) = \sum_{-\nu}^{\nu} a_\lambda e^{i\lambda t} \qquad (\bar{a}_\lambda = a_{-\lambda})$$

with constant coefficients a_λ.

But

$$\sum_{s=0}^{k-1} e^{i\lambda\frac{t+2s\pi}{k}} = e^{\frac{i\lambda t}{k}} \sum_{s=0}^{k-1} e^{i\frac{\lambda 2s\pi}{k}}$$

$$= \begin{cases} ke^{i\frac{\lambda t}{k}} = ke^{i\mu t} & \text{for } \dfrac{\lambda}{k} = \mu \text{ an integer} \\[2ex] 0 & \text{for } \dfrac{\lambda}{k} = \mu \text{ not an integer} \end{cases}$$

and, accordingly, the sum

$$\sum_{s=0}^{k-1} \tau_\nu\left(\frac{t+2s\pi}{k}\right) = \sum_{-\nu}^{\nu} a_\lambda \sum_{s=0}^{k-1} e^{i\lambda\frac{t+2s\pi}{k}}$$

is a trigonometric polynomial of order ν. But then K_ν is also a trigonometric polynomial of order ν.

It follows from (8) that

$$T_\nu - f = (-1)^{l+1} \int_0^{2\pi} \tau_\nu(t) \Delta_{x,t}^l f(x, \mathbf{y})\, dt, \tag{11}$$

so that by applying the generalized Minkowski inequality we obtain the inequality

$$\|T_\nu - f\|_{L_p(\mathscr{E}_*)} \leq \int_0^{2\pi} |\tau_\nu(t)|\, \|\Delta_{x,t}^l f(x, \mathbf{y})\|_{L_p(\mathscr{E}_*)}\, dt \tag{12}$$

We have the following theorem, reducing to an inequality analogous to 5.2.1 (7).

5.3.2. Theorem. *Suppose that* $1 \leqq p \leqq \infty$, $\mathscr{E} = \mathbb{R}_1 \times \mathscr{E}_1' \subset \mathbb{R}_n$, *and that the function* $f = f(x, \mathbf{y})$ $(x \in \mathbb{R}_1, \mathbf{y} \in \mathscr{E}_1)$ *is defined on* $\mathscr{E}$, *has period* 2π *relative to* x *for almost all* $\mathbf{y} \in \mathscr{E}'$, *and belongs to the class* $L_p(\mathscr{E}_*)$, $\mathscr{E}_* = [0, 2\pi] \times \mathscr{E}_{\mathscr{E}}'$. *Moreover suppose that* f *has on* $\mathscr{E}$ *a generalized derivative* $f_x^{(\varrho)} = \dfrac{\overset{*}{\partial^\varrho} f}{\partial x^\varrho}$ *of order* $\varrho (f_x^{(0)} = f)$. *Finally suppose that the even nonnegative trigonometric polynomials* $\tau_\nu(t)$ *of order* ν *satisfy along with the condition* 5.3.1 (4) *the further condition*

$$\int_0^\pi \tau_\nu(t) t^\varrho dt \leqq \frac{a_\varrho}{(\nu+1)^\varrho}, \tag{1}$$

where the constants a_ϱ *do not depend on* $\nu = 0, 1, 2, \ldots$. (*Such polynomials may be obtained relative to formula* 5.3.1 (6) *with the appropriate choice of* σ *and* λ.)

Then the function $T_\nu(x, \mathbf{y})$ *defined by equations* 5.3.1 (8) (*trigonometric polynomial of order* ν *relative to* x) *approximates* f *in the metric of* $L_p(\mathscr{E}_*)$ *with the following estimate*:

$$\|f - T_\nu\|_{L_p(\mathscr{E}_*)} \leqq b_\varrho \frac{\overset{k}{\omega}{}^*_{x, L_p(\mathscr{E})}\left(f_x^{(\varrho)}, \dfrac{\pi}{\nu+1}\right)}{(\nu+1)^\varrho}, \tag{2}$$

$$(\nu = 0, 1, \ldots)$$

where the b_ϱ *are constants not depending on* ϱ.

Proof. We already know that the trigonometric polynomials $d_\nu(t)$ defined by the relations 5.3.1 (6), (7) satisfy the conditions 5.3.1 (4). We verify that for $\nu \geqq 1$ they satisfy also inequality (1) for some constant a_{p+1}, under the hypothesis that $2\sigma - \varrho \geqq 3$. With this there will be established the existence of polynomials satisfying the conditions of the theorem. Indeed,

$$\int_0^\pi d_\nu(t)\, t^\varrho dt \leqq \frac{c_1}{a_\nu} \int_0^\pi \left(\frac{\sin \frac{\lambda t}{2}}{t}\right)^{2\sigma} t^\varrho dt \leqq \frac{c_2}{\lambda^\varrho} \int_0^\infty \frac{(\sin u)^{2\sigma}}{u^{2\sigma-\varrho}} du$$

$$\leqq \frac{c_3}{\lambda^\varrho} \leqq \frac{a_\varrho}{(\nu+1)^\varrho} \quad (\nu = 1, 2, \ldots),$$

where the last inequality follows from 5.3.1 (7).

We recall that the inequality

$$(3)\qquad \|\Delta_{x,t}^{\varrho+k} f(x, y)\|_{L_p^*(\mathscr{E})} \leqq |t|^\varrho \| \Delta_{x,t}^k f_x^{(\varrho)}(x, y)\|_{L_p^*(\mathscr{E})} \leqq |t|^\varrho \omega_*^k(|t|)$$

holds, where

$$\omega_*^k(\delta) = \overset{k}{\omega}{}^*_{x, L_p(\mathscr{E})}(f, \delta).$$

We note further that the inequality

$$(4)\qquad \omega_*^k(t) \leqq c(\nu+1)^k t^k \omega_*^k\left(\frac{\pi}{(\nu+1)}\right) \left(\frac{\pi}{\nu} \leqq t\right).$$

holds, which is proved in analogy with inequality 4.2 (8).

Now we apply inequality 5.3.1 (12) for $l = \varrho + k$, taking into account (3) and (4):

$$\|f - \tau_\nu\|_{L_p^*} \leqq \int_{-\pi}^{\pi} \tau_\nu(t) \, \|\Delta_{x,t}^{\varrho+k}(x, y)\|_{L_p^*} dt$$

$$\leqq \int_{-\pi}^{\pi} \tau_\nu(t) \, |t|^\varrho \omega_*^k(|t|) \, dt = 2 \int_0^{\pi} \tau_\nu(t) \, t^\varrho \omega_*^k(t) \, dt$$

$$= 2 \int_0^{\frac{\pi}{\nu+1}} \tau_\nu(t) \, t^\varrho \omega_*^k(t) \, dt + 2 \int_{\frac{\pi}{\nu+1}}^{\pi} \tau_\nu(t) \, t^\varrho \omega_*^k(t) \, dt$$

$$\leqq 2\omega_*^k\left(\frac{\pi}{\nu+1}\right) \left[\left(\frac{\pi}{\nu+1}\right)^\varrho + \frac{2}{\pi} (\nu+1)^k \int_{\frac{\pi}{\nu+1}}^{\pi} \tau_\nu(t) \, t^{\varrho+k} dt\right]$$

$$\leqq 2\omega_*^k\left(\frac{\pi}{\nu+1}\right) \left[\left(\frac{\pi}{\nu+1}\right)^\varrho + \frac{2}{\pi} \frac{a_\varrho + k}{(\nu+1)^\varrho}\right] = \frac{b_\varrho}{(\nu+1)^\varrho} \, \omega_*^k\left(\frac{\pi}{\nu+1}\right),$$

where

$$b_\varrho = 2\left(\pi^\varrho + \frac{2}{\pi} a_{\varrho+1}\right).$$

This proves the theorem.

Remark 1. For the trigonometric polynomial T_ν in question equation 5.3.1 (11) is satisfied, so that

$$T_\nu^{(\varrho)}(x, y) = f^\varrho(x, y) + (-1)^{l+1} \int_0^{2\pi} \tau_\nu(t) \, \Delta_{x,t}^l f_x^{(\varrho)}(x, y) \, dt.$$

and we have obtained an equation analogous to equation 5.2.2 (1). Reasoning as in 5.2.2. for $l = \varrho + k$, it is easy to obtain an inequality analogous to 5.2.2 (5):

$$(5) \qquad \overset{k}{\omega}{}^*_{x,L_p(\mathscr{E})}(T_\nu^{(\varrho)}, \delta) \leqq c\overset{k}{\omega}{}^*_{x,L_p(\mathscr{E})}(f^{(\varrho)}, \delta),$$

where the constant c does not depend on the factor standing beside it.

Remark 2. If the periodic function $f(x, \mathbf{y})$ is such that its mean value over the period is equal to zero, i.e.

$$\int_0^{2\pi} f(u, \mathbf{y})\, du = 0,$$

then $T_0 = 0$ (see 5.3.1 (8)), so that inequality (2) for $\nu = 0$ reduces to the following inequality:

$$(6) \qquad \|f\|_{L_p(\mathscr{E}_*)} \leqq b_\varrho \overset{k}{\omega}{}^*_{x,L_p(\mathscr{E})}(f_x^{(\varrho)}, \pi).$$

5.3.3. As in 5.2.3, we may define functions $T_{\nu_1,\ldots,\nu_m}(x_1, \ldots, x_n)$, analogous to the $g_{\nu_1,\ldots,\nu_m}(\mathbf{x})$, given on the measurable set $\mathscr{E} = \mathbb{R}_m \times \mathscr{E}_1 \subset \mathbb{R}_n$, which are trigonometric polynomials for almost all $\mathbf{y} = (x_{m+1}, \ldots, x_n) \in \mathscr{E}'$ relative to the variables $x_1, \ldots, x_m$, of orders $\nu_1, \ldots, \nu_m$ respectively. As in 5.2.3 we will admit that some of the ν_k, $k = 1, \ldots, m$, may be equal to ∞.

Suppose that $\mathbf{r} = (r_1, \ldots, r_m) > 0$. We define even nonnegative trigonometric polynomials $\tau_{\nu,r_j}(t)$ of order ν, satisfying the conditions of Theorem 5.3.2 for $\varrho = r_1, \ldots, \varrho = r_m$ respectively. Thus, they satisfy the conditions

$$(1) \qquad \int_0^{2\pi} \tau_{\nu,r_j}(t)\, dt = 1,$$

$$(2) \qquad \int_0^{2\pi} \tau_{\nu,r_j}(t)\, dt \leqq \frac{a_{r_j}}{(\nu+1)^{r_j}} \qquad (j = 1, \ldots, m;\ \nu = 1, 2, \ldots).$$

Further we define trigonometric polynomials K_{ν,r_j} (kernels) of orders ν, by the formulas

$$K_{\nu,r_j}(t) = \sum_{q=1}^{l_j} \frac{d_q}{q} \sum_{s=0}^{q-1} \tau_{\nu,r_j}\left(\frac{t + 2s\pi}{q}\right) \qquad (k = 1, \ldots, m)$$

(see 5.3.1), where $l_j \geqq r + k$.

Further we put

$$T_{\nu_1,\infty,\ldots,\infty}(\boldsymbol{x}) = \int_0^{2\pi} K_{\nu_1 r_1}(u)\, f(x_1+u, x_2, \ldots, x_m; \boldsymbol{y})\, du,$$

. . . .

$$T_{\nu_1,\ldots,\nu_m}(\boldsymbol{x})$$

$$= \int_0^{2\pi} \cdots \int_0^{2\pi} K_{\nu_1 r_1}(u_1) \cdots K_{\nu_m r_m}(u_m)\, f(x_1+u_1, \ldots, x_m+u_m; \boldsymbol{y})\, du_1 \cdots du_m.$$

For the indicated family $T_{\nu_1,\ldots,\nu_m}$ of functions f one may formulate and prove a theorem analogous to Theorem 5.2.4, a generalization of Jackson's theorem.

In particular, it follows from it that if $f \in H^{r}_{\boldsymbol{u}p}{}^{*}(\mathscr{E})$, then

(3) $$E^*_{x_1\nu}(f)_p \leqq \frac{c\|f\|^{r}_{h^*_{x_1,p}}}{(\nu+1)^r},$$

where c does not depend on the factor standing beside it.

5.4. Inverse Theorems of the Theory of Approximations

In this section we will explain a schema, following which we may obtain inverse theorems of the theory of approximations, indicating that a function belongs to one or another class, if estimates of its approximations are known.

We prove a general theorem, at the basis of which there lies the inverse theorem of the theory of approximations (for trigonometric polynomials and entire functions of exponential type), due to S. N. Bernšteĭn[1].

5.4.1. Theorem. *Suppose that* $\mathbb{R}_n$ *is an n-dimensional space of points* $\boldsymbol{x} = (\boldsymbol{u}, \boldsymbol{y})$, $\boldsymbol{u} = (x_1, \ldots, x_m)$, $\boldsymbol{y} = (x_{m+1}, \ldots, x_n)$ *and* $\mathbb{R}_m = (\boldsymbol{u}, 0)$ *is an m-dimensional subspace of it* ($1 \leqq m \leqq n$). *Suppose further that* $r > 0$, k *is a natural number,* $1 \leqq p \leqq \infty$, *and* $\mathfrak{M}_\nu$ *is a linear set of functions, defined on an open set* $g \subset \mathbb{R}_n$ *and depending on a parameter* $\nu \geqq 1$, *while*

(1) $$\mathfrak{M}_\nu \subset \mathfrak{M}_{\nu'} \quad (\nu < \nu').$$

We suppose that each function $\tau_\nu \in \mathfrak{M}_\nu$ *has the following properties:* τ_ν *has on* g *all derivatives relative to* $\boldsymbol{u}$ *of orders less than* $r+k$, *and the following*

[1] S. N. Bernšteĭn [1], pages 11–104.

inequalities hold:

(2) $$\|\tau_\nu^{(s)}\|_{L_p(g)} \leqq c\nu^{|s|}\|\tau_\nu\|_{L_p(g)},$$

$$s = (s_1, \ldots, s_m, 0, \ldots, 0), \ |s| < r + k,$$

where the constant c does not depend on ν.

Suppose moreover for a given function $f \in L_p(g)$ *that there exists a family of functions* $\tau_\nu \in \mathfrak{M}_\nu$, *depending on* ν, *such that*

(3) $$\|f - \tau_\nu\|_{L_p(g)} \leqq \frac{K}{\nu^r} \quad (\nu \geqq 1),$$

where K does not depend on ν.

Then $f \in H^r_{up}(g)$ (see 4.3.3) *and the inequalities*

(4) $$\|f^{(\varrho)}\|_{L_p(g)} \leqq A\,(\|f\|_{L_p(g)} + K)$$

are satisfied for all the derivatives $f^{(\varrho)}$ *of f of order* $\varrho < r$ *and*

(5) $$\|f\|_{H^r_{up}(g)} \leqq A\,(\|f\|_{L_p(g)} + K),$$

where A does not depend on the factor standing beside it.

Remark 1. The functions τ_ν may further be considered to be periodic in x_1 of period 2π, defined on $g = \mathscr{E} = \mathbb{R}_1 \times \mathscr{E}'$, then in (1)—(5) we have to replace $L_p(g)$, $H^r_{up}(g)$ by $L_{p*}(\mathscr{E})$, $H^r_{up*}(\mathscr{E})$.

Remark 2. We may suppose that ν runs through values $\nu = \nu(s)$, depending on $s = 0, 1, \ldots$, and satisfying the conditions

$$1)\ \nu(s) \geqq 1,$$

$$2)\ \nu(s) \to \infty \ (s \to \infty),$$

$$3)\ \frac{\nu(s+1)}{\nu(s)} \leqq \Lambda < \infty \ (s = 0, 1, \ldots),$$

where Λ does not depend on s. In particular, we may suppose that $\nu(s) = a^s$, $a > 1$.

Indeed, suppose that

$$\|f - \tau_{\nu(s)}\|_{L_p(g)} \leqq \frac{K}{\nu(s)^r} \quad (s = 0, 1, \ldots)$$

and $\nu_0 = \min \nu(s)$, $s = 0, 1, \ldots$.

If $1 \leqq \nu \leqq \nu_0$, then we put $\tau_\nu = 0$. Then $\|f - \tau_\nu\| = \|f\| \leqq \frac{\|f\|\nu_0^r}{\nu^r}$. If on the other hand $\nu > \nu_0$, then we choose s so that

$$\nu(s) \leqq \nu < \nu(s+1).$$

Since $\tau_{\nu(s)} \subset \mathfrak{M}_{\nu(s)} \subset \mathfrak{M}_\nu$, then we may suppose that $\tau_{\nu(s)} = \tau_\nu$, so that

$$\|f - \tau_\nu\|_{L_p(g)} \leqq \frac{K}{\nu(s)_r} \leqq \frac{K}{\nu^r}\left(\frac{\nu(s+1)}{\nu(s)}\right)^r \leqq \frac{KA^r}{\nu^r}, \qquad \nu \geqq 1.$$

Thus

$$\|f - \tau_\nu\| \leqq \frac{K_1}{\nu^r}, \quad \nu \geqq 1,$$

where

$$K_1 = \|f\|\nu_0^r + KA,$$

and inequality (3) is satisfied for all $\nu \geqq 1$, as required by the theorem. The conclusion of the theorem (see (4), (5)) does not change from the replacement of K by K_1, because $\|f\| + K_1 \ll \|f\| + K$.

Proof of Theorem 5.4.1. In view of (3) the function f may be represented in the form of a series

$$f = \sum_0^\infty Q_j, \tag{6}$$

where

$$Q_0 = \tau_1 = \tau_{2^0}, \ Q_j = \tau_{2^j} - \tau_{2^{j-1}} \ (j = 1, 2, \ldots),$$

converging in the sense of $L_p(g)$. Here $(\|\cdot\|_{L_p(g)} = \|\cdot\|)$,

$$\|Q_0\| = \|\tau_1\| \leqq K + \|f\|, \tag{7}$$

$$\|Q_j\| \leqq \|\tau_{2^j} - f\| + \|f - \tau_{2^{j-1}}\|$$

$$\leqq \frac{K}{2^{jr}} + \frac{K}{2^{(j-1)r}} = \frac{c_1K}{2^{jr}} \quad (j = 1, 2, \ldots).$$

We choose any derivative of f of order ϱ, mixed or nonmixed:

$$f^{(\varrho)} = \sum_0^\infty Q_j^{(\varrho)}, \ |\varrho| = \varrho. \tag{8}$$

Since the sets $\mathfrak{M}_{2^j}$ are linear and $\mathfrak{M}_{2^{j-1}} \subset \mathfrak{M}_{2^j}$ $(j = 1, 2, \ldots)$, then $Q_j \in \mathfrak{M}_{2^j}$, and on the basis of the estimate (2) we obtain the inequalities

$$\|Q_0^{(\varrho)}\| \leqq c\|Q_0\| \leqq c_1(\|f\| + K),$$

$$\|Q_j^{(\varrho)}\| \leqq c2^{\varrho j}\|Q_j\| \leqq \frac{c_1K}{2^{j(r-\varrho)}}, \tag{9}$$

which show that the formal term-by-term differentiation (8) of the series (6) is legitimate for $\varrho < r$ and that the series (8) converges in the sense of $L_p(g)$ to $f^{(\varrho)}$ (see Lemma 4.4.7). In addition $f \in W^{\varrho}_{up}(g)$, and inequality (4) holds.

Suppose given a vector $\boldsymbol{h} = (h_1, \ldots, h_m, 0, \ldots, 0) \in \mathbb{R}_m$. Choose a natural number N so that

$$(10) \qquad \frac{1}{2^{n+1}} < |\boldsymbol{h}| \leq \frac{1}{2^N}, \quad |\boldsymbol{h}|^2 = \sum_{j=1}^{m} h_j^2.$$

Consider the kth difference of the function $f^{(\varrho)}$, corresponding to the shift $\boldsymbol{h}$. Taking account of the equation

$$\Delta_h \varphi(\boldsymbol{x}) = |\boldsymbol{h}| \int_0^1 \frac{\partial \varphi}{\partial \boldsymbol{h}} (\boldsymbol{x} + t\boldsymbol{h})\, dt,$$

we get

$$(11) \quad \Delta_{\boldsymbol{h}}^k f^{(\varrho)}(\boldsymbol{x}) = \sum_0^N |\boldsymbol{h}|^k \int_0^1 \cdots \int_0^1 \frac{\partial^k}{\partial \boldsymbol{h}^k} Q_j^{(\varrho)}\big(\boldsymbol{x} + \boldsymbol{h}(u_1 + \cdots + u_k)\big)\, du_1 \cdots du_k$$
$$+ \sum_{N+1}^{\infty} \Delta_h^k Q_j^{(\varrho)}(\boldsymbol{x}),$$

where $\boldsymbol{x} \in g_{kh}$. Obviously,

$$(12) \qquad \|\Delta_{\boldsymbol{h}}^k f^{(\varrho)}(\boldsymbol{x})\|_{L_p(g_{k\boldsymbol{h}})} \leq |\boldsymbol{h}|^k \sum_0^N \left\| \frac{\partial^k}{\partial \boldsymbol{h}^k} Q_j^{(\varrho)} \right\|_{L_p(g_{k\boldsymbol{h}})}$$
$$+ 2^k \sum_{N+1}^{\infty} \|Q_j^{(\varrho)}\|_{L_p(g)} = J_1 + J_2.$$

Recalling that the derivative $\dfrac{\partial^k}{\partial \boldsymbol{h}^k}$ is a finite linear combination of ordinary derivatives with respect to the coordinates $x_1, \ldots, x_m$, and using inequalities (2), (9), (10), we get

$$(13) \qquad I_1 \leq |\boldsymbol{h}|^k c_4(\|f\| + K) \sum_0^N 2^{[k-(r-\varrho)]j}$$
$$\leq c_5 |\boldsymbol{h}|^k (\|f\| + K) 2^{[k-(r-\varrho)]N} \leq c_6(\|f\| + K)|\boldsymbol{h}|^{r-\varrho}.$$

It is important to note that we are supposing that $k > r - \varrho$. If we had supposed that $k = r - \varrho$, then the sum

$$\sum_0^N 2^{[k-(r-\varrho)]j} = N + 1$$

would not have been of order $2^{[k-(r-\varrho)]N} = 1$.

Further

(14) $$I_2 \leqq c_7 K \sum_{N+1}^{\infty} \frac{1}{2^{(r-\varrho)j}} \leqq c_8 K |\boldsymbol{h}|^{r-\varrho}.$$

(5) follows from (12)—(14) and (4) for $\varrho = 0$.

5.4.2. Theorem (inverse to 5.2.1 (8))[1]. *Suppose $r > 0$, $1 \leqq p \leqq \infty$, $1 \leqq m \leqq n$, $\mathscr{E} = \mathbb{R}_m \times \mathscr{E}'$ and $f \in L_p(\mathscr{E})$.*

If for the best approximation to f in the metric of $L_p(\mathscr{E})$ by means of entire functions of exponential spherical type ν the inequality

(1) $$E_{u\nu}(f)_{L_p(\mathscr{E})} \leqq \frac{K}{\nu^r} \quad (\nu \geqq 1),$$

is satisfied, where K does not depend on ν (ν may run through the values $\nu = \nu(s)$, satisfying the conditions of Remark 2 in 5.4.1, $s = 0, 1, \ldots$), *then $f \in H_p^r(\mathscr{E})$ and*

(2) $$\|f\|_{H_p^r(\mathscr{E})} \leqq A(\|f\|_{L_p(\mathscr{E})} + K),$$

(3) $$\|f\|_{L_p(\mathscr{E})} \leqq A(\|f\|_{L_p(\mathscr{E})} + K), \quad (|\boldsymbol{\varrho}| = 0, 1, \ldots, \bar{r})$$

where A does not depend on the factor standing beside it.

Proof. From the hypothesis it follows that there exists a family of functions $g_\nu(\boldsymbol{u}, \boldsymbol{y})$ $(\boldsymbol{u} \in \mathbb{R}_m, \boldsymbol{y} \in \mathscr{E}')$ of exponential spherical type ν relative to $\boldsymbol{u}$ (for almost all $\boldsymbol{y} \in \mathscr{E}'$), such that

$$\|f - g_\nu\|_{L_p(\mathscr{E})} \leqq 2E_\nu(f)_{L_p(\mathscr{E})} \leqq \frac{2K}{\nu^r}.$$

But then the assertion of the theorem follows directly from Theorem 5.4.1, if one recalls that the g_ν are also functions of exponential type ν relative to each of the variables $x_1, \ldots, x_m$, and therefore they satisfy the inequality (see 3.3.2 (9))

$$\|g_\nu^{(\boldsymbol{k})}\|_{L_p(\mathscr{E})} \leqq \nu^{|\boldsymbol{k}|} \|_{L_p(\mathscr{E})},$$

for any derivative of order $\boldsymbol{k} = (k, \ldots, k_m, 0, \ldots, 0)$. The case when $\nu = \nu(s)$ runs through the values described in Remark 2 in 5.4.1 reduces, according to that same remark, to the case of a continuously varying ν.

5.4.3. Theorem (Inverse to 5.3.3 (3))[2]. *Suppose that $r > 0$, $1 \leqq p \leqq \infty$, $\mathscr{E} = \mathbb{R}_1 \times \mathscr{E}' \subset \mathbb{R}_n$ and that the function $f(\boldsymbol{x}) = f(x_1, \boldsymbol{y})$ is, for almost all $\boldsymbol{y} \in \mathscr{E}'$), periodic of period 2π relative to the variable x_1, and that it belongs to $L_{\underset{p}{*}}(\mathscr{E})$.*

---

[1] For $m = 1$, $p = \infty$, S. N. Bernšteĭn [1], pp. 431—432; for $m = n = 1$, $1 \leqq p \leqq \infty$, N. I. Ahiezer [1]; $m = 1$, $1 \leqq p \leqq \infty$, S. M. Nikol'skiĭ [3].

[2] See the remark at the end of the book to 5.4.

If for the best approximation to f in the metric of $L_{\underset{p}{*}}(\mathscr{E})$ by means of the functions $T_\nu(x_1, \mathbf{y})$, being for almost all $\mathbf{y} \in \mathscr{E}'$ trigonometric polynomials of order ν, the inequality

$$E_{x_1\nu}*(f)_{L_p(\mathscr{E})} \leqq \frac{K}{(\nu+1)^r} \qquad (\nu = 0, 1, \ldots), \tag{1}$$

is satisfied, then $f \in H^r_{x_1p}*(\mathscr{E})$ and

$$\|f^{(\varrho)}_{x_1}\|_{L_p^*(\mathscr{E})} \leqq AK \qquad (1 \leqq \varrho < r), \tag{2}$$

$$\|f\|_{h^r_{*\,x_1p}} = Mf \leqq AK, \tag{3}$$

where A does not depend on K.

In the hypotheses of the theorem it suffices to suppose that the numbers ν run through the values described in Remark 2 in 5.4.1.

This theorem is analogous to Theorem 5.4.2. But historically it appeared earlier than the latter[1], first of all for $p = \infty$, $n = 1$. Its hypothesis (1) for $\nu = 0$ contains an inequality for the best approximation of f by means of a constant, which reduces to inequalities (2), (3) with right sides depending only on K (cf. 5.4.2 (2), (3)).

5.4.4. Harmonic polynomials. Consider the harmonic polynomial

$$\Phi_n(\varrho, \theta) = \sum_0^n \varrho^k(a_k \cos k\theta + b_k \sin k\theta) \tag{1}$$

$$(0 \leqq \varrho \leqq 1, -\infty < \theta < \infty).$$

Write

$$\|\Phi\|_{L_p(\sigma\mathrm{R})} = \int_{\mathrm{R}}^1 \int_0^{2\pi} |\Phi(\varrho, \theta)|^p \varrho \, d\varrho \, d\theta \qquad (0 \leqq \mathrm{R} \leqq 1). \tag{2}$$

Theorem. *The following inequalities hold*[2]*:*

$$\left\| \frac{\partial^l \Phi_n}{\partial \varrho^l} \right\|_{L_p(\sigma\mathrm{R})} \leqq c_{\mathrm{R}} n^l \|\Phi_n\|_{L_p(\sigma\mathrm{R})}, \tag{3}$$

$$\left\| \varrho^l \frac{\partial^l \Phi_n}{\partial \varrho^l} \right\|_{L_p(\sigma\mathrm{R})} \leqq c n^l \|\Phi_n\|_{L_p(\sigma_0)}, \tag{4}$$

$$\left\| \frac{\partial^l \Phi_n}{\partial \theta^l} \right\|_{L_p(\sigma\mathrm{R})} \leqq n^l \|\Phi_n\|_{L_p(\sigma\mathrm{R})}$$

$$(n, l = 1, 2, \ldots, 1 \leqq p \leqq \infty). \tag{5}$$

---

[1] S. N. Bernšteĭn [1], pages 11–104. See the remark at the end of the book to 5.4.

[2] A. L. Šaginjan [1] for $p = \infty$, Ja. S. Bugrov [2] for the general case.

Proof. Put

$$x(x-1)\cdots(x-l+1)=\sum_{i=1}^{l}\lambda_i x^i,$$

where the λ_i are thus numbers depending on i and l. Then

$$\frac{\partial^l \Phi_n}{\partial \varrho^l}=\frac{1}{\varrho^l}\sum_{k=1}^{n}k(k-1)\cdots(k-l+1)\,\varrho^k(a_k\cos k\theta+b_k\sin k\theta)$$

$$=\frac{1}{\varrho^l}\sum_{k=1}^{n}\sum_{i=1}^{l}\lambda_i k^i\varrho^k(a_k\cos k\theta+b_k\sin k\theta)=\frac{1}{\varrho^l}\sum_{i=1}^{l}\Psi_n^{(i)}(\varrho,\theta),$$

where

$$\Psi_n^{(i)}(\varrho,\theta)=\frac{\lambda_i}{\varrho^l}\sum_{k=1}^{n}k^i\varrho^k(a_k\cos k\theta+b_k\sin k\theta).$$

We have

$$\Psi_n^{(i)}(\varrho,\theta)=\frac{\lambda_i}{\pi\varrho^l}\int_0^{2\pi}\Phi_n(\varrho,u+\theta)\sum_{k=1}^{n}k^l\cos ku\,du \tag{6}$$

$$=\frac{2\lambda_i n^l}{\pi\varrho^l}\int_0^{2\pi}\cos nu\Phi_n(\varrho,u+\theta)\,\chi_n^{(i)}(u)\,du,$$

where $\chi_n^{(i)}(u)=\sum_0^{n-1}{}'\left(1-\frac{k}{n}\right)^i\cos ku$, the prime denoting that the term corresponding to $k=0$ is equal to $\frac{1}{2}$. Here we need to recall that $ku=\frac{1}{2}(\cos(n+k)u+\cos(n-k)u)$ and that the terms $\cos(n+k)u$ obtained in this way for $k>0$ are orthogonal on the segment $(0,2\pi)$ of variation of u to the function $\Phi_n(\varrho,\theta+u)$, which is a trigonometric polynomial of order n relative to u.

Applying Abel's transformation, we get

$$\chi_n^{(i)}(u)=\sum_{k=0}^{n-3}(k+1)\Delta^2\left(1-\frac{k}{u}\right)^i F_k(u)$$

$$+\frac{1}{n^{i-1}}F_{n-1}(u)+\frac{n-1}{n^i}(2^i-2)F_{n-2}(u),$$

where $F_k(u)$ is the Féjer kernel (see (2.2.2 (1)) $\Delta^2\mu_k=\mu_k-2\mu_{k+1}+\mu_{k+2}$.

It is essential to note that $F_k(u)\geqq 0$ and $\Delta^2\left(1-\frac{k}{n}\right)^i\geqq 0$, as a consequence of which $\chi_n^{(i)}(u)\geqq 0$ and $\frac{1}{\pi}\int_0^{2\pi}|\chi_n^{(j)}(u)|\,du=1$ $(i=1,2,\ldots)$.

Applying the generalized Minkowski inequality to (6), we get

$$\|\Psi_n^{(i)}\|_{L_p(\sigma\mathbb{R})} \leqq 2|\lambda_i| n^i \mathbb{R}^{-l} \|\Phi_n\|_{L_p(\sigma\mathbb{R})} \frac{1}{\pi} \int_0^{2\pi} |\chi_n^{(i)}(u)|\, du \leqq c_{\mathbb{R}} n^i \|\Phi_n\|_{L_p(\sigma\mathbb{R})},$$

from which follows (3) and (4). Inequality (5) follows from the fact that $\Phi_n(\varrho, \theta)$ is a trigonometric polynomial relative to θ of order n.

5.5. Direct and Inverse Theorems on Best Approximations. Equivalent H-Norms

In this section we compare the direct and inverse theorems proved above on best approximations. We shall see that functions of the classes H are completely characterized by the behavior of their best approximations. As throughout this Chapter, $\mathscr{E} = \mathbb{R}_m \times \mathscr{E}' \subset \mathbb{R}_n$.

A best approximation of a function f measurable on $\mathscr{E}$ by means of entire functions of exponential spherical type ν relative to $\boldsymbol{u}$, according to 5.2.1 (7), satisfies the inequality

(1) $$E_\nu(f) = E_{u\nu}(f)_{L_p(\mathscr{E})} \leqq \frac{c}{\nu^\varrho} \Omega^k_{\mathbb{R}_m}\left(f^\varrho, \frac{1}{\nu}\right) \leqq \frac{c}{\nu^\varrho} \sum_{|s|=\varrho} \Omega^k_{\mathbb{R}_m}\left(f^{(s)}, \frac{1}{\nu}\right),$$

if, of course, its right side has a meaning. Thus,

(2) $$E_\nu(f) = o(\nu^{-\varrho}) \quad (\nu \to \infty)$$

for $f \in W^\varrho_{up}(\mathscr{E})$ and p finite ($1 \leqq p < \infty$), and for $p = \infty$ if the derivatives $f^{(s)}(|s| = \varrho)$ are *uniformly continuous on* $\mathscr{E}$ *in the direction of* $\mathbb{R}_m$, which means that for any $\varepsilon > 0$ there exists a $\delta > 0$ such that

$$|f^{(s)}(\boldsymbol{x} + \boldsymbol{h}) - f^{(s)}(\boldsymbol{x})| < \varepsilon, \quad (|\boldsymbol{h}| < \delta,\ \boldsymbol{h} \in \mathbb{R}_m).$$

As the examples presented below in (5.5.5) show, the estimate (1) does not invert for $\varrho > 0$, i.e. from the fact that (2) is satisfied for $f \in L_p(\mathscr{E})$ it does not follow that $f \in W^\varrho_{up}(\mathscr{E})$.

On the other hand, in the case $\varrho = 0$ it does invert. Indeed, the following two theorems (of Weierstrass type) hold.

5.5.1. Theorem. *Suppose that* $1 \leqq p < \infty$. *For a function* f *to lie in* $L_p(\mathscr{E})$, *it is necessary and sufficient that there should exist a family of functions* $g_\nu \in L_p(\mathscr{E})$ *of entire exponential spherical type* ν *relative to* $\boldsymbol{u}$ *and such that*

(1) $$\|f - g_\nu\|_{L_p(\mathscr{E})} \to 0 \quad (\nu \to \infty).$$

The necessity follows from 5.5 (1) for $\varrho = 0$. The sufficiency is trivial.

5.5.2. Theorem[1]. *For a function f to be bounded and uniformly continuous on $\mathscr{E}$ in the direction of $\mathbb{R}_m$, it is necessary and sufficient that there exist a family of uniformly bounded functions g_ν on $\mathscr{E}$ of entire exponential spherical type ν relative to $\boldsymbol{u}$, such that*

$$\lim_{\nu\to\infty} g_\nu(\boldsymbol{x}) = f(\boldsymbol{x}) \tag{1}$$

uniformly on $\mathscr{E}$.

Proof. Again the necessity follows from 5.5 (1) for $\varrho = 0$. We shall prove sufficiency. Since the g_ν are bounded and since the convergence in (1) is uniform, then f is bounded and there exists a constant λ such that for all the ν in question and $\boldsymbol{x} \in \mathscr{E}$

$$|g_\nu(\boldsymbol{x})| \leqq \lambda.$$

Therefore for $\boldsymbol{h} \in \mathbb{R}_m$

$$|g_\nu(\boldsymbol{x}+\boldsymbol{h}) - g_\nu(\boldsymbol{x})| \leqq |\boldsymbol{h}| \sup_x \left| \frac{\partial}{\partial \boldsymbol{h}} g_\nu(\boldsymbol{x}) \right| \leqq |\boldsymbol{h}|\, \nu \sup_x |g_\nu(\boldsymbol{x})| \leqq \lambda\nu|\boldsymbol{h}|,$$

i.e. the g_ν are uniformly continuous on $\mathscr{E}$ in the direction of $\mathbb{R}_m$. Therefore f as well is uniformly continuous on $\mathscr{E}$ in the direction of $\mathbb{R}_m$.

5.5.3. Consider the norms

$${}^j\|f\|_{H^r_{\boldsymbol{u}p(\mathscr{E})}} = {}^j\|f\| = \|f\| + {}^j\|f\|_h, \quad \|\cdot\| = \|\cdot\|_{L_p(\mathscr{E})},$$

and the classes ${}^jH$, ${}^jh$, corresponding to them, where

$${}^j\|f\|_h = {}^jM_f \quad (j = 1, 2, 3, 4)$$

are the smallest constants M for which the following inequalities are respectively satisfied (see 4.3.3):

$$\Omega^k_{\mathbb{R}_m}(f^{(\boldsymbol{s})}, \delta)_{L_p(\mathscr{E})} \leqq M\delta^{r-\varrho} \quad (|\boldsymbol{s}| = \varrho), \tag{1}$$

$$\Omega^k_{\mathbb{R}_m}(f^{\varrho}, \delta)_{L_p(\mathscr{E})} \leqq M\delta^{r-\varrho}, \tag{2}$$

$$\|\Delta^k_{\boldsymbol{h}} f^{(\boldsymbol{s})}(\boldsymbol{x})\|_{L_p(\mathscr{E})} \leqq M|\boldsymbol{h}|^{r-\varrho} \quad (|\boldsymbol{s}| = \varrho), \tag{3}$$

$$\|\Delta^k_{\boldsymbol{h}} f^{\varrho}_{\boldsymbol{h}}(\boldsymbol{x})\|_{L_p(\mathscr{E})} \leqq M|\boldsymbol{h}|^{r-\varrho} \tag{4}$$

---

[1] S. N. Bernšteĭn [2], page 37, for $n = 1$.

($\varrho \geqq 0$, $k > r - \varrho > 0$) and $\boldsymbol{h} \in \mathbb{R}_m$. We introduce further the norm

$$(5) \qquad {}^5\|f\|_{\boldsymbol{h}} = \sup_{\nu>0} \nu^r E_\nu(f),$$

where $E_\nu(f) = E_{u\nu}(f)$ is the best approximation of the function f in the metric of $L_p(\mathscr{E})$ by entire functions of spherical type ν relative to $\boldsymbol{u}$. Here ν may also run through the values $\nu(s) = a^s$, $a > 1$ ($s = 0, 1, 2, \ldots$).

Moreover,

$$(6) \qquad {}^6\|f\|_H = \sup_{s=0,1,\ldots} a^{sr}\|Q_{a^s}\|,$$

where we are supposing that the function f is representable in the form of a series converging to it in the metric of $L_p(\mathscr{E})$:

$$(7) \qquad f = \sum_{s=0}^{\infty} Q_{a^s}(\boldsymbol{x}),$$

whose terms are entire functions of spherical type a^s relative to $\boldsymbol{u}$, while the norm (6) is finite. We note that the norm of f does not enter explicitly into (6).

For $j = 1, 2, 3, 4$, we may also consider variations of the constants ${}^jM_f$, which we shall denote by ${}^jM'_f$. These are the smallest constants in the respective inequalities (1)—(4), when $\delta \leqq \eta$ or $|\boldsymbol{h}| < \eta$, where η is a given arbitrary positive number. We will denote the corresponding classes by ${}^jH'$, ${}^jh'$ and the norms as follows:

$${}^j\|f\|_{H'} = \|f\| + {}^j\|f\|_{h'}.$$

Our object will be to prove that all the classes ${}^jH$, ${}^lH'$ (but in general not the ${}^jh$, ${}^lh'$) are equivalent to one another. Each of them may be chosen with any independent system of admissible parameters k, ϱ, η, α. However, the constants of the corresponding imbeddings depend on these parameters (along with r, n, m).

What has been said gives a basis in what follows to use a single notation $\|f\|_{H^r_p(\mathscr{E})}$ for all the norms ${}^j\|\cdot\|$, ${}^j\|\cdot\|_{H'}$, dropping the j and the prime. As to the norms ${}^j\|\cdot\|_h$, ${}^j\|\cdot\|_{h'}$, for them these notations are generally speaking essential. Along the way we shall obtain some imbeddings for the classes h, interesting in themselves.

It follows directly from the definitions of the moduli of continuity entering into (1), (2), that the equivalences

$$(8) \qquad {}^1H \rightleftarrows {}^3H,\ {}^3H \rightleftarrows {}^4H,$$

hold, if the classes being compared are taken for one and the same pairs k, ϱ. This will be the case, if in (8) one replaces H by h, H' or h' (in comparisons with the same η). We will show below that ${}^1H \rightleftharpoons {}^2H \rightleftharpoons {}^1H' \rightleftharpoons {}^2H'$, and in addition these classes may be chosen independently with any admissible k, ϱ, and also any $\eta > 0$. Then, in view of (8) it is possible to adjoin the classes ${}^jH$, ${}^jH'$ for $j = 3, 4$ to this chain as well. These classes thus also may be each chosen with various admissible k, ϱ, η. The cases $j = 3, 4$ may be regarded as having been treated.

Suppose given an $\eta > 0$ and admissible $k, \varrho (\varrho \geqq 0, k > r - \varrho > 0)$, From (1) and (2) it follows that

(9) $$ {}^1h' \to {}^2h'. $$

Further

(10) $$ {}^2h' \to {}^5h', $$

where we are for the present supposing

(11) $$ {}^5\|f\|_{h'} = \sup_{\nu \geqq \frac{1}{\eta}} \nu^r E_\nu(f). $$

Indeed, in view of 5.5 (1), for $f \in {}^2h'$

$$ E_\nu(f) \leqq \frac{c}{\nu^\varrho} \, {}^2M_f \frac{1}{\nu^{r-\varrho}} = \frac{c\, {}^2M_f}{\nu^r} \quad \left(\nu > \frac{1}{\eta}\right), $$

so that

$$ {}^5\|f\|_{h'} \leqq c_1 {}^2M_f. $$

If now $f \in {}^5H'$, then $f \in L_p(\mathscr{E})$ and the norm (11) is finite for f.

Then in view of the inverse theorem 5.4.2, $f \in {}^1H$ and

(12) $$ {}^5H' \to {}^1H \to {}^1H'. $$

The second imbedding is trivial. In (9) and (10) one may obviously replace h by H, which along with (12) yields

(13) $$ \begin{array}{c} \qquad\qquad\qquad\qquad\qquad\qquad {}^2H \to {}^2H' \\ \qquad\qquad\qquad\qquad\qquad \nearrow \\ {}'H' \to {}^2H' \to {}^5H' \to {}^1H \to {}^1H', \end{array} $$

i.e. the classes entering into this chain are equivalent. Since ${}^5H'$ does not depend on k, ϱ, then one may suppose that in this chain the pairs $k\,\varrho$, in ${}^jH$ and ${}^jH'$ ($j = 1, 2$) are chosen arbitrarily and independently. The number η in ${}^1H$ and ${}^2H'$ also may be chosen independently, because ${}^1H$ does not depend on η.

The task of showing that the variants of the norm ${}^5\|\cdot\|$ are equivalent to one another still faces us.

It follows from 5.5 (1) that for $f \in {}^1H$

$$E_\nu(f) \leqq \frac{c\,{}^1M_f}{\nu^r} \quad (\nu > 0),$$

so that

$${}^1M_f \gg \sup_{\nu>0} \nu^r E_\nu(f) \geqq \sup_{s=0,1,\ldots} a^{sr} E_{a^s}(f).$$

But if the right side in these relations is finite, $a > 1$ and $f \in L_p(\mathscr{E})$, then by the inverse theorem 5.4.2 (see also Remark 2 to 5.4.2) $f \in {}^1H$. This proves the equivalence.

Further, suppose that $f \in {}^5H$ and that the g_ν are functions of entire spherical type ν relative to $\boldsymbol{u}$ such that

$$\|f - g_\nu\| \leqq 2E_\nu(f) \leqq \frac{2\,{}^5\|f\|_h}{\nu^r}.$$

Suppose given an $a > 0$. Put

$$Q_1 = Q_{a^0} = g_{a^0},\ Q_{a^s} = g_{a^s} - g_{a^{s-1}} \quad (s = 1, 2, \ldots).$$

Then we obtain a representation of the function f in the form of a series tending to it in the sense of $L_p(\mathscr{E})$:

$$f = \sum_{s=0}^{\infty} Q_{a^s} \tag{14}$$

with the estimate

$$\|Q_1\| \leqq \|f\| + {}^5\|f\|_h \ll {}^5\|f\|_H,$$

$$\|Q_{a^s}\| \leqq \|g_{a^s} - f\| + \|f - g_{a^{s-1}}\| \ll \frac{{}^5\|f\|_h}{a^{rs}},$$

where the constant in the last inequality depends on a.

Thus,

$${}^6\|f\| = \sup_{s=0,1,\ldots} a^{rs}\|Q_{a^s}\| \ll {}^5\|f\|_H.$$

Conversely, from the fact that the function $f \in {}^6H$ it follows that it is representable in the form of a series (14). Therefore

$$E_{a^N}(f) \leqq \left\| f - \sum_0^{N-1} Q_{a^s} \right\| = \sum_N^{\infty} \|Q_{a^s}\| \leqq {}^6\|f\|_H \sum_N^{\infty} a^{-rs} \ll \frac{{}^6\|f\|_H}{a^{Nr}},$$

i.e.

$${}^5\|f\|_h \sup_N a^{Nr} E_{a^N}(f) \ll {}^6\|f\|_H.$$

Moreover,

$$\|f\| \leqq \sum \|Q_{a^s}\| \leqq {}^6\|f\|_H \sum a^{-rs} \ll {}^6\|f\|_H,$$

and we have proved that

$${}^5H \to {}^6H \to {}^5H,$$

i.e.

$${}^5H = {}^6H.$$

The results just obtained, in particular, contain the following theorem.

5.5.4. Theorem. *For a function f defined on $\mathscr{E} = \mathbb{R}_m \times \mathscr{E}'$ to belong to one of the classes ${}^jH_p^r(\mathscr{E})$ $(j = 1, 2, 3, 4)$ or ${}^jH_p^{\prime r}(\mathscr{E})$ $(j = 1, 2, 3, 4)$, it is necessary and sufficient that its best approximation by means of entire functions of exponential spherical type ν relative to $\boldsymbol{u}$ satisfy the inequality*

$$E_{u\nu}(f)_{L_p(\mathscr{E})} \leqq \frac{c}{\nu^r},$$

where c does not depend on $\nu > 0$ or $\nu = a^s$ $(s = 0, 1, \ldots; \nu > 0, a > 1)$.

5.5.5. *Example 1.* It is well known that if a real function $f(x)$ of period 2π belongs to the space L_2, then it decomposes into a Fourier series

$$f(x) = \frac{a_0}{2} + \sum_1^\infty (a_k \cos kx + b_k \sin kx), \tag{1}$$

converging to it in the sense of $L_2^* = L_2(0, 2\pi)$, where

$$\begin{Bmatrix} a_k \\ b_k \end{Bmatrix} = \frac{1}{\pi} \int_0^{2\pi} f(t) \begin{Bmatrix} \cos kt \\ \sin kt \end{Bmatrix} dt \quad (k = 0, 1, \ldots). \tag{2}$$

In addition

$$\frac{1}{\pi} \int_0^{2\pi} f^2 dx = \frac{a_0^2}{2} + \sum_{k=1}^\infty (a_k^2 + b_k^2). \tag{3}$$

Conversely, if a series of arbitrary real numbers a_k, b_k on the right side of (3) converges, then the series (1) converges in the L_2^* sense to some function $f \in L_2^*$, and equations (2) hold.

As a consequence of the well-known orthogonal properties of trigonometric functions, the square of the best approximation by means of trigonometric polynomials of order $n - 1$ (in the L_2^* sense) of the function $f \in L_2^*$ defined by the series (1) is equal to

$$(4) \quad E_n(f)_{L_2^*} = \min_{\gamma_k, \delta_k} \int_0^{2\pi} \left[f(x) - \frac{\gamma_0}{2} - \sum_1^{n-1} (\gamma_k \cos kx + \delta_k \sin kx) \right]^2 dx$$

$$= \int_0^{2\pi} \left[f(x) - \frac{a_0}{2} - \sum_1^{n-1} (a_k \cos kx + b_k \sin kx) \right]^2 dx$$

$$= \int_0^{2\pi} \left[\sum_n^{\infty} (a_k \cos kx + b_k \sin kx) \right]^2 dx = \pi \sum_n^{\infty} (a_k^2 + b_k^2).$$

If the function f belongs to $W_{2}^{1}{}^{*}$, i.e. it is absolutely continuous and its derivative (existing almost everywhere) $f' \in L_2^*$, then its Fourier coefficients a_k, b_k may, by integrating by parts, be presented in the form

$$(5) \qquad a_k = -\frac{\beta_k}{k}, \; b_k = \frac{\alpha_k}{k} \quad (k = 1, 2),$$

where α_k, β_k are the Fourier coefficients of the derivative f', for which the series

$$(6) \qquad \sum_1^{\infty} \alpha_k^2 + \beta_k^2 = \sum_1^{\infty} k^2(a_k^2 + b_k^2)$$

converges. Conversely, f will belong to $W_{2}^{1}{}^{*}$ if it is representable in the form of a series (1) converging to it in the sense of L_2^*, while $\sum_1^{\infty} k^2(a_k^2 + b_k^2) < \infty$.

The best approximation of a function $f \in W_{2}^{1}{}^{*}$, by means of trigonometric polynomials of the $(n-1)^{st}$ order, obeys the inequality

$$(7) \quad E_{n-1}(f)_{L_2^*}^2 = \pi \sum_n^{\infty} \frac{\alpha_k^2 + \beta_k^2}{k^2} \leqq \frac{\pi}{n^2} \sum_n^{\infty} (\alpha_k^2 + \beta_k^2) = o(n^{-2}) \; n \to \infty,$$

which is consistent with the general theory (periodic analogue of formula 5.5 (1)).

In order to verify the fact that, on the contrary, it does not follow from (7) that f belongs to the class $W_{2}^{1}{}^{*}$, we consider the function

$$\varphi(x) = \sum_1^{\infty} \frac{\cos kx}{k^{3/2} \sqrt{\ln k}}.$$

Obviously

$$E_n(f)^2_{L_2^*} = \pi \sum_n^\infty \frac{1}{k^3 \ln k} \leqq \pi \frac{1}{\ln n} \sum_n^\infty \frac{1}{k^3} = o(n^{-2}) \quad (n \to \infty).$$

On the other hand, $f \notin W^1_{\frac{*}{2}}$, since the series

$$\sum_1^\infty \frac{1}{k \ln k},$$

corresponding to the series (6), diverges.

Example 2. The function of period 2π given by

$$f(x) = \sum_1^\infty \frac{\sin kx}{k^2 \ln k}$$

obviously is continuous and has a best approximation using trigonometric polynomials of the $(n-1)^{st}$ order in the metric of C (or L_∞), satisfying the inequality

$$E_{n-1}(f)_C \leqq \sum_2^\infty \frac{1}{k^2 \ln k} \leqq \frac{c}{n \ln n} = \frac{o(1)}{n} \quad (n \to \infty).$$

At the same time the termwise differentiated series

$$f'(x) = \sum_2^\infty \frac{\cos kx}{k \ln k}, \tag{8}$$

in view of the fact that its coefficients decrease monotonically to zero, converges uniformly on $[\varepsilon, 2\pi - \varepsilon]$ for any $\varepsilon > 0$ (see Zygmund, [1], 2.6). Thus its sum is continuous on the interval $(0, 2\pi)$ and is equal to the derivative $f'(x)$. In addition the series (8) is the Fourier series for $f'(x)$, since $\sum \frac{1}{k^2 \ln^2 k} < \infty$. In such a case f' is discontinuous at the point $x = 0$, because if f' were continuous everywhere, then its nth Féjer sum at $x = 0$ would tend to $f'(0)$. At the same time the Féjer sum, as the arithmetic mean of the first $n + 1$ Fourier sums, tends at $x = 0$ along with these sums to ∞.

5.5.6. The anisotropic case. We will start from the estimate proved in 5.2.4 (5):

$$\|f - g_\nu\|_{L_p(\mathscr{E})} \leqq c \sum_{j=1}^m \frac{\omega^{k_j}_{x_j}\left(f^{(\varrho_j)}_{x_j}, \frac{1}{\nu_j}\right)_{L_p(\mathscr{E})}}{\nu_j^{\varrho_j}}$$

$$(\mathscr{E} = \mathbb{R}_m \times \mathscr{E}', \nu_j > 0). \tag{1}$$

From it, for the best approximation of $f \in W^r_{up}(\mathscr{E})$ using entire functions g_ν of exponential type $\boldsymbol{v} = (\nu_1, \ldots, \nu_m)$ relative to $\boldsymbol{u} = (x_1, \ldots, x_m)$, there follows the inequality

$$E_\nu(f)_{L_p(\mathscr{E})} = \sum_{j=1}^m \frac{o(1)}{\nu_j^{r_j}} \qquad (\nu_j \to 0) \tag{2}$$

for $1 \leqq p \leqq \infty$ or for $p = \infty$, if the partial derivatives $f^{(r_j)}_{x_j}$ are respectively uniformly continuous on $\mathscr{E}$ in the directions x_j.

If $f \in H^r_p(\mathscr{E})$, then it follows from (1) that

$$E_\nu(f)_{L_p(\mathscr{E})} \leqq \|f - g_\nu\|_{L_p(\mathscr{E})} \leqq c\|f\|_{h^r_p(\mathscr{E})} \sum_1^m \frac{1}{\nu_j^{r_j}}. \tag{3}$$

In particular, if one replaces the ν_j in this inequality by $\nu^{l/r_j} (\nu > 0)$ respectively, we obtain (dropping the $L_p(\mathscr{E})$)

$$\nu E_{\nu^{1/r_1}, \ldots, \nu^{1/r_m}}(f) \leqq c_1 \|f\|_{h^r_p(\mathscr{E})} \quad (\nu > 0). \tag{4}$$

Suppose given $a > 1$. Introduce the norms

$${}^j\|\cdot\|_H = \|\cdot\| + {}^j\|\cdot\|_h \ (j = 1, 2, 3), \ \|\cdot\| = \|\cdot\|_{L_p(\mathscr{E})}, \tag{5}$$

where

$${}^1\|f\|_h = \|f\|_{h^r_p(\mathscr{E})}, \tag{6}$$

$${}^2\|f\|_h = \sup_{\nu > 0} \nu E_{\nu^{1/r_1}, \ldots, \nu^{1/r_m}}(f), \tag{7}$$

$${}^3\|f\|_h = \sup_{s = 0, 1, \ldots} a^s E_{a^{s/r_1}, \ldots, a^{s/r_m}}(f), \tag{8}$$

Moreover, put

$${}^4\|f\|_H = \sup_{s = 0, 1, \ldots} a^s \|Q_s\|, \tag{9}$$

where the last norm (not containing $\|f\|$ explicitly) must be understood in the sense that f is representable in the form of a series converging to it in the metric of $L_p(\mathscr{E})$:

$$f = \sum_0^\infty Q_s,$$

whose terms are functions of entire type a^{s/r_j} relative to $x_j (j = 1, \ldots, m)$, such that the norm (9) is finite.

All four norms ${}^j\|\cdot\|_H$ (but not the ${}^j\|\cdot\|_h$) and the classes ${}^jH$ corresponding to them, are equivalent. Below this assertion is proved for $j = 1, 2, 3$.

The equivalence $^3H = {}^4H$ will follow (see 5.6.1) as a special case of the corresponding proposition for the B-classes.

One has the obvious inequality

$$E_{x_j a^{s/r_j}}(f) \leq E_{a^{s/r_1}, \ldots, a^{s/r_m}}(f),$$

whose the left side is the best approximation of f in the metric of $L_p(\mathscr{E})$ in terms of functions of entire eүponential type a^{s/r_j} relative to one variable x_j. Therefore, if $f \in {}^3H$, then

$$\sup_s a^s E_{x_j a^{s/r_j}}(f) \leq \sup_s a^s E_{a^{s/r_1}, \ldots, a^{s/r_m}}(f) = {}^3\|f\|_h.$$

But then, recalling further that $f \in L_p(\mathscr{E})$, we find that $f \in H^{r_j}_{x_j p}(\mathscr{E})$ (see 5.5.3) and

$$\|f\|_{H^{r_j}_{x_j p}(\mathscr{E})} \ll {}^3\|f\|_H \quad (j = 1, \ldots, m).$$

Therefore $f \in H^r_p(\mathscr{E}) = {}^1H$ and

$${}^1\|f\|_H \ll {}^3\|f\|_H.$$

We have proved that

$${}^1H \to {}^2H \to {}^3H \to {}^1H,$$

i.e. these classes are equivalent.

The results just obtained contain in particular the following theorem.

5.5.7. Theorem[1]. *For the function f to lie in the class $H^r_p(\mathscr{E})$, it is necessary and sufficient that the following inequality be satisfied:*

$$(1) \qquad E_\nu(f) \leq \sum_1^m \frac{1}{\nu_j^{r_j}} \quad (\nu_j > 0).$$

Inequality (1) necessarily follows from 5.5.6 (3). Conversely, if it is satisfied for arbitrary independent $\nu_j > 0$, then it is so much the more satisfied for ν_j of the form $\nu_j = \nu^{1/r_j}$ $(j = 1, \ldots, m)$. But then the upper bound in 5.5.6 (7) is finite.

Of course, the classes $H^{r_j}_{x_j p}$ defining H^r_p may be considered in the several variants described in 5.5.3 $(m = 1, \mathbb{R}_m = \mathbb{R}_{x_j})$.

5.5.8. All the theorems proved in this section carry over without change to the periodic case. Here, throughout the arguments it is necessary to replace H and W by $H*$, $W*$ respectively and the entire functions g_ν by trigonometric polynomials. In doing this it is necessary to take account of what was said in 5.3.

---

[1] S. N. Bernšteĭn [2], pp. 421–426, $p = \infty$, S. M. Nikol'skiĭ [1], $1 \leq p < \infty$.

5.6. Definition of B-Classes with the Aid of Best Approximations. Equivalent Norms

Suppose that $\mathscr{E} = \mathbb{R}_m \times \mathscr{E}' \subset \mathbb{R}_n$, $r > 0$, k and ϱ admissible integers, (satisfying the inequalities $\varrho \geqq 0$, $k > r - \varrho > 0$), $1 \leqq p \leqq \infty$, $1 \leqq \theta < \infty$, $a > 1$, and the function f is measurable on $\mathscr{E}$.

The basic object will be the proof of the fact that the norms

$$ {}^{j}\|f\|_{B^r_{\boldsymbol{u}p\theta}(\mathscr{E})} = {}^{j}\|f\|_B = \|f\| + {}^{j}\|f\|_b \quad (j = 1, \dots, 5), $$

where $\|\cdot\| = \|\cdot\|_{L_p(\mathscr{E})}$,

$$ (1) \qquad {}^{1}\|f\|_b = \sum_{|\boldsymbol{s}|=\varrho} \left(\int_0^\infty t^{-1-\theta(r-\varrho)} \Omega^k (f^{(\boldsymbol{s})}, t)^\theta \, dt \right)^{1/\theta}, $$

$$ (2) \qquad {}^{2}\|f\|_b = \left(\int_0^\infty t^{-1-\theta(r-\varrho)} \Omega^k (f^\varrho, t)^\theta \, dt \right)^{1/\theta}, $$

$$ (3) \qquad {}^{3}\|f\|_b = \sum_{|\boldsymbol{s}|=\varrho} \left(\int_{\mathbb{R}_m} |\boldsymbol{u}|^{-m-\theta(r-\varrho)} \|\Delta^k_{\boldsymbol{u}} f^{(\boldsymbol{s})}(\boldsymbol{x})\|^\theta_{L_p(\mathscr{E})} \, d\boldsymbol{u} \right)^{1/\theta}, $$

$$ (4) \qquad {}^{4}\|f\|_b = \left(\int_{\mathbb{R}_m} |\boldsymbol{u}|^{-m-\theta(r-\varrho)} \|\Delta^k_{\boldsymbol{u}} f^\varrho_{\boldsymbol{u}}(\boldsymbol{x})\|^\theta_{L_p(\mathscr{E})} \, d\boldsymbol{u} \right)^{1/\theta}, $$

$$ (5) \qquad {}^{5}\|f\|_b = \left\{ \sum_{l=0}^\infty a^{lr\theta} E^\theta_{\boldsymbol{u}a^l}(f)_{L_p(\mathscr{E})} \right\}^{1/\theta}, $$

are equivalent, and moreover equivalent to the norm (not containing $\|f\|$ explicitly)

$$ (6) \qquad {}^{6}\|f\|_B = \left\{ \sum_{l=0}^\infty a^{lr\theta} \|Q_{\boldsymbol{u}a^l}\|^\theta_{L_p(\mathscr{E})} \right\}^{1/\theta} \quad (\alpha > 1). $$

This last expression needs to be understood in the sense that f may be represented in the form of a series converging to it in the sense of $L_p(\mathscr{E})$:

$$ (7) \qquad f = \sum_{l=0}^\infty Q_{\boldsymbol{u}a^l}(\boldsymbol{x}), $$

whose terms are entire exponential spherical of type a^l relative to $\boldsymbol{u} \in \mathbb{R}_m$, so that the norm ${}^{6}\|f\|_B$ is finite.

Here $f^{(\boldsymbol{s})}$ denotes any derivative of f of order $\boldsymbol{s} = (s_1, \ldots, s_m)$, $|\boldsymbol{s}| = \varrho$, relative to the variables $u_1, \ldots, u_m$, and $f_{\boldsymbol{u}}^{\varrho}$ the derivative in the direction $\boldsymbol{u} \in \mathbb{R}_m$ of order ϱ,

$$\Omega^k(f^{(\boldsymbol{s})}, \delta) = \Omega^k_{\mathbb{R}_m}(f^{(\boldsymbol{s})}, \delta)_{L_p(\mathscr{E})} = \sup_{|\boldsymbol{u}|<\delta} \|\Delta^k_{\boldsymbol{u}} f^{(\boldsymbol{s})}(\boldsymbol{x})\|_{p(\mathscr{E})},$$

$$\Omega^k(f^{\varrho}, \delta) = \Omega^k_{\mathbb{R}_m}(f^{\varrho}, \delta)_{L_p(\mathscr{E})} = \sup_{\substack{|\boldsymbol{u}|=1 \\ \boldsymbol{u}\in\mathbb{R}_m}} \sup_{|t|<\delta} \|\Delta^k_{t\boldsymbol{u}} f^{\varrho}_{\boldsymbol{u}}(\boldsymbol{x})\|_{L_p(\mathscr{E})}.$$

Further we introduce the norm ${}^j\|f\|_{B'} = \|f\| + {}^j\|f\|_{b'}$ $(j = 1, 2, 3, 4)$. These are the same norms as ${}^j\|\ \|_B$, ${}^j\|\ \|_b$, respectively, but in them, by definition, the integration is carried out relative to $t \in [0, \eta]$ or relative to $\boldsymbol{u}$ with $|\boldsymbol{u}| < \eta$.

It will be proved that these norms are equivalent to the preceding ones (without the primes), but with constants depending on η. We need to keep in mind that each of the classes just enumerated depends further on the admissible pair (k, ϱ). It will be shown that any two such classes, corresponding to different pairs, are also equivalent, with constants depending on these pairs.

We note that equivalence of the norm (5) with one of the other norms for the clases $B^r_{\boldsymbol{u}p\theta}(\mathscr{E})$ corresponds to the assertion of Theorem 5.5.4, giving in the terms of best approximations necessary and sufficient conditions for the function f to belong to the class $H^r_{\boldsymbol{u}p}(\mathscr{E})$. It follows from (5) that $B^r_{\boldsymbol{u}p\infty}(\mathscr{E}) = H^r_{up}(\mathscr{E})$.

The classes corresponding to the indicated norms will for short be denoted by ${}^jB$, ${}^jb$ $(j = 1, \ldots, 6)$, ${}^jB'$, ${}^jb'$ $(j = 1, \ldots, 4)$. We need to keep in mind that the seminorms ${}^jb$, ${}^lb'$, themselves, generally speaking, are not equivalent. But if to add to them $\|f\| = \|f\|_{Lp(\mathscr{E})}$, we obtain sums, the norms ${}^jB$, ${}^jB'$ which are equivalent.

Below we shall prove a series of imbeddings, from which the assertions on equivalence made above will follow. This imbeddings present interest in themselves as well. Certain of them are true for pairs (k, ϱ) which may not be admissible.

We have for the present, for one and the same not necessarily admissible pair of natural numbers k, ϱ,

$$(8) \qquad {}^1b \to {}^1b' \to {}^3b' \to {}^4b'.$$

The first and second imbeddings are obvious, and the third follows from the relations

$$\|\Delta^k_{\boldsymbol{u}} f^{\varrho}_{\boldsymbol{u}}(\boldsymbol{x})\| = \left\| \Delta^k_{\boldsymbol{u}} \sum_{|\boldsymbol{s}|=\varrho} f^{(\boldsymbol{s})} \left(\frac{\boldsymbol{u}}{|\boldsymbol{u}|}\right)^{\boldsymbol{s}} \right\| \leqq \sum_{|\boldsymbol{s}|=\varrho} \|\Delta^k_{\boldsymbol{u}} f^{(\boldsymbol{s})}\|.$$

Analogously, also for one and the same not necessarily admissible pair k, ϱ:

$$(9) \qquad {}^1b \to {}^1b' \to {}^2b' \to {}^4b'.$$

Now suppose that $f \in {}^4B'$ for some not necessarily admissible pair k, ϱ.

For each $\nu > 0$ there exists an entire function g_ν of spherical type ν relative to $\boldsymbol{u} \in \mathbb{R}_m$ such that (5.2.1 (6))

$$(10) \qquad g_\nu - f = (-1)^{j-1} \int_{\mathbb{R}_m} g(|\boldsymbol{u}|)\, \Delta_{\boldsymbol{u}/\nu}^{k+\varrho} f(\boldsymbol{x})\, d\boldsymbol{u},$$

and then

$$E_{a^j}(f) \le \|g_{a^j} - f\| = \left\| \int_{\mathbb{R}_m} g(|\boldsymbol{u}|) \Delta_{a^{-j}\boldsymbol{u}}^{k+\varrho} f(\boldsymbol{x}) d\boldsymbol{u} \right\|$$

$$= c \left\| \int_0^\infty \int_{|\xi|=1} g(t) \Delta_{a^{-j}t\xi}^{k+\varrho} f(\boldsymbol{x})\, t^{m-1} d\xi\, dt \right\|.$$

Therefore (explanation below)

$$(11) \qquad {}^5\|f\|_b = \left\{ \sum_{j=0}^\infty a^{jr\theta} E_{a^j}^\theta(f) \right\}^{1/\theta} \le a^r \left\{ \int_{-1}^\infty a^{jr\theta} E_{a^j}^\theta(f)\, dj \right\}$$

$$\ll \left\{ \int_{-1}^\infty a^{jr\theta} \left\| \int_0^\infty \int_{|\xi|=1} g(t) \Delta_{a^{-j}t\xi}^{k+\varrho} f(x) t^{m-1}\, d\xi\, dt \right\|^\theta dj \right\}^{1/\theta}$$

$$\le \int_0^\infty t^{m-1} g(t) \left\{ \int_{-1}^\infty a^{jr\theta} \left\| \int_{|\xi|=1} |\Delta_{a^{-j}t\xi}^{k+\varrho} f(\boldsymbol{x})| d\xi \right\|^\theta dj \right\}^{1/\theta} dt$$

$$\ll \int_0^\infty t^{m-1+r} g(t) \left\{ \int_0^{at} \int_{|\xi|=1} v^{-r\theta-1} \|\Delta_{v\xi}^{k+\varrho} f(\boldsymbol{x})\|^\theta d\xi\, dv \right\}^{1/\theta} dt$$

$$\ll \int_0^\infty t^{m-1+r} g(t)\, dt \left\{ \int_{|\boldsymbol{u}|<\eta} |\boldsymbol{u}|^{-m-(r-\varrho)\theta} \|\Delta_{\boldsymbol{u}}^k f^\varrho\|^\theta\, du \right.$$

$$\left. + \int_\eta^\infty v^{-r\theta-1}\, dv \|f\|^\theta \right\}^{1/\theta} \ll {}^4\|f\|_{b'} + \eta^{-r}\|f\| \ll {}^4\|f\|_{B'}.$$

In the fourth relation (inequality) we have applied the generalized Minkowski inequality: first the norm $\|\cdot\|$ is brought under the sign of the integral with respect to j, and then norm relative to j under the integral sign relative to t. In the fifth relation the j in the integral is replaced by v by means of the substitution $a^{-j}t = v$.

If $\eta = \infty$, then

(12) $${}^5\|f\|_b \ll {}^4\|f\|_b,$$

i.e.

(13) $${}^4B' \to {}^5B,$$

(14) $${}^4b \to {}^5b.$$

In what follows we shall make use only of the imbedding (13). But the imbedding (14) is interesting in itself.

Now suppose that $f \in {}^5B$. We denote by g_{a^l} a function of entire spherical degree a^l relative to $\boldsymbol{u}$, such that

$$\|f - g_{a^l}\| \leqq \boldsymbol{v} E_{a^l}(f) \ (l = 0, 1, \ldots),$$

and put

$$Q_{a^0} = g_{a^0}, Q_{a^l} = g_{a^l} - g_{a^{l-1}} \ (l = 1, 2, \ldots).$$

Then in the sense of $L_p(\mathscr{E})$

$$f = \sum_{l=0}^{\infty} \Phi_{a^l},$$

because it follows from the finiteness of ${}^5\|\cdot\|_b$ that $E_{a^l}(f) \to 0$ as $l \to \infty$. Further

$$\|Q_{a^0}\| \leqq \|f\| + 2E_{a^0}(f),$$
$$\|Q_{a^l}\| \leqq \|g_{a^l} - f\| + \|f - g_{a^{l-1}}\| \leqq 4E_{a^{l-1}}(f),$$

because $E_{a^l}(f)$ does not increase when l increases. Hence

$$\begin{aligned} {}^6\|f\|_B &\ll \left\{ \left(\|f\| + 2E_{a^0}(f)\right)^\theta + \sum_{l=1}^{\infty} a^{lr\theta} E_{a^{l-1}}(f)^\theta \right\}^{1/\theta} \\ &\ll \|f\| + \left\{ \sum_{l=1}^{\infty} a^{lr\theta} E_{a^l}(f)^\theta \right\}^{1/\theta} = {}^5\|f\|_B, \end{aligned}$$

and we have proved that

(15) $${}^5B \to {}^6B.$$

Now suppose that $f \in {}^6B$ and that f is representable in the form (7). Suppose given any admissible natural numbers k, ϱ. For any $\boldsymbol{u} \in \mathbb{R}_m$, integer vector $\boldsymbol{s} = (s_1, \ldots, s_m, 0, \ldots, 0)$ with $|\boldsymbol{s}| = \varrho$ and natural number N,

$$\Delta_{\boldsymbol{u}}^k f^{(\boldsymbol{s})}(\boldsymbol{x}) = \sum_{l=0}^{N-1} \Delta_{\boldsymbol{u}}^k Q_{a^l}^{(\boldsymbol{s})}(\boldsymbol{x}) + \sum_{l=N}^{\infty} \Delta_{\boldsymbol{u}}^k Q_{a^l}^{(\boldsymbol{s})}(\boldsymbol{x}),$$

$$\|\Delta_{\boldsymbol{u}}^k f^\varrho\| \leqq |\boldsymbol{u}|^c \sum_{l=0}^{N} a^{l(\varrho+k)} \|Q_{a^l}\| + 2^k \sum_{l=N}^{\infty} a^{l\varrho} \|Q_{a^l}\|.$$

Hence

$$\Omega^k(f^{(s)}, a^{-N}) = \sup_{|u|<a^{-N},\, u\in \mathbb{R}_m} \|\Delta_u^k f^{(s)}(x)\| \ll a^{-Nk} \sum_{l=0}^{N} a^{l(\varrho+k)}\|Q_{a^l}\| + \sum_{l=N}^{\infty} a^{l\varrho}\|Q_{a^l}\|.$$

We shall estimate ${}^1\|f\|_{b'}$. We have

$$(16)\qquad \int_0^1 t^{-1-\theta(r-\varrho)}\Omega^k(f^{(s)}, t)^\theta dt = \ln a \int_0^\infty a^{\theta(r-\varrho)n}\Omega^k(f^{(s)}, a^{-n})^\theta\, dn$$

$$= \ln a \sum_{N=0}^{\infty} \int_N^{N+1} a^{\theta(r-\varrho)n}\Omega^c(f^{(s)}, a^{-n})^\theta\, dn$$

$$\ll \sum_{N=0}^{\infty} a^{\theta(r-\varrho)N}\Omega^k(f^{(s)}, a^{-N})^\theta = J_1 + J_2,$$

where (explanation below)

$$(17)\qquad J_1 = \sum_{N=0}^{\infty} a^{\theta(r-\varrho-k)N}\left(\sum_{l=0}^{N} a^{l(\varrho+k)}\|Q_{a^l}\|\right)^\theta \ll \sum_{l=0}^{\infty} a^{r\theta l}\|Q_{a^l}\|^\theta,$$

$$(18)\qquad J_2 = \sum_{N=0}^{\infty} a^{\theta(r-\varrho)N}\left(\sum_{l=N} a^{l\varrho}\|Q_{a^l}\|\right)^\theta \ll \sum_{l=0}^{\infty} a^{lr\theta}\|Q_{a^l}\|^\theta.$$

The inequalities $\ll$ are justified as follows. If $a > 1$, $0 < \delta < \beta$, $b_l \geqq 0$ $(l = 0, 1, \ldots)$, then

$$(19)\qquad \sum_{N=0}^{\infty} a^{-\theta\beta N}\left(\sum_{l=0}^{N} b_l\right)^\theta = \sum_{N=0}^{\infty} a^{-\theta\beta N}\left(\sum_{l=0}^{N} a^{(\beta-\delta)l}a^{(\delta-\beta)l}b_l\right)^\theta$$

$$\ll \sum_{N=0}^{\infty} a^{-\theta\delta N}\sum_{l=0}^{N} a^{(\delta-\beta)\theta l}b_l = \sum_{l=0}^{\infty} a^{(\delta-\beta)\theta l}b_l^\theta \sum_{N=l}^{\infty} a^{-\theta\delta N} \ll \sum_{l=0}^{\infty} a^{-\beta\theta l}b_l^\theta,$$

$$(20)\qquad \sum_{N=0}^{\infty} a^{\theta\beta N}\left(\sum_{l=N}^{\infty} b_l\right)^\theta = \sum_{N=0}^{\infty} a^{\theta\beta s}\left(\sum_{l=N}^{\infty} a^{(\delta-\beta)}\, a^{(\beta-\delta)l}b_l\right)^\theta$$

$$\ll \sum_{N=0}^{\infty} a^{\theta\beta N}\left(\sum_{l=N}^{\infty} a^{(\delta-\beta)\theta' l}\right)^{\theta/\theta'}\left(\sum_{l=N}^{\infty} a^{(\beta-\delta)\theta l}b_l^\theta\right)$$

$$\ll \sum_{N=0}^{\infty} a^{\theta\delta N}\sum_{l=N}^{\infty} a^{(\beta-\delta)\theta l}b_l^\theta = \sum_{l=0}^{\infty} a^{(\beta-\delta)\theta l}b_l^\theta \sum_{N=1}^{l} a^{\theta\delta N} \ll \sum_{l=0}^{\infty} a^{\beta\theta l}b_l^\theta;$$

In this formula the expression $A \ll B$ has to be understood in the sense $A \leq cB$, where c is a constant depending on a and δ, but not on the b_l.

The inequality (17) is obtained from (19), if one puts $\beta = k - r + \varrho (> 0)$, $b_l = a^{l(\varrho+k)}\|Q_{a^l}\|$, and inequality (18) from (20), if one puts

$$\beta = r - \varrho, \quad b_l = a^{l\varrho}\|Q_{a^l}\|.$$

The application of these two inequalities requires the assumption of admissibility of the pairs k, ϱ, i.e. the condition $k > r - \varrho > 0$ must hold.

We have proved that

$$\left(\int_0^\eta t^{-\theta(r-\varrho)-1}\Omega^k(f^{(s)}, t)^\theta \, dt\right)^{1/\theta} \ll {}^6\|f\|_B. \tag{21}$$

Further, putting $\dfrac{1}{\theta} + \dfrac{1}{\theta'} = 1$, we get

$$\|f\| \leqq \sum_0^\infty \|Q_{a^l}\| = \sum_0^\infty a^{-lr} a^{lr} \|Q_{a^l}\| \leqq \left(\sum_0^\infty a^{-lr\theta'}\right)^{1/\theta'} {}^6\|f\|_B = c\,{}^6\|f\|_B. \tag{22}$$

Therefore it follows from (21) and (22) that, for any admissible pair k, ϱ,

$${}^6B \to {}^1B'. \tag{23}$$

Finally, using (7), it follows that ($|s| = \varrho < r$)

$$\|f^{(s)}\| \ll \sum_{l=0}^\infty a^{l\varrho}\|Q_{a^l}\| = \sum_{l=0}^\infty a^{-l(r-\varrho)} a^{lr}\|Q_{a^l}\| \ll \left(\sum_0^\infty a^{\theta l r}\|Q_{a^l}\|^\theta\right)^{1/\theta} = {}^6\|f\|_B, \tag{24}$$

Then, since the function $t^{-\theta(r-\varrho)-1}$ is integrable on $(1, \infty)$ and $|s| = \varrho$

$$\Omega^k(f^{(s)}, t) \ll \|f^{(s)}\|,$$

then

$$\int_\eta^\infty t^{-\theta(r-\varrho)-1}\Omega^k(f^{(s)}, t)\, dt \ll {}^6\|f\|_B,$$

so that we have the imbedding, stronger than (23),

$${}^6B \to {}^1B, \tag{25}$$

true for any admissible pair k, ϱ.

Now suppose that k, ϱ is an admissible pair. Gathering (8), (9), (13), (15), (25) together, we get

$$\begin{array}{ccc} & \nearrow {}^1b' \to {}^3b' \searrow & \\ {}^1b & & {}^4b', \\ & \searrow {}^1b' \to {}^2b' \nearrow & \end{array} \quad {}^4B' \to {}^5B \to {}^6B \to {}^1B.$$

Since here we may always replace b by B (because this only means that the corresponding inequality does not change if one adds $\|f\|$ to both sides), then

$$ {}^1B \to {}^1B' \begin{smallmatrix} \nearrow {}^3B' \searrow \\ \searrow {}^2B' \nearrow \end{smallmatrix} {}^4B' \to {}^5B \to {}^6B \to {}^1B. $$

On the other hand, it is obvious (the chains (8), (9) are true if one everywhere drops the primes), that

$$ {}^1B \begin{smallmatrix} \nearrow {}^3B \searrow \\ \searrow {}^2B \nearrow \end{smallmatrix} {}^4B \to {}^4B' \to {}^1B. $$

This shows that all the classes entering into both chains are equivalent. For another admissible pair k', ϱ' we again obtain the equivalence of these classes, and since the class ${}^5B$, as well as ${}^6B$, does not depend on the (admissible) k, ϱ, then, evidently, all the indicated classes ${}^jB$ $(j = 1, ..., 6)$ and ${}^jB'$ $(j = 1, \ldots, 4)$ are equivalent to one another independently of what pairs k, ϱ or parameter $\eta > 0$ define them. Of course, the imbedding constants arising here depend in general on k, ϱ, η, a. We note further that the classes ${}^5B$, ${}^6B$ remain equivalent when the quantity $a > 1$ varies. This follows for example from the fact that they are equivalent (but with constants depending on a) to the class ${}^1B$, which does not depend on a.

Remark. Suppose that $f \in {}^1B$. We define for f functions g_ν by means of equation (10). It is easy to see that the g_ν are obtained from f by means of a linear operation $g_\nu = A_\nu(f)$ (see 5.2.1 (4)). From the chain of inequalities (11), which we need to consider starting with the third term, and from the further estimates (see (12)), one obtains the inequality

$$ \left(\sum_{j=0}^{\infty} a^{jr\theta} \|g_{a^j} - f\|^\theta \right)^{1/\theta} \ll {}^4\|f\| \ll {}^1\|f\|. $$

Therefore, if one puts

$$ Q_{a^0} = g_{a^0}, \quad Q_{a^i} = g_{a^i} - g_{a^{i-1}} \quad (i = 1, 2, ...) $$

and takes into account the fact that

$$ \|Q_{a^i}\| \ll \|g_{a^i} - f\| + \|f - g_{a^{i-1}}\|, $$

then it is easy to obtain the inequality

$$ {}^6\|f\|_B = \left(\sum_{i=0}^{\infty} a^{ir\theta} \|Q_{a^i}\|^\theta \right)^{1/\theta} \ll {}^1\|f\|_B. $$

This calculation has been carried out in order to emphasize the fact that if we introduce for the functions $f \in B^r_{p\theta}(\mathscr{E})$ a norm of the type ${}^6\|\cdot\|_B$, then it is always possible to suppose that in addition the functions Q_s are obtained from f by means of completely defined linear operations (5.2.1 (4)). It is important to note further that for a given $r_0 > 0$, for all $r < r_0$ these operations for each s may be taken as one and the same.

5.6.1. The anisotropic case. Suppose given a function $f \in B^r_{p\theta}(\mathscr{E})$, where

$$\boldsymbol{p} = (p_1, \ldots, p_m), \boldsymbol{\theta} = (\theta_1, \ldots, \theta_m), \boldsymbol{r} = (r_1, \ldots, r_m),$$

$$1 \leqq m \leqq n, 1 \leqq p_j, \theta_j \leqq \infty, r_j > 0, a > 1, \mathscr{E} = \mathbb{R}_m \times \mathscr{E}'.$$

We shall define for it a family of functions $g_{\nu_1, \ldots, \nu_m}$ of entire exponential type ν_j relative to x_j, by formulas 5.2.4 (1), where $0 < \nu_j \leqq \infty$, and introduce a constant $a > 1$. We shall prove that there hold inequalities which generalize inequalities 5.2.4 (2) to the case of finite $\boldsymbol{\theta}$:

(1)
$$\left(\sum_{s=0}^{\infty} a^{\theta_1 s}\|f - g_{a^{s/r_1}, \infty, \ldots, \infty}\|^{\theta_1}_{L_{p_1}(\mathscr{E})}\right)^{1/\theta_1} \leqq c\|f\|_{b^r_{p\theta}(\mathscr{E})},$$

$$\left(\sum_{s=0}^{\infty} a^{\theta_2 s}\|g_{a^{s/r_1}, \infty, \ldots, \infty} - g_{a^{s/r_1}, a^{s/r_2}, \infty, \ldots, \infty}\|^{\theta}_{L_{p_2}(\mathscr{E})}\right)^{1/\theta_2} \leqq c\|f\|_{b^r_{p\theta}(\mathscr{E})},$$

. . . .

$$\left(\sum_{s=0}^{\infty} a_m^{\theta_m s}\|g_{a^{s/r_1}, \ldots, a^{s/r_{m-1}}, \infty} - g_{a^{s/r_1}, \ldots, a^{s/r_m}}\|^{\theta_m}_{L_{p_m}(\mathscr{E})}\right)^{1/\theta_m} \leqq c\|f\|_{b^r_{p\theta}(\mathscr{E})}.$$

For $\theta_j = \infty$ the corresponding j^{th} inequality has the form

$$a^s\|g_{a^{s/r_1}, \ldots, a^{s/r_{j-1}}, \infty, \ldots, \infty} - g_{a^{s/r_1}, \ldots, a^{s/r_j}, \infty, \ldots, \infty}\|_{L_{p_j}(\mathscr{E})} \leqq c\|f\|_{h^r_p(\mathscr{E})}.$$

This follows directly from 5.2.4 (2). If now θ_j is finite, then ($r_j - \varrho_j > 0$, $\varrho_j \geqq 0$, see 5.2.4 (2) and 5.6)

$$\sum_{s=0}^{\infty} a^{\theta_j s}\|g_{a^{s/r_1}, \ldots, a^{s/r_{j-1}}, \infty, \ldots, \infty} - g_{a^{s/r_1}, \ldots, a^{s/r_j}, \infty, \ldots, \infty}\|^{\theta_j}_{L_{p_j}(\mathscr{E})}$$

$$\leqq \sum_{s=0}^{\infty} a^{\theta_j \frac{r_j - \varrho_j}{r_j} s} \omega^{k_j}_{x_j}(f^{(\varrho_j)}_{x_j}, a^{-s/r_j})^{\theta_j}_{L_{p_j}(\mathscr{E})} \ll \int_0^1 t^{-(r_j - \varrho_j)\theta_j - 1} \omega^{k_j}_{x_j}(f^{(\varrho_j)}_{x_j}, t)^{\theta_j} dt \ll \|f\|^{\theta_j}_{b^r_{p\theta}(\mathscr{E})}$$

and we have proved (1).

Now suppose that $p = p_1 = \cdots = p_n$, $\theta = \theta_1 = \cdots = \theta_m$.

We introduce the norms ($\|f\| = \|f\|_{L_p(\mathscr{E})}$):

(2)
$$^j\|f\|_B = \|f\| + {}^j\|f\|_b \quad (j = 1, 2, 3).$$

We suppose that ${}^1b = b^r_{p\theta}(\mathscr{E})$, i.e. that it is already a known class,

(3) $$ {}^2\|f\|_b = \left(\sum_{s=0}^{\infty} a^{\theta s} E_{a^{s/r_1},\ldots,a^{s/r_m}}(f)^\theta_{L_p(\mathscr{E})}\right)^{1/\theta} $$

and

(4) $$ {}^3\|f\|_B = \left(\sum_{s=0}^{\infty} a^{\theta s} \|Q_s\|^\theta\right)^{1/\theta}, $$

where we are supposing that f is representable in the form of a series

(5) $$ f = \sum_{s=0}^{\infty} Q_s $$

converging to it in the sense of $L_p(\mathscr{E})$, whose terms Q_s are functions of entire type a^{s/r_j} relative to the respective x_j, $j = 1, \ldots, m$.

The norms (2) (of the classes B), but not the ${}^j\|\cdot\|_b$, are equivalent.

Indeed, suppose that $f \in {}^1B = B^r_{p\theta}(\mathscr{E})$

(6) $$ {}^2\|f\|_b \leq \left(\sum_{s=0}^{\infty} a^{\theta s} \|f - g_{a^{s/r_1},\ldots,a^{s/r_m}}\|^\theta_{L_p(\mathscr{E})}\right)^{1/\theta} \ll {}^1\|f\|_B $$

(the middle part of (6) does not exceed the sum of the left sides of inequalities (1) for equal p_j and equal θ_j).

On the other hand,

(7) $$ {}^2\|f\|_B = \|f\|_{L_p(\mathscr{E})} + \left(\sum_{0}^{\infty} a^{\theta s} E_{x_j a^{s/r_j}}(f)^\theta\right)^{1/\theta} \gg \|f\|_{B^{r_j}_{x_j p\theta}(\mathscr{E})}, $$

where the second inequality is true in view of the equivalence of the norms corresponding to the seminorms 5.6 (1) and 5.6 (5).

From (6) and (7) it follows that ${}^1B = {}^2B$.

We turn to the proof of the equivalence ${}^1B = {}^3B$. Suppose that $f \in {}^1B$. We define for f a family of entire functions $g_s = g_{a^{s/r_1},\ldots,a^{s/r_m}}$ ($a > 1$, $s = 0, 1, \ldots$), for which (6) holds:

(8) $$ \left(\sum_{s=0}^{\infty} a^{\theta s} \|f - g_s\|^\theta\right)^{1/\theta} \ll {}^1\|f\|_B. $$

Hence in particular it follows that

$$ \|f - g_0\| \ll {}^1\|f\|_B \text{ and } \|g_0\| \ll {}^1\|f\|_B. $$

Put

(9) $$ Q_0 = g_0, \quad Q_s = g_s - g_{s-1} \quad (s = 1, 2, \ldots). $$

From the convergence of the series entering into (8) it follows that the function is representable in the form of a series (5) converging to it in the sense of $L_p(\mathscr{E})$.

Further,

$$^3\|f\|_B = \left(\sum_{s=0}^{\infty} a^{\theta s}\|Q_s\|^\theta\right)^{1/\theta} \leqq \|Q_0\| + \left(\sum_{s=1}^{\infty} a^{\theta s}\|g_s - f\|^\theta\right)^{1/\theta}$$

$$+ \left(\sum_{s=1}^{\infty} a^{\theta s}\|g_{s-1} - f\|^\theta\right)^{1/\theta} \leqq 3\|f\|_{1_B}.$$

Finally, if $f \in {}^3B$, then f is representable in the form of a series (5) with finite norm (4). But Q_s for each f is entire of type a^{s/r_j} relative to x_j, so that $f \in B^r_{x_j p}(\mathscr{E})$ (see 5.6 (6), replace a^r by a and put $m = 1$, $\mathbb{R}_m = \mathbb{R}_{x_j}$) and

$$\|f\|_{B^{r_j}_{x_j p}(\mathscr{E})} \ll {}^3\|f\|_B \quad (j = 1, \ldots, m).$$

Thus $f \in B^r_p(\mathscr{E})$ and

$$^1\|f\| = \|f\|_{B^r_p(\mathscr{E})} \ll {}^3\|f\|_B.$$

We have proved that ${}^1B = {}^3B$.

In conclusion we emphasize that the norms of the classes ${}^1B = B^r_p(\mathscr{E})$ are expressed (4.3.4) in terms of the norms $B^{r_j}_{x_j p}(\mathscr{E})$ $(j = 1, \ldots, m)$, which may be conceived of in any of the equivalent forms decribed in 5.6.6 $(m = 1, \mathbb{R}_m = \mathbb{R}_{x_j})$.

We note that we have been supposing throughout here that θ_j and θ may be equal to ∞, so that in particular we have proved that, in the notations of subsection 5.5.6, ${}^3H = {}^4H$.

5.6.2. We shall show the equivalance of the classes

$$B^{r,\ldots,r}_{p\theta}(\mathscr{E}) = B^r_{p\theta}(\mathscr{E}) \quad (1 \leqq \theta \leqq \infty). \tag{1}$$

The first of them we will denote by B, the second by B'. We choose a number a so that $a^{1/r} \geqq \sqrt{m}$. Then $\sqrt{m}a^{s/r} \leqq a^{s+1/r}$ $(s = 0, 1, \ldots, m)$. We note that the entire function $Q_{a^{s/r},\ldots,a^{s/r}}$, of type $a^{s/r}$ relative to each variable x_j $(j = 1, \ldots, m)$, is at the same time spherical of type $\sqrt{m}\,a^{s/r}$ relative to $\boldsymbol{u}$, and all the more of spherical type $a^{(s+1)/r}$ relative to $\boldsymbol{u}$:

$$Q_{a^{s/r},\ldots,a^{s/r}} = Q_{\boldsymbol{u}a^{(s+1)/r}}.$$

Now suppose that $f \in B$. Then

$$f = \sum_{s=0}^{\infty} Q_{a^{s/r},\ldots,a^{s/r}} = \sum_{s=0}^{\infty} Q_{ua^{(s+1)/r}}$$

and

$$\|f\|_B = \left(\sum_{s=0}^{\infty} a^{\theta s}\|Q_{ua^{(s+r)/r}}\|_p^\theta\right)^{1/\theta}$$

$$= \frac{1}{a}\left(\sum_{s=0}^{\infty} a^{\theta(s+1)}\|Q_{ua^{(s+1)/r}}\|_p^\theta\right)^{1/\theta} \leqq \frac{1}{a}\left(\sum_{s=0}^{\infty} a^{\theta s}\|Q_{ua^{s/r}}\|_p^\theta\right)^{1/\theta} = \frac{1}{a}\|f\|_{B'},$$

where we have put $Q_{ua^0} = Q_{u1} \equiv 0$.

Thus we have proved that if the function $f \in B$, then it is representable in the form of a series

$$f \in \sum_0^\infty Q_{ua^{s/r}}$$

of entire functions of spherical type $a^{s/r}$ relative to $\boldsymbol{u}$, such that

$$\|f\|_{B'} \ll \|f\|_B.$$

I.e., we have proved that $B \to B'$. The reverse imbedding is trivial, and we have proved (1).

5.6.3. Theorem[1]. *Suppose that* $f \in B^{\boldsymbol{r}}_{p\theta}(\mathscr{E})$ *and that* $\boldsymbol{l} = (l_1, \ldots, l_m)$ — *is an integer nonnegative vector* ($l_j \geqq 0$) *such that*

(1) $$\varkappa = 1 - \sum_{j=1}^m \frac{l_j}{r_j} > 0.$$

Then there exists a derivative

(2) $$f^{(\boldsymbol{l})} \in B^{\varkappa \boldsymbol{r}}_{p\theta}(\mathscr{E})$$

and

(3) $$\|f^{(\boldsymbol{l})}\|_{B^{\varkappa\boldsymbol{r}}_{p\theta}(\mathscr{E})} \leqq c\|f\|_{B^{\boldsymbol{r}}_{p\theta}(\mathscr{E})},$$

where c *does not depend on* f.

The theorem is no longer true if $\varrho = \varkappa\boldsymbol{r}$ is replaced by $\varrho + \varepsilon$, where $\varepsilon > 0$ (see 7.5). Moreover, it is generally speaking not true for $\varkappa = 0$ (see the Remark to 5.6.3).

Proof. In view of the hypothesis of the theorem

$$f = \sum_{s=0}^\infty Q_{a^{s/r_1}, \ldots, a^{s/r_m}} = \sum_0^\infty Q_s \quad (a > 1),$$

[1] See the remarks at the end of the book to subsections 5.6.2—5.6.3.

where the terms of the series are entire functions of exponential type a^{s/r_j} relative to x_j $(j = 1, \ldots, m)$, while

$$\|f\|_B = \left(\sum_0^\infty a^{\theta s}\|Q_s\|^\theta\right)^{1/\theta}$$

$$(B = B^r_{p\theta}(\mathscr{E}),\ \|\cdot\|_{L_p(\mathscr{E})},\quad a > 1).$$

We have for the time being formally

$$f^{(\boldsymbol{k})} = \sum_0^\infty Q_s^{(\boldsymbol{k})}, \tag{4}$$

where $\boldsymbol{k}$ is any one of the vectors $(l_1, 0, \ldots, 0)$, $(l_1, l_2, 0, \ldots, 0)$, $\ldots$, $\boldsymbol{l} = (l_1, \ldots, l_n)$. We note that $\|Q_s^{(\boldsymbol{k})}\| \leqq a^{s \sum_1^m \frac{l_j}{r_j}} \|Q_s\| = a^{s(1-\varkappa)}\|Q_s\|$.

Hence

$$\left(\sum_0^\infty a^{s\varkappa\theta}\|Q_s^{(\boldsymbol{k})}\|^\theta\right)^{1/\theta} \leqq \left(\sum_0^\infty a^{s\theta}\|Q_s\|^\theta\right)^{1/\theta} = \|f\|_B. \tag{5}$$

It follows from (5) that the series (4) converges in the sense of L_p, so that, on the basis of Lemma 4.4.7, term-by-term differentiation in (4), in the generalized sense, is legitimate.

We note that $Q_s^{(\boldsymbol{l})}$, just as is Q_s, is entire of type a^{s/r_j} relative to x_j $(j = 1, \ldots, m)$. If one puts $a^\varkappa = b$ $(b > 1)$, then inequality (5) for $\boldsymbol{k} = \boldsymbol{l}$ may be written in the form

$$\left(\sum_0^\infty b^{s\theta}\|Q_s^{(\boldsymbol{l})}\|^\theta\right)^{1/\theta} \leqq \|f\|_B,$$

where $Q_s^{(\boldsymbol{l})}$ is entire of type $b^{s/r_j\varkappa}$ relative to x_j. In such a case $f^{(\boldsymbol{l})} \in B^{\varkappa \boldsymbol{r}}_{p\theta}(\mathscr{E})$, and inequality (3) is satisfied.

We wish to go into this somewhat further. Suppose that we wanted to differentiate the derivative $f^{(\boldsymbol{l})}$ of the theorem still $\boldsymbol{l}' = (l_1', \ldots, l_m')$ "times" more. According to that theorem this is possible, if the quantity

$$\varkappa' = 1 - \sum_{j=1}^m \frac{l_j'}{r_j\varkappa} > 0.$$

Hence

$$\varkappa\varkappa' = \varkappa - \sum_{j=1}^m \frac{l_j'}{r_j} = 1 - \sum_{j=1}^m \frac{l_j + l_j'}{r_j} = \varkappa_* > 0.$$

But the quantity $\varkappa_*$ is in its turn a constant like the $\varkappa$ appearing in our theorem, if one replaces $\boldsymbol{l}$ in it by $\boldsymbol{l} + \boldsymbol{l}'$.

In this sense the theorem we have just proved has a transitive character.

5.6.4. *Example.* Below we present an example showing that the seminorms ${}^3b$ and ${}^3b'$ are generally speaking not equivalent (see 5.6 (3), (4)). We restrict ourselves to the one-dimensional case:

$$m = 1, r = 1 - \frac{1}{p} < 1, \quad \varrho = 0, \quad k = 1, \theta = p.$$

Suppose that $f_N(x)$ is an even function, given by

$$f_N(x) = \begin{cases} \dfrac{x}{N}, & 0 \leqq x \leqq N, \\ 1, & N < x. \end{cases}$$

Then

$${}^3\|f_N\|_p^b = 2\int_0^\infty dh \int_{-\infty}^\infty \left|\frac{f_N(x+h) - f_N(x)}{h}\right|^p dx \geqq 2\int_0^N dx \int_0^{N-x} \frac{dh}{N^p} = N^{2-p},$$

$$\frac{1}{2}\,{}^3\|f_N\|_{b'}^p = \int_0^1 dh \left\{\int_0^\infty + \int_{-\infty}^{-h} + \int_{-h}^0\right\} dx = J_1 + J_2 + J_3 = O(N^{1-p}),$$

because

$$J_2 = J_1 \leqq \int_0^{N-1} dx \int_0^1 \frac{dh}{N^p} + \int_{N-1}^N dx \left\{\int_0^{N-x} \frac{dh}{N^p} + \int_{N-x}^1 \left|\frac{1 - \frac{x}{N}}{h}\right|^p dh\right\}$$
$$= O(N^{1-p}),$$

$$J_3 = \int_0^1 dh \int_{-h}^0 \left|\frac{\frac{x+h}{N} + \frac{x}{N}}{h}\right|^p dx = O(N^{1-p}).$$

Hence it is clear that there does not exist a constant c such that for all $N > 0$ the inequality ${}^3\|f_N\|_b \leqq c\,{}^3\|f_N\|_{b'}$ is satisfied.

5.6.5. Continuity relative to a shift. Theorem. *As* $h \to 0$

(1) $\quad \|f(x+h) - f(x)\|_W \to 0 \; (f \in W = W_p^l(\mathbb{R}_n),\ 1 \leqq p < \infty,\ l \geqq 0).$

(2) $\quad \|f(x+h) - f(x)\|_B \to 0 \; (f \in B = B_p^r(\mathbb{R}_n),\ 1 \leqq p, \theta < \infty,\ r \geqq 0).$

Assertion (1) does not hold for $p = \infty$, nor (2) for $\theta = \infty$ ($B^{\boldsymbol{r}}_{p\infty} = H^{\boldsymbol{r}}_p$; see 7.4.1. farther on). For $p = \infty$ and $1 \leqq \theta < \infty$ (2) remains true.

Proof. In the case: $\boldsymbol{l} = \boldsymbol{0}$ ($W^{\boldsymbol{0}}_p = L_p(\mathbb{R}_n)$) property (1) is a well-known fact (not true, however, for $p = \infty$).

The general case reduces to it as well, because $\|f\|_W$ is the sum of the norm f and $\dfrac{\partial^{l_j} f}{\partial x_j^{l_j}}$ in $L_p(\mathbb{R}_n)$ ($j = 1, \ldots, n$). For a function $f \in B$ we have the representation

$$f = \sum_0^\infty Q_s,$$

$$\|f\|_B = \left\{\sum_0^\infty 2^{s\theta} \|Q_s\|_p^\theta\right\}^{1/\theta},$$

where the Q_s are entire functions of type $2^{s/r_j}$ relative to x_j. Hence

$$\|f(\boldsymbol{x} + \boldsymbol{h}) - f(\boldsymbol{x})\|_B \leqq \left\{\sum_0^{N-1} 2^{s\theta} \|Q_s(\boldsymbol{x} + \boldsymbol{h}) - Q_s(\boldsymbol{x})\|_p^\theta\right\}^{1/\theta}$$

$$+ 2\left\{\sum_N^\infty 2^{s\theta} \|Q_s\|_p^\theta\right\}^{1/\theta} < \varepsilon + \varepsilon = 2\varepsilon, \quad |\boldsymbol{h}| < \delta;$$

one now takes N sufficiently large and thereupon chooses a sufficiently small δ.

Remark. In (1) and (2) one may replace p by $\boldsymbol{p} = (p_1, \ldots, p_n)$ ($1 \leqq p_j < \infty$), because these relations are true in particular for the classes $W^{l_j}_{x_j, p_j}(\mathbb{R}_n)$, $B^{r_j}_{x_j, p_j}\mathbb{R}_n)$, $j = 1, \ldots, n$.

5.6.6. Under the condition that $1 \leqq \theta, p < \infty$, $g \subset \mathbb{R}_n$ is an open set, $g^N = g\,(\mathbb{R}_n - V_N)$, where V_N is a solid sphere with center at the origin of radius N, and $f \in B^r_{p\theta}(g) = B(g)$, we have

$$(1) \qquad \|f\|_{B(g^N)} \to 0 \quad (N \to \infty).$$

This is clear from the definition of the norm $\|\cdot\|_B$, for example in the form 4.3.4 (2) ($\varrho = 0$, $k \geqq 2$):

$$^2\|f\|_{b^r_{p\theta}(g^N)} = \left(\int_0^\infty t^{-1-\theta r} \Omega^k_{\mathbb{R}_m}(f, t)^\theta_{L_p(g^N)}\right)^{1/\theta} \to 0 \quad (N \to \infty).$$

Indeed, $f \in L_p(g)$, so that $\Omega^k_{\mathbb{R}_m}(f, t)_{L_p(g^N)}$ is finite for any t and tends monotonically to zero as $N \to \infty$. One may then apply Lebesgue's theorem on the passage to the limit under the integral sign.

Chapter 6

Imbedding Theorems for Different Metrics and Dimensions

6.1. Introduction

We begin by presenting the formulation of the imbedding theorem of S. L. Sobolev [3], with later complements due to V. I. Kondrašov [1] and V. P. Il'in [2][1]. As applied to the space $\mathbb{R}_n$ and to its coordinate subspace $\mathbb{R}_m (1 \leqq m \leqq n)$, this theorem reads:

If the function $f \in W_p^l(\mathbb{R}_n)$ and

(1) $$0 \leqq \varrho = l - \frac{n}{p} + \frac{m}{p'}, \quad 1 \leqq p < p' < \infty,$$

then[2]

(2) $$W_p^l(\mathbb{R}_n) \to W_{p'}^{[\varrho]}(\mathbb{R}_m),$$

where $[\varrho]$ *is the integer part of* ϱ. *This means that there exists a trace of the function* $f|_{\mathbb{R}_m} = \varphi$, *lying in the class* $W_{p'}^{(\varrho)}(\mathbb{R}_m)$, *and that the inequality*

(3) $$\|\varphi\|_{W_{p'}^{[\varrho]}(\mathbb{R}_m)} \leqq c\|f\|_{W_p^l(\mathbb{R}_n)}$$

is satisfied, where c does not depend on f.

The concept of the trace of f will be explained in what follows. For the time being we will say that in every case, if f is continuous on $\mathbb{R}_n$, then its *trace* on $\mathbb{R}_m$ is the function $\varphi = f|_{\mathbb{R}_m}$ induced by the function f on $\mathbb{R}_m$.

In particular, for $m = n$ it follows from (2) that one has the "pure" imbedding of different metrics:

(4) $$W_p^l(\mathbb{R}_n) \to W_{p'}^{[\varrho]}(\mathbb{R}_n),$$

---

[1] See the remark to 6.1 at the end of the book.

[2] The general theorem of S. L. Sobolev may be written in the form of formula (2), where we need to replace $\mathbb{R}_n$, $\mathbb{R}_m$ respectively by g, $\Lambda_m = \mathbb{R}_m g$ and suppose that g is star-shaped relative to some n-dimensional solid sphere.

asserting that if $f \in W_p^l(\mathbb{R}_n)$, then $f \in W_{p'}^{[\varrho]}(\mathbb{R}_n)$ and

$$(5) \qquad \|f\|_{W_{p'}^{[\varrho]}(\mathbb{R}_n)} \leqq c\|f\|_{W_p^l(\mathbb{R}_n)}$$

under the condition that (1) is satisfied (for $m = n$).

The imbedding theorems of S. L. Sobolev will be proved in Chapter 9.

In the present chapter we intend to consider the questions just mentioned for the classes $B_{p\theta}^r(\mathbb{R}_n)$, in particular (for $\theta = \infty$) the classes $H_p^r(\mathbb{R}_n)$. Besides, from the theorems obtained in this chapter, it will follow in particular that the theorems formulated above, in the case when ϱ is a positive noninteger real number, will follow. And then, as we shall see, they are true under wider conditions: $1 \leqq p \leqq p' \leqq \infty$.

We shall present a now already characteristic theorem on imbeddings of different metrics, which, in particular, will be obtained in this chapter:

$$(6) \qquad B_{p\theta}^r(\mathbb{R}_n) \to B_{p'\theta}^\varrho(\mathbb{R}_n),$$

if

$$(7) \qquad 1 \leqq p < p' \leqq \infty, \quad 1 \leqq \theta \leqq \infty, \quad \varrho = r - n\left(\frac{1}{p} - \frac{1}{p'}\right) > 0.$$

Thus, if the function f belongs to the left class (6), then it belongs to the right class, and moreover the inequality

$$(8) \qquad \|f\|_{B_{p'}^\varrho(\mathbb{R}_n)} \leqq c\|f\|_{B_{p\theta}^r(\mathbb{R}_n)}$$

is satisfied, where c does not depend on f.

The characteristic (direct) imbedding theorem for different dimensions which will be proved in this chapter is written as follows:

$$(9) \qquad B_{p\theta}^r(\mathbb{R}_n) \to B_{p\theta}^\varrho(\mathbb{R}_m),$$

where

$$(10) \qquad 1 \leqq p, \theta \leqq \infty, \quad 1 \leqq m < n, \quad \varrho = r - \frac{n-m}{p} > 0.$$

It asserts that under the hypotheses (10), if one is given on $\mathbb{R}_n$ a function f of the class $B_{p\theta}^r(\mathbb{R}_n)$, then it has a trace φ on $\mathbb{R}_m$, lying in the class $B_{p\theta}^\varrho(\mathbb{R}_m)$, and that the inequality

$$(11) \qquad \|\varphi\|_{B_{p\theta}^\varrho(\mathbb{R}_m)} \leqq c\|f\|_{B_{p\theta}^r(\mathbb{R}_n)}$$

is satisfied, where c does not depend on f.

Inequality (11) is important in the applications. It exhibits the definite (stable) dependence of the norms of the traces of functions f on the norms of the f themselves.

Imbedding theorems for different dimensions for the classes $B^r_{p\theta}$ are characterized by the fact that they are completely inversible. As an example we present a theorem inverse to theorem (9). It is written as follows:

$$B^{\varrho}_{p\theta}\mathbb{R}(_m) \to B^r_{p\theta}(\mathbb{R}_n) \tag{12}$$

(under conditions (10)), and reads as follows: *to each function φ defined on $\mathbb{R}_m$ and belonging to the class $B^{\varrho}_{p\theta}(\mathbb{R}_m)$, it is possible to bring into correspondence its extension to $\mathbb{R}_n$—a function $f \in B^r_{p\theta}(\mathbb{R}_n)$ — such that $f|_{\mathbb{R}_m} = \varphi$*

$$f|_{\mathbb{R}_m} = \varphi \textit{ and } \|f\|_{B^r_{p\theta}(\mathbb{R}_n)} \leqq c\|\varphi\|_{B^{\varrho}_{p\theta}(\mathbb{R}_m)}, \tag{13}$$

where c does not depend on φ.

The more general imbedding theorems for different dimensions which the reader will find in the present chapter are also respectively completely inversible. This shows in particular that these theorems cannot be improved. As to the imbedding theorems for different metrics, they are also nonimprovable in the terms in which they are expressed. This is proved in the following chapter. There the reader will find some interesting so-called transitive properties of imbedding theorems.

We begin this chapter by establishing the simplest connections among the classes W, H, B, expressed in terms of imbeddings.

Here we note only the following connections:

$$H^{r+\varepsilon}_p \to W^r_p \to H^r_p \quad (\varepsilon > 0, r = 0, 1, \ldots), \tag{14}$$

the second of which is already known to us.

From (14), (6) and (9) it follows, for $\varrho = l - \frac{n}{p} + \frac{m}{p'} > 0$ noninteger:

$$W^l_p(\mathbb{R}_n) \to H^l_p(\mathbb{R}_n) \to H_{p'}^{l-n\left(\frac{1}{p}-\frac{1}{p'}\right)}(\mathbb{R}_n) \to H^{\varrho}_{p'}(\mathbb{R}_m) \to W^{[\varrho]}_{p'}(\mathbb{R}_m),$$

i.e. (5).

For the classes $H^r_p(\mathbb{R}_n)$ the imbedding theorems of different metrics and dimensions, just as the inverse imbedding theorems for different dimensions, were proved by S. M. Nikol'skiĭ [3] by methods of approximation by entire functions of exponential type. They were generalized by O. V. Besov [2, 3] to the classes $B^r_{p\theta}(\mathbb{R}_n)$ ($H^r_p = B^r_{p\infty}$) introduced by him. Besov also based himself on the method of approximations by entire functions of exponential type. Certain theorems for imbeddings of different metrics for the one-dimensional classes H^r_p were known to Hardy and Littlewood [1]. For the more general classes $H_p^r(\mathbb{R}_n)$ (the p_j being in general different), an imbedding theorem for different dimensions was proved by S. M. Nikol'skiĭ [10] by the method of approximation. Then it was generalized to the classes $B_{p\theta}^r$ by V. P. Il'in and V. A. Solonnikov [1, 2], but by quite different methods.

In the proofs below we shall base ourselves on the methods of approximation, including the consideration of the theorem indicated above for the general classes $B_{p\theta}^r$[1].

6.2. Connections Among the Classes B, H, W

The functions of these classes will be considered on a cylindrical measurable set $\mathscr{E} = \mathscr{E}_m \times \mathscr{E}'$ $\left(1 \leqq m \leqq n,\ \boldsymbol{u} = (x_1, \ldots, x_m),\ \boldsymbol{w} = (x_{m+1}, \ldots, x_n)\right)$.

We put for short

$$B_{\boldsymbol{u}p\theta}^r(\mathscr{E}) = B_{p\theta}^r,\ H_{\boldsymbol{u}p}^r(\mathscr{E}) = H_p^r,\quad W_{\boldsymbol{u}p}^r(\mathscr{E}) = W_p^r,$$

$$\|f\|_{L_p(\mathscr{E})} = \|f\|,\quad r > 0,\quad 1 \leqq \theta \leqq \infty.$$

The following imbeddings hold, where $r = \bar{r} + \alpha$, $0 < \alpha \leqq 1$, $\bar{r}$ an integer:

$$B_{p1}^r \to B_{p\theta}^r \to B_{p\theta'}^r \to B_{p\infty}^r = H_p^r\ (1 \leqq \theta < \theta' \leqq \infty), \tag{1}$$

$$W_p^r \to H_p^r \quad (r = 1, 2, \ldots), \tag{2}$$

$$H_p^{r+\varepsilon} \to B_{p\theta}^r \to H_p^r \qquad (\varepsilon > 0), \tag{3}$$

$$H_p^r \to W_p^\varrho \qquad (\varrho = 0, 1, \ldots, \bar{r}), \tag{4}$$

$$B_{p\theta}^{r+\varepsilon} \to B_{p\theta}^r \qquad (\varepsilon > 0). \tag{5}$$

The imbeddings (1) show that as θ increases the classes $B_{p\theta}^r$ widen. The proof of (1) follows directly from the fact that (see 5.6 (6), (7)) the function $f \in B_{p\theta}^r$ may be defined as the sum of the series

$$f = \sum_{s=0}^{\infty} Q_s, \tag{6}$$

converging to it in the sense of L_p. Here the terms are functions Q_s which are entire of spherical type $a^s (a > 1)$ relative to $\boldsymbol{u}$ such that

$$\|f\|_{B_{p\theta}^r} = \left(\sum_{s=0}^{\infty} a^{sr\theta} \|Q_s\|^\theta\right)^{1/\theta} \quad (a > 1). \tag{7}$$

But as θ increases the right side of (7) decreases (see 3.3.3). From the chain (1) it is also clear that for fixed r and p the "worst" class is the class H_p^r and the "best" B_{p1}^r.

[1] See T. I. Amanov [3].

The imbeddings (3), from which it follows that

$$B^{r+\varepsilon}_{p\theta'} \to B^r_{p\theta} \to B^{r-\varepsilon}_{p\theta''}$$

for arbitrary $1 \leqq \theta', \theta, \theta'' \leqq \infty$, no matter how small the positive ε is, show that the class $B^r_{p\theta}$ depends more essentially on r than it does on θ. The second imbedding (3) already has been proved in (1). Suppose that $f \in H^{r+\varepsilon}_p$. Then

$$\|f\|_{H^{r+\varepsilon}_p} = \sup_s a^{s(r+\varepsilon)} \|Q_s\| = M < \infty,$$

so that

$$\|f\|_{B^r_{p\theta}} \leqq \left\{\sum_{s=0}^{\infty} \left(a^{sr} \frac{M}{a^{s(r+\varepsilon)}}\right)^\theta\right\}^{1/\theta} \leqq cM,$$

where c does not depend on M. Hence the first imbedding in (3) follows. The imbedding (5) follows from the fact that the right side of (7) grows along with r. We need to keep in mind that for any $r_0 > 0$, for all $r < r_0$ the functions Q_s may be considered as one and the same (see the remark at the end of § 5.6).

The imbedding (2) follows from the inequalities ($\boldsymbol{h} \in \mathbb{R}_m$):

$$\|\Delta^k_{\boldsymbol{h}} f^{(\bar{r})} \boldsymbol{x})\| \leqq 2^{k-1} \|\Delta_{\boldsymbol{h}} f^{(\bar{r})}(\boldsymbol{x})\| \leqq 2^{k-1} |\boldsymbol{h}| \left\| \frac{\partial}{\partial \boldsymbol{h}} f^{(\bar{r})}(\boldsymbol{x}) \right\|$$

$$\ll |\boldsymbol{h}| \sum \|f^{(r)}\| \quad (\bar{r} = r - 1),$$

where the sum is extended over all derivatives $f^{(r)}$ of f of order r.

It follows from (1) and (4) that

(8) $$B^r_{p\theta} \to H^r_p \to W^\varrho_p \quad (\varrho = 0, 1, \ldots, \bar{r}).$$

For the anisotropy classes

$$B^{\boldsymbol{r}}_{up\theta}(\mathscr{E}) = B^{\boldsymbol{r}}_{p\theta}, \quad H^{\boldsymbol{r}}_{up}(\mathscr{E}) = H^{\boldsymbol{r}}_{up}, \quad W^{\boldsymbol{r}}_{up}(\mathscr{E}) = W^{\boldsymbol{r}}_p$$

the following imbeddings hold ($\boldsymbol{p} = (p_1, \ldots, p_n)$):

(9) $$B^{\boldsymbol{r}}_{p1} \to B^{\boldsymbol{r}}_{p\theta} \to B^{\boldsymbol{r}}_{p\theta'} \to B^{\boldsymbol{r}}_{p\infty} = H^{\boldsymbol{r}}_p \quad 1 \leqq \theta < \theta' < \infty,$$

(10) $$W^{\boldsymbol{\varrho}}_p \to H^{\boldsymbol{\varrho}}_p \quad (\boldsymbol{\varrho} \text{ — an integer vector})$$

(11) $$H^{\boldsymbol{r}+\boldsymbol{\varepsilon}}_p \to B^{\boldsymbol{r}}_{p\theta} \to H^{\boldsymbol{r}}_p \quad (\boldsymbol{\varepsilon} > \boldsymbol{0}, \text{ i.e. } \varepsilon_j > 0),$$

(12) $$B^{\boldsymbol{r}+\boldsymbol{\varepsilon}}_{\boldsymbol{p}\theta} \to B^{\boldsymbol{r}}_{\boldsymbol{p}\theta} \quad (\boldsymbol{r} > \boldsymbol{0}),$$

(13) $$B^{\boldsymbol{r}}_{p\theta} \to H^{\boldsymbol{r}}_p \to W^{\boldsymbol{\varrho}}_p \quad (\boldsymbol{\varrho} < \boldsymbol{r}, \boldsymbol{\varrho} \text{ an integer vector})$$

They are analogous to the imbeddings (1)—(5), (8), and follow immediately from them. If $p = p_1 = \cdots = p_n$, then $\mathbf{p}$ will be replaced by p throughout.

6.3. Imbedding of Different Metrics

The following imbedding holds[1]:

$$B^{\mathbf{r}}_{p\theta}(\mathbb{R}_n) \to B^{\mathbf{r}'}_{p'\theta}(\mathbb{R}_n), \tag{1}$$

if the following conditions are satisfied:

$$1 \leqq p < p' \leqq \infty, \quad 1 \leqq \theta \leqq \infty, \tag{2}$$

$$\varkappa = 1 - \left(\frac{1}{p} - \frac{1}{p'}\right) \sum_1^n \frac{1}{r_j} > 0, \tag{3}$$

$$\mathbf{r}' = \varkappa \mathbf{r}. \tag{4}$$

Here we are supposing that $\mathbf{r} > 0$.

In particular if we take into account that $B^r_p(\mathbb{R}_n) = B^{r,\ldots,r}_p(\mathbb{R}_n)$ (see 5.6.2), we have

$$B^{r}_{p\theta}(\mathbb{R}_n) \to B^{r'}_{p\theta}(\mathbb{R}_n) \tag{1'}$$

under the conditions

$$1 \leqq p < p' \leqq \infty, \tag{2'}$$

$$\varkappa = 1 - \left(\frac{1}{p} - \frac{1}{p'}\right) \frac{n}{r} > 0, \tag{3'}$$

$$r' = \varkappa r. \tag{4'}$$

For example, for $p' = \infty$ and

$$r' = r - n\left(\frac{1}{p} - \frac{1}{\infty}\right) = r - \frac{n}{p} > 0,$$

$$B^r_{p\theta}(\mathbb{R}_n) \to B^{r'}_{\infty\theta}(\mathbb{R}_n) \to H^{r'}_{\infty}(\mathbb{R}_n)$$

and, accordingly, if the function $f \in B^r_{p\theta}(\mathbb{R}_n)$, then it is continuous and bounded on $\mathbb{R}_n$ along with its partial derivatives of order less than r'. Morevoer, if, for example, $r' = \varrho + \alpha$ where ϱ is an integer and $0 < \alpha < 1$, then the derivatives $f^{(\varrho)}$ of order ϱ satisfy on $\mathbb{R}_n$ a Hölder condition of order α.

---

[1] S. M. Nikol'skiĭ [3], the case $H^{\mathbf{r}}_p(\mathbb{R}_n) = B^{\mathbf{r}}_{p\infty}(\mathbb{R}_n)$; O. V. Besov [2, 3], the case $1 \leqq \theta < \infty$; for certain one-dimensional classes H^r_p, Hardy and Littlewood [1].

Let us prove (1). Suppose that B and B' denote respectively the first and second classes in (1) and that $\|\cdot\|_p = \|\cdot\|_{L_p(\mathbb{R}_m)}$. Suppose given a function $f \in B$ and a number $a > 1$. The function may be represented in the form of a series

(5) $$f = \sum_{s=0}^{\infty} Q_s,$$

whose terms Q_s are entire of order a^{r/s_j} relative to x_j ($j = 1, \ldots, n$), and in addition

(6) $$\|f\|_B = \left(\sum_0^\infty a^{\theta s}\|Q_s\|_p^\theta\right)^{1/\theta} < \infty \quad (a > 1).$$

For the function Q_s the inequality of different metrics is satisfied:

$$\|Q_s\|_{p'} \leqq 2^n a^{(1-\varkappa)s}\|Q_s\|_p,$$

so that

$$\left(\sum_0^\infty a^{\theta\varkappa s}\|Q_s\|_{p'}^\theta\right)^{1/\theta} \ll \left(\sum_0^\infty a^{\theta s}\|Q_s\|_p^\theta\right)^{1/\theta} = \|f\|_B.$$

But if one puts $a^\varkappa = b$ ($b > 1$) then we obtain the inequality

(7) $$\left(\sum_0^\infty b^{\theta s}(\|Q_s\|_{p'})^\theta\right)^{1/\theta} \ll \|f\|_B,$$

where the Q_s are entire of type b^{s/r'_j} relative to x_j. It follows from this that the series (5) converges in the metric of $L_{p'}$, and in fact to f, because it already converges to f in the metric of L_p (see 1.3.7)). Moreover, it follows from (7) that $f \in B'$ and that the left side of (7) is $\|f\|_{B'}$. We have proved that

$$\|f\|_{B'} \ll \|f\|_B,$$

and the imbedding (1) is proved.

In the case at hand the conditions (2')—(4') are equivalent to the following:

$$r, r' > 0, \quad 1 \leqq p < p' \leqq \infty, \quad r - \frac{n}{p} = r' - \frac{n}{p'}.$$

The quantity $r - \dfrac{n}{p}$ enters essentially into them. This has to be an invariant, in order for the imbedding to be guaranteed. In the general case for this topic see 7.1.

We note that in (1) it is not possible to replace $\mathbb{R}_n$ by $\mathbb{R}_n \times \mathscr{E}'$ since in this case there is no inequality similar to (7). And indeed that is in essence not possible, as one may discover on examples.

6.4. Trace of a Function

A function f, belonging to one or another of the classes $B(\mathbb{R}_n)$, $W(\mathbb{R}_n)$, is defined on $\mathbb{R}_n$ only up to a set of n-dimensional measure zero, or, as we will also say, up to equivalency relative to $\mathbb{R}_n$ or in the sense of $\mathbb{R}_n$. Therefore the trace of the function f,

$$(1) \qquad f|_{\mathbb{R}_m} = \varphi = \varphi(x_1, \ldots, x_m),$$

on an arbitrary subspace $\mathbb{R}_m \subset \mathbb{R}_n (m < n)$ does not make sense if this is understood literally.

Below we give a definition of the trace of a function f on $\mathbb{R}_m$, leading to a function φ unique up to equivalence relative to $\mathbb{R}_m$.

We shall denote each point $\boldsymbol{x} \in \mathbb{R}_n$ in the form of a pair $\boldsymbol{x} = (\boldsymbol{u}, \boldsymbol{w})$, where $\boldsymbol{u} = (x_1, \ldots, x_m)$, $\boldsymbol{w} = (x_{m+1}, \ldots, x_n)$, and suppose that $\mathbb{R}_m(\boldsymbol{w})$ is the m-dimensional subspace of points $(\boldsymbol{u}, \boldsymbol{w})$, where $\boldsymbol{w}$ is fixed, and $\boldsymbol{u}$ runs through all possible values. In particular, suppose that $\mathbb{R}_m(0) = \mathbb{R}_m$.

Suppose that $f(\boldsymbol{x})$ is a measurable function on $\mathbb{R}_n$.

We will say that the function

$$(2) \qquad \varphi = \varphi(\boldsymbol{u}) = f|_{\mathbb{R}_m}$$

is the trace of f on $\mathbb{R}_m$, if f may be altered on a set of n-dimensional zero in such a way that after this, for some p with $1 \leqq p \leqq \infty$ the following properties will be satisfied:

$$(1) \qquad f(\boldsymbol{u}, \boldsymbol{w}) \in L_p(\mathbb{R}_m(\boldsymbol{w})), \ |\boldsymbol{w}| < \delta,$$

$$(2) \qquad f(\boldsymbol{u}, 0) = \varphi(\boldsymbol{u}),$$

$$(3) \qquad \|f(\boldsymbol{u}, \boldsymbol{w}) - \varphi(\boldsymbol{u})\|_{L_p(\mathbb{R}_m(\boldsymbol{w}))} \to 0, \ |\boldsymbol{w}| \to 0,$$

where δ is sufficiently small.

We shall show that the trace of f on $\mathbb{R}_m$ defined in this may is unique up to equivalency in the sense of $\mathbb{R}_m$.

In fact, suppose we succeeded in finding two variants f_1 and f_2 of the function f on a set of n-dimensional measure zero and numbers p_1 and p_2 $(1 \leqq p_1 \leqq p_2 \leqq \infty)$ such that for each set f_1, φ_1, p_1 and f_2, φ_2, p_2 the relations (1)—(3) are satisfied. Suppose that $g \subset \mathbb{R}_m$ is any bounded open set. Then

$$(3) \qquad \|\varphi_2(\boldsymbol{u}) - \varphi_1(\boldsymbol{u})\|_{L_{p_1}(g)} \leqq \|\varphi_1(\boldsymbol{u}) - f_1(\boldsymbol{u}, \boldsymbol{w})\|_{L_{p_1}(g)}$$
$$+ \|f_1(\boldsymbol{u}, \boldsymbol{w}) - f_2(\boldsymbol{u}, \boldsymbol{w})\|_{L_{p_1}(g)} + c\|f_2(\boldsymbol{u}, \boldsymbol{w}) - \varphi_2(\boldsymbol{u})\|_{L_{p_2}(g)},$$

where c is a constant depending on the measure of g. The functions f_1 and f_2 are equivalent in the sense of $\mathbb{R}_n$, so that

$$\iint_{\mathbb{R}_n} |f_1 - f_2|^{p_1} du\, dw = 0.$$

By Fubini's theorem, for almost all w

$$\int_{\mathbb{R}_m(w)} |f_1 - f_2|^{p_1}\, du = 0.$$

But from the set of points w for which this equation is satisfied it is always possible to select a sequence $w_1, w_2, \ldots$ with $|w_k| \to 0$. The right side of (3), when w runs through this sequence, tends to zero. But then the left side is equal to zero. Since g is arbitrary in $\mathbb{R}_m$, then $\varphi_1 = \varphi_2$ on $\mathbb{R}_m$.

It is not hard to see that if the function f is not only measurable in $\mathbb{R}_n$, but also continuous in an n-dimensional neighborhood of $\mathbb{R}_m$, then its trace φ coincides with the trace of f on $\mathbb{R}_m$ in the usual understanding of this word. We note further that if for two measurable functions f_1, f_2 and for some p the operation of taking the trace (2) is possible, then it is possible also for any linear combination

$$c_1 f_1 + c_2 f_2,$$

where c_1 and c_2 are any real numbers.

If in fact we denote the operation of taking the trace by

$$\varphi = A(f) = f|_{\mathbb{R}_m}, \tag{4}$$

then for the f_1, f_2 and c_1, c_2 we have the equation

$$A(c_1 f_1 + c_2 f_2) = c_1 A(f_1) + c_2 A(f_2).$$

Thus the set of all measurable functions f for which the operation (4) is possible for some p is linear, and (4) is a linear operation (operator) defined on them. As will be clear in what follows, the functions of the classes $B^r_{p\theta}$ and W^r_p, for corresponding values of the parameters p, r have traces on $\mathbb{R}_m$ in the sense indicated above.

Suppose that the region $g \subset \mathbb{R}_n$ and $g' \subset \bar{g}$, so that in particular g' may be the boundary of g. Suppose further that a class of functions $\mathfrak{M}$ is defined on g'. Suppose given a function f on g, and suppose that it has a trace on g',

$$\varphi = f|_{g'},$$

lying in $\mathfrak{M}$. Then we will not only write $\varphi = f|_{g'} \in \mathfrak{M}$, but also $f \in \mathfrak{M}$.

6.5. Imbeddings of Different Dimensions

We have the following imbedding[1]:

$$B^{\boldsymbol{r}}_{p\theta}(\mathbb{R}_n) \to B^{\boldsymbol{r}'}_{p\theta}(\mathbb{R}_m) \tag{1}$$

under the hypotheses

$$0 \leqq m < n, \quad 1 \leqq p, \theta \leqq \infty, \tag{2}$$

$$\varkappa = 1 - \frac{1}{p} \sum_{j=m+1}^{n} \frac{1}{r_j} > 0, \tag{3}$$

$$\boldsymbol{r}' = (r'_1, \ldots, r'_m), \quad r'_j = \varkappa r_j. \tag{4}$$

Here $\mathbb{R}_m$ denotes an m-dimensional subspace of points

$$\boldsymbol{x} = (\boldsymbol{u}, \boldsymbol{y}) = (x_1, \ldots, x_m, \quad x_{m+1}, \ldots, x_n),$$

$\boldsymbol{u} = (x_1, \ldots, x_m)$, $\boldsymbol{y} = (x_{m+1}, \ldots, x_n)$, where $\boldsymbol{y}$ is fixed. Suppose that B, B' are correspondingly the first and second classes in (1) and $\|\cdot\|^m = \|\cdot\|_{L_p(\mathbb{R}_m)}$. The imbedding (1) asserts that each function $f \in B$ has a trace

$$f|_{\mathbb{R}_m} = \varphi \in B',$$

and that the inequality [2]

$$\|\varphi\|_{B'} \leqq c\|f\|_B$$

is satisfied, where c does not depend on f.

In case $m = 0$ one supposes that

$$B^{\boldsymbol{r}'}_{p\theta}(\mathbb{R}_0) = B^{\boldsymbol{r}'}_{\infty,\theta}(\mathbb{R}_n).$$

Thus, in this case we are dealing with imbedding in different metrics (from p to $p' = \infty$), which was already proved in 6.2.

If we suppose that $B^r_p(\mathbb{R}_n) = B^{r,\ldots,r}_p(\mathbb{R}_n)$, $B^r_p(\mathbb{R}_m) = B^{r,\ldots,r}_p(\mathbb{R}_m)$, then it follows in particular from (1) that

$$B^r_{p\theta}(\mathbb{R}_n) \to B^{r'}_{p\theta}(\mathbb{R}_m) \tag{1'}$$

---

[1] S. M. Nikol'skiĭ, [3], $H^r_p = B^r_{p\infty}$; O. V. Besov, [2, 3], the case $1 \leqq \theta < \infty$.

[2] Under certain stipulations there is a more precise inequality $\|f\|_{b'} \leqq c\,\|f\|_b$ (see 7.2 (10) and (11)).

under the conditions

(2′) $$0 \leqq m < n, \quad 1 \leqq p, \theta \leqq \infty,$$

(3′) $$\varkappa = 1 - \frac{n-m}{rp} > 0,$$

(4′) $$r' = r\varkappa = r - \frac{n-m}{p}.$$

We turn to the proof for $1 \leqq m < n$. The function $f \in B^r_{p\theta}(\mathbb{R}_n)$ may be represented in the form of a series

(5) $$f = \sum_0^\infty Q_s$$

of functions which are entire of type $a^{s/r_j}(a > 1)$ relative to x_j $(j = 1, \ldots, n)$, with the norm

(6) $$\|f\|_B = \left(\sum_0^\infty a^{s\theta}(\|Q_s\|^n)^\theta\right)^{1/\theta}.$$

For Q_s we apply the estimate (3.4.2 (1)):

$$\|Q_s\|^m \leqq 2^{n-m} a^{s(1-\varkappa)} \|Q_s\|^n,$$

so that

$$\left(\sum_0^\infty a^{\theta\varkappa s}(\|Q_s\|^m)^\theta\right)^{1/\theta} \ll \left(\sum_0^\infty a^{s\theta}(\|Q^s\|^n)^\theta\right)^{1/\theta} = \|f\|_B.$$

Putting $a^\varkappa = b(b > 1)$, we get

(7) $$\left(\sum_0^\infty b^{\theta s}(\|Q_s\|^m)^\theta\right)^{1/\theta} \ll \|f\|_B \quad \left(Q_s = Q_{b^{s/r'_1}, \ldots, b^{s/r'_m}}\right).$$

This inequality shows in particular that the series (5) converges, for any fixed $\boldsymbol{y}$ in the sense of $L_p(\mathbb{R}_m)$, relative to $\boldsymbol{u} = (x_1, \ldots, x_m)$ to some function $f_1(\boldsymbol{x}) = f_1(\boldsymbol{u}, \boldsymbol{y}) \in L_p(\mathbb{R}_m)$. But then $f_1 = f$ almost everywhere in the sense of n-dimensional measure (see 1.3.9).

In view of inequality (7), $f_1(\boldsymbol{u}, \boldsymbol{y}) \in B'$ *for* any $\boldsymbol{y}$ and

(8) $$\|f_1(\boldsymbol{u}, \boldsymbol{y})\|_{B'} \ll \|f\|_B.$$

The constant in this inequality does not depend on $\boldsymbol{y}$.

If it is proved that $f_1(\boldsymbol{u}, \boldsymbol{y})$ is the trace of f on $\mathbb{R}_m$ for any $\boldsymbol{y}$, then along with inequality (7) this leads to the required imbedding (1). Since (see 6.1 (14))

$$B = B^{r}_{p\theta}(\mathbb{R}_n) \to H^{r}_{p}(\mathbb{R}_n),$$

then

(9) $$\|Q_s\|^n \ll a^{-s}.$$

The increment of $Q_s(\boldsymbol{x})$ is in its turn an entire function of type a^{s/r_j} relative to x_j ($j = 1, \ldots, n$), so that on the basis of 3.4.2 (1), 3.3.2 (7) and 4.4.4 (2)

$$\|\Delta_{x_j h} Q_s\|^m \leqq 2^{n-m} a^{s(1-\varkappa)} \|\Delta_{x_j h} Q_s\|^n \leqq 2^{n-m} a^{s\left(1-\varkappa+\frac{1}{r_j}\right)} \|Q_s\|^n |h|.$$

We have the inequality

$$\|\Delta^{x_j h} f_1\|^m \leqq \sigma'_\mu + \sigma''_\mu,$$

where

$$\sigma'_\mu \leqq \sum_0^{\mu-1} \|\Delta_{x_j h} Q_s\|^m, \quad \sigma''_\mu = \sum_\mu^\infty \|\Delta_{x_j h} Q_s\|^m.$$

Suppose given a number h with $|h| \leqq 1$. Choose an integer μ such that

(10) $$a^{-\mu/r_j} < |h| \leqq a^{-(\mu-1)/r_j}.$$

Then (see (9))

(11) $$\sigma'_\mu \leqq 2^{n-m} \sum_0^{\mu-1} a^{s\left(1-\varkappa+\frac{1}{r_j}\right)} \|Q_s\|^n \, |h| \ll |h| \sum_0^{\mu-1} \frac{1}{a^{s\delta}},$$

where

$$\delta = \varkappa - \frac{1}{r_j}.$$

If $\delta < 0$, in other words if $r'_j = r_j \varkappa < 1$, then

$$\sigma'_\mu \ll |h| \, |a^{-\mu\delta}| \ll |h|^{r'_j}.$$

If $\delta > 0$, i.e. if $1 < r'_j$, then

$$\sigma'_\mu \ll |h|.$$

For $\delta = 0$, i.e. $r'_j = 1$,

$$\sigma'_\mu \ll |h| \, \mu \ll |h| |\ln |h||.$$

On the other hand,

$$(12)\qquad \sigma''_\mu \leqq 2 \sum_\mu^\infty \|Q_s\|^m \ll \sum_\mu^\infty a^{s(1-\varkappa)}\|Q_s\|^n \ll \sum_\mu^\infty \frac{1}{a^{\varkappa s}} \ll a^{-\varkappa\mu} \ll |h|^{r'_j}.$$

From the estimates just obtained, it follows, obviously, that[1]

$$(13)\qquad \|\Delta_{x_j h} f_1(\boldsymbol{x})\|^m = \begin{cases} O\left(|h|^{r'_j}\right) & 0 < r'_j < 1. \\ O\left(|h|\,|\ln|h||\right), & r'_j = 1, \\ O\left(|h|\right), & r'_j > 1. \end{cases}$$

The right sides of (13) tend to zero along with h, so that $f_1(\boldsymbol{u}, \boldsymbol{y})$ is the trace of $f(\boldsymbol{u}, \boldsymbol{y})$ for any $\boldsymbol{y}$.

We emphasize that the initial function $f(x) = f(\boldsymbol{u}, \boldsymbol{y})$ was known up to a set of n-dimensional measure zero, so that to consider it as a function of $\boldsymbol{u}$ for fixed $\boldsymbol{y}$ has no meaning. Above we indicated a method of obtaining the trace of a function $f \in B$. For this we need to decompose f into a series (5) with a finite norm (6) and to fix $\boldsymbol{y}$ in its terms Q_s. Then the resulting series of functions of $\boldsymbol{u}$ converges in the sense of $L_p(\mathbb{R}_m)$, and in fact to the trace $f_1(\boldsymbol{u}, \boldsymbol{y})$ of the function f.

Usually in inequalities of the type of (13), one writes f in place of f_1, understanding this in the sense that f may be altered on a set of n-dimensional measure zero in such a way that after this, for any $\boldsymbol{y}$, (13) holds and moreover with a constant not depending on $\boldsymbol{y}$.

6.5.1. *Remark.* The imbedding 6.5 (1) remains true under the same hypotheses 6.5 (2)—(4), if one replaces $\mathbb{R}_n$, $\mathbb{R}_m$ in it respectively by measurable cylindrical sets $\mathscr{E}_n = \mathbb{R}_n \times \mathscr{E}'$, $\mathscr{E}_m = \mathbb{R}_m \times \mathscr{E}'$, where as before $\mathbb{R}_m \subset \mathbb{R}_n$ and $\boldsymbol{z} = (\boldsymbol{x}, \boldsymbol{w})$, $\boldsymbol{x} \in \mathbb{R}_n$, $\boldsymbol{w} \in \mathscr{E}'$.

Indeed, for finite p, for almost any $\boldsymbol{w}$, the inequality corresponding to 6.5 (1) is true, where the constant c is independent not only of $\boldsymbol{y}$ but also of $\boldsymbol{w}$.

We raise both of its sides to the power p, integrate with respect to $\boldsymbol{w}$, and then raise to the power $1/p$. As a result we obtain the needed inequality. For $p = \infty$ this assertion is trivial.

6.5.2. Inequalities 6.5 (13) are interesting for their own sake. They indicate for functions of the class $H_p^r(\mathbb{R}_m)$ the order of convergence of their traces in the mean. This order cannot be improved (see 7.6).

It is not hard to show that the inequality

$$(1)\qquad \|\Delta_{x_j h}^k f_1(\boldsymbol{x})^m\| = O(|h|) \quad (r'_j = 1,\, k > 1)$$

[1] L. D. Kudrjavcev [2], Part 1.

holds (already without the ln), complementing the second inequality 6.5 (13).

Since

$$W_p^{\boldsymbol{r}}(\mathbb{R}_n) \to H_p^{\boldsymbol{r}}(\mathbb{R}_n)$$

($\boldsymbol{r}$ an integer vector),

then the estimates 6.5 (13) are applicable also to $W_p^{\boldsymbol{r}}(\mathbb{R}_n)$[1]. In this case as well they cannot be improved in the sense that it is not possible in their right sides to replace the power of $|h|$ indicated there by a larger one. However for each individual function $f \in W_p^{\boldsymbol{r}}(\mathbb{R}_n)$ the following estimates hold:

$$(2) \qquad \|\Delta_{x_j h} f(\boldsymbol{x})\|^m = \begin{cases} o\left(|h|^{r'_j}\right) \ (|h| \to 0), & 0 < r'_j < 1, \\ o\,(|h|\,|\ln|h||)\ (|h| \to 0), & r'_j = 1, \\ O\,(|h|)\ (|h| < 1), & r'_j > 1, \end{cases}$$

improving for $r'_j \leqq 1$ the estimates 6.5 (13).

Indeed (see 5.6.1 (9) and 5.2.4 (5)), in this case

$$\|Q_s\|^n \leqq \|g_{a^{s/r_1},\ldots,a^{s/r_n}} - f\| + \|f - g_{a^{(s-1)/r_1},\ldots,a^{(s-1)/r_n}}\| = o\left(a^{-s}\right) \quad (s \to \infty),$$

and then the inequalities 6.5 (11), (12) are replaced by in equalities of the following sort:

$$\sigma'_\mu \ll o\left(|h|^{r_j}\right) \ (r'_j < 1),\ \sigma'_\mu \ll o\,(|h|\,|\ln|h||) \ (r'_j = 1),\ \sigma''_\mu \ll o\left(|h|^{r'_j}\right).$$

On the other hand the estimate $O(|h|)$ cannot be improved. This is easily verified on the example $f(x, y) = g(x)\, g(y)$, where $g(x) \in L_p(\mathbb{R}_1)$ is entire of type 1 and such that $g'(0) \neq 0$, and $\mathbb{R}_m = \mathbb{R}_n$.

6.6. The Simplest Inverse Theorem on Imbedding of Different Dimensions

Suppose $1 \leqq m < n$ and that $\mathbb{R}_m$ is a coordinate subspace of the points $(x_1, \ldots, x_m, 0, \ldots, 0) \in \mathbb{R}_n$. For definiteness we will suppose that it consists of points $(\boldsymbol{u}, \boldsymbol{0}) = (x_1, \ldots, x_m, 0, \ldots, 0)$.

In § 6.5 we proved the theorem:

$$(1) \qquad B_{p\theta}^{\boldsymbol{r}}(\mathbb{R}_n) \to B_{p\theta}^{\boldsymbol{r}'}(\mathbb{R}_m)$$

---

[1] These estimates for the classes W_p^l $(l = 1, 2, \ldots)$ were obtained directly by V. I. Kondrašov [1].

under the hypothesis that

(2) $$r' = r\varkappa = r - \frac{n-m}{p} > 0.$$

Below we shall prove that there is a theorem which inverts it completely[1]:

(3) $$B^{r'}_{p\theta}(\mathbb{R}_m) \to B^{r}_{p\theta}(\mathbb{R}_n)$$

under condition (2). For an explanation see 6.1. (12).

We denote by B, B' respectively the first and second classes (1). Suppose given an arbitrary function $\varphi \in B'$. It may be represented in the form of a series converging to it in the sense of $L_p(\mathbb{R}_m)$:

$$\varphi(u) = \sum_{s=0}^{\infty} Q_{a^{s/r}},$$

where the $Q_{a^{s/r}}$ are entire of spherical type $a^{s/r} (a > 1)$

(4) $$\|\varphi\|_{B'} = \left\{\sum_0^\infty a^{\varkappa s\theta}(\|Q_s\|^m)^\theta\right\}^{1/\theta}, \quad \|\cdot\|^m = \|\cdot\|_{L_p(\mathbb{R}_m)}.$$

Put

$$F_\nu(t) = \left(\frac{\sin\frac{\nu t}{2}}{\frac{\nu t}{2}}\right)^2 \qquad (\nu > 0).$$

This is an entire function of one variable t, of type ν, such that

$$\|F_\nu\|_{L_p(\mathbb{R}_1)} = \left(\int\left(\frac{\sin\frac{\nu t}{2}}{\frac{\nu t}{2}}\right)^{2p} d\left(\frac{\nu t}{2}\right)\right)^{1/p}\left(\frac{2}{\nu}\right)^{1/p} = \frac{c_p}{\nu^{1/p}} \ (\nu > 0).$$

We introduce a new function of $x \in \mathbb{R}_n$, defined by the series

(5) $$f(x) = \sum_0^\infty q_{a^{s/r}}(x),$$

where

(6) $$q_{a^{s/r}}(x) = Q_{a^{s/r}}(u) \prod_{m+1}^{n} F_{a^{s/r}}(x_j).$$

---

[1] S. M. Nikol'skiĭ [5], the case $H^r_p = B^r_{p\theta}$; O. V. Besov [2, 3], the case $1 \leqq \theta < \infty$.

Obviously (see (2)),

$$\|q_{a^{s/r}}\|^n = \|Q_{a^{s/r}}\|^m \, a^{-(1-\varkappa)s} c_p^{n-m}.$$

In such a case

$$(7) \quad \left(\sum_0^\infty a^{s\theta}(\|q_{a^{s/r}}\|^n)^\theta\right)^{1/\theta} = c_p^{n-m}\left(\sum_0^\infty a^{\varkappa s\theta}(\|Q_{a^{s/r}}\|^m)^\theta\right)^{1/\theta} = c_p^{n-m}\|\varphi\|_{B'}.$$

Since the functions $q_{a^{s/r}}$ are entire of exponential type $a^{s/r}$ relative to each variable $x_1, \ldots, x_n$, then in view of 5.6.1 (4), (5) the left side of (7) is the norm of f in the sense of $B_{p\theta}^{r,\ldots,r}(\mathbb{R}_n)$. But $B_{p\theta}^{r,\ldots,r}(\mathbb{R}_n) \to B_{p\theta}^r(\mathbb{R}_n) = B$, and we have proved that $f \in B$ $\|f\|_B \ll \|\varphi\|_{B'}$.

The function f is defined by the series (5), converging to it in the sense of $L_p(\mathbb{R}_n)$. But this series, for each $\boldsymbol{y} = (x_{m+1}, \ldots, x_n)$ converges also in the sense of $L_p(\mathbb{R}_m)$ to some function $f_1(\boldsymbol{x})$, which can differ from f only on a set of n-dimensional measure zero (1.3.9). Obviously,

$$f_1(\boldsymbol{u}, \boldsymbol{0}) = \varphi(\boldsymbol{u})$$

in the sense of $\mathbb{R}_m$, i.e. for almost all $\boldsymbol{u}$ in the sense of m-dimensional measure. Further, taking into account the facts that $F_\nu(0) = 1$, $F_\nu(t)$ is bounded relative to ν and t and $\sum_0^\infty \|Q_{a^{s/r}}\|^n < \infty$, we get

$$\|f_1(\boldsymbol{u}, \boldsymbol{y}) - f_1(\boldsymbol{u}, \boldsymbol{0})\|^m \leq \sum_0^\infty \left| \prod_{m+1}^n F_{a^{s/r}}(x_j) - 1 \right| \|Q_{a^{s/r}}\|^m \to 0$$

$$(\boldsymbol{y} = (x_{m+1}, \ldots, x_n) \to \boldsymbol{0}).$$

This argument shows that φ is the trace of f (6.3). With this the proposition (3) is completely proved.

We note that the class $B = B_{p\theta}^r(\mathbb{R}_n)$ is a Banach space. If

$$r' = r - \frac{n-m}{p} > 0.$$

then, by (1) for functions $f \in B$ the operation of taking the trace

$$Af = f|_{\mathbb{R}_m} = \varphi$$

on $\mathbb{R}_m \subset \mathbb{R}_n$ $(1 \leq m < n)$ has a meaning. This operation is linear. Moreover, by (1), it maps B into B' boundedly. In view of the invertibility of the imbedding (1), one sees further that this mapping is not into B', but onto B'. Above we saw that in its turn B' can be mapped onto some part of B by means of some bounded linear operator. This last is not unique; it is possible to indicate an infinite set of such operators.

In the language of functional analysis a bounded linear operator mapping a Banach space B onto a Banach space B' is said to be continuously invertible[1].

In the following sections we shall prove an imbedding theorem more general than 6.5, in which one is dealing with the boundary properties not only of the function f itself on a subspace $\mathbb{R}_m \subset \mathbb{R}_n$, but also those of certain of its partial derivatives. We will thereupon completely invert this theorem.

6.7. General Imbedding Theorem for Different Dimensions

Theorem[2]. *Suppose that* $f \in B^r_{p\theta}(\mathbb{R}_n)$, $0 \leqq m < n$ *and that for some vector* $\lambda = (\lambda_{m+1}, \ldots, \lambda_n)$ *with nonnegative integer coordinates the inequality*

$$(1) \qquad \varrho_i^{(\lambda)} = r_i \left(1 - \sum_{j=m+1}^{n} \frac{\lambda_j}{r_j} - \frac{1}{p} \sum_{j=m+1}^{n} \frac{1}{r_j}\right) > 0$$

is satisfied. Suppose moreover that

$$\psi = \frac{\partial^{\lambda_{m+1}+\cdots+\lambda_n} f}{\partial x_{m+1}^{\lambda_{m+1}} \cdots \partial x_n^{\lambda_n}}.$$

Denote by $\mathbb{R}_m$ *some* m*-dimensional subspace of* $\mathbb{R}_n$, *gotten by fixing a vector* $\mathbf{y} = (x_{m+1}, \ldots, x_n)$, *and put* $\varrho(\lambda) = (\varrho_1^{(\lambda)}, \ldots, \varrho_n^{(\lambda)})$.

Then

$$\psi|_{R_m} = \varphi \in B^{\varrho(\lambda)}_{p\theta}(\mathbb{R}_m),$$

and the inequality

$$\|\varphi\|_{B^{\varrho(\lambda)}_{p\theta}(\mathbb{R}_m)} \leqq c \|f\|_{B^r_{p\theta}(\mathbb{R}_n)}$$

is satisfied with a constant c *not depending on* f *and* $\mathbf{y}$.

Proof. It follows from (1) that

$$\sum_{j=m+1}^{n} \frac{\lambda_j}{r_j} < 1,$$

[1] F. Hausdorff [1], supplement.

[2] S. M. Nikolskiĭ [5], the case $H^r_p = B^r_{p\infty}$; O. V. Besov [2, 3], the case $1 \leqq \theta < \infty$.

so that on the basis of Theorem 5.6.3.

$$\psi \in B^{\mathbf{r}'}_{p\theta}(\mathbb{R}_n),$$

$$\mathbf{r}' = (r'_1, \ldots, r'_n), r'_i = r^i\left(1 - \sum_{k=m+1}^{n} \frac{\lambda_k}{r_k}\right), \quad i = 1, \ldots, n,$$

and the inequality

$$\|\psi\|_{B^{\mathbf{r}'}_{p\theta}(\mathbb{R}_n)} \ll \|f\|_{B^{\mathbf{r}}_{p\theta}(\mathbb{R}_n)}$$

is satisfied.

In order to see to which class the trace ψ on $\mathbb{R}_m$ belongs, we make use of the imbedding theorem 6.5 (1). It is applicable, because

$$\varkappa = 1 - \frac{1}{p}\sum_{m+1}^{n} \frac{1}{r'_j} = \frac{1 - \sum\limits_{m+1}^{n} \frac{\lambda_j}{r_j} - \frac{1}{p}\sum\limits_{m+1}^{n} \frac{1}{r_j}}{1 - \sum\limits_{m+1}^{n} \frac{\lambda_j}{r_j}} > 0,$$

so that the assertion of the theorem holds.

6.8. General Inverse Imbedding Theorem

Theorem[1]. *Suppose given a vector* $\mathbf{r} = (r_1, \ldots, r_n) > \mathbf{0}$ *and all possible vectors*

(1) $$\boldsymbol{\lambda} = (\lambda_{m+1}, \ldots, \lambda_n)$$

with nonnegative integer coordinates, for which the vectors $\boldsymbol{\varrho}(\boldsymbol{\lambda}) = (\varrho_1^{(\lambda)}, \ldots, \varrho_n^{(\lambda)})$ *defined by formula 6.7 (1) are positive.*

Suppose moreover that to each vector ($\boldsymbol{\lambda}$) *there has been assigned a function*

(2) $$\varphi_{(\lambda)}(x_1, \ldots, x_m) \in B^{\rho(\lambda)}_{p\theta}(\mathbb{R}_m).$$

Then it is possible to construct on $\mathbb{R}_n$ *a function* $f \in B^r_{p\theta}(\mathbb{R}_n)$ *such that*

(3) $$\|f\|_{B^{\mathbf{r}}_{p\theta}(\mathbb{R}_n)} \leq c \sum_{\lambda} \|\varphi_{(\lambda)}\|_{B^{\rho(\lambda)}_{p\theta}(\mathbb{R}_m)},$$

where c does not depend on $\varphi(\lambda)$. *The sum in* (3) *is extended over all possible vectors* $\boldsymbol{\lambda}$ *indicated above, and the* $\varphi_{(\lambda)}$ *are the traces of the partial deriva-*

---

[1] S. M. Nikol'skiĭ [5], the case $H^r_p = B^r_{p\infty}$; O. V. Besov [2, 3], the case $1 \leq \theta < \infty$.

tives of the function f:

(4) $$\frac{\partial^{\lambda_{m+1}+\cdots+\lambda_n} f}{\partial x_{m+1}^{\lambda_{m+1}} \cdots \partial x_n^{\lambda_n}}\Bigg|_{\mathbb{R}_m} = \varphi_{(\lambda)}.$$

Proof. Put

(5) $$r_i^{(\lambda)} = r_i \left(1 - \sum_{m+1}^{n} \frac{\lambda_j}{r_j}\right) \quad (i = 1, \ldots, n),$$

(6) $$\varkappa^{(\lambda)} = 1 - \frac{1}{p} \sum_{m+1}^{n} \frac{1}{r_i^{(\lambda)}}.$$

Then obviously

(7) $$\varrho_i^{(\lambda)} = r_i^{(\lambda)} \varkappa^{(\lambda)}.$$

Put $\varphi_{(\lambda)} \in B_{p\theta}^{\varrho(\lambda)}(\mathbb{R}_m) = B^{(\lambda)}$. Then

$$\varphi_{(\lambda)} = \sum_{0}^{\infty} Q_{s(\lambda)},$$

where the $Q_{s(\lambda)}$ are entire of type $2^{s/r_i(\lambda)}$ relative to the $x_i (i = 1, \ldots, m)$ and

(8) $$\|\varphi_{(\lambda)}\|_{B^{(\lambda)}} = \left(\sum_{0}^{\infty} 2^{\theta\varkappa(\lambda)s} (\|Q_s\|^m)^\theta\right)^{1/\theta},$$

$$(\|\cdot\|^m = \|\cdot\|_{L_p(\mathbb{R}_m)}).$$

We introduce trigonometric polynomials $T_\nu(x)$, $\nu = 1, \ldots, l$, where l denotes the largest of the numbers λ_j occuring in the various vectors $\lambda = (\lambda_{m+1}, \ldots, \lambda_n)$ being considered. Suppose that these polynomials have the following properties: the function

$$\Phi_\nu(x) = \frac{T_\nu(x)}{x^2}$$

is entire, and, moreover,

(9) $$\Phi_\nu^{(\nu)}(0) = \frac{d^\nu}{dx_\nu} \Phi_\nu(x)\Bigg|_{x=0} = 1, \; \Phi_\nu^{(k)}(0) = 0$$

$$(k = 0, \ldots, \nu - 1, \nu + 1, \ldots, l).$$

We are not concerned with the magnitude of the degree of the trigonometric polynomial, so that conditions indicated above do not uniquely define it. We will suppose that we have chosen fully defined polynomials, having degree $\mu(\nu)$. Then $\Phi_\nu(x)$ is entire of type $\mu(\nu)$, and $\Phi_\nu\left(\frac{k}{\mu(\nu)}x\right)$ is entire of type k. Obviously, further,

$$\left\|\Phi_\nu\left(\frac{k}{\mu(\nu)}x\right)\right\|_{L_p(\mathbb{R}_1)} = \left(\int_{-\infty}^{\infty}\left|\Phi_\nu\left(\frac{k}{\mu(\nu)}x\right)\right|^p dx\right)^{1/p} \tag{10}$$

$$= \left(\frac{\mu(\nu)}{k}\right)^{1/p}\left(\int_{-\infty}^{\infty}|\Phi_\nu(u)|^p\,du\right)^{1/p} = \frac{A_\nu}{k^{1/p}},$$

where A_ν depends only on ν.

We define functions $f_{(\lambda)}(x_1, \ldots, x_n)$ corresponding to the various vectors λ by means of the series

$$f_{(\lambda)}(x_1, \ldots, x_n) \tag{11}$$

$$= \sum_{s=0}^{\infty} Q_{s(\lambda)} \prod_{j=m+1}^{n}\left(\frac{\mu(\lambda_j)}{2^{s/r_j(\lambda)}}\right)^{\lambda_j}\Phi_{\lambda_j}\left(2^{s/r_j(\lambda)}\frac{x_j}{\mu(\lambda_j)}\right) = \sum_{s=0}^{\infty}\mathbb{R}_{s(\lambda)},$$

where, obviously, the $\mathbb{R}_{s(\lambda)}$ are of type $2^{s/r_i(\lambda)}$ relative to the $x_i (i = 1, \ldots, n)$.

Taking account of (5)—(8), (10), we have

$$\|\mathbb{R}_{s(\lambda)}\|^m \leqslant \frac{\|Q_{s(\lambda)}\|^m}{2^{s\left[\sum\limits_{m+1}^{n}\frac{\lambda_j}{r_j(\lambda)} + \frac{1}{p}\sum\limits_{m+1}^{n}\frac{1}{r_j(\lambda)}\right]}} = \frac{\|Q_{s(\lambda)}\|^m\, 2^{s\varkappa(\lambda)}}{2^{s\frac{r_i(\lambda)}{r_i}}},$$

or

$$2^{s\frac{r_i(\lambda)}{r_i}}\|\mathbb{R}_{s(\lambda)}\|^n \leqslant \|Q_{s(\lambda)}\|^m\, 2^{s\varkappa(\lambda)},$$

so that, in view of (8) and further recalling that

$$\frac{r_i(\lambda)}{r_i} = 1 - \sum_{m+1}^{n}\frac{\lambda_j}{r_j}$$

in fact does not depend on i, we obtain the inequality

$$\|f_{(\lambda)}\|_{B^{r}_{p\theta}(\mathbb{R}_n)} \leqslant \|\varphi_{(\lambda)}\|_{B^{(\lambda)}}.$$

We note further that in view of the properties of the function Φ_ν, if the vector λ is admissible, i.e. satisfies conditions 6.7 (1), then we have for $f_{(\lambda)}$ the following equation:

$$\left.\frac{\partial^{\lambda_{m+1}+\cdots+\lambda_n} f_{(\lambda)}}{\partial x_{m+1}^{\lambda_{m+1}} \cdots \partial x_n^{\lambda_n}}\right|_{\mathbb{R}_m} = \varphi_{(\lambda)}(x_1, \ldots, x_n). \tag{12}$$

Indeed, if one formally differentiates the series (11) termwise relative to $x_{m+1}, \ldots, x_n$, respectively $\lambda_{m+1}, \ldots, \lambda_n$, times, then we obtain

$$\frac{\partial^{\lambda_{m+1}+\cdots+\lambda_n}}{\partial x_{m+1}^{\lambda_{m+1}} \cdots \partial x_n^{\lambda_n}} f = \tag{13}$$

$$= \sum_{s=0}^{\infty} Q_{s(\lambda)} \prod_{j=m+1}^{n} \Phi_{\lambda_j}^{(\lambda_j)}\left(2^{s/r_j(\lambda)} \frac{x_j}{\mu_j(\lambda)}\right) = \sum_{s=0}^{\infty} \mu_{s(\lambda)}.$$

From the estimates we have presented for $\mathbb{R}_{s(\lambda)}$ it follows that at any stage of the differentiation one obtains series converging in the sense of $L_p(\mathbb{R}_n)$, so that equation (13) really holds in the sense of convergence in $L_p(\mathbb{R}_n)$ (see Lemma 4.4.7). Further, in view of the boundedness of the Φ_{λ_j} the derivatives $\Phi_{\lambda_j}^{(\lambda_i)}$, are bounded as well, so that

$$\left\|\sum_{0}^{\infty} (\mu_{s(\lambda)} - Q_{s(\lambda)})\right\|_{L_p(\mathbb{R}_m)} \tag{14}$$

$$\leq \sum_{0}^{N} \|Q_{s(\lambda)}\|_{L_p(\mathbb{R}_m)} \left|\Phi_{\lambda_j}^{(\lambda_j)}\left(2^{s/r_j(\lambda)} \frac{x_j}{\mu_j(\lambda)}\right) - 1\right| + 2c \sum_{N+1}^{\infty} \|Q_{s(\lambda)}\|_{L_p(\mathbb{R}_m)}.$$

Now suppose given an $\varepsilon > 0$. Choose N sufficiently large that the second term in the right side of (14) is less than ε, and now select δ so small that for $|x_j| < \delta$, $j = m + 1, \ldots, n$, the first term is less than ε.

If now $\{\lambda'\}$ is another admissible vector $(\lambda'_{m+1}, \ldots, \lambda'_n)$, then, using similar arguments, we get

$$\left.\frac{\partial^{\lambda'_{m+1}+\cdots+\lambda'_n} f_{(\lambda)}}{\partial x_{n+1}^{\lambda'_{n+1}} \cdots \partial x_n^{\lambda'_n}}\right|_{\mathbb{R}_m} = 0.$$

In such a case the function

$$f = \sum_{\lambda} f_\lambda,$$

where the summation is extended over all possible admissible vectors λ, satisfies all the requirements of the theorem.

6.9. Generalization of the Imbedding Theorem for Different Metrics

Below we shall give a generalization of the imbedding theorem (6.3 (1) to the case of the classes $B^r_{p\theta}(\mathbb{R}_n)$.

Theorem[1]. *Suppose that for the numbers considered below the inequalities* $(r_j > 0)$

(1) $$1 \leqq p_j \leqq p' \leqq \infty,$$

(2) $$\varkappa' = 1 - \sum_{l=1}^{n}\left(\frac{1}{p_l} - \frac{1}{p'}\right)\frac{1}{r_l} > 0,$$

(3) $$\varkappa_i = 1 - \sum_{l=1}^{n}\left(\frac{1}{p_l} - \frac{1}{p_i}\right)\frac{1}{r_l} > 0 \quad (i = 1, \ldots, n)$$

are satisfied, and

(4) $$\varrho_i = \frac{r_i\varkappa'}{\varkappa_i}\,[2].$$

Suppose further that $\mathbf{r} = (r_1, \ldots, r_n)$, $\boldsymbol{\varrho} = (\varrho_1, \ldots, \varrho_n)$, $\mathbf{p} = (p_1, \ldots, p_n)$. *Then one has the imbedding*

(5) $$B^{\mathbf{r}}_{\mathbf{p}\theta}(\mathbb{R}_n) \to B^{\varrho}_{\mathbf{p}'\theta}(\mathbb{R}_n).$$

From (5) for $p = p_1 = \cdots = p_n$ it follows that 6.3 (1) holds with $\varrho = r'$. We note further that $\varkappa' > 0$ implies that $\varkappa_i > 0$ for all i, since $p' \geqq p_i$.

Proof. Select a family of functions $g_\nu = g_{\nu_1,\ldots,\nu_n}$ $(0 < \nu_j \leqq \infty;$ $j = 1, \ldots, n)$ of entire exponential type ν_j relative to the variable x_j, defined by the last equation in 5.2.4 (1) for $m = n$.

Put

$$\nu_k = \nu_k(s) = 2^{s/\varrho_k} \ (k = 1, \ldots, n;\ s = 0, 1, \ldots, \text{and } s = \infty)$$

and

(6) $$Q_0 = g_{\nu(0)}, \quad Q_s = g_{\nu(s)} - g_{\nu(s-1)} \quad (s = 1, 2, \ldots).$$

Obviously,

(7) $$Q_s = \sum_{i=1}^{n} Q_s^{(i)} \quad (s = 1, 2, \ldots),$$

---

[1] S. M. Nikol'skiĭ [10], the case $H^r_p = B^r_{p\theta}$; V. P. Il'in and V. A. Solonnikov [1, 2], the case $1 \leqq \theta < \infty$ (by the method of the theory of approximation, T. I. Amonov [3]).

[2] It is possible in this theorem to start from the hypothesis that all the $\varrho_j > 0$, since for that i for which p_i takes on its least value one has $\varkappa_i > 0$. But then $\varkappa' > 0$ as well, so that the remaining $\varkappa_i > 0$.

where

$$(8)\quad Q_s^{(i)} = g_{\nu_1(s),\ldots,\nu_i(s),\nu_{i+1}(s-1),\ldots,\nu_n(s-1)} - g_{\nu_1(s),\ldots,\nu_{i-1}(s),\nu_i(s-1),\ldots,\nu_n(s-1)}.$$

We have

$$(9)\qquad \|Q_s\|_{p'} \leq \sum_{i=1}^{n} \|Q_s^{(i)}\|_{p'} \quad (s = 1, 2, \ldots)$$

$$(\|\cdot\|_{p'} = \|\cdot\|_{L_{p'}(\mathbb{R}_n)}).$$

We apply to each ith term of this sum the inequality of different metrics (3.3.5):

$$(10)\qquad \|Q_s^{(i)}\|_{p'} \leq 2^n 2^{s\left(\frac{1}{p_i}-\frac{1}{p'}\right)\sum_1^n \frac{1}{\varrho_j}} \|Q_s^{(i)}\|_{p_i}$$

$$= 2^n 2^{-s\left[\frac{r_i}{\varrho_i}-\left(\frac{1}{p_i}-\frac{1}{p'}\right)\sum_{j=1}^n \frac{1}{\varrho_j}\right]} 2^{s\frac{r_i}{\varrho_i}} \|Q_s^{(i)}\|_{p_i}$$

$$(i = 1, \ldots, n).$$

Now we choose numbers ϱ_j so that the expressions in brackets will be equal to unity:

$$(11)\qquad 1 = \frac{r_i}{\varrho_i} - \left(\frac{1}{p_i} - \frac{1}{p'}\right) \sum_{l=1}^{n} \frac{1}{\varrho_l} \quad (i = 1, \ldots, n).$$

Dividing all equations by r_i, replacing i by l and summing on l, we get

$$(12)\qquad \sum_{l=1}^{n} \frac{1}{r_l} = \left(1 - \sum_{l=1}^{n} \frac{\frac{1}{p_l} - \frac{1}{p'}}{r_l}\right) \sum_{l=1}^{n} \frac{1}{\varrho_l}.$$

Eliminating the sum from (11) and (12), we get

$$(13)\qquad \varrho_i = r_i \frac{\varkappa}{\varkappa_i} \quad (i = 1, \ldots, n).$$

Therefore summation of (10) on i leads (see (7)) to the inequality

$$2^s \|Q_s\|_{p'} \leq 2^n \sum_{i=1}^{n} 2^{s\frac{r_i}{\varrho_i}} \|Q_s^{(i)}\|_{p_i},$$

from which it follows that (explanation below)

(14)
$$\left\{\sum_1^\infty 2^{\theta s}\|Q_s\|_{p'}^\theta\right\}^{1/\theta} \ll \sum_{i=1}^n \left\{\sum_{s=1}^\infty 2^{\theta s\frac{r_i}{\varrho_i}}\|Q_s^{(i)}\|_{p_i}^\theta\right\}^{1/\theta}$$
$$\ll \sum_{i=1}^n \left(\sum_{s=1}^\infty 2^{\theta s\frac{\alpha_i}{\varrho_i}}\omega_{x_i}^k\left(f_{x_i}^{\bar r_i}, 2^{-\frac{s}{\varrho_i}}\right)_{p_i}^\theta\right)^{1/\theta}$$
$$\leqq \sum_{i=1}^n \left(\int_0^\infty 2^{\theta s\frac{\alpha_i}{\varrho_i}}\omega_{x_i}^k\left(f_{x_i}^{\bar r_i}, 2^{-\frac{s}{\varrho_i}}\right)_{p_i}^\theta ds\right)^{1/\theta}$$
$$\ll \sum_{i=1}^n \left(\int_0^1 \frac{\omega_{x_i}^k\left(f_{x_j}^{\bar r}, t\right)_{p_i}^\theta}{t^{1+\theta\alpha_i}}\,dt\right)^{1/p} \ll \|f\|_{B_{\mathbf{p}\theta}^{\mathbf{r}}(\mathbb{R}_n)}.$$

The second inequality in (14) follows from the fact that if $\nu_1, \ldots, \nu_n$ and ν_n' are arbitrary numbers and $\nu_n \leqq \nu_n'$, then in view of 5.2.4 (2)

$$\left\|g_{\nu_1,\ldots,\nu_n} - g_{\nu_1,\ldots,\nu_{n-1},\nu_n'}\right\|_{p_n} \leqq \|g_{\nu_1,\ldots,\nu_n} - f\|_{p_n}$$
$$+ \left\|f - g_{\nu_1,\ldots,\nu_{n-1},\nu_n'}\right\|_{p_n} \leqq \frac{2c\omega_{x_n}^k\left(f_{x_n}^{\bar r_n}, \frac{1}{\nu_n}\right)_{p_n}}{\nu_r^{\bar r_n}} \quad (r_n - \bar r_n = \alpha_n).$$

Further, since $f \in L_{p_i}$, then also $Q_0 \in L_{p_i}$ (see the integral representation 5.2.4 (1)). Moreover $Q_0 \in L_{p'}$ since $p_i \leqq p'$ and

(15)
$$\|Q_0\|_{p'} \leqq \|f\|_{p_i} \leqq \|f\|_{B_{\mathbf{p}\theta}^{\mathbf{r}}(\mathbb{R}_n)}.$$

It follows in particular from (14) and (15) that the series

(16)
$$\sum_0^\infty Q_s$$

converges in the sense of L_{p_i}. It converges *a fortiori* in the sense of L_{p_i} to f, because

$$\left\|f - \sum_0^N Q_s\right\|_{p_i} = \|f - g_{\nu(N)}\|_{p_i} \ll \frac{\omega_{x_i}^k\left(f_{x_i}^{\bar r_i}, \nu_i(N)^{-1}\right)_{p_i}}{\nu_i^{\bar r_i}}$$
$$\ll \frac{1}{\nu_i(N)^{r_i}} \to 0 \quad (N \to \infty),$$

since $B_{\mathbf{p}\theta}^{\mathbf{r}} \to H_{\mathbf{p}}^{\mathbf{r}} \to H_{x_ip_i}^{r_i}$.

Thus, the series (16) converges to f, inequalities (14) and (15) hold, and the Q_s are entire of type $2^{s/\varrho_i}$ relative to the x_i, $i = 1, \ldots, n$. Accordingly, $f \in B^{\varrho}_{p'}(\mathbb{R}_n)$ and the imbedding (5) holds.

6.9.1. Suppose that instead of the number p' (see 6.9) we are given a vector $\boldsymbol{p}' = (p'_1, \ldots, p'_n)$ such that $p'_i \geqq p_j$, $i, j = 1, \ldots, n$. Put

$$\varkappa'_i = 1 - \sum_{l=1}^{n} \left(\frac{1}{p_l} - \frac{1}{p'_i}\right)\frac{1}{r_l} > 0 \tag{1}$$

and

$$r'_i = \frac{r_i \varkappa'_i}{\varkappa_i}, \tag{2}$$

where the $\varkappa_i$ are defined as in 6.9 (3).

Then, in virtue of Theorem 6.9,

$$B^{\boldsymbol{r}}_{\boldsymbol{p}\theta}(\mathbb{R}_n) \to B^{r'_i}_{x_i p'_i \theta}(\mathbb{R}_n) \quad (i = 1, \ldots, n), \tag{3}$$

so that

$$B^{\boldsymbol{r}}_{\boldsymbol{p}\theta}(\mathbb{R}_n) \to B^{\boldsymbol{r}'}_{\boldsymbol{p}'\theta}(\mathbb{R}_n). \tag{4}$$

6.10. Supplementary Information

6.10.1. The theorems obtained in this chapter carry over automatically to the periodic case. Their formulations remain true, if the symbols W, H, and B in them are replaced respectively by W_*, $H*$, $B*$.

In the proofs in the periodic case the role of the entire functions of exponential type is now already played, of course, by the trigonometric polynomials. In our exposition it was essential that the entire functions of exponential type possess certain properties. For them the following inequalities must be valid: 1) inequalities for the derivatives (of the type of the Bernšteĭn inequalities), 2) inequalities for different metrics, and 3) inequalities of different dimensions. The trigonometric polynomials have similar properties. Moreover, for periodic and nonperiodic functions one may, as we know (see 5.2.1 (6) and 5.3.1 (11)), construct analogous methods of approximating them by trigonometric polynomials and correspondingly by functions of exponential type. We have made use of these methods in expounding the theory in the nonperiodic case.

6.10.2. It is possible to indicate a method of obtaining general systems of functions which are not analytic, but satisfying inequalities which are quite similar to the inequalities discussed above. Suppose (O. V. Besov)

$$(1) \qquad \boldsymbol{h} = (h_1, \ldots, h_n), \ h_i > 0, \ \boldsymbol{y}:\boldsymbol{h} = \left(\frac{y_1}{h_1}, \ldots, \frac{y_n}{h_n}\right),$$

$$\varphi_{\boldsymbol{h}}(\boldsymbol{x}) = \int\limits_{\mathbb{R}_n} \prod_{i=1}^{n} \frac{1}{h_i} \chi(\boldsymbol{y}:\boldsymbol{h}) \, \varphi(\boldsymbol{x} + \boldsymbol{y}) \, d\boldsymbol{y},$$

where the function $\chi(\boldsymbol{y})$ is infinitely differentiable on $\mathbb{R}_n$ with support within the first orthant, and

$$(2) \qquad \int\limits_{\mathbb{R}_n} \chi(\boldsymbol{y}) d\boldsymbol{y} = 1.$$

We shall call the function $\varphi_{\boldsymbol{h}}(x)$ the *mean for* $\varphi(\boldsymbol{x})$ *with the vector step* $\boldsymbol{h} = (h_1, \ldots, h_n)$.

For the mean functions the following inequality holds:

$$(3) \qquad \|D^{\alpha} \varphi_h(\boldsymbol{x})\|_{L_q(\mathbb{R}_m)} \leq c_1 \left(\prod_1^n h_i^{-\alpha_i}\right) \left(\prod_1^n h_i^{-\frac{1}{p}}\right) \left(\prod_1^m h_i^{\frac{1}{q}}\right) \|\varphi\|_{L_p(\mathbb{R}_n)}.$$

Inequality (3) is to a certain degree[1] analogous to the corresponding estlmates for entire functions of finite degrees $\nu_i = l/h_i$, which makes it possible without essential changes to carry the theory that has been presented to the case of approximation by mean functions $\varphi_{\boldsymbol{h}}$ (or second means $\varphi_{\boldsymbol{hh}} = (\varphi_{\boldsymbol{h}})_{\boldsymbol{h}}$ taking in place of

$$g(t) = \mu \left(\frac{\sin \frac{t}{\lambda}}{t} \right)^{\lambda}$$

in 5.2.1 (5) a smooth function $\xi(t)$ with compact support.

In this way one may for example arrive at the integral representation of V. P. Il'in [6] of a function in terms of its differences, obtained by him using different arguments. We note that in the construction (1) of the mean function $\varphi_{\boldsymbol{h}}(\boldsymbol{x})$ there took part only values of the function $\varphi(\boldsymbol{x} + \boldsymbol{y})$ for points of the neighborhood of the point $\boldsymbol{y} = \boldsymbol{0}$ lying in the first orthant. This guarantees the possibility of setting up a corresponding "local" theory.

---

[1] There is something of a difference in that in the right side of (3) there is a φ inside the norm sign, and not $\varphi_{\boldsymbol{h}}$. The way out of this situtation may be found in the fact that inequality (3) is applied for $\varphi_{\boldsymbol{h}}$, and then $\varphi_{\boldsymbol{hh}}$ appears in the left side instead of $\varphi_{\boldsymbol{h}}$.

We shall prove inequality (3). It is obviously possible to suppose that $\alpha = 0$. Using the Hölder inequality for the three functions

$$|\chi|^{1-\varepsilon}, \ |\varphi(\boldsymbol{x}+\boldsymbol{y})|^{\frac{q-p}{q}}, \ |\chi|^{\varepsilon}\,|\varphi(\boldsymbol{x}+\boldsymbol{y})|^{\frac{p}{q}} \quad (\varepsilon > 0)$$

with the exponents $\lambda_1 = \dfrac{p}{p-1}$, $\lambda_2 = \dfrac{pq}{q-p}$, $\lambda_3 = q$ we have[1]

$$\left| \int_{\mathbb{R}_n} \prod_{i=1}^{n} \frac{1}{h_i}\, \chi(\boldsymbol{y}:\boldsymbol{h})\, \varphi(\boldsymbol{x}+\boldsymbol{y}) d\boldsymbol{y} \right|$$

$$\leqq \left(\prod_{i=1}^{n} \frac{1}{h_i} \right) \left(\int_{\mathbb{R}_n} |\chi(\boldsymbol{y}:\boldsymbol{h})|^{\frac{1-\varepsilon}{p-1}p} d\boldsymbol{y} \right)^{1-\frac{1}{p}}$$

$$\times \|\varphi\|_{L_p(\mathbb{R}_n)}^{1-\frac{p}{q}} \left(\int_{\mathbb{R}_n} |\chi(\boldsymbol{y}:\boldsymbol{h})|^{\varepsilon q} |\varphi(\boldsymbol{x}+\boldsymbol{y})|^p d\boldsymbol{y} \right)^{\frac{1}{q}},$$

so that

$$\|\varphi_{\boldsymbol{h}}\|_{L_q(\mathbb{R}_m)} \leqq c_1 \left(\prod_{i=1}^{1} h_i \right)^{-\frac{1}{p}} \|\varphi\|_{L_p(\mathbb{R}_n)}^{1-\frac{p}{q}}$$

$$\times \operatorname*{sup\,vrai}_{x_{m+1},\ldots,x_n} \left\{ \int_{\mathbb{R}_n} |\chi(\boldsymbol{y}:\boldsymbol{h})|^{\varepsilon q} \int_{\mathbb{R}_m} |\varphi(\boldsymbol{x}+\boldsymbol{y})|^p \, dx_1 \ldots dx_n d\boldsymbol{y} \right\}^{1/q}$$

$$\leqq c \left(\prod_{1}^{n} h_i^{-\frac{1}{p}} \right) \left(\prod_{1}^{m} h_i^{\frac{1}{q}} \right) \|\varphi\|_{L_p(\mathbb{R}_n)}.$$

6.10.3. It is useful to keep the following lemma in mind.

Lemma *Suppose that on* $\mathbb{R}_n = \mathbb{R}_m \times \mathbb{R}_{n-m}$, $(\boldsymbol{x} = (\boldsymbol{u}, \boldsymbol{w}), \boldsymbol{u} \in \mathbb{R}_m, \boldsymbol{w} \in \mathbb{R}_{n-m})$ *we are given two functions* $f \in L_p(\mathbb{R}_n)$ $(1 \leqq p \leqq \infty)$ *and* $f*$, *and a sequence of functions* f_k, $k = 1, 2, \ldots$, *continuous on* $\mathbb{R}_n$, *such that the following properties are satisfied:*

1) $$\|f_k - f\|_{L_p(\mathbb{R}_n)} \to 0 (k \to \infty);$$

2) $$\|f_k(\boldsymbol{u}, \boldsymbol{w}) - f_*(\boldsymbol{u}, \boldsymbol{w})\|_{L_p(\mathbb{R}_m)} \to 0 \ (k \to \infty).$$

uniformly relative to $\boldsymbol{w}(|\boldsymbol{w}| < a)$;

3) $$\|f_k(\boldsymbol{u}, \boldsymbol{w}) - f_k(\boldsymbol{u}, \boldsymbol{w}')\|_{L_p(\mathbb{R}_m)} \to 0$$
$$(|\boldsymbol{w} - \boldsymbol{w}'| \to 0, \ |\boldsymbol{w}|, |\boldsymbol{w}'| < a)$$

uniformly relative to k $= 1, 2, \ldots$

---

[1] For the exponents λ_1, λ_2, λ_3 we have $1 \leqq \lambda_i \leqq \infty$ and $\lambda_1^{-1} + \lambda_2^{-1} + \lambda_3^{-1} = 1$.

Then f_*, for an arbitrary fixed $\boldsymbol{w}(|\boldsymbol{w}| < a)$ is the trace of the function f on the corresponding m-dimensional space $\mathbb{R}_m(\boldsymbol{w})$.

Proof. From properties 1), 2) it follows in view of Lemma 1.3.9 that f and f_* are equivalent on $\mathbb{R}_n$; $f = f_*$ almost everywhere on $\mathbb{R}_n$. Further, for the indicated $\boldsymbol{w}$, $\boldsymbol{w}'$

$$\|f_*(\boldsymbol{u}, \boldsymbol{w}) - f_*(\boldsymbol{u}, \boldsymbol{w}')\|_{L_p(\mathbb{R}_m)} \leqq \|f_*(\boldsymbol{u}, \boldsymbol{w}) - f_k(\boldsymbol{u}, \boldsymbol{w})\|_{L_p(\mathbb{R}_m)}$$

$$+ \|f_k(\boldsymbol{u}, \boldsymbol{w}) - f_k(\boldsymbol{u}, \boldsymbol{w}')\|_{L_p(\mathbb{R}_m)} + \|f_k(\boldsymbol{u}, \boldsymbol{w}') - f_*(\boldsymbol{u}, \boldsymbol{w}')\|_{L_p(\mathbb{R}_m)}$$

$$< \varepsilon + \varepsilon + \varepsilon = 3\varepsilon \ (k > k_0, |\boldsymbol{w} - \boldsymbol{w}'| < \delta)$$

for sufficently small δ and large k_0. This is possible in view of properties 1), 2), 3).

In order to make clear what might be the significance of the lemma just presented, we turn to the theory of imbedding of different dimensions, for simplicity restricting ourselves to the isotropic case. Using this lemma it is easy to verify that it is sufficient to prove the trace lemma only for continuous, or even only for infinitely differentiable functions of the corresponding class, as it will automatically be true for all functions of this class. Let us explain what this means.

Let

$$B = B^r_{p\theta}(\mathbb{R}_n),\ B' = B^\varrho_{p\theta}(\mathbb{R}_m(\boldsymbol{w}))\ \left(\varrho = r - \frac{n-m}{p} > 0,\ 1 \leqq m < n\right)$$

and let $\mathfrak{M} \subset B$ be a set of continuous functions dense in B (in the metric of B)[1]. Suppose further that the inequalities

$$\|f\|_{B'} \leqq c\|f\|_n, \tag{1}$$

$$\|f(\boldsymbol{u}, \boldsymbol{w}) - f(\boldsymbol{u}, \boldsymbol{w}')\|_{L_p(\mathbb{R}_m)} \leqq \|f\|_B\, \lambda(|\boldsymbol{w} - \boldsymbol{w}'|), \tag{2}$$

$$\left(\lambda(\delta) \to 0,\ \delta \to 0\right),$$

have been proved, where c does not depend on $\boldsymbol{w}$ and on the indicated f, while the function $\lambda(\delta)$ does not depend on f, $\boldsymbol{w}$, $\boldsymbol{w}'$. Then these inequalities, with the same constant c and function $\lambda(\delta)$, hold for all $f \in B$. Indeed, suppose that $f_k \in \mathfrak{M}$ $(k = 1, 2, \ldots)$ and $\|f_k - f\|_B \to 0$ $(k \to \infty)$.

$$\|f_k\|_{B'} \leqq c\|f_k\|, \tag{3}$$

$$\|f_k(\boldsymbol{u}, \boldsymbol{w}) - f_k(\boldsymbol{u}, \boldsymbol{w}')\|_{L_p(\mathbb{R}_m)} \leqq K\lambda(|\boldsymbol{w} - \boldsymbol{w}'|), \tag{4}$$

---

[1] In this discussion it is possible to replace B by $W^l_p(\mathbb{R}_n)$ $(l = 1, 2, \ldots)$.

where the constant K does not depend on k. It already follows from (3) that

$$\|f_k - f_l\|_{B'} \leqq c\|f_k - f_l\|_B,$$

so that as a consequence of the completeness of B', for any $\boldsymbol{w}$ there exists a function $f_*(\boldsymbol{x}) = f_*(\boldsymbol{u},\boldsymbol{w})$ such that

$$\|f_k - f_*\|_{L_p(\mathbb{R}_m)} \leqq \|f_k - f_*\|_{B'} \leqq c\|f_k - f_*\|_B \to 0, \quad k \to \infty, \tag{5}$$

$$\|f_*(\boldsymbol{u}, \boldsymbol{w}') - f_*(\boldsymbol{u}, \boldsymbol{w})\|_{L_p(\mathbb{R}_m)} \leqq K\lambda(|\boldsymbol{w} - \boldsymbol{w}'|),$$

$$\|f_*\|_{B'} \leqq c\|f\|_B.$$

Thus, for f, f_*, f_k conditions 1)—3) of the lemma are satisfied, so that f^*, for any $\boldsymbol{w}$, is the trace of f on $\mathbb{R}_m(\boldsymbol{w})$. This proves inequalities (1), (2) for an arbitrary function $f \in B$. Here we need to recall that the constant K in (4), for sufficiently large k, l, may be chosen to differ by arbitrarily little from $\|f\|_B$.

A similar argument may be carried through in the case of an inverse imbedding theorem. Suppose that $\mathfrak{M}' \subset B'$ is a set of continuous functions dense in B', and suppose that to each continuous function $\varphi \in \mathfrak{M}'$ defined on $\mathbb{R}_m = \mathbb{R}_m(0)$ there has been assigned a continuous function $A\varphi = f(x) \in B$, defined on $\mathbb{R}_n$ and such that the trace of f on $\mathbb{R}_m$ is φ, and the inequality[1]

$$\|f\|_B \leqq c\|\varphi\|_{B'} \tag{6}$$

is satisfied, where c does not depend on $\varphi \in \mathfrak{M}'$. Suppose given any function $\varphi \in B'$, and suppose that $\varphi_k \in \mathfrak{M}'$, $\|\varphi - \varphi_k\|_{B'} \to 0$ $(k \to \infty)$, $A\varphi_k = f_k$. Then

$$\|f_k - f_l\|_B \leqq c\|\varphi_k - \varphi_l\|_{B'} \to 0 \quad (k, l \to \infty),$$

and there exists an $f \in B$ (B is complete) such that $\|f - f_k\|_B \to 0$. Obviously for the functions φ, f inequality (6) is satisfied, and with the same constant c.

We note that for a finite θ the set $\mathfrak{M}$ of entire functions $f \in L_p(\mathbb{R}_n)$ $(1 \leqq p \leqq \infty)$ of exponential spherical types (all types) is dense in any $B = B^r_{p\theta}(\mathbb{R}_n)$ (in the metric of B). Indeed it is dense in any $B = B^r_{p\theta}(\mathbb{R}_n)$. $1 \leqq \theta < \infty$, because if $f \in B$, then (see 6.2 (6))

$$f = \sum_0^\infty Q_s, \ \|f\|_B = \left(\sum_0^\infty a^{sr\theta}\|Q_s\|^\theta\right)^{1/\theta}$$

[1] Once again B can be replaced here by $W^l_p(\mathbb{R}_n)$ $(l = 1, 2, \ldots)$. The corresponding theorem on extension from $\mathbb{R}_m$ to $\mathbb{R}_n$ is proved in 9.5.2.

and

$$\|f - f_k\|_B = \left(\sum_{k+1}^{\infty} a^{sr\theta} \|Q_s\|^\theta\right) \to 0 \quad (k \to \infty),$$

where

$$f_k = \sum_0^k Q_s \in \mathfrak{M}.$$

For $1 \leqq p < \infty$, the set $\mathfrak{M}$ is also dense in $W_p^l(\mathbb{R}_n)$ $(l = 0, 1, 2, \ldots)$, as follows from the estimate 5.2.2 (4).

Of course it follows from what has been said that the set of all infinitely differentiable functions of the class $B_{p\theta}^r(\mathbb{R}_n)$ $(1 \leqq \theta < \infty)$ or of $W_p^l(\mathbb{R}_n)$ is dense in the corresponding class, because it contains a set of functions of exponential types belonging to $L_p(\mathbb{R}_n)$.

Chapter 7

Transitivity and Unimprovability of Imbedding Theorems. Compactness

7.1. Transitive Properties of Imbedding Theorems[1]

Suppose given a system of numbers

$$(1) \qquad \boldsymbol{r} = (r_1, \ldots, r_n) > \boldsymbol{0}, \quad \boldsymbol{p} = (p_1, \ldots, p_n) \quad (1 \leqq p_l \leqq \infty)$$

and numbers p', p'', satisfying the inequalities

$$(2) \qquad p_l \leqq p' < p'' \leqq \infty .$$

If the following conditions are satisfied:

$$(3) \qquad \varrho_i' = \frac{r_i \varkappa}{\varkappa_i},$$

$$(4) \qquad \varkappa = 1 - \sum_{l=1}^{n} \frac{\frac{1}{p_l} - \frac{1}{p'}}{r_l} > 0,$$

$$(5) \qquad \varkappa_i = 1 - \sum_{l=1}^{n} \frac{\frac{1}{p_l} - \frac{1}{p_i}}{r_l} > 0 \qquad (i = 1, \ldots, n),$$

then the imbedding theorem (6.8) holds:

$$B^{\boldsymbol{r}}_{\boldsymbol{p}\theta}(\mathbb{R}_n) \to B^{\boldsymbol{\varrho}'}_{p'\theta}(\mathbb{R}_n)$$

accomplishing the transition from the system of numbers (1) to the system of numbers

$$\boldsymbol{\varrho}' = (\varrho_1', \ldots, \varrho_n'), \, p' .$$

We need to keep in mind that $\varkappa \leqq \varkappa_i$, so that inequality (5) is a consequence of inequality (4).

---

[1] S. M. Nikol'skiĭ [3, 10].

But now we may assume the class $B^{\varrho'}_{p'\theta}(\mathbb{R}_n)$ to be the initial class, and in the presence of the inequality

$$\varkappa' = 1 - \left(\frac{1}{p'} - \frac{1}{p''}\right) \sum_1^n \frac{1}{\varrho'_k} > 0$$

conclude that there holds a further imbedding of classes

$$B^{(\varrho')}_{p'\theta}(\mathbb{R}_n) \to B^{(\varrho'')}_{p''\theta}(\mathbb{R}_n),$$

where

$$\varrho'' = (\varrho''_1, \ldots, \varrho''_n) = \varkappa' \varrho'.$$

Thus we have transformed the system $(\boldsymbol{r}, \boldsymbol{p})$ into a system (ϱ', p'), which in its turn we have transformed into a system (ϱ'', p''). We need to keep in mind that ϱ' is defined in terms of $\boldsymbol{r}, \boldsymbol{p}$ and p', and ϱ'' in terms of ϱ', p' and p''. It is worth noting that these transformations bear a *transitive* character: the passage from the first system to the second, and then from the second to the third, may be replaced by one passage from the first system to the third.

Indeed,

$$\varrho''_k = \frac{r_k \varkappa \varkappa'}{\varkappa_k} \quad (k = 1, \ldots, n),$$

where we have assumed that

(6) $$\varkappa, \varkappa', \varkappa_k > 0 \quad (k = 1, \ldots, n).$$

On the other hand, suppose that $p_k \leq p''$ $(k = 1, \ldots, n)$, and suppose that we have the inequalities

(7) $$\varkappa'', \varkappa_k > 0 \quad (k = 1, \ldots, n),$$

where

$$\varkappa'' = 1 - \sum_{l=1}^n \frac{\frac{1}{p_l} - \frac{1}{p''}}{r_l}.$$

Then we have the imbedding

$$B^{\boldsymbol{r}}_{\boldsymbol{p}}(\mathbb{R}_n) \to B^{(\varrho''_*)}_{p''\theta}$$

i.e. a passage from $(\boldsymbol{r}, \boldsymbol{p})$ directly to (ϱ''_*, p''), where

$$\varrho''_* = \{\varrho''_{*1}, \ldots, \varrho''_{*n}\}$$

and

$$\varrho''_{*k} = \frac{r_k \varkappa''}{\varkappa_k} \quad (k = 1, \ldots, n).$$

But it is easy to calculate that

$$\varkappa'' = \varkappa\varkappa', \tag{8}$$

so that

$$\varrho_*'' = \varrho''.$$

Moreover, in view of the inequality $p' < p''$, it is obvious that $\varkappa' > \varkappa''$, i.e. $\varkappa' > 0$. But then, in view of (8), $\varkappa > 0$, and the transitivity is proved.

The transitivity of the imbedding theorems for different dimensions,

$$B_{p\theta}^{\boldsymbol{r}}(\mathbb{R}_n) \to B_{p\theta}^{\boldsymbol{r}'}(\mathbb{R}_{m_1}) \to B_{p\theta}^{\boldsymbol{r}''}(\mathbb{R}_{m_2})$$

$$(1 \leqq m_2 < m_1 < n),$$

where

$$r_i' = r_i\varkappa' \ (i = 1, \ldots, m_1),$$

$$\varkappa' = 1 - \frac{1}{p} \sum_{m_1+1}^{n} \frac{1}{r_j} > 0,$$

$$r_i'' = r_i'\varkappa'' \ (i = 1, \ldots, m),$$

$$\varkappa'' = 1 - \frac{1}{p} \sum_{m_2+1}^{n} \frac{1}{r_j'},$$

follows from the easily verified equation

$$\varkappa = 1 - \frac{1}{p} \sum_{m_2+1}^{n} \frac{1}{r_j} = \varkappa'\varkappa''.$$

7.2. Inequalities with a Parameter ε. Multiplicative Inequalities

Suppose given a function $f(\boldsymbol{x}) \in H_p^{\boldsymbol{r}}(\mathbb{R}_n) = H_p^{\boldsymbol{r}}$ and a positive vector $\varepsilon = (\varepsilon_1, \ldots, \varepsilon_n)$. Put $F(\boldsymbol{x}) = f(\varepsilon_1 x_1, \ldots, \varepsilon_n x_n) = f_\varepsilon(\boldsymbol{x})$.

Obviously $(k_j > r_j - \varrho_j > 0)$

$$\frac{\left\|\Delta_h^{k_j} F_{x_j}^{(\varrho_j)}\right\|_p}{h^{r_j-\varrho}} = \frac{\varepsilon_j^{r_j}\left\|\Delta_{x_j\varepsilon_j h}^{k_j} f_{x_j}^{(\varrho_j)}(\varepsilon_1 x_1, \ldots, \varepsilon_n x_n)\right\|_p}{(\varepsilon_j h)^{r_j-\varrho_j}}$$

$$= \frac{\varepsilon_j^{r_j}\varepsilon^{-\frac{1}{p}}\left\|\Delta_{x_j\varepsilon_j h}^{k} f_{x_j}^{(\varrho_j)}(\boldsymbol{x})\right\|_p}{(\varepsilon_j h)^{r_j-\varrho_j}} \quad (\varepsilon^\alpha = \varepsilon_1^\alpha \ldots \varepsilon_n^\alpha).$$

Taking the upper bound of both sides of the inequality relative to h, we get

(1) $$\|f_\varepsilon(\boldsymbol{x})_{h^{r_j}_{x_jp}}\| = \varepsilon_j^{r_j}\varepsilon^{-\frac{1}{p}}\|f\|_{h^{r_j}_{x_jp}},$$

for any $\varepsilon > 0$.

Consider further the seminorm $b^{r_j}_{x_jp} = b^{r_j}_{x_jp\theta}(\mathbb{R}_n)$, $1 \leqq \theta \leqq \infty$:

(2) $$\|f\|_{b^{r_j}_{x_jp}} = \left(\int_0^\infty t^{1-\theta(r_j-\varrho_j)}\,\Omega^{k_j}_{x_j}\left(f^{(\varrho_j)}_{x_j}, t\right)^\theta_{L_p(\mathbb{R}_n)}dt\right)^{1/\theta}.$$

Here we may also insert the function f_ε and carry out a change of variables in the integral under the sign Ω. As a result we obtain an equation, analogous to (1):

(3) $$\|f_\varepsilon(\boldsymbol{x})\|_{b^{r_j}_{x_jp}} = \varepsilon_j^{r}\varepsilon^{-\frac{1}{p}}\|f\|_{b^{r_j}_{x_jp}}.$$

It is thus true for any θ $(1 \leqq \theta \leqq \infty)$.

Further, it is obvious that

(4) $$\|f_\varepsilon\|_p = \varepsilon^{-\frac{1}{p}}\|f\|_p,$$

so that

(5) $$\|f_\varepsilon\|_{B^r_p} = \varepsilon^{-\frac{1}{p}}\left\{\|f\|_p + \sum_{j=1}^n \varepsilon_j^{r_j}\|f\|_{B^{r_j}_{x_jp}}\right\}.$$

If we put $f_\varepsilon(\boldsymbol{x}) = f(\varepsilon\boldsymbol{x})$, for functions f belonging to isotropy classes, where ε is now a positive scalar, and argue as above, we get

(6) $$\|f(\varepsilon\boldsymbol{x})\|_p = \varepsilon^{-\frac{n}{p}}\|f\|_p,\ \|f(\varepsilon\boldsymbol{x})\|_{h^r_p} = \varepsilon^{r-\frac{n}{p}}\|f\|_{h^r_p},$$

$$\|f(\varepsilon\boldsymbol{x})\|_{b^r_p} = \varepsilon^{r-\frac{n}{p}}\|f\|_{b^r_p}.$$

We present some examples of the application of formulas (3)—(6). To the imbedding $B^r_p \to B^\varrho_p$ $(0 < \varrho < r)$ there is connected the inequality

(7) $$\|f\|_{B^\varrho_p} \leqq c\left(\|f\|_p + \|f\|_{b^r_p}\right),$$

from which, in view of (6), one obtains the inequality

(7′) $$\|f\|_{B^\varrho_p} \leqq c\left(\varepsilon^{-\varrho}\|f\|_p + \varepsilon^{r-\varrho}\|f\|_{b^r_p}\right)$$

with an arbitrary parameter ε.

Conversely, (7) follows from (7') on putting $\varepsilon = 1$. In the applications inequality (7') is employed when it is desired that a certain one of the terms of its right hand side should be sufficiently small. Minimizing the right side of (7') with respect to ε, we obtain the inequality

$$(7'') \qquad \|f\|_{b_p^\varrho} \leqq c\left[\left(\frac{r-\varrho}{\varrho}\right)^{\varrho/r} + \left(\frac{\varrho}{r-\varrho}\right)^{1-\varrho/r}\right] \|f\|^{1-\frac{\varrho}{p}} \left(\|f\|_{b_p^r}\right)^{\varrho/r},$$

which is called the *multiplicative inequality*. From (7''), obviously, (7') follows.

We shall consider further the following inequalities, connected with imbeddings of different dimensions and metrics:

$$(8) \qquad \|f\|_{B_p^\varrho(\mathbb{R}_m)} \leqq c\left(\|f\|_{L_p(\mathbb{R}_n)} + \|f\|_{b_p^r(\mathbb{R}_n)}\right)$$

$$\left(1 \leqq m < n,\ \varrho = r - \frac{n-m}{p} > 0\right),$$

$$(9) \qquad \|f\|_{B_{p'}^{r'}} \leqq c\left(\|f\|_{L_p(R_n)} + \|f\|_{b_p^r(\mathbb{R}_n)}\right),$$

$$r' = r - \left(\frac{1}{p} - \frac{1}{p'}\right) n > 0,$$

where $\mathbb{R}_m$ is the subspace of points $(x_1, \ldots, x_m, x_{m+1}, \ldots, x_n)$ with an arbitrary fixed $\boldsymbol{y} = (x_{m+1}, \ldots, x_n)$, and c does not depend on f and φ. If in these inequalities one replaces f by f_ε, and then take the ε out from under the norm by using (6), then one obtains respectively

$$\|f\|_{b_p^\varrho(\mathbb{R}_m)} \leqq c\left(\varepsilon^{-r}\|f\|_{L_p(\mathbb{R}_n)} + \|f\|_{b_p^r(\mathbb{R}_n)}\right),$$

$$\|f\|_{b_{p'}^{r'}(\mathbb{R}_n)} \leqq c\left(\varepsilon^{-r}\|f\|_{L_p(\mathbb{R}_n)} + \|f\|_{b_p^r(\mathbb{R}_n)}\right).$$

Passing to the limit as $\varepsilon \to \infty$, we obtain the inequalities

$$(10) \qquad \|f\|_{b_p^\varrho(\mathbb{R}_m)} \leqq c\, \|f\|_{b_p^r(\mathbb{R}_n)},$$

$$(11) \qquad \|f\|_{b_{p'}^{r'}(\mathbb{R}_n)} \leqq c\, \|f\|_{b_p^r(\mathbb{R}_n)}.$$

These improve inequalities (8), (9), since the same c enters into them, but they do not contain the term $\|f\|_{L_p(\mathbb{R}_n)}$, which is assumed to be finite. However if $\|f\|_{L_p(\mathbb{R}_n)} = \infty$, then inequalities (10), (11), generally speaking, are not true. So, for $r - \varrho \geqq 1$ the polynomial

$$P_l(\boldsymbol{x}) = \sum_{|\boldsymbol{k}| \leqq l} a^{\boldsymbol{k}} \boldsymbol{x}^{\boldsymbol{k}},$$

where $l = \bar{r}$ if r is not an integer and $l = \bar{r} + 1$ if r is an integer, makes the right side of inequality (10) equal to zero, while its left side in general is not equal to zero. For $r - \varrho < 0$ inequalities (10) may be satisfied without the norm being finite (see the remark to (7.2)).

It is possible, in the spirit of formulas (10)—(11), to give a sharpening of the theorem on mixed derivatives. For example, for $W_p^2 (1 < p < \infty)$ we have the inequality (see 9.2.2)

$$\left\| \frac{\partial^2 u}{\partial x_1 \, \partial x_2} \right\| \leqq c \left(\left\| \frac{\partial^2 u}{\partial x_1^2} \right\| + \left\| \frac{\partial^2 u}{\partial x_2^2} \right\| + \|u\| \right),$$

so that

$$\left\| \frac{\partial^2 u}{\partial x_1 \, \partial x_2} \right\| \leqq c \left(\left\| \frac{\partial^2 u}{\partial x_1^2} \right\| + \left\| \frac{\partial^2 u}{\partial x_2^2} \right\| + \varepsilon^{-2} \|u\| \right).$$

After passage to the limit as $\varepsilon \to 0$ we obtain the inequality

$$\left\| \frac{\partial^2 u}{\partial x_1 \, \partial x_2} \right\| \leqq c \left\| \frac{\partial^2 u}{\partial x_1^2} \right\| + \left\| \frac{\partial^2 u}{\partial x_2^2} \right\| \Bigg),$$

valid under the condition that $\|u\| < \infty$.

Similar sharpenings do not always hold. For example, in inequality (7) the first term on its right side cannot be dropped, as is clear from the inequality (7′) which is equivalent to it. If the first term in the latter were dropped, then on passing to the limit as $\varepsilon \to 0$ we would find that the left side is equal to zero. This is however possible only if f is a polynomial.

We shall further consider an example relating to the anisotropic case.

In the inequality of different dimensions

$$\|f\|_{b_{x_j p}^{\varrho_j}(\mathbb{R}_m)} \leqq c \left(\|f\|_{L_p} + \|f\|_{b_p^{\mathbf{r}}(\mathbb{R}_n)} \right), \tag{12}$$

$$\varkappa = 1 - \frac{1}{p} \sum_{m+1}^{n} \frac{1}{r_j} > 0, \quad \varrho_j = \varkappa r_j, \tag{13}$$

$$(j = 1, \ldots, m)$$

the first term in the right side is superfluous. In fact choosing $j = 1$ for convenience and substituting f_ε in (12), we find on the basis of (3) that

$$\varepsilon_1^{\varrho_1} (\varepsilon_1 \ldots \varepsilon_m)^{-1/p} \|f\|_{b_{x_1 p}^{\varrho_1}(\mathbb{R}_m)} \leqq c (\varepsilon_1 \ldots \varepsilon_n)^{-1/p} \left\{ \{\|f\|_{L_p(\mathbb{R}_n)} + \sum_{j=1}^{u} \varepsilon_j^{r_j} \|f\|_{b_{x_j p}^{r_j}(\mathbb{R}_n)} \right\}.$$

Cancelling the $(\varepsilon_1 \ldots \varepsilon_m)^{-1/p}$, we pass to the limit as $\varepsilon_j \to 0$ only when $j = 2, \ldots, m$, and put $\varepsilon_j = \varepsilon_1^{r_1/r_j}$ when $j = m + 1, \ldots, n$. Then

$$\|f\|_{b^{\varrho_1}_{x_1 p}(\mathbb{R}_m)} \leqq c \left\{ \varepsilon_1^{-r_1} \|f\|_{L_p(\mathbb{R}_n)} + \sum_{j=m+1}^{n} \|f\|_{b^{r_j}_{x_j p}(\mathbb{R}_n)} \right\}.$$

Passage to the limit as $\varepsilon_1 \to \infty$ leads to the inequality (12), but already without the first term in the right side for $j = 1$. But we can do the same thing for any $j = 1, \ldots, m$. Summing on j, we obtain (if $\|f\|_{L_p} < \infty$) the inequality

$$\|f\|_{b^{\varrho}_{p}(\mathbb{R}_m)} \leqq c_1 \|f\|_{b^{r}_{p}(\mathbb{R}_n)},$$

which sharpens the corresponding imbedding theorem for different dimensions.

7.3. Boundary Functions in $H_p^{\mathbf{r}}$. Unimprovability of Imbedding Theorems

We will write $\varepsilon = (\varepsilon_1, \ldots, \varepsilon_n) > 0$, if for all j we have ε_j nonnegative, and at least one ε_j is positive. We will call the function f a *boundary function* in the class $H_{\mathbf{p}}^{\mathbf{r}}$, if it belongs to $H_{\mathbf{p}}^{\mathbf{r}}$ and does not belong to $H_{\mathbf{p}}^{\mathbf{r}+\varepsilon}$ for any vector $\varepsilon > \mathbf{0}$.

We will consider the class $H_{\mathbf{p}}^{\mathbf{r}}(\mathbb{R}_n)$, where $\mathbf{r} = (r_1, \ldots, r_n) > \mathbf{0}$, $\mathbf{p} = (p_1, \ldots, p_n)$, $1 \leqq p_j \leqq \infty$, $j = 1, \ldots, n$.

As always, if $p = p_1 = \cdots = p_n$, then in place of the vector $\mathbf{p}$ we will speak of the number p, and in place of $H_{\mathbf{p}}^{\mathbf{r}}$ we shall write $H_p^{\mathbf{r}}$. On the vector $\mathbf{p}$ we impose the condition

$$(1) \qquad \varkappa_j = \varkappa_j(\mathbf{p}) = 1 - \sum_{l=1}^{n} \left(\frac{1}{p_l} - \frac{1}{p_j} \right) \frac{1}{r_l} > 0 \quad (j = 1, \ldots, n).$$

In particular,

$$\varkappa_j(p) = 1 \quad (j = 1, \ldots, n),$$

and in the case of the classes $H_p^{\mathbf{r}}(\mathbb{R}_n)$ the condition (1) is automatically satisfied.

We note that

$$\sum_{l=1}^{n} \frac{\varkappa_j}{r_j} = \sum_j \frac{1}{r_j} + \sum_j \sum_l \left(\frac{1}{p_l} - \frac{1}{p_j} \right) \frac{1}{r_l r_j} = \sum_j \frac{1}{r_j}.$$

Put

(2) $$F(t) = \left(\frac{\sin\frac{t}{2}}{\frac{t}{2}}\right)^2$$

and

(3) $$\psi(\boldsymbol{x}) = \psi_{\boldsymbol{p},\boldsymbol{r}}(a, \boldsymbol{x}) = \sum_{s=0}^{\infty} \frac{\prod_{j=1}^{n} F\left(a^{\frac{s\varkappa_j}{r_j}} x_j\right)}{a^{s\left(1-\sum_{l=1}^{n}\frac{1}{p_l r_l}\right)}}$$

$$\left(a > 1,\ \varkappa_j = \varkappa_j(\boldsymbol{p})\right).$$

In particular,

(4) $$\psi_{\boldsymbol{p},\boldsymbol{r}}(a, \boldsymbol{x}) = \sum_{s=0}^{\infty} \frac{\prod_{j=1}^{n} F\left(a^{\frac{s}{r_j}} x_j\right)}{a^{s\left(1-\frac{1}{p}\sum_{l}^{n}\frac{1}{r_l}\right)}}.$$

We shall show that $\psi_{\boldsymbol{p},\boldsymbol{r}}(a, \boldsymbol{x}) \in H_{\boldsymbol{p}}^{\boldsymbol{r}}(\mathbb{R}_n)$. Indeed, suppose that Q_s is the sth term of the series (3). Since F is an entire function of type 1 of one variable, then Q_s is an entire function of type $\nu_j(s) = a^{s\frac{\varkappa_j}{r_j}}$ relative to x_j, and in addition

(5) $$\|Q_s\|_{L_{p_j}(\mathbb{R}_n)} \sim a^{-s\varkappa_j} \quad (s = 0, 1, \ldots),$$

because

(6) $$1 - \sum_{1}^{n} \frac{1}{p_l r_l} + \frac{1}{p_i} \sum_{1}^{n} \frac{\varkappa_l}{r_l} = 1 - \sum_{1}^{n} \left(\frac{1}{p_l} - \frac{1}{p_i}\right)\frac{1}{r_l} = \varkappa_i.$$

Accordingly[1],

(7) $$\nu_i(s)^{r_i}\|Q_s\|_{L_{p_i}(\mathbb{R}_n)} = \left(a^{s\frac{\varkappa_i}{r_i}}\right)^{r_i} \|Q_s\|_{L_{p_i}(\mathbb{R}_n)} \sim 1 \ (s = 0, 1, \ldots).$$

Thus the left side of (7) is bounded for $\nu_j(s)$ running through an increasing progression. This shows (see 5.5.3 (6)) that $\psi \in H_{x_i p_i}^{r_i}(\mathbb{R}_n)$ for any $i = 1, \ldots, n$, i.e. $\psi \in H_{\boldsymbol{p}}^{\boldsymbol{r}}(\mathbb{R}_n)$.

But below (see 7.4) it will be shown that in any case for a sufficiently large $a > 1$ the function $\psi_{\boldsymbol{p},\boldsymbol{r}}$ not only belongs to $H_{\boldsymbol{p}}^{\boldsymbol{r}}(\mathbb{R}_n)$, but

---

[1] By definition $a_s \cap b_s$ $(s \in e)$ if there exist positive constants c_1, c_2, not depending on $s \in e$, such that $c_1 a_s \leqq b_s \leqq c_2 a_s$ $(s \in e)$.

is also a boundary function in this class. We will meanwhile make certain deductions arising from this.

Suppose given a number $p' \geqq p_j$ $(j = 1, \ldots, n)$, which may in particular be equal to ∞, such that

$$\varkappa = 1 - \sum_1^n \left(\frac{1}{p_l} - \frac{1}{p'}\right)\frac{1}{r_l} > 0.$$

Then automatically $\varkappa_j > 0$, $j = 1, \ldots, n$. As in the imbedding theorem for different metrics we define the numbers

(9) $$\varrho_i = \frac{r_i \varkappa}{\varkappa_i} \quad (i = 1, \ldots, n).$$

If we put

(8) $$b = a^{\varkappa}, \quad a^{s\frac{\varkappa_j}{r_j}} = b^{s/\varrho_j} \qquad (j = 1, \ldots, n),$$

then we get

(9) $$\psi(\boldsymbol{x}) = \psi_{p',\varrho}(b, \boldsymbol{x}) = \sum_{s=0}^{\infty} \frac{F\left(b^{s\frac{1}{r_j}} x_j\right)}{b^{s\left(1 - \frac{1}{p'}\Sigma \frac{1}{\varrho_l}\right)}}.$$

Indeed,

$$\varkappa\left(1 - \frac{1}{p'}\sum_l \frac{1}{\varrho_l}\right) = \varkappa - \frac{1}{p'}\sum_l \frac{\varkappa_l}{r_l} = 1 - \sum_l \left(\frac{1}{p_l} - \frac{1}{p'}\right)\frac{1}{r_l}$$
$$- \frac{1}{p'}\sum_1 \frac{1}{r_l} = 1 - \sum \frac{1}{p_l r_l}.$$

The equations (3) and (9) exhibit the fact that ψ is at the same time a function $\psi_{\boldsymbol{p},\boldsymbol{r}}(a, \boldsymbol{x})$ and a function $\psi_{p',\varrho}(b, \boldsymbol{x})$ where b and a are connected by equation (8). But if a is sufficiently large, then $\psi_{\boldsymbol{p},\boldsymbol{r}} \in H_{\boldsymbol{p}}^{\boldsymbol{r}}(\mathbb{R}_n)$ and $\psi_{p',\varrho} \in H_{p'}^{\varrho}(\mathbb{R}_n)$, which is in accordance with the imbedding theorem. But $\psi_{p',\varrho}$ is a boundary function in the class $H_{p'}^{\varrho}(\mathbb{R}_n)$. It does not belong to any class $H_{p'}^{\varrho+\varepsilon}(\mathbb{R}_n)$, where $\varepsilon > 0$. This shows the the imbedding $H_{\boldsymbol{p}}^{\boldsymbol{r}}(\mathbb{R}_n) \to H_{p'}^{\varrho+\varepsilon}(\mathbb{R}_n)$ $(\varepsilon > 0)$ is false. But then the imbedding $B_{\boldsymbol{p}\theta}^{\boldsymbol{r}}(\mathbb{R}_n) \to B_{p'}^{\varrho+\varepsilon}(\mathbb{R}_n)$, is also false, because if one supposed that it were true one would have

$$H_p^{r+\frac{1}{2}\varepsilon}(\mathbb{R}_n) \to B_{\boldsymbol{p}\theta}^{\boldsymbol{r}}(\mathbb{R}_n) \to B_{p\theta}^{\varrho+\varepsilon}(\mathbb{R}_n) \to H_{p'}^{\varrho+\varepsilon}(\mathbb{R}_n),$$

which, as we have proved, is not possible. Now let us start from the function $\psi_{p,\boldsymbol{r}}(a, \boldsymbol{x})$ (see (3)). Suppose that

(10) $$\varkappa = 1 - \frac{1}{p}\sum_{m+1}^{n} \frac{1}{r_l} > 0 \quad (1 \leqq m < n).$$

We will suppose that the vector $\varrho = (\varrho_1, \ldots, \varrho_m)$, now already m-dimensional, is defined as in the imbedding theorem for different dimensions, by the equations

$$\varrho_j = r_j \varkappa \ (j = 1, \ldots, m),$$

Put $x = (u, y)$, $u = (x_1, \ldots, x_m)$, $y = (x_{m+1}, \ldots, x_n)$, $\psi(x) = \psi(u, y)$.

Denote by $\mathbb{R}_m$ the coordinate subspace of points $(u, 0)$. The trace of ψ on $\mathbb{R}_m$ is the function $(F(0) = 1)$

$$\psi(u, 0) = \sum_{s=0}^{\infty} \frac{\prod_{j=1}^{m} F\left(a^{\frac{s}{r_j}} x_j\right)}{a^{s\left(1-\frac{1}{p}\sum_1^n \frac{1}{r_l}\right)}} = \sum_{s=0}^{\infty} \frac{\prod_{j=1}^{m} F\left(b^{\frac{s}{\varrho_j}} x_j\right)}{b^{s\left(1-\frac{1}{p}\sum_1^m \frac{1}{\varrho_l}\right)}} = \psi_{p,\varrho}(u) \qquad (b = a^{\varkappa})$$

because

$$\varkappa\left(1 - \frac{1}{p}\sum_1^m \frac{1}{\varrho_l}\right) = \varkappa - \frac{1}{p}\sum_1^m \frac{1}{r_l} = 1 - \frac{1}{p}\sum_1^n \frac{1}{r_l}.$$

From (11) we see that the trace of $\psi_{p,r}(x)$ on $\mathbb{R}_m$ is $\psi_{p,\varrho}(u)$. In addition $\psi_{p,r}(x) \in H_p^r(\mathbb{R}_n)$, $\psi_{p,\varrho}(u) \in H_p^\varrho(\mathbb{R}_m)$, which is in conformity with the imbedding theorem for different dimensions. But ψ_p^ϱ is a boundary function in $H_p^\varrho \mathbb{R}_m)$ and does not belong to $H_p^{\varrho+\varepsilon}(\mathbb{R}_m)$ $(\varepsilon > 0)$, so that the imbedding $H_p^r(\mathbb{R}_n) \to H_p^{\varrho+\varepsilon}(\mathbb{R}_m)$ it false. Reasoning as above we arrive at the conclusion that the imbedding $B_{p\theta}^r(\mathbb{R}_n) \to B_{p\theta}^{\varrho+\varepsilon}(\mathbb{R}_m)$ is false as well. This shows that the imbedding theorem for different metrics is not improvable in the indicated sense. However it can be improved in the terms of wider classes. For example, A. S. Džafarov [1] obtained an improvement of the imbedding theorems for the classes H_p^r, considering the wider classes $H_p^{r,s}(H_p^{r,0} = H_p^r)$ of functions f, which, for example, for $n = 1$, $r < 1$, are defined as follows: $f \in H_p^{r,s}$ if $f \in L_p$ and

$$\|f(x + h) - f(x)\| \leq M|h|^r \left|\ln \frac{1}{|h|}\right|^s.$$

We have seen that the conclusion on the impossibility of the indicated imbedding reduced to the proof of the impossibility of the inequality accompanying it. However it remains to this point not clear as to whether there exists in the class $B_{p\theta}^r(\mathbb{R}_m)$ a function which does not belong to $B_{p\theta}^{\varrho+\varepsilon}(\mathbb{R}_m)$. It will be proved in 7.6 that such a function does exist.

7.4. More on Boundary Functions in H_p^r

We turn to the proof of the fact that $\psi = \psi_{pr}(a, x)$ (7.3 (3)) for sufficiently large a is a boundary function in $H_p^r(\mathbb{R}_n)$.

We note (see 7.3 (2)) that the function $F(t)$ has the following properties: for each natural number l it is possible to indicate numbers c, δ, depending on l, such that

1) the derivative $F^{(l)}(t)$ preserves its sign on $(0, \delta)$;
2) on $(0, \delta)$ the following inequality holds:

(1) $$|F^{(l)}(t)| \geqq ct.$$

The first property follows from the analyticity of F. The second follows from the fact that

$$F(t) = a_0 + a_2 t^2 + a_4 t^4 + \ldots,$$

where $a_i \neq 0$ for arbitrary $i = 0, 2, 4, \ldots,$.

Put

(2) $$\gamma = 1 - \sum_1^n \frac{1}{p_l r_l}$$

and note that

(3) $$\frac{1}{p_i} \sum_{j=1}^n \frac{\varkappa_j}{r_j} - \varkappa_i = \frac{1}{p_i} \sum_l \frac{1}{r_l} - \left(1 - \sum_l \left(\frac{1}{p_l} - \frac{1}{p_i}\right) \frac{1}{r_l}\right) = -\gamma.$$

Suppose given a sequence

(4) $$h = h_\mu = \frac{\delta}{2} a^{-\mu \frac{\varkappa_1}{r_1}} \quad (\mu = 0, 1, \ldots),$$

where δ is the number spoken of above, chosen for $l = \bar{r}_1 + 2$ ($r_1 = \bar{r}_1 + \alpha$, $\bar{r}_1$ an integer and $0 < \alpha \leqq 1$).

Our function may be written in the form

$$\psi = \sum_0^\infty Q_s, \quad Q_s = a^{-s\gamma} \prod_{j=1}^n F\left(a^{s \frac{\varkappa_j}{r_j}} x_j\right).$$

Our object is the estimation from below of the norm $\Delta_{x_1 h}^2 \psi_{x_1}^{(\bar{r}_1)}(\boldsymbol{x})$ in the metric $L_{p_1}(\mathbb{R}_n)$, where $\psi_{x_1}^{(\bar{r}_1)}$ denotes the derivative of ψ relative to x_1 of order $\bar{r}_1$.

We have

(5) $$\Delta_{x_1 h}^2 \psi_{x_1}^{(\bar{r}_1)} = \sum_{s \leqq \mu} \Delta_{x_1 h}^2 Q_{s x_1}^{(\bar{r}_1)} + \sum_{s > \mu} = s(h) + \sigma(h),$$

(6) $$\|\sigma(h)\|_{L_{p_1}(\mathbb{R}_n)} \leqq 4 \sum_{s > \mu} \|Q_{s x_1}^{(\bar{r}_1)}\|_{L_{p_1}(\mathbb{R}_n)} \leqq 4 \sum_{s > \mu} a^{-s\varkappa_1\left(1 - \frac{\bar{r}_1}{r_1}\right)}$$

$$= 4a^{-(\mu+1)\frac{\varkappa_1}{r_1}\alpha} \sum_0^\infty a^{-s\frac{\varkappa_1}{r_1}\alpha} = \left(\frac{2}{\delta} h\right)^\alpha \frac{4}{a^{\frac{\varkappa_1}{r_1}\alpha}\left(1 - a^{-\frac{\varkappa_1}{r_1}\alpha}\right)}.$$

In the second inequality we have made use of the estimate 7.3 (5). In the last equation we carried out a replacement by $h = h_\mu$ according to formula (4). We did not calculate the constant on h^α in vain; now it is clear that it tends to zero as $a \to \infty$.

On the other hand (see below for explanation)

$$(7) \quad \|s(h)\|_{L_{p_1}(\mathbb{R}_n)} = \left\| \sum_{s \leq \mu} a^{-s\gamma} \left(ha^{s\frac{\varkappa_1}{r_1}}\right)^2 a^{s\bar{r}_1 \frac{\varkappa_1}{r_1}} F^{(\bar{r}_1+2)}\left(a^{s\frac{\varkappa_1}{r_1}}(x_1 + \theta h)\right) \right.$$

$$\left. \times \prod_{j=2}^{n} F\left(a^{s\frac{\varkappa_j}{r_j}} x_j\right) \right\|_{L_{p_1}(\mathbb{R}_n)} \geq \|\cdot\|_{L_{p_1}[(0,h)\times\mathbb{R}_{n-1}]} \geq \left(ha^{\mu\frac{\varkappa_1}{r_1}}\right)^2 a^{-\mu\left(\gamma - \bar{r}_1\frac{\varkappa_1}{r_1}\right)}$$

$$\times \left\| F_{x_1}^{(\bar{r}_1+2)}\left(a^{\mu\frac{\varkappa_1}{r_1}}(x_1 + \theta h)\right) \prod_{j=2}^{n} F\left(a^{\mu\frac{\varkappa_j}{r_j}} x_j\right) \right\|_{L_{p_1}[(0,h)\times\mathbb{R}_{n-1}]}$$

$$\geq c_1 \left(\frac{\delta}{2}\right)^2 a^{-\mu\left(\gamma - \varkappa_1 + \frac{\alpha\varkappa_1}{r_1}\right)} \left(\int_0^h \left| ca^{\mu\frac{\varkappa_1}{r_1}} x_1 \right|^{p_1} dx_1\right)^{1/p_1} a^{-\mu\frac{1}{p_1}\sum_2^n \frac{\varkappa_j}{r_j}}$$

$$= c_2 \left(\frac{\delta}{2}\right)^2 \left(ha^{\mu\frac{\varkappa_1}{r_1}}\right)^{1+\frac{1}{p_1}} \left(ha^{\mu\frac{\varkappa_1}{r_1}}\right)^{-\alpha} h^\alpha = c_2 \left(\frac{\delta}{2}\right)^{3-\alpha+\frac{1}{p_1}} h^\alpha.$$

In the second relation (inequality) the region of integration $\mathbb{R}_n$ is replaced by its subset $(0, h) \times \mathbb{R}_{n-1}$ consisting of the points $\boldsymbol{x}$ where $0 < x_1 < h$, $-\infty < x_j < \infty$ for $j = 2, \ldots, n$. In such a case, for $s \leq \mu$, it follows from (4) that $a^{s\frac{\varkappa_1}{r_1}}(x_1 + \theta h) \leq a^{\mu\frac{\varkappa_1}{r_1}} 2h \leq \delta$. Hence the functions $F_{x_1}^{(\bar{r}_1+2)}\left(a^{s\frac{\varkappa_1}{r_1}}(x_1 + \theta h)\right)$ preserve their sign. Since further $F \geq 0$, then the norm at which we have arrived can only decrease if one drops one term in the sum corresponding to $s = \mu$. This explains the passage from the third to the fourth term. The passage from the fourth to the fifth term is realized in view of (4) and inequality (1) for $l = \bar{r}_1 + 2$. In integrating with respect to $\mathbb{R}_{n-1}$ we need to take into account the fact that

$$\left(\int |F(Nx)|^p dx\right)^{1/p} = \frac{1}{N^{1/p}} \left(\int |F(u)|^p du\right)^{1/p} = c_l N^{-1/p}.$$

The passage from the fifth to the sixth term is based on the application of (3). Finally, in the last equation we apply (4). It is essential to note that the constants c, c_1, c_2 in (7) are independent not only of h and μ, but also of a. On the other hand, as we already noted above, the constant on h^α in inequality (6) may be made arbitrarily small for sufficiently large a. Accordingly it follows from (5), (6), and (7) that for sufficiently large a we have the inequality

(8) $$\|\Delta^2_{x,h}\psi^{\bar{r}_1}_{x_1}\|_{L_p(\mathbb{R}_n)} \geqq \|s(h)\|_{L_p(\mathbb{R}_n)} - \|\sigma(h)\|_{L_p(\mathbb{R}_n)} \geqq c(a)h^\alpha,$$

where h runs through the sequence (4), decreasing to zero. But then the function ψ cannot belong to the class $H^{r_1+\varepsilon}_{p_1x_1}(\mathbb{R}_n)$ $(\varepsilon > 0)$. Indeed, suppose that $\psi \in H^{r_1+\varepsilon}_{p_1x_1}(\mathbb{R}_n)$, and let $0 < \eta < \min\{\varepsilon, 1\}$. Then also $\psi \in H^{r_1+\eta}_{p_1x_1}(\mathbb{R}_n)$ and in addition $r_1 + \eta - \bar{r}_1 = \alpha + \eta < 2 = k$, so that the inequality

$$\|\Delta^2_{x_1h}\psi^{\bar{r}_1}_{x_1}\|_{L_{p_1}(\mathbb{R}_n)} \leqq M|h|^{\alpha+\eta}$$

must be satisfied for all h, which contradicts (8). Analogously one proves that $\psi \notin H^{r_i+\varepsilon}_{p_ix_1}(\mathbb{R}_n)$ for any $i = 1, \ldots, n$, if $\varepsilon > 0$.

We have proved that the function $\psi_{p,r}(a, x)$ for sufficiently large a does not belong to any class $H^{r+\varepsilon}_p(\mathbb{R}_n)$, where $\varepsilon > 0$.

7.4.1. For a boundary function ψ in $H^r_p(\mathbb{R}_n) = H(r > 0)$ the norm $\|\psi(x + h) - \psi(x)\|$ does not tend to zero as $h \to 0$. Indeed, suppose that $r_1 > 0$ and $r_1 = \bar{r}_1 + \alpha$, where $\bar{r}_1$ is an integer and $0 < \alpha < 1$. In view of 7.4 (8), for real $h > 0$, running through some sequence tending to zero $(\|\cdot\|_{L_p(\mathbb{R}_n)} = \|\cdot\|_p)$:

(1) $$\|\Delta_{x_1h}\psi\|_H \geqq \sup_{k>0} \frac{\|\Delta_{x_1k}\Delta_{x_1h}\psi^{\bar{r}_1}_{x_1}\|_p}{k^\alpha} \geqq \frac{\|\Delta^2_{x_1h}\psi^{\bar{r}_1}_{x_1}\|_p}{h^\alpha} \geqq m > 0.$$

For $a = 1$ it would be necessary to operate in (1) with the second difference relative to k (instead of the first), which leads to the necessity of proving inequality 7.4 (8) for the third difference (instead of the second). This is done analogously.

7.5. Unimprovability of Inequalities for Mixed Derivatives

In 5.6.3 we proved the inequality

(1) $$\|f^{(l)}\|_{B^\varrho_{p\theta}(\mathbb{R}_n)} \leqq c\|f\|_{B^r_{p\theta}(\mathbb{R}_n)}$$

under the hypothesis that

(2) $$\varrho = \varkappa r, \quad \varkappa = 1 - \sum_1^n \frac{l_k}{r_k} > 0.$$

It stops being true if ϱ is replaced by $\varrho + \varepsilon$ ($\varepsilon > 0$). This may also be proved by considering the boundary function

$$\psi = \psi_{p,r} = \sum_{s=0}^{n} \frac{\prod_{j=1}^{n} F(a^{s/r_j} x_j)}{a^{s\left(1-\frac{1}{p}\Sigma\frac{1}{r_i}\right)}} \quad (a > 1).$$

Its derivative

$$\psi^{(l)}(x) = \sum_{s} \frac{\prod_{j=1}^{n} F^{(l_j)}(b^{s/\varrho_j} x_j)}{b^{s\left(1-\frac{1}{p}\Sigma\frac{1}{\varrho_i}\right)}} \quad (a^{\varkappa} = b),$$

although not a special case of the family of boundary functions we considered above, is nevertheless a boundary function in the class $H_p^{\varrho}(\mathbb{R}_n)$, and this is proved in a way quite analogous to what was done in 7.4 where we need to suppose that $p = p_1 = \cdots = p_n$. The circumstance that now there are different functions $F^{(l_j)}$, under the sign $\prod$ does not have any essential significance.

This proves our assertion for H-classes[1], and then for B-classes also.

7.6. Another Proof of the Unimprovability of Imbedding Theorems

We shall consider the question as it relates to the general theory of functional spaces. Suppose that E_1 and E_2 are Banach spaces (i.e. normed complete linear spaces). The following theorem holds.

Theorem[2]. *If the bounded linear operator maps E_1 in a 1 — 1 way onto E_2, then its inverse operator A^{-1} is (obviously) linear and, mapping E_2 onto E_1, is in its turn bounded.*

Suppose that the Banach spaces E_1 and E_2 have a nonempty intersection E_1E_2. We shall assign to elements $x \in E_1E_2$ the norm

(1) $$\|x\|_{E_1E_2} = \|x\|_{E_1} + \|x\|_{E_2}.$$

With this E_1E_2 becomes a normed space.

Theorem 2. *If E_1E_2 is a complete space, i.e. a Banach space, and if there does not exist a constant $c > 0$ such that*

$$\|x\|_{E_2} \leqq c\|x\|_{E_1}$$

or all $x \in E_1E_2$, then there exists an element in E_1 which does not belong to E_2.

---

[1] S. M. Nikol'skiĭ [2], the case $p = \infty$.

[2] See Hausdorff [1] or the Russian edition [2] of Hausdorff's Mengenlehre, the Supplement, or else the German original [3] of that Supplement. See also Banach [1], X, §1, Theorem 10.

Proof. Suppose that in fact this is not so, i.e. $E_1 \subset E_2$. Each element $\boldsymbol{x}$ of the Banach space E_1E_2 may be considered as a mapping into $\boldsymbol{x}$ itself, but belonging to E_1. This operation is linear and bounded:

$$\|\boldsymbol{x}\|_{E_1} \leqq \|\boldsymbol{x}\|_{E_1} + \|\boldsymbol{x}\|_{E_2} = \|\boldsymbol{x}\|_{E_1E_2},$$

and maps E_1E_2 onto E_1. But then on the basis of Theorem 1 there must exist a constant c such that

$$\|\boldsymbol{x}\|_{E_1} + \|\boldsymbol{x}\|_{E_2} \leqq c\|\boldsymbol{x}\|_{E_1}$$

or

$$\|\boldsymbol{x}\|_{E_2} \leqq c\|\boldsymbol{x}\|_{E_1}, \quad \boldsymbol{x} \in E_1E_2,$$

and we have arrived at a contradiction with the hypothesis of the theorem.

The application of Theorem 2 requires verification of the completeness of E_1E_2.

If $E_1 = B_{\boldsymbol{p}}^{\boldsymbol{r}}(\mathbb{R}_n)$, $E_2 = B_{p'}^{\varrho}(\mathbb{R}_n)$ (here $B_{\boldsymbol{p}} = B_{\boldsymbol{p}\theta}$), then completeness of E_1E_2 holds, because then the fact that $\|f_k - f_l\|_{E_1E_2} \to 0$, $k, l \to \infty$, implies, as a consequence of the completeness of E_1 and E_2, the existence of $f \in E_1$ and $F \in E_2$ such that $\|f - f_k\|_{E_1} \to 0$, $\|F - f_k\|_{E_2} \to 0$ $(k \to \infty)$ But then also

$$\|f - f_k\|_{L_{p_i}(\mathbb{R}_n)} \to 0, \ \|F - f_k\|_{L_{p_i}(\mathbb{R}_n)} \to 0 \quad (i = 1, \ldots, n).$$

Hence (see 1.3.9) $f = F$ almost everywhere, and we have proved the existence of an $f \in E_1E_2$ such that

$$\|f - f_k\|_{E_1E_2} \to 0.$$

Suppose that $\boldsymbol{r} = (r_1, \ldots, r_n) > \boldsymbol{0}$, $\boldsymbol{p} = (p_1, \ldots, p_n)$, $1 \leqq p_j \leqq \infty$, $\varkappa_j = \varkappa_j(\boldsymbol{r}, \boldsymbol{p}) > 0$ (see 7.3 (1)).

We introduce a function $F(t)$ of one variable, finite and infinitely differentiable.

Its norms in the metrics of the $B_p^{\varrho}(\mathbb{R}_1)$ $(0 < \varrho \leqq l)$ are positive; otherwise it would be zero.

We construct a family of functions $\left(\text{see 7.4 (2)}, \ \gamma = 1 - \sum_{j=1}^{n} \frac{1}{p_j r_j}\right)$

$$\Phi_N = \Phi_{N,\boldsymbol{p},\boldsymbol{r}}(\boldsymbol{x}) = \frac{1}{N^\gamma} \prod_{j=1}^{n} F\left(N^{\frac{\varkappa_j}{r_j}} x_j\right), \tag{2}$$

depending on a parameter $N > 0$.

On the basis of formulas 7.2 (1) and (6)

(3)
$$\|\Phi_{N,\mathbf{p},\mathbf{r}}\|_{p_i} \sim N^{-\frac{1}{p_i}\sum_{i=1}^{n}\frac{\varkappa_j}{r_j}-\gamma} = N^{-\varkappa_i} \quad (N > 0),$$

$$\|\Phi_{N,\mathbf{p},\mathbf{r}}\|_{b^{r_i}_{x_ip_i}} \sim N^{\frac{r_i\varkappa_i}{r_i}} N^{-\varkappa_i} = N^0 = 1 \quad (i = 1, \dots, n)$$

$$\left(b^{r_i}_{x_ip_i} \equiv b^{r_i}_{x_ip_i\theta}\right).$$

Fix attention on a definite i and suppose given numbers $p^* \geqq p_i$, $r^* \geqq r_j$, where at least one of these inequalities is strict. We calculate for comparison the norms

$$\|\Phi_{N,\mathbf{p},\mathbf{r}}\|_{p_*} \sim N^{-\frac{1}{p_*}\sum_1^n\frac{\varkappa_j}{r_j}-\gamma} = N^{-(\varkappa_j-\varepsilon)},$$

$$\|\Phi_{N,\mathbf{p},\mathbf{r}}\|_{b^{r^*}_{x_ip_*}} \sim N^{-(\varkappa_j-\varepsilon)} N^{\frac{r^*\varkappa_i}{r_i}} = N^{\left(\frac{r^*}{r_i}-1\right)\varkappa_j+\varepsilon}$$

$$(\varepsilon > 0, \quad N > 0).$$

It is essential to note that there ε is a positive number, so that

(4)
$$\|\Phi_{N,\mathbf{p},\mathbf{r}}\|_{b^{r^*}_{x_ip_*}} \to \infty \ (N \to \infty).$$

Thus, to each pair of vectors $\mathbf{p}$, $\mathbf{r}$ satisfying the conditions indicated above we have brought into correspondence a family of functions $\Phi(N, \mathbf{p}, \mathbf{r})$, whose norms

$$\|\Phi(N, \mathbf{p}, \mathbf{r})\|_{B^{\mathbf{r}}_{\mathbf{p}}(\mathbb{R}_n)} \leqq c < \infty \ (N > 0)$$

are bounded, and at the same time, for each i, the property (4) is satisfied, given only that $r^* \geqq r_i$, $p^* \geqq p_i$ and one of these inequalities is strict.

We shall call the family $\Phi(N, \mathbf{p}, \mathbf{r})$ a *boundary family of functions in the class* $B^{\mathbf{r}}_{\mathbf{p}}(\mathbb{R}_n)$.

Recall the imbedding

$$B^{\mathbf{r}}_{\mathbf{p}\theta}(\mathbb{R}_n) \to B^{\boldsymbol{\varrho}}_{p'\theta}(\mathbb{R}_n)$$

proved in 6.9 under the conditions $1 \leqq p_j \leqq p' \leqq \infty$, $\varkappa > 0$ which imply also that $\varkappa_j > 0$ (see 7.1 (4), (5)), and

(5)
$$\varrho_j = \frac{r_j\varkappa}{\varkappa_j} \quad (j = 1, \dots, n)$$

This imbedding ceases to be valid if one increases at least one of the components ϱ_j or the number p', or does both of these things. Indeed, if one takes $N_1 = N^{\varkappa}$, then (explanation below)

$$(6) \qquad \Phi_{N,p,r} = \frac{1}{N^\gamma} \prod_{j=1}^{n} F\left(N^{\frac{\varkappa_j}{r_j}} x_j\right) = \frac{1}{N_1^{\gamma_1}} \prod_{j=1}^{n} F(N_1^{1/\varrho_j} x_j) = \Phi_{N_1,p,\varrho}$$

$$(N_1 = N^{\varkappa}).$$

Here

$$\gamma_1 = 1 - \frac{1}{p'} \sum_1^n \frac{1}{\varrho_l} = \gamma(\varrho, p'),$$

because

$$\varkappa\gamma_1 = \varkappa - \frac{1}{p'} \sum \frac{\varkappa_j}{r_j} = 1 - \sum \frac{1}{p_l r_l} + \frac{1}{p'} \sum \frac{1}{r_l} - \frac{1}{p'} \sum \frac{1}{r_l}$$

$$- \frac{1}{p'} \sum \frac{1}{r_l} + \frac{1}{p'} \sum_j \sum_l \left(\frac{1}{p_i} - \frac{1}{p_j}\right) \frac{1}{r_i r_j} = 1 - \sum \frac{1}{p_l r_l} = \gamma.$$

Further,

$$\varkappa_j(\varrho, p') = 1 - \sum \left(\frac{1}{p'} - \frac{1}{p'}\right) \frac{1}{r_l} = 1.$$

This proves equations (6).

Thus the family of functions Φ_N is a boundary class simultaneously in the classes B_p^r and $B_{p'}^{\varrho}$, and the norms Φ_N in the metrics of these classes are uniformly bounded relative to N. However, the norms Φ_N are not bounded in the metric of $B_{p'+\eta}^{\varrho+\varepsilon}$. But then there does not exist a constant c independent of N and such that

$$\|\Phi_N\|_{B_{p'+\eta}^{\varrho+\varepsilon}} \leqq c \|\Phi_N\|_{B_p^r},$$

and we have proved our assertion.

In view of Theorem 2 it follows in such a case that for arbitrary $\varepsilon > 0$, $\eta \geqq 0$, where one of the inequalities is strict, there exists in the class $B_p^r(\mathbb{R}_n)$ a function not belonging to $B_{p'+\eta}^{\varrho+\varepsilon}(\mathbb{R}_n)$. In particular, there exists in the class $B_p^r(\mathbb{R}_n)$ a function not belonging to $B_{p+\eta}^{r+\varepsilon}(\mathbb{R}_n)$.

The imbedding (proved in 6.5)

$$B_p^r(\mathbb{R}_n) \to B_p^{\varrho}(\mathbb{R}_m),$$

$$1 \leqq m < n, \quad \varrho_j = \varkappa r_j, \quad j = 1, \ldots, m,$$

$$\varkappa = 1 - \frac{1}{p} \sum_{m+1}^{n} \frac{1}{r_l},$$

ceases being valid if in it one replaces ϱ, p by $\varrho^* \geqq \varrho$, $p^* \geqq p$, where one of the inequalities is strict.

Indeed, there exists a function $\varphi \in B_p^\varrho(\mathbb{R}_m)$ but not belonging to $B_{p+\eta}^{\varrho+\varepsilon}$. On the basis of the extension theorem φ may be extended from $\mathbb{R}_m$ to $\mathbb{R}_n$ in such a way that the extended function $f \in B_p^{\boldsymbol{r}}(\mathbb{R}_n)$. Since $f|_{\mathbb{R}_m} = \varphi$, then f is an example of a function $f \in B_p^{\boldsymbol{r}}(\mathbb{R}_n)$ whose trace on $\mathbb{R}_m$ does not belong to $B_{p+\eta}^{\varrho+\varepsilon}(\mathbb{R}_m)$.

From what has just been said it follows in passing that the extension theorem

$$B_p^\varrho(\mathbb{R}_m) \to B_p^{\boldsymbol{r}}(\mathbb{R}_n)$$

also cannot be improved in the terms of the classes being considered. However it does not mean that theorem cannot be improved in other terms. For example, in Chapter 9 it will be proved that under the same connection between $\boldsymbol{r}$ and ϱ, n and m there hold the mutually inverse imbeddings

$$B_p^\varrho(\mathbb{R}_m) \rightleftharpoons L_p^{\boldsymbol{r}}(\mathbb{R}_n),$$

where the class $L_p^{\boldsymbol{r}}$ for $p \neq 2$ is not equivalent to $B_p^{\boldsymbol{r}}(\mathbb{R}_n)$.

Suppose further given a family (a boundary family in $B_p^{\boldsymbol{r}}(\mathbb{R}_n)$)

$$\Phi_N = \Phi_{N,p,\boldsymbol{r}} = \frac{1}{N^\gamma} \prod_{j=1}^n F(N^{1/r}x_j).$$

In the case at hand

$$\gamma = 1 - \frac{1}{p} \sum_1^n \frac{1}{r_l} > 0, \quad \varkappa_j(p, \boldsymbol{r}) = 1$$

$$(j = 1, \ldots, n).$$

Suppose further that $F(t)$ moreover has compact support and is infinitely differentiable, and that it has the Taylor expansion

$$F(t) = 1 + a_1 t + \cdots + a_{l+1} t^{l+1} + \mathbb{R}_{l+1}$$

with coefficients at the odd or even places unequal to zero. Then, as is easily seen, it is easy to indicate a positive number δ and a constant B such that

$$|F^{(k)}(t)| > Bt \; (k = 0, 1, \ldots, l, |t| < \delta).$$

Suppose that $\mathbb{R}_m$ is the subspace of points $(x_1, \ldots, x_m, \ldots, 0) = (\boldsymbol{u}, \boldsymbol{0})$, $\boldsymbol{u} = (x_1, \ldots, x_m)$. Then

$$\Phi_N(\boldsymbol{u}, \boldsymbol{0}) = \frac{1}{N_1^{\gamma_1}} \prod_{j=1}^m F\left(N_1^{\frac{1}{\varrho_j}} x_j\right) = \Phi_{N_1,p,\varrho} \quad (N_1 = N^\varkappa),$$

where

$$\gamma_1 = \gamma_1(p, \varrho) = 1 - \frac{1}{p} \sum_1^m \frac{1}{\varrho_l}.$$

(Recall that $\gamma = \varkappa\gamma_1$.)

Suppose that $h > 0$ and $i = m + 1, \ldots, n$. Consider the increment

$$\Delta_{x_i h} \Phi_N(u, 0) = \frac{1}{N_1^{\gamma_1}} \prod_{j=1}^m F\left(N_1^{\frac{1}{\varrho_j}} x_j\right)\left[F\left(N_1^{\frac{1}{\varrho_i}} h\right) - F(0)\right]$$

$$= \Phi_{N_1, p, \varrho}(u)\left[F\left(N_1^{\frac{1}{\varrho_j}} h\right) - F(0)\right].$$

The function F is not identically zero, so that there exists a $\delta > 0$ such that $|F(\delta) - F(0)| = K > 0$. We will consider values of h and N_1 connected by the equation $\delta = N_1^{\frac{1}{\varrho_i}} h$. In view of the first estimate (3) we have (in our case $\varkappa_i = 1$)

$$\|\Delta_{x_i h} \Phi_N(u, 0)\|_{L_p(R_m)} = \|\Phi_{N_1, p, \varrho}\|_{L_p(R_m)} K \gg \frac{1}{N} \gg |h|^{\varrho_i}.$$

This estimate from below shows that the first inequality ((6.5 (13)) ($\varrho_i = r_i'$) introduced earlier is attained and moreover not only for the class $H_p^r(\mathbb{R}_n)$, but also for $B_{p\theta}^r(\mathbb{R})$.

7.7. Theorems on Compactness

Theorem. *Suppose given a sequence of functions f_l, having one of the following properties:*

(1) a) $\|f_l\|_{L_p(\mathbb{R}_n)} \leqq M, \quad \|f_l\|_{w_p^r(\mathbb{R}_n)} \leqq N$

(r an integer vektor, $1 < p < \infty$)

(2) b) $\|f_l\|_{L_p(\mathbb{R}_n)} \leqq M, \quad \|f_l\|_{b_{p\theta}^r} \leqq N$

$$(1 \leqq p, \ \theta \leqq \infty).$$

Then it is possible to select a subsequence $\{f_{l_k}\}$ and a function f, satisfying[1] *respectively conditions (1), (2), such that for any numbers r_j' with*

---

[1] The function f satisfies (1) or (2) with the same constant N, if one understands the norm in one and the same sense. In case b) below the proof is carried through for the variant ${}^4\|\cdot\|_B$ of this norm (see 5.6, $\varrho = \bar{r}$.)

$0 < r'_j < r_j$ $(j = 1, \ldots, n)$ *one has*

(3) $$\|f_{l_k} - f\|_{H_p^{r'}(g)} \to 0 \ (l_k \to \infty)$$

on any bounded region $g \subset \mathbb{R}_n$.

The proof of this theorem will be based on the following lemma from functional analysis.

Lemma. *Suppose that the same set of elements* x *is normed by two norms* $\|\cdot\|$ *and* $\|\cdot\|_*$, *and that the resulting normed spaces* E *and* E_* *are complete and* $\|x\|_* \leqq c\|x\|$, *where the constant* c *does not depend on* x.

Suppose given in E *a bounded set* F *and a sequence of operators* $A_n(x)$, $n = 1, 2, \ldots$, *mapping* E *into* E^*, *defined by the equations*

$$A_n(x) = x - U_n(x)$$

and satisfying the conditions:

1) *the operators* $y = U_n(x)$ $(x \in E, y \in E_*)$ *are completely continuous* (linearity of the U_n is not required)

2) $$\sup_{x \in F} \|A_n(x)\| = \eta_n \to 0 \ (n \to \infty).$$

Then the set F *is compact in* E_*.

Proof. Suppose given any sequence of elements $x_1, x_2, \ldots$, lying in F. It is bounded and in view of property 1) one may select a subsequence $x_1^{(1)}, x_2^{(1)}, \ldots$ from it for which $U_1(x_k^{(1)})$ $(k = 1, 2, \ldots)$ converges in E_*. In turn one may select from this sequence a subsequence $x_1^{(2)}, x_2^{(2)}, \ldots$ for which $U_2(x_k^{(2)})$ $(k = 1, 2, \ldots)$ converges in E_*. Carrying out this process without limit and selecting the diagonal sequence $z_1 = x_1^{(1)}$, $z_2 = x_2^{(2)}, \ldots$, we find that $U_n(z_k)$ converges in E_* for $k \to \infty$ and any n. Now suppose given $\varepsilon > 0$. In view of condition 2), for some $n = N$ the inequality

$$\|A_N(x)\|_* \leqq c\|A_N(x)\| < \varepsilon$$

is satisfied for all $x \in F$. If p and q exceed a sufficiently large number, then

$$\|z_p - z_q\|_* \leqq \|A_N(z_p)\|_* + \|U_N(z_p) - U_N(z_q)\|_* + \|A_N(z_Q)\|_* < 3\varepsilon,$$

and the compactness of F in E_* is proved.

Proof of the Theorem. Suppose that $K = M + N$. Consider first case b) with $\theta = \infty$, i.e. the case of the class $H_p^r = H_p^r(\mathbb{R}_n)$.

Suppose that $\mathfrak{M}$ is the set of all functions f for which inequality (2) is satisfied for $\theta = \infty$. For each of them one has the decomposition 5.5.3 (6),(7):

$$f = \sum_{s=0}^{\infty} Q_s,$$

where the

$$Q_s = Q_{a^{s/r_1},\ldots,a^{s/r_n}} \quad (a > 1)$$

are entire functions of exponential type a^{s/r_j} relative to the x_j $(j = 1,\ldots,n)$ respectively and

$$\sup_s a^s \|Q_s\|_p = \|f\|_{H_p^r} \leqq cK.$$

Suppose given a number γ satisfying the inequalities $0 < \gamma < 1$, and put

$$T_m(f) = T_m = \sum_0^{m-1} Q_s, \quad a^\gamma = b.$$

Then $(\|\cdot\|_p = \|\cdot\|_{L_p(\mathbb{R}_n)})$

$$\|f - T_m\|_{H_p^{\gamma r}} = \sup_{s \geqq m} b^s \left\| Q_{b^{s/\gamma r_1},\ldots,b^{s/\gamma r_n}} \right\|_p$$

$$= \sup_{s \geqq m} a^{\gamma s} \|Q_s\|_p \leqq \frac{1}{a^{(1-\gamma)m}} \sup_s a^s \|Q_s\|_p \leqq \frac{cK}{a^{(1-\gamma)m}}.$$

Moreover,

$$\|f\|_{H_p^{\gamma r}} \leqq c\|f\|_{H_p^r} \leqq cK$$

(see 6.2 (3)).

We will consider functions f from the space

$$E = H_p^{\gamma r} = H_p^{\gamma r}(\mathbb{R}_n)$$

also as elements of the space

$$E_* = H_p^{\gamma r}(g) \; (g \subset \mathbb{R}_n),$$

while, obviously,

$$\|f\|_{E_*} \leqq \|f\|_E.$$

We have

$$f = T_m(f) + f - T_m(f)),$$

where for $f \in \mathfrak{M}$

$$\|f - T_m(f)\|_E \leqq \frac{cK}{a^{(1-\gamma)m}} \to 0 \quad (m \to \infty).$$

Further

$$\|T_m(f)\|_p \leqq \|T_m(f)\|_{H_p^{\gamma r}} \leqq \|f\|_E = \frac{cK}{a^{(1-\gamma)m}}.$$

Therefore the image of any sphere E under the transformation T_m is a set of functions $T_m(f)$ of exponential type a^{m/r_j} relative to x^l and bounded in the sense of $L_p = L_p(\mathbb{R}_n)$. In such a case that set is compact on any bounded set $g \subset \mathbb{R}_n$ in the sense of the metric $c^l(g)$ (see 3.3.6[1]) for any natural number l. Accordingly it is compact also in the sense of $E_* = H_p^{\gamma r}(g)$. We have proved that $T_m(f)$ is a completely continuous operator (in general nonlinear).

As a consequence of the lemma proved above, $\mathfrak{M}$ is compact in $H_p^{\gamma r}(g)$. Since this reasoning is valid for any γ with $0 < \gamma < 1$, then $\mathfrak{M}$ is compact in the sense of $H_p^{\gamma r}$ for any of the indicated γ. We choose a definite sequence of numbers γ_k, tending monotonically to 1, and suppose given an arbitrary sequence of functions f_l of $F(\subset \mathfrak{M})$. In view of what has been proved and the completeness of $H_p^{\gamma_1 r}$ (see 4.7) it is possible to select a subsequence $\{f_{l_k^1}\}$ from it which converges in the metric of $H_p^{\gamma_1 r}$ to some function $f \in H_p^{\gamma_1 r}$. From the subsequence just obtained one may in turn select a subsequence $\{f_{l_k^2}\}$ which converges in the metric of $H_p^{\gamma_2 r}$ to some function $f \in H_p^{\gamma_2 r}$, evidently to the same one. Continuing this process without limit and taking the diagonal sequence, which we denote by $\{f_{l_k}\}$, we find that $f_{l_k} \to f$ in the metric of $H_p^{\gamma_s r}$ for any s. But then the convergence holds also in the sense of the metric $H_p^{r'}$, where $r'_j < r_j$ $(j = 1, \ldots, n)$, as follows from 6.2 (3).

We have proved (3) in the case b) for $\theta = \infty$. The remaining cases a) and b) with $1 \leqq \theta \leqq \infty$ reduce to this case, because $W_p^r, B_{p\theta}^r \to H_p^r$. But still remains for us to prove the more precise fact that the limit function f belongs respectively to $W_{p\theta}^r$, $B_{p\theta}^r$ and that the following inequalities are respectively satisfied:

$$\|f\|_{w_p^r}, \|f\|_{b_{p\theta}^r} \leqq N.$$

The inequality $\|f\| \leqq M$ follows from (1), (2), (3).

As always, we will suppose that $r_j = \bar{r}_j + \alpha_j$, where $\bar{r}_j$ is an integer and $0 < \alpha_j \leqq 1$. Suppose that $f_{x_j}^{\bar{r}_j}$ denotes the partial derivative of f of order $\bar{r}_j$ with respect to x_j $(\bar{r}_j < r'_j < r_j)$.

Then (6.2 (3))

$$(4) \qquad \|f_{l_k} - f\|_p, \left\|f_{l_k x_j}^{\bar{r}_j} - f_{x_j}^{\bar{r}_j}\right\|_p \leqq c\|f_{l_k} - f\|_{H_p^{r'}} \to 0 \quad (l_k \to \infty).$$

---

[1] From the boundedness (in the sense of L_p) of the functions $T_m(f_k)$ $(k = 1, 2, \ldots)$ follows the boundedness of their derivatives of any given order. The application of 3.3.6 not only to the functions, but also to their derivatives to order l inclusive, and the diagonal process, leads to compactness not only in the sense of $c(g)$ but also in the sense of $c^l(g)$.

In case b) the functions f_{l_k} are subordinate to inequality (2), so that

$$(5)\qquad \left(\int_{-\infty}^{\infty} |u|^{-1-\theta\alpha_j} \left\| \Delta^2_{x_j u} f^{\bar{r}_j}_{l_k x_j}(\boldsymbol{x}) \right\|_p^\theta du\right)^{1/\theta} = m_j^{(k)} \quad (1 \leqq \theta < \infty),$$

$$\left\| \Delta^2_{x_j u} f^{\bar{r}_j}_{l_k x_j}(\boldsymbol{x}) \right\|_p \leqq m_j^{(k)} |u|^{\alpha_j} \quad (\theta = \infty),$$

where

$$\sum_{j=1}^{n} m_j^{(k)} \leqq N \quad (k = 1, 2, \ldots).$$

Passing to the limit as $k \to \infty$ in (5), we find on the basis of (4) that

$$m_j = \left(\int_{-\infty}^{\infty} |u|^{-1-\theta\alpha_j} \left\| \Delta^2_{x_j u} f^{\bar{r}_j}_{x_j}(\boldsymbol{x}) \right\|_p^\theta du\right)^{1/\theta} \leqq \overline{\lim_{k\to\infty}}\, m_j^{(k)} \quad (1 \leqq \theta < \infty),$$

$$\left\| \Delta^2_{x_j u} f^{\bar{r}_j}_{l_k x_j}(\boldsymbol{x}) \right\|_p \leqq \overline{\lim}\, m_j^{(k)} |u|^{\alpha_j} \quad (\theta = \infty),$$

so that $(f \in L_p)$ $f \in B^{\boldsymbol{r}}_{p\theta}$,

$$\|f\|_{b^{\boldsymbol{r}}_{p\theta}} = \sum_{j=1}^{n} m_j \leqq N.$$

In case a) the functions f_{l_k} are subordinate to the inequalities

$$(6)\qquad \left\| \frac{\Delta_{x_j u} f^{\bar{r}_j}_{l_k x_j}}{u} \right\|_p \leqq \left\| f^{r_j}_{l_k x_j} \right\|_p \leqq m_j^{(k)},$$

where

$$\sum_{j=1}^{n} m_j^{(k)} \leqq N.$$

Passing to the limit in (6) as $k \to \infty$, we get

$$m_j = \left\| \frac{\Delta_{x_j u} f^{\bar{r}_j}_{x_j}}{u} \right\| \leqq \overline{\lim_{k\to\infty}}\, m_j^{(k)},$$

and since further $f \in L_p$, then (see 4.8) $f \in W^{\boldsymbol{r}}_p$ and

$$\boldsymbol{r}\ \|f\|_{w^{\boldsymbol{r}}_p} = \sum_{j=1}^{n} m_j \leqq N.$$

Remark. In the theorem just proved $W^{\boldsymbol{r}}_p$, $B^{\boldsymbol{r}}_{p\theta}$ may be replaced respectively by W^r_p, $B^r_{p\theta}$. Then in (3) we have to replace $\boldsymbol{r}'$ by r', $0 < r' < r$. The case of W^r_p and similar cases, which can be proved by analogy, find an application in the theory of variational methods. For the applications it is very essential that an inequality of type (1) implies the same inequality for the limit function, with the same constant. In this theorem it is also possible to replace the classes considered there by the corresponding periodic classes.

7.7.1. Theorem. *For a set* $\mathfrak{M}$ *of functions* $f \in L_p = L_p(g)$, *where* $g \subset \mathbb{R}_n$ *is any region, to be compact, it is necessary and sufficient that it should be*: 1) *bounded in* L_p, 2) *equicontinuous relative to a shift in* L_p:

$$\Lambda(\delta) = \sup_{f \in \mathfrak{M}} \omega(\delta, f)_p \to 0 \quad (\delta \to 0),$$

$$\omega(\delta, f)_p = \sup_{|\boldsymbol{h}| < \delta} \|f(\boldsymbol{x} + \boldsymbol{h}) - f(\boldsymbol{x})\|_p \quad (f = 0 \text{ on } \mathbb{R}_n - g),$$

and that 3) *the norm of a function* $f \in \mathfrak{M}$ *should decrease uniformly in norm in* L_p *at infinity*:

$$\sup_{f \in \mathfrak{M}} \|f\|_{L_p(|\boldsymbol{x}| > N, \boldsymbol{x} \in g)} \to 0 \ (N \to \infty).$$

This theorem is proved in the book of S. L. Sobolev [4], Ch. I, §4.3. For a bounded region g property 3) evidently drops out. For $p = \infty$ in general the theorem ceases to be true. In this case the norm of the shift of a particular function in general does not tend to zero as $\boldsymbol{h} \to \boldsymbol{0}$.

7.7.2. Theorem. *For a set* $\mathfrak{M}$ *of functions* $f \in W = W_p^{\boldsymbol{l}}(\mathbb{R}_n)$ $(1 \leqq p < \infty, \boldsymbol{l} \geqq \boldsymbol{0})$ *to be compact in* W, *it is necessary and sufficient that* $\mathfrak{M}$ *should be equicontinuous relative to the shift*:

$$\Lambda(\delta) = \sup_{f \in \mathfrak{M}} \sup_{|\boldsymbol{h}| < \delta} \|f(\boldsymbol{x} + \boldsymbol{h}) - f(\boldsymbol{x})\|_W \to 0 \quad (\delta \to 0), \tag{1}$$

and that the functions $f \in \mathfrak{M}$ *should uniformly decrease in norm at infinity*:

$$\sup_{f \in \mathfrak{M}} \|f\|_{L_p(|\boldsymbol{x}| > N)} \to 0 \ (N \to \infty). \tag{2}$$

In this formulation W *may be replaced by* $B = B_{p\theta}^{\boldsymbol{r}}(\mathbb{R}_n)$ $(1 \leqq p, \theta < \infty, \boldsymbol{r} \geqq \boldsymbol{0})$.

Proof. We shall consider the space W; however W may be replaced by B throughout. Suppose that $\mathfrak{M}$ is compact in W. Then it is compact in L_p, so that (see 7.7.1) property (2) is satisfied. By the general criterion for compactness (Hausdorff [1]), for a given $\varepsilon > 0$ it is possible to indicate a finite system of functions f_j, $j = 1, \ldots, N$, such that for any function $f \in \mathfrak{M}$ there exists a j, depending on f, for which

$$\|f - f_j\|_W < \varepsilon.$$

It is also possible to indicate δ and N so that the following inequalities are satisfied (see 5.6.5):

$$\|f_j(\boldsymbol{x}+\boldsymbol{h}) - f_j(\boldsymbol{x})\|_W < \varepsilon,\ \|f_j\|_{L_p(|\boldsymbol{x}|>N)} < \varepsilon,\ |\boldsymbol{h}| < \delta$$

for all $j = 1, \ldots, n$. But then for any $f \in \mathfrak{M}$, for the corresponding j

$$\|f(\boldsymbol{x}+\boldsymbol{h}) - f(\boldsymbol{x})\|_W \leqq \|f(\boldsymbol{x}+\boldsymbol{h}) - f_j(\boldsymbol{x}+\boldsymbol{h}\|_W$$
$$+ \|f_j(\boldsymbol{x}+\boldsymbol{h}) - f_j(\boldsymbol{x})\|_W + \|f_j(\boldsymbol{x}) - f(\boldsymbol{x})\|_W < 3\varepsilon \quad (|\boldsymbol{h}| < \delta),$$

if δ is sufficiently small, and we have proved (1). The necessity of the conditions of the theorem is proved.

Suppose conversely that $\mathfrak{M}$ is a bounded set in L_p which satisfies conditions (1) and (2). Then on the basis of 7.7.1 it is compact in $L_p(\|\cdot\|_W \gg \|\cdot\|_{L_p})$. We now introduce a new concept—the *modulus of continuity of* $f \in W$:

$$\omega(t) = \omega(f, t) = \sup_{|\boldsymbol{h}|<t} \|f(\boldsymbol{x}+\boldsymbol{h}) - f(\boldsymbol{x})\|_W.$$

It satisfies the conditions

(3) $$0 \leqq \omega(\delta_2) - \omega(\delta_1) \leqq \omega(\delta_2 - \delta_1)\ (0 < \delta_1 < \delta_2),$$
$$\omega(l\delta) \leqq (l+1)\ \omega\ (\delta)\quad (l, \delta > 0).$$

This is proved exactly as it was for the modulus of continuity of f in L_p (see 4.2). It follows from (3) that for the function $\Lambda(\delta)$ (see (1)) the following inequality is also satisfied:

(4) $$\Lambda(l\delta) \leqq (l+1)\Lambda(\delta)\quad (l, \delta > 0).$$

We introduce further the function of one variable

$$K_k(t) = a_k \left(\frac{\sin kt}{t}\right)^\lambda \quad (k > 1),$$

which is entire exponential of type $k\lambda$, where $\lambda > n+1$ is an even natural number and the constant a_k is determined from the equation

$$1 = \int\limits_{\mathbb{R}_n} K_k(|u|)\, d\boldsymbol{u} = a_k \varkappa_n \int\limits_0^\infty \left(\frac{\sin kt}{t}\right)^\lambda t^{n-1} dt$$
$$= k^{\lambda-n} a_k \varkappa_n \int\limits_0^\infty \left(\frac{\sin t}{t}\right)^\lambda t^{n-1} dt = ck^{\lambda-n} a_k,$$

where $\varkappa_n$ is the surface area of the unit sphere in $\mathbb{R}_n$ and c does not depend on c and a_k. Hence it follows that

$$a_k = O(k^{n-\lambda}) \quad (k > 1).$$

Put

$$U_k f = \int K_k(|\boldsymbol{u}|)\, f(\boldsymbol{x} + \boldsymbol{u})\, d\boldsymbol{u},$$

so that

$$\|U_k f\|_p \leqq \|K_k\|_L \|f\|_p. \tag{5}$$

For $f \in \mathfrak{M}$

$$f - U_k f = \int K_k(|\boldsymbol{u}|)\, [f(\boldsymbol{x}) - f(\boldsymbol{x} + \boldsymbol{u})]\, d\boldsymbol{u},$$

so that

$$\begin{aligned} \|f - U_k f\|_W &\leqq \int K_k(|\boldsymbol{u}|) \| f(\boldsymbol{x}) - f(\boldsymbol{x} + \boldsymbol{u})\|_{\boldsymbol{x},W}\, d\boldsymbol{u} \\ &\leqq \int K_k(|\boldsymbol{u}|)\, \Lambda(|\boldsymbol{u}|) d\boldsymbol{u} \leqq \int\limits_{|\boldsymbol{u}|<\delta} K_k(|\boldsymbol{u}|)\, \Lambda(|\boldsymbol{u}|)\, d\boldsymbol{u} \\ &\quad + \int\limits_{|\boldsymbol{u}|>\delta} K_k(|\boldsymbol{u}|)\, \Lambda\left(\frac{|\boldsymbol{u}|}{\delta}\,\delta\right) d\boldsymbol{u} \\ &\leqq \Lambda(\delta) + \Lambda(\delta) \int\limits_{|\boldsymbol{u}|>\delta} K_k(|\boldsymbol{u}|) \left(1 + \frac{|\boldsymbol{u}|}{\delta}\right) d\boldsymbol{u} \\ &< \varepsilon + \varepsilon = 2\varepsilon \quad (k, k_0), \end{aligned} \tag{6}$$

where k_0 is sufficiently large.

Fix on an $\varepsilon > 0$. Then choose δ so that $\Lambda(\delta) < \varepsilon$. With that δ fixed, we observe that

$$\begin{aligned} \int\limits_{|\boldsymbol{u}|>\delta} K_k(|\boldsymbol{u}|) \left(1 + \frac{|\boldsymbol{u}|}{\delta}\right) d\boldsymbol{u} &\ll k^{n-\lambda} \int\limits_{\delta}^{\infty} \left(\frac{\sin kt}{t}\right)^{\lambda} \left(1 + \frac{t}{\delta}\right) t^{n-1} dt \\ &\ll k^{n-\lambda} \int\limits_{\delta}^{\infty} \left(1 + \frac{1}{\delta}\right) t^{n-\lambda-1} dt = c_\delta k^{n-\lambda} \to 0 \quad (k \to \infty). \end{aligned} \tag{7}$$

Hence for k sufficiently large, say $k > k_0$, we find from (6) and (7) that

$$\|f - U_k f\| < 2\varepsilon.$$

We have thus proved that

$$\sup_{f \in \mathfrak{M}} \|f - U_k f\|_W \to 0 \quad (k \to \infty). \tag{8}$$

Now suppose given a sequence of functions $f_l \in \mathfrak{M}$. It is compact in L_p, so that it is possible to extract from it a subsequence, which we shall again denote by $\{f_l\}$, converging to some function $f \in L_p$. For any fixed k (see (5))

$$U_k f_l \to U_k f \quad (l \to \infty)$$

in L_p. But then this is so in W as well, since for a fixed k the functions $U_k f_l$ $(l = 1, 2, \ldots)$ are of entire exponential spherical type $k\lambda$ (see 3.6.2 and Lemma 7.7.3 below).

In view of (8), for any $\varepsilon > 0$ it is possible to choose k so that

$$\|f_l - U_k f_l\|_W < \varepsilon \quad \text{(for all} \quad l = 1, 2, \ldots).$$

Accordingly, the sequence $\{f_l\}$ has the property that for any $\varepsilon > 0$ it is possible to indicate a k for which

$$f_l = U_k f_l + (f_l - U_k f_l),$$

where the first term converges as $l \to \infty$ in the sense of W, and the second, in the norm of W, does not exceed ε for any $l = 1, 2, \ldots$. But then as a consequence of the completeness of W

$$f_l \to f \quad (l \to \infty)$$

in W. The theorem is proved.

7.7.3. Lemma. *In the notation of Theorem 7.7.2, the following inequalities hold:*

$$\|g_\nu\|_W \leqq \left(1 + \sum_1^n \nu_j^{l_j}\right) \|g_\nu\|_{L_p},$$

$$\|g_\nu\|_B \leqq c \left(1 + \sum_1^n \nu_j^{r_j}\right) \|g_\nu\|_{L_p}.$$

Here c is a constant not depending on the factor standing next to it, g_ν is an entire function of exponential type $\boldsymbol{\nu} = (\nu_1, \ldots, \nu_n) \geqq \mathbf{0}$. In (2) B may be replaced by $H = H_p^{\mathbf{r}}(\mathbb{R}_n)$.

Thus, if the sequence g_ν^l of entire functions of one and the same type converge to some function g_ν (see 3.5) in the sense of L_p, then it converges in the sense of W, H, and B as well.

Proof. Inequality (1) converges directly from the definition of W and the Bernšteĭn inequality 3.2.2 (9). The function $g = g_\nu$ is entire of type ν_j relative to x_j, and accordingly of type $2^s > 1 + \nu_j$, where s is the smallest natural number for which that inequality is satisfied. Put

$$g_{2^0} = g_2 = \cdots = g_{2^{s-1}} = 0, \; g_{2^s} = g .$$

Then (see 5.6.6 (6))

$$g = g_{2^0} + \sum_1^s (g_{2^j} - g_{2^{j-1}}),$$

$$\|g\|_{B^{r_j}_{x_jp\theta}} = 2^{sr_j}\|g\|_{L_p} \leqq 2^{r_j}(1+\nu_j)^{r_j}\|g\|_p \leqq c(1+\nu_j^{r_j})\|g\|_p$$

$$(j = 1, \ldots, m),$$

from which (2) follows. In these considerations one may obviously replace B by H.

7.7.4. Theorem 7.7.2 remains valid and is proved in exactly the same way, if one replaces W in it by $H = H_p^{\boldsymbol{r}}(\mathbb{R}_n)$ ($\boldsymbol{r} \geqq \boldsymbol{0}$, $1 \leqq p < \infty$), and then supposes that for each function $f \in \mathfrak{M}$ one has

(1) $$\|f(\boldsymbol{x}+\boldsymbol{h}) - f(\boldsymbol{x})\|_H \to 0 \quad (|\boldsymbol{h}| \to 0),$$

which does not in general hold.

In the case $p = \infty$ one has the following theorem.

7.7.5. Theorem. *Suppose given a set $\mathfrak{M} \subset H = H_\infty^{\boldsymbol{r}}(\mathbb{R}_n)$ ($\boldsymbol{r} \geqq \boldsymbol{0}$) of functions f, each of which belongs further to the class $\tilde{C} = \tilde{C}(\mathbb{R}_n)$ of functions continuous on $\mathbb{R}_n$ and having finite limits at the point $\boldsymbol{x} = \infty$. Then for each function $f \in \mathfrak{M}$ one obviously has 7.7.4 (1). Suppose moreover that $\mathfrak{M}$ is bounded in C.*

For $\mathfrak{M}$ to be compact in H, it is necessary and sufficient that that the condition

$$\Lambda(\delta) = \sup_{f\in\mathfrak{M}} \sup_{|\boldsymbol{h}|<\delta} \|f(\boldsymbol{x}+\boldsymbol{h}) - f(\boldsymbol{x})\|_H \to 0 \quad (\delta \to 0)$$

be satisfied, and that for any $\varepsilon > 0$ there exist an $N > 0$ such that

(1) $$|f(\boldsymbol{x}) - f(\boldsymbol{x}')| < \varepsilon$$

for any pair $\boldsymbol{x}, \boldsymbol{x}'$ satisfying the inequalities $|\boldsymbol{x}|, |\boldsymbol{x}'| > N$ for all $f \in \mathfrak{M}$.

The proof is also exactly the same as the proof of 7.7.2, if one observes that the following assertion holds: for a set $\mathfrak{M} \subset \tilde{C}$ of functions to be compact in $\tilde{C}$, it is necessary and sufficient that: 1) it is bounded, 2) it is equicontinuous on $\mathbb{R}_n$, and 3) that for any $\varepsilon > 0$ there is an N such that property (1) holds.

This last assertion is easily obtained from Arzelà's theorem. That conditions 1) and 2) should be satisfied on an arbitrary sphere $|\boldsymbol{x}| \leqq N$ is necessary and sufficient for the compactness of $\mathfrak{M}$ on that sphere.

Chapter 8

Integral Representations and Isomorphism of Isotropy Classes

8.1. Bessel-MacDonald Kernels

The Fourier-transform of the function $(1 + |\boldsymbol{x}|^2)^{-r/2}$, for sufficiently large $r > 0$, may be obtained effectively, since it is a function of $|\boldsymbol{x}|$, and the well-known formula[1]

$$\widehat{(1 + |\boldsymbol{x}|^2)^{-r/2}} = \frac{1}{(2\pi)^{n/2}} \int \frac{e^{iu\xi} d\xi}{(1 + |\xi|^2)^{r/2}}$$

$$= \frac{1}{|\boldsymbol{u}|^{\frac{n-r}{2}}} \int_0^\infty \frac{\varrho^{n/2}}{(1 + \varrho^2)^{r/2}} I_{\frac{n-2}{2}} (|\boldsymbol{u}|\varrho) d\varrho\,,$$

where I_μ is the Bessel Function of order μ, is applicable to it.

This integral (of Hankel type) is computed, for example, in the book of Titchmarsh[2], where we need to suppose that $\mu + 1 = \frac{r}{2}$, $\nu + 1 = \frac{n}{2}$, which yields

$$\widehat{(1 + |\boldsymbol{x}|^2)^{-r/2}} = \frac{1}{2^{\frac{r-2}{2}} \Gamma\left(\frac{r}{2}\right)} \frac{K_{\frac{n-r}{2}}(|\boldsymbol{x}|)}{|\boldsymbol{x}|^{\frac{n-r}{2}}} = G_r(|\boldsymbol{x}|)\,, \tag{1}$$

$$K_\nu(z) = K_{-\nu}(z) = \frac{1}{2}\left(\frac{z}{2}\right)^\nu \int_0^\infty \xi^{-\nu-1} e^{-\xi - \frac{z^2}{4\xi}} d\xi\,. \tag{2}$$

The function $K_\nu(x)$ is called the MacDonald function of order ν, or the modified Bessel function of order ν.

---

[1] Bochner [1], Theorem 5.6.

[2] Titchmarsh [1], 7.11.6, see also Watson [1], § 13.6 (2), and N. Ja. Sonin [1].

For the kernel $K_\nu(x)$, as a function of one variable x, a number of asymptotic estimates are known. Here we shall present them without proof, referring to the book of Watson [1], which we shall refer to as W. The following asymptotic equations hold:

$$K_\nu(x) = \left(\frac{\pi}{2x}\right)^{1/2} e^{-x}\left(1 + O\left(\frac{1}{x}\right)\right) \quad (1 < x)$$

(W 7.2.3 (1)),

$$K_0(x) = \ln\frac{1}{x} + O(1) \quad (0 < x < 1)$$

(W 3.7.1 (14)),

$$K_n(x) = \frac{1}{2}\frac{(n-1)!}{\left(\frac{1}{2}x\right)^n} + O\left(\frac{1}{x^{n-2}}\right) \quad (0 < x < 1,\ n \neq 0 \text{ an integer})$$

(W 3.7.1 (15)),

$$K_\nu(x) = \frac{\pi}{2\sin|\nu|\pi\Gamma(-|\nu|+1)}\left(\frac{1}{2}x\right)^{-|\nu|} + O(x^{-|\nu|})$$

(W 3.7 (6), 3.1 (8)).

For our objectives it will be fully sufficient to keep in mind that from the estimates presented above it follows that

$$|K_\nu(x)| \leqq \frac{ce^{-x}}{x^{1/2}} \quad (1 < x),$$

$$|K_0(x)| \leqq c\left(\ln\frac{1}{x} + 1\right) \quad (0 < x < 1), \tag{3}$$

$$|K_\nu(x)| \leqq \frac{c}{x^{|\nu|}} \quad (0 < x < 1,\ \nu \neq 0 \text{ arbitrary}),$$

where c depends on ν but not on x.

Inequalities (3) may be obtained directly by estimating the integral

$$\Phi(\nu, x) = \int_0^\infty \xi^{-\nu-1} e^{-\xi-\frac{x^2}{4\xi}} d\xi. \tag{4}$$

In the integral (4) one may regard the parameter $\nu = \lambda + i\mu$ as complex. If we recall that

$$|\xi^{-\nu}| = |\xi^{-\lambda}|, \tag{5}$$

then the estimates (3) remain true, on replacing ν by λ in them, also for complex ν. We note that the integral has only two singularities, $\xi = \infty$ and $\xi = 0$, and that the integrand is continuous relative to (ξ, x, ν) $(\xi > 0)$ for arbitrary real x and complex ν. Moreover the integral converges uniformly relative to the indicated x, ν in a sufficiently small neighborhood of any indicated point x_0, ν_0. This shows that $\Phi(\nu, x)$ is continuous relative to ν, x. Similar facts hold also for the integral when it is formally differentiated with respect to ν. This shows that the function $\Phi(\nu, x)$ has a derivative $\frac{\partial}{\partial \nu} \Phi(\nu, x)$ with respect to ν, continuous with respect to (ν, x). Thus $\Phi(\nu, x)$ is analytic with respect to ν.

In equation (1) its left side, if it is considered as a generalized function, has a meaning for any complex r. The right side, expressed using the integral (2), also has a meaning as an ordinary function of $(r, \boldsymbol{x})$, for any complex number r with $\operatorname{Re} r > 0$ and point $\boldsymbol{x} \neq \boldsymbol{0}$ in $\mathbb{R}_n$. Moreover, $G_r(|\boldsymbol{x}|)$ is continuous relative to the indicated $(r, \boldsymbol{x})$, as well as its derivative relative to r. Thus it is analytic relative to r.

From the estimates (3) and equation (1) it follows that

$$|G_r|(|\boldsymbol{x}|)| \leq c_r \begin{cases} \dfrac{e^{-|\boldsymbol{x}|}}{|\boldsymbol{x}|^{\frac{n-r+1}{2}}} & (|\boldsymbol{x}|) > 1,\ n,\ r \text{ arbitrary}), \\ \ln \dfrac{1}{|\boldsymbol{x}|} + 1 & (|\boldsymbol{x}| < 1,\ n - r = 0), \\ \dfrac{1}{|\boldsymbol{x}|^{n-r}} & (|\boldsymbol{x}|) < 1,\ n - r > 0), \\ 1 & (|\boldsymbol{x}|) < 1,\ n - r < 0), \end{cases} \tag{6}$$

where $c_r > 0$ is a continuous function of r.

In these inequalities we have regarded r as real. They are in view of (5) valid also when r is regarded in their left sides as complex and $r = \lambda + i\mu$ is replaced throughout in their right sides by λ.

It is easy to see from (6) that $G_r(|\boldsymbol{x}|) \in L(\mathbb{R}_n) = L$. From what has been said it follows that equation (1) is indeed true for any complex r, if $\operatorname{Re} r = \lambda > 0$. Indeed, suppose $\varphi \in S$. Then the function

$$\widetilde{\left((1 + |\boldsymbol{x}|^2)^{-r/2}, \varphi\right)} = \left((1 + |\boldsymbol{x}|^2)^{-r/2}, \hat{\varphi}\right) = \psi(r)$$

is, as is easily verified, an analytic function of r. On the other hand, using estimates (6), it is directly verified that the function $G_r(|\boldsymbol{x}|)$ cannot uniformly exceed a summable function[1] uniformly for r satisfying the inequality $|r - r_0| < \delta$ $(\lambda_0 > \delta > 0)$. And since $G_r(|\boldsymbol{x}|)\,\varphi(\boldsymbol{x})$ is continuous in $(r, \boldsymbol{x})$ for $\boldsymbol{x} \neq \boldsymbol{0}$ and φ is bounded, then, according to Weierstrass' rule, the function

$$\psi_1(r) = \big(G_r(|\boldsymbol{x}|), \varphi(\boldsymbol{x})\big) = \int G_r(|\boldsymbol{x}|)\,\varphi(\boldsymbol{x})\,d\boldsymbol{x}$$

is a continuous function in r with compact support $(\lambda > 0)$. Using the estimates (3), (6) the analogous fact[2] is established for the derivative

$$\frac{d}{dr}\psi_1(r) = \left(\frac{d}{dr}G_r(|\boldsymbol{x}|), \varphi(\boldsymbol{x})\right).$$

This shows that $\psi_1(r)$ is analytic for $\lambda > 0$. Moreover it is equal to $\psi(r)$ for sufficiently large real r, and accordingly for any complex r with $\lambda > 0$ as well, for any $\varphi \in S$. This implies equation (1). We shall show that for the derivatives of $G_r(|\boldsymbol{x}|)$ of order $\boldsymbol{s} = (s_1, \ldots, s_n)$ the following estimates hold:

$$|D^{\boldsymbol{s}}G_r(|\boldsymbol{x}|)|$$

$$\leqq c \begin{cases} \dfrac{e^{-|\boldsymbol{x}|}}{|\boldsymbol{x}|^{\frac{n-r+1}{2}}} & (|\boldsymbol{x}| > 1,\ n, r, \boldsymbol{s} \text{ arbitrary}), \\ \ln\dfrac{1}{|\boldsymbol{x}|} + 1 & (|\boldsymbol{x}| < 1,\ n - r + |\boldsymbol{s}| = 0,\ \boldsymbol{s} \text{ even}) \\ \dfrac{1}{|\boldsymbol{x}|^{n-r+|\boldsymbol{s}|}} & (|\boldsymbol{x}| < 1,\ n - r + |\boldsymbol{s}| > 0 \text{ and } n - r + |\boldsymbol{s}| = 0, \text{ and } |\boldsymbol{s}| \text{ is odd}), \\ 1 & (|\boldsymbol{x}| < 1,\ n - r + |\boldsymbol{s}| < 0), \end{cases} \tag{7}$$

where c depends continuously on n, r, and $\boldsymbol{s}$, but not on $\boldsymbol{x}$.

We note that by induction it is easily verified that

$$D^{\boldsymbol{s}} e^{-\frac{|\boldsymbol{x}|^2}{4\xi}} = e^{-\frac{|\boldsymbol{x}|^2}{4\xi}} \sum \frac{A_{\boldsymbol{k},l}\boldsymbol{x}^{\boldsymbol{k}}}{\xi^l} \quad (2l - |\boldsymbol{k}| \leqq |\boldsymbol{s}|;\ |\boldsymbol{k}| \leqq l \leqq |\boldsymbol{s}|, \tag{8}$$

where $D^{\boldsymbol{s}}$ is the differentiation operator of order $\boldsymbol{s} = (s_1, \ldots, s_n)$, $\boldsymbol{x}^{\boldsymbol{k}} = x_1^{k_1} \ldots x_n^{k_n}$, the $\boldsymbol{k} = (k_1, \ldots, k_n)$ are integer nonnegative vectors, the $A_{\boldsymbol{k},l}$

---

[1] The constants c_r in inequality (6) are bounded for the indicated values of r.
[2] In 9.4 we shall consider the analogous anisotropic case in detail.

are constants, and the sum is extended over the pairs $\boldsymbol{k}$, l satisfying the inequalities indicated in the parentheses.

Thus

$$|D^s G_r(|\boldsymbol{x}|)| \ll \left| D^s \int_0^\infty \xi^{\frac{n-r}{2}-1} e^{-\xi - \frac{|\boldsymbol{x}|^2}{4\xi}} d\xi \right| \tag{9}$$

$$\ll \sum \left| \boldsymbol{x}^{\boldsymbol{k}} \int_0^\infty \xi^{-\frac{n-r+2l}{2}-1} e^{-\xi - \frac{|\boldsymbol{x}|^2}{4\xi}} d\xi \right| \ll \sum |\boldsymbol{x}^{\boldsymbol{k}} G_{r-2l}(|\boldsymbol{x}|)|,$$

where the sums are extended over the pairs $\boldsymbol{k}$, l indicated in (8).

If $|\boldsymbol{x}| > 1$, then in view of the first estimate in (6)

$$|D^s G_r(|\boldsymbol{x}|)| \ll \sum \frac{|\boldsymbol{x}^{\boldsymbol{k}}| \, e^{-|\boldsymbol{x}|}}{|\boldsymbol{x}|^{\frac{n-(r-2l)+1}{2}}} \ll \sum \frac{e^{-|\boldsymbol{x}|}}{|\boldsymbol{x}|^{\frac{n-r+1}{2}+l-|\boldsymbol{k}|}} \ll \frac{e^{-|\boldsymbol{x}|}}{|\boldsymbol{x}|^{\frac{n-r+1}{2}}},$$

because $l - |\boldsymbol{k}| \geqq 0$. We have proved the first inequality in (7).

Now suppose that $|\boldsymbol{x}| < 1$. If, moreover, $n - r + 2l > 0$, then in view of the third estimate in (6)

$$|\boldsymbol{x}^{\boldsymbol{k}} G_{r-2l}(|\boldsymbol{x}|)| \ll \frac{|\boldsymbol{x}^{\boldsymbol{k}}|}{|\boldsymbol{x}|^{n-r+2l}} = \frac{1}{|\boldsymbol{x}|^{n-r+2l-|\boldsymbol{k}|}} \ll \frac{1}{|\boldsymbol{x}|^{n-r+|\boldsymbol{s}|}}, \tag{10}$$

because $2l - |\boldsymbol{k}| \leqq |\boldsymbol{s}|$.

Further, if $n - r + 2l < 0$, then in view of the fourth estimate in (6)

$$\boldsymbol{x}^{\boldsymbol{k}} G_{r-2l}(|\boldsymbol{x}|)| \ll |\boldsymbol{x}^{\boldsymbol{k}}| \ll 1. \tag{11}$$

Now if for some l (just one) $n - r + 2l = 0$, then in view of the second estimate in (6)

$$|\boldsymbol{x}^{\boldsymbol{k}} G_{r-2l}(|\boldsymbol{x}|) \,| \ll |\boldsymbol{x}^{\boldsymbol{k}}| \ln \frac{1}{|\boldsymbol{x}|}. \tag{12}$$

Further, if $n - r + |\boldsymbol{s}| > 0$, the right side of (10) is larger than the right sides of (11) and (12) ($|\boldsymbol{k}| \geqq 0$). We have proved the third estimate for $n - r + |\boldsymbol{s}| \geqq 0$. Now if $n - r + |\boldsymbol{s}| = 0$ and $|\boldsymbol{s}|$ is odd, then there is no natural number l for which $n - r + 2l = 0$. In this case the estimate (12) does not arise, and estimates (10) and (11) yield 1. Thus the third estimate in (7) is proved completely. Now if $n - r + |\boldsymbol{s}| = 0$ with $|\boldsymbol{s}|$ even, then the estimate (12) may indeed arise. Thus the second inequality in (7) is proved.

Finally, if $n - r + |\boldsymbol{s}| < 0$, then the right sides of (10), (11), and, for $|\boldsymbol{k}| > 0$, (12), are estimated by unity. It remains only to investigate the case (12) for $\boldsymbol{k} = \mathbf{0}$. But this is impossible, since, if $n - r + |\boldsymbol{s}| < 0 = n - r + 2l$, it follows that $|\boldsymbol{s}| < 2l$, which contradicts the fact that for $|\boldsymbol{k}| = 0$ we must have satisfied along with this the inequality $2l - |\boldsymbol{k}| \leqq |\boldsymbol{s}|$, i.e. $2l \leqq |\boldsymbol{s}|$. This proves the last inequality in (7).

It is easy to see from inequalities (7) that for any $r > 0$ and natural number n, $G_r(|\boldsymbol{x}|)$ belongs to $L(\mathbb{R}_n) = L$, so that for the functions $f \in L_p(\mathbb{R}_n) = L_p$ $(1 \leqq p \leqq \infty)$ the convolution

$$(13) \quad F(\boldsymbol{x}) = \frac{1}{(2\pi)^{n/2}} \int G_r(|\boldsymbol{x} - \boldsymbol{u}|)\, f(\boldsymbol{u})\, d\boldsymbol{u} = \widehat{(1 + |\boldsymbol{u}|^2)^{-r/2} \tilde{f}} = I_r f$$

has meaning. In addition, obviously, $F \in L_p$. Indeed, the function F possesses, as we shall see, significantly better properties.

8.2. The Isomorphism Classes W_p^l

We will say that the Banach spaces E_1 and E_2 are *isomorphic*, if there exist a linear operator A, mapping E_1 onto E_2 in a $1 - 1$ manner, and two positive constants c_1 and c_2, not depending on $\boldsymbol{x} \in E_1$, such that

$$(1) \qquad c_1 \|\boldsymbol{x}\|_{E_1} \leqq \|A(\boldsymbol{x})\|_{E_2} \leqq c_2 \|\boldsymbol{x}\|_{E_1}$$

for all $\boldsymbol{x} \in E_1$.

Concerning the operator A we shall say that it realizes the isomorphism of E_1 and E_2:

$$(2) \qquad A(E_1) = E_2.$$

Then the inverse operator A^{-1}, obviously, exists, is linear, and in its turn realizes the isomorphism

$$A^{-1}(E_2) = E_1.$$

We shall show that the operation I_l, for a natural number l, realizes the isomorphism

$$(3) \qquad I_l(L_p) = W_p^l$$

$$(1 < p < \infty;\ W_p^l = W_p^l(\mathbb{R}_n),\ L_p = W_p^l,\ l = 0, 1, \ldots).$$

Suppose that $F \in W_p^l$. Then

$$\widehat{(iu_j)^l \tilde{F}} = \frac{\partial^l F}{\partial x_j^l} \in L_p \quad (j = 1, \ldots, n),$$

and in view of the fact that $(i^3 \operatorname{sign} u_j)^l$ a Marcinkiewicz multiplier (see 1.5.5, Example 1, and 1.5.4.1),

$$\|\widetilde{|u_j|^l \tilde{F}}\|_p \leqq c_1 \left\| \frac{\partial^l F}{\partial x_j^l} \right\|_p .$$

Therefore, taking account as well of the fact that $F \in L_p$, we get

$$\left\| \widetilde{\left(1 + \sum_{j=1}^{n} |u_j|^l\right) \tilde{F}} \right\|_p \leqq c_2 \|F\|_{W_p^l} .$$

But the function $(1 + |\boldsymbol{u}|^2)^{l/2} \left(1 + \sum_1^n |u_j|\right)^{-l}$ is a Marcinkiewicz multiplier (1.5.5, Example 7), so that

$$f = \widetilde{(1 + |\boldsymbol{u}|^2)^{l/2} \tilde{F}} \in L_p$$

and

(4) $$\|f\|_p \leqq c_3 \|F\|_{W_p^l} .$$

Now suppose that $f \in L_p$. Then $\tilde{F} = \tilde{G}_l \tilde{f} = (1 + |\boldsymbol{\lambda}|^2)^{-l/2} \tilde{f}$ and, accordingly (1.5 (10)),

$$\widetilde{F^{(\boldsymbol{k})}} = (i\boldsymbol{\lambda})^{\boldsymbol{k}} (1 + |\boldsymbol{\lambda}|^2)^{-l/2} \tilde{f} .$$

But for $|\boldsymbol{k}| = l$ the function

$$(i\boldsymbol{\lambda})^{\boldsymbol{k}} (1 + |\boldsymbol{\lambda}|^2)^{-l/2}$$

is a Marcinkiewicz multiplier (see 1.5.5, Example 5). Hence

(5) $$\|F^{(\boldsymbol{k})}\|_p \leqq c_4 \|f\|_p .$$

But also (8.1 (13)) $\|F\|_p \leqq c_5 \|f\|_p$, so that $F \in W_p^l$ and

$$\|F\|_{W_p^l} \leqq c \|f\|_p .$$

We have proved that the operation I_l realizes the isomorphism (3). In what follows we will prove that it can serve as a means for the definition and realization of isomorphisms of other classes of differentiable functions.

8.3. Properties of the Bessel-MacDonald Kernel

Below it is proved that the Bessel-MacDonald kernel $G_r(|\boldsymbol{x}|)$ satisfies for $r > 0$ the estimate

$$(1) \qquad \Lambda_1 = \int \left| \Delta h^2_{hx_j} \frac{\partial^2 G_r(|\boldsymbol{x}|)}{\partial x_j^s} \right| d\boldsymbol{x} \leqq M_r |h|^\alpha .$$

Here $-\infty < h < \infty$, $j = 1, \ldots, n$; $s = \bar{r}$, $r = \bar{r} + \alpha$, where $\bar{r}$ is an integer and $0 < \alpha \leqq 1$. Since $G_r(|\boldsymbol{x}|) \in L = L(R_n)$ for $r > 0$, then it follows from (1) that (see the definition of the classes H_p^r in 4.3.3)

$$(2) \qquad G_r(|\boldsymbol{x}|) \in H_1^r = H_1^r(\mathbb{R}_n)$$

and

$$(3) \qquad \|G_r(|\boldsymbol{x}|)\|_{H_1^r} = \|G_r(|\boldsymbol{x}|)\|_L + M_r ,$$

where M_r is the smallest constant for which the inequality (1) is satisfied.

Put $\boldsymbol{u} = (u_j, \boldsymbol{u}^j)$, $\boldsymbol{u}^j = (u_1, \ldots, u_{j-1}, u_{j+1}, \ldots, u_n)$,

$$g^{(s)}(\boldsymbol{x}) = \frac{\partial^s G_r(|\boldsymbol{x}|)}{\partial x_j^s}, \quad \Delta_h^2 \varphi(t) = \varphi(t+h) - 2\varphi(t) + \varphi(t-h).$$

We shall make use of the four estimates in 8.1 (7). We shall denote these by 1), 2), 3), 4).

By 1)—3)

$$\Lambda \leqq 4 \int |g^{(s)}(\boldsymbol{u})|\, d\boldsymbol{u}$$

$$\ll \int\limits_{|\boldsymbol{u}|<1} \left(\ln \frac{1}{|\boldsymbol{u}|} + \frac{1}{|\boldsymbol{u}|^{n-r+s}} \right) d\boldsymbol{u} + \int\limits_{|\boldsymbol{u}|\geqq 1} e^{-\frac{|\boldsymbol{u}|}{2}}\, d\boldsymbol{u} \leqq c < \infty ,$$

because $n - (r - s) = n - \alpha < n$. Accordingly, for $|h| \geqq 1$

$$(4) \qquad \Lambda \leqq c \leqq c|h|^\alpha .$$

We turn now to the case $|h| < 1$. For definiteness we suppose that $0 < h < 1$. We have

$$(5) \qquad \Lambda = \Lambda_1 + \Lambda_2 ,$$

where Λ_1 is an integral the same as Λ, but taken over the ball $|\boldsymbol{u}| < 4h$ instead of over the whole space. Then

$$(6) \qquad \Lambda_1 \leqq 4 \int\limits_{|\boldsymbol{u}|<2h} |g^{(s)}(\boldsymbol{u})|\, d\boldsymbol{u} \ll \int\limits_{|\boldsymbol{u}|<2h} \frac{d\boldsymbol{u}}{|\boldsymbol{u}|^{n+s-r}} \ll \int_0^{2h} \varrho^{\alpha-1} d\varrho \ll h^\alpha$$

in view of the estimate (3).

However there remains the case

$$n - r + s = 0, \qquad s \text{ even.}$$

Since $0 < r - s \leqq 1$, then this can be so only if $n = 1$, $s = r - 1$ is even, i.e. $\alpha = r - s = 1$.

The needed estimate is then obtained as follows (the integrals are one-dimensional):

$$\Delta_1 \leqq \int\limits_{|u|<4h} |g^{(s)}(u+h) - g^{(s)}(u)|\, du + \int\limits_{|u|<4h} |g^{(s)}(u) - g^{(s)}(u-h)|\, du$$

$$\ll 2 \int\limits_{|u|<5h} |g^{(s)}(u+h) - g^{(s)}(u)|\, du$$

$$= 2 \int\limits_{-5h}^{-h} |g^{(s)}(u+h) - g^{(s)}(u)|\, du + 2 \int\limits_{-h}^{0} + 2 \int\limits_{0}^{5h} = \Delta_1^{(1)} + \Delta_1^{(2)} + \Delta_1^{(3)},$$

where

$$\Delta_1^{(3)} = 2 \int\limits_0^{5h} du \left| \int\limits_u^{u+h} g^{s+1}(t)\, dt \right| \ll \int\limits_0^{5h} du \int\limits_u^{u+h} \frac{dt}{t}$$

$$= \int\limits_0^{5h} \ln\left(1 + \frac{h}{u}\right) du = h \int\limits_0^5 \ln(1+t)\, dt \ll h,$$

and analogously

$$\Delta_1^{(1)} \ll h.$$

Further, taking account of the fact that $G_r(|u|)$ is an even function, so that for $s = r - 1$ even the function $g^{(s)}(u)$ is also even, we find

$$\Delta_1^{(2)} = 2 \int\limits_{-h}^{0} |g^{(s)}(u+h) - g^{(s)}(u)\, du = 2 \int\limits_{-h}^{0} |g^{(s)}(u+h) - g^{(s)}(-u)|\, du$$

$$= 2 \int\limits_0^h |g^{(s)}(u) - g^{(s)}(h-u)|\, du = 2 \int\limits_0^h du \left| \int\limits_{h-u}^{u} g^{(s+1)}(t)\, dt \right|$$

$$\ll \int\limits_0^h du \left| \int\limits_{h-u}^{u} \frac{dt}{t} \right| = 2 \int\limits_0^h \left| \ln \frac{u}{h-u} \right| du = 2h \int\limits_0^1 \left| \ln \frac{t}{1-t} \right| dt \ll h.$$

Thus (6) is completely proved. We turn to the estimate

$$\Lambda_2 = \int\limits_{|\boldsymbol{u}|>4h} \Delta^2_{h,x_j} g^{(s)}| d\boldsymbol{u} = \int\limits_{|\boldsymbol{u}|>4h} \left| \int\limits_0^h \int\limits_0^h \frac{\partial^2 g^{(s)}}{\partial x_2^j} (u_j + v + t, \boldsymbol{u}^j)\, dv\, dt \right| d\boldsymbol{u}$$

$$= \int\limits_{\substack{|\boldsymbol{u}|>4h \\ u_j>0}} + \int\limits_{\substack{|\boldsymbol{u}|>4h \\ -2h<u_j<0}} + \int\limits_{\substack{|\boldsymbol{u}|>4h \\ h_j<-2h}}^{\infty} = \Lambda_2^{(1)} + \Lambda_2^{(2)} + \Lambda_2^{(3)},$$

where, in view of 3) and recalling that $n + s - r + 2 = n - \alpha + 2 \geqq n + 1 > 0$,

$$\Lambda_2^{(1)}, \Lambda_2^{(3)} \ll h^2 \int\limits_{\substack{|\boldsymbol{u}|<2h \\ u_j>0}} \frac{d\boldsymbol{u}}{|\boldsymbol{u}|^{n+s-r+2}} \ll h^2 \int\limits_{2h} \varrho^{\alpha-3} d\varrho \ll g h^\alpha. \tag{7}$$

Taking into account the fact that, for $|\boldsymbol{u}| > 4h$, $-2h < u_j < 0$, $|\boldsymbol{u}_j| \geqq |\boldsymbol{u}| - |u_j| \geqq 4h - 2h = 2h$, we get

$$\Lambda_2^{(2)} \ll h^3 \int\limits_{|\boldsymbol{u}_j|>2h} \frac{d\boldsymbol{u}^j}{|\boldsymbol{u}^j|^{n+s-r+2}} \ll h^3 \int\limits_{2h}^{\infty} \frac{\varrho^{n-2}}{\varrho^{n+s-r+2}}\, d\varrho \ll h^3 \int\limits_{2h}^{\infty} \varrho^{\alpha-4}\, d\varrho \ll h^\alpha. \tag{8}$$

(1) now follows from (4), (6), (7), and (8).

8.4. Estimate of the Best Approximation for $I_r f$

Suppose that the function $f \in L_p = L_p(\mathbb{R}_n)$, $r > 0$ and (8.1 (13))

$$F = I_r f = \frac{1}{(2\pi)^{n/2}} \int G_r(|\boldsymbol{u}|)\, f(\boldsymbol{x} - \boldsymbol{u})\, d\boldsymbol{u}. \tag{1}$$

Suppose further that $w_\nu \in L$, $\lambda_\nu \in L_p$ are arbitrary functions of exponential spherical type ν. Thus, $\omega_\nu \in S\mathfrak{M}_{\nu 1}$, $\lambda_\nu \in S\mathfrak{M}_{\nu p}$. Put

$$F(\boldsymbol{x}) - \Omega_\nu(\boldsymbol{x}) = \frac{1}{(2\pi)^{n/2}} \int [G_r(|\boldsymbol{u}|) - \omega_\nu(\boldsymbol{u})]\, [f(\boldsymbol{x} - \boldsymbol{u}) - \lambda_\nu(\boldsymbol{x} - \boldsymbol{u})]\, d\boldsymbol{u}.$$

Obviously, $\Omega_\nu \in S\mathfrak{M}_{\nu p}$ (see 3.6.2) and

$$\|F - \Omega_\nu\|_p \leqq \frac{1}{(2\pi)^{n/2}} \|G_r(|\boldsymbol{x}|) - \omega_\nu(\boldsymbol{x})\|_L \|f - \lambda_\nu\|_p.$$

Therefore, taking account of the fact that the function $G_r(|\boldsymbol{x}|) \in H_1^r$ (see 8.3) and that accordingly its best approximation in the metric of L by means of entire functions of spherical degree ν has order $Q(\nu^{-r})$ (see 5.5.4), we will have

$$E_\nu(F)_p \leqq \frac{1}{(2\pi)^{n/2}} E_\nu(G_r(|\boldsymbol{x}|))_L \, E_\nu(f)_p = \frac{b_r}{\nu^r} E_\nu(f)_p, \tag{2}$$

where $E_\nu(\varphi)_p$, $E_\nu(\varphi)_L$ denote the best approximations of φ by means of entire functions of spherical type ν in the metrics of L_p and L respectively, and the constant b_r does not depend on the factor next to it.

Again suppose that $f \in L_p$, and that moreover the Fourier transform $\tilde{f}$ (in general a generalized function) is equal to zero on the sphere v_ν with center at the origin and of radius ν (see 3.2.6 (5)):

$$\tilde{f} = 0 \text{ on } v_\nu. \tag{3}$$

Then (see 3.2.6 (6)), if $0 < \lambda < \nu$, the convolution of any function $\omega_\lambda \in S\mathfrak{M}_{\nu 1}$ with f is equal to zero:

$$\omega_\lambda * f = \frac{1}{(2\pi)^{n/2}} \int \omega_\lambda(\boldsymbol{u}) \, f(\boldsymbol{x} - \boldsymbol{u}) \, d\boldsymbol{u} = 0,$$

so that

$$F(\boldsymbol{x}) = I_r f = \frac{1}{(2\pi)^{n/2}} \int [G_r(|\boldsymbol{u}|) - \omega_\lambda(\boldsymbol{u})] \, f(\boldsymbol{x} - \boldsymbol{u}) \, d\boldsymbol{u}$$

and

$$\|F\|_p \leqq \frac{1}{(2\pi)^{n/2}} \|G_r - \omega_\lambda\|_L \, \|f\|_p.$$

But then, taking the lower bound relative to ω_λ, we obtain the inequality

$$\|F\|_p \leqq \frac{1}{(2\pi)^{n/2}} E_\lambda(G_r)_L \, \|f\|_p = \frac{b_r \|f\|_p}{\lambda^r},$$

valid for all $\lambda < \nu$, so that

$$\|I_r f\|_p = \|F\|_p \leqq \frac{b_r \|f\|_p}{\nu^r} \quad (\nu > 0, \, (\tilde{f})_{v_\nu} = 0), \tag{4}$$

where b_r is the constant entering into inequality (2), not depending on $\nu > 0$ and the f in question.

8.5. Multipliers Equal to Unity on a Region

By definition the generalized function $f \in S'$ is equal to zero on an open set $g \subset \mathbb{R}_n$, if for any function φ in g with compact support we have

$$(f, \varphi) = 0.$$

If in addition f not only belongs to S', but is also locally summable on g, then almost everywhere

$$f(\boldsymbol{x}) = 0 \text{ on } g.$$

Indeed, suppose that $\sigma \subset g$ is any ball. There exists (see 1.4.2) a sequence of functions φ_N with compact support in σ, for which one has bounded convergence:

$$\lim_{N \to \infty} \varphi_N(\boldsymbol{x}) = \operatorname{sign} f(\boldsymbol{x}) \text{ almost everywhere on } \sigma.$$

Therefore in view of the Lebesgue theorem

$$0 = (f, \varphi_N) = \int_\sigma f(\boldsymbol{x}) \, \varphi_N(\boldsymbol{x}) d\boldsymbol{x} \to \int_\sigma |f(\boldsymbol{x})| \; d\boldsymbol{x} \;\; (N \to \infty),$$

i.e. $f(\boldsymbol{x}) = 0$ on σ almost everywhere and accordingly on g as well.

If $f_1, f_2 \in S'$ and $f_1 - f_2 = 0$ on an open set g, then it is natural to say that $f_1 = f_2$ on g.

8.5.1. Lemma. *Suppose that μ is a multiplicator in $L_p (1 \leqq p \leqq \infty$; $\hat{\mu} \in L$ for $p = \infty$; see 1.5.1, 1.5.1.1), equal to unity on the open set $g \subset \mathbb{R}_n$. Then for $f \in L_p$ and, more generally, for a function f regular in the sense of L_p,*

$$(1) \qquad \widetilde{K * f} = \mu \tilde{f} = \tilde{f} \quad \text{on} \quad g (K = \hat{\mu}).$$

Proof. For $\varphi \in S$ with support in g and a C^∞ function f with compact support,

$$(2) \qquad (\mu \tilde{f}, \varphi) = (\mu, \tilde{f} \varphi) = (1, \tilde{f} \varphi) = (\tilde{f}, \varphi).$$

Here we need to recall that by definition μ is an ordinary measurable function by the hypothesis of the lemma, equal to unity on g, so that the second term in (2) is a Lebesgue integral. Moreover, by the hypothesis of the lemma $\mu(\boldsymbol{x}) = 1$ on g, and $\tilde{f}\varphi$ has its support in g, which proves the second equation.

If $f \in L_p$, then there exists a sequence of C^∞ functions f_l such that $f_l \to f$, $\mu \tilde{f}_l \to \mu \tilde{f}$ weakly. Putting f_l in the place of f in (2) and passing to the limit as $l \to \infty$, we again obtain (2), but now already for $f \in L_p$.

If now f is regular in the sense of L_p, then for $\varphi \in S$ with support in g, for sufficiently large ϱ we find that

$$(\widetilde{K*f}, \varphi) = (\widetilde{I_{-\varrho}(K*I_\varrho f)}, \varphi) = (\widetilde{K*I_\varrho f}, (1 + |\lambda|^2)^{\varrho/2}\varphi)$$

$$= (\widetilde{I_\varrho f}, (1 + |\lambda|^2)^{\varrho/2}\varphi) = (\tilde{f}, \varphi),$$

i.e. (1).

8.5.2. Lemma. *Suppose that the multiplicator* $\mu = \mu_N = 1$ *on* $\Delta_N = \{|x_j| < N; j = 1, \ldots, n\}$. *Then, if* $N' < N$ *and the function* $\omega_{N'} \in \mathfrak{M}_{N'p}$ *(i.e. it is entire of exponential type* N' *relative to all the variables and lies in* L_p*), then*

$$K*\omega_{N'} = \widehat{\mu\tilde{\omega}_{N'}} = \omega_{N'} \ (K = \tilde{\mu}). \tag{1}$$

Proof. Suppose that $\varepsilon > 0$ and $N' + \varepsilon < N$. Since ψ_ε is exponential of type ε, $\psi_\varepsilon w_{N'} \in \mathfrak{M}_{N+\varepsilon,p}$. Moreover, $\psi_\varepsilon w_{N'} \in S$, because $\psi_\varepsilon \in S$, and $\psi_{N'}$ along with all of its derivatives is bounded ($w_{N'}$ is of polynomial growth). Hence

$$(\widetilde{\mu\psi_\varepsilon\omega_{N'}}, \varphi) = (\mu, \widetilde{\psi_\varepsilon\omega_{N'}}\varphi) = (1, \widetilde{\psi_\varepsilon\omega_{N'}}\varphi) = (\widetilde{\psi_\varepsilon\omega_{N'}}, \varphi),$$

where the second equation holds because the support of $\widetilde{\psi_\varepsilon\omega_{N'}}\varphi$ belongs to Δ_N. Accordingly,

$$\widehat{\mu\widetilde{\psi_\varepsilon\omega_{N'}}} = \psi_\varepsilon\omega_{N'}. \tag{2}$$

Passing to the limit in (2) in the weak sense as $\varepsilon \to 0$, we obtain (1). For the right side of (1) this follows from 1.5.8 (6). As to the left side, then we need to recall that

$$\|\psi_\varepsilon\omega_{N'} - \omega_{N'}\|_p^p = \int |(\psi_\varepsilon(\boldsymbol{x}) - 1)\omega_{N'}(\boldsymbol{x})|^p \, d\boldsymbol{x} \to 0 \quad (\varepsilon \to 0)$$

by the Lebesgue theorem. Hence in view of the fact that μ is a multiplicator, the left side of (2) tends to the left side of (1) not only weakly, but even in L_p.

8.6. de la Vallée Poussin Sums of a Regular Function

In the theory of the Fourier integral the kernel

$$\frac{\sin Nt}{t} = \int_0^N \cos nt \, dn \tag{1}$$

for integer N corresponds to the trigonometric polynomial

$$(2)\qquad D_N^*(t) = \frac{1}{2} + \sum_{n=1}^{N} \cos nt = \frac{\sin\left(N + \frac{1}{2}\right)t}{2\sin\frac{t}{2}} \qquad (N = 0, 1, \ldots),$$

called the Dirichlet kernel of order N.

The arithmetic mean

$$(3)\qquad \nu_N^* = \nu_N^*(t) = \frac{D_{N+1}^* + \cdots + D_{2N}^*}{N}$$

$$= \frac{1}{2} + \sum_{0}^{N} \cos kt + \frac{1}{N} \sum_{N+1}^{2N} (2N + 1 - k) \cos kt$$

$$= \frac{\cos(N+1)t - \cos(2N+1)t}{4N \sin^2 \frac{t}{2}}$$

is called *the kernel of de la Vallée Poussin*[1]. We shall say that it has order N.

The important properties of the de la Vallée Poussin kernel consist in the following:

1*) ν_N^* is an even trigonometric polynomial of order $2N$;

2*) The Fourier coeffcients ν_N^* with indices $k = 0, 1, \ldots, N$ are equal to unity;

$$3^*)\ \frac{1}{\pi} \int_{-\pi}^{\pi} \nu_N^*(t) dt = 1;$$

$$4^*)\ \frac{1}{\pi} \int_{-\pi}^{\pi} |\nu_N^*(t)|\, dt = \frac{1}{2N\pi} \int_{0}^{\pi} \frac{|\cos(N+1)t - \cos(2N+1)t|}{\sin^2 \frac{t}{2}} dt$$

$$\leq \frac{\pi}{N} \int_{0}^{\pi} \frac{\left|\sin \frac{N}{2} t \sin\left(\frac{3N}{2} + 1\right)t\right|}{t^2} dt$$

[1] de la Vallée Poussin [1].

$$\leqq \frac{\pi}{N}\int_0^{\pi} \frac{\left|\sin\frac{N}{2}t\sin\frac{3N}{2}t\right| dt}{t^2} + \frac{\pi}{N}\int_0^{\pi} \frac{\left|\sin\frac{N}{2}t\right|}{t}\, dt$$

$$\leqq \pi\int_0^{N\pi} \frac{\left|\sin\frac{u}{2}\sin\frac{3}{2}u\right|}{u^2}\, du + \frac{\pi^2}{2} < A < \infty,$$

where A does not depend on $N \geqq 1$.

Below we shall consider the appropriate analogue of the de la Vallée Poussin kernel in the case of Fourier integrals in the n-dimensional case.

We begin by considering an ordinary function $g(\boldsymbol{x})$, measurable and bounded on $\mathbb{R} = \mathbb{R}_n$ and such that its Fourier transform $\tilde{g}$ is in its turn an ordinary bounded function. Suppose further that $\boldsymbol{\lambda} = (\lambda_1, \ldots, \lambda_n)$ is a vector parameter which can vary on the n-dimensional interval $\Omega_a = \{a < \lambda_j < 2a;\, j = 1, \ldots, n\}$, where $a > 0$. The following equation holds:

$$\widetilde{\int_{\Omega_a} g(\lambda_1 x_1, \ldots, \lambda_n x_n) d\boldsymbol{\lambda}} = \int_{\Omega_a} \widetilde{g(\lambda_1 x_1, \ldots, \lambda_n x_n)} d\boldsymbol{\lambda}. \tag{4}$$

Indeed, if $\varphi \in S$, then

$$\left(\widetilde{\int_{\Omega_a} g(\lambda_1 x_1, \ldots, \lambda_n x_n) d\boldsymbol{\lambda}}, \varphi\right) = \int_{\Omega_a} d\boldsymbol{\lambda} \int \widetilde{g(\lambda_1 x_1, \ldots, \lambda_n x_n)}\varphi(\boldsymbol{x}) d\boldsymbol{x}$$

$$= \int \left(\int_{\Omega_a} \widetilde{g(\lambda_1 x_1, \ldots, \lambda_n x_n)} d\boldsymbol{\lambda}\right) \varphi(\boldsymbol{x})\, d\boldsymbol{x} = \left(\int_{\Omega_a} \widetilde{g(\lambda_1 x_1, \ldots, \lambda_n x_n)} d\lambda, \varphi(\boldsymbol{x})\right).$$

All the equations here are obvious. All that has to be clarified is the fact that $\widetilde{g(\lambda_1 x_1, \ldots, \lambda_n x_n)}$ is for $\boldsymbol{\lambda} \in \Omega_a$ an ordinary bounded function. But this follows from the equations

$$\widetilde{g(\lambda_1 x_1, \ldots, \lambda_n x_n)} = \frac{1}{(2\pi)^{n/2}} \int g(\lambda_1 u_1, \ldots, \lambda_n u_n) e^{-i\boldsymbol{x}\boldsymbol{u}}\, d\boldsymbol{u}$$

$$= \frac{1}{\prod_1^n \lambda_j (2\pi)^{n/2}} \int g(\boldsymbol{u}) e^{-i\sum_{j=1}^{n} \frac{x_j u_j}{\lambda_j}}\, d\boldsymbol{u}$$

and the hypothesis that $\tilde{g}$ is an ordinary bounded measurable function.

The analogue of the de la Vallée Poussin kernel is defined using an equation analogous to (3):

(5)
$$V_N(\boldsymbol{t}) = \frac{1}{N^n} \int\limits_{\Omega_N} \prod_{j=1}^{n} \frac{\sin \lambda_j t_j}{t_j}\, d\lambda$$

$$= \frac{1}{N^n} \prod_{j=1}^{2N} \int\limits_N^{2N} \frac{\sin \nu t_j}{t_j}\, d\nu = \frac{1}{N^n} \prod_{j=1}^{n} \frac{\cos Nt_j - \cos 2Nt_j}{t_j^2}.$$

The kernel V_N satisfies properties analogous to the properties 1*)–4*):

1) $V_N(\boldsymbol{z})$ is an entire function of exponential type of degree $2N$ relative to each of the variables z_j, $j = 1, \ldots, n$, bounded and summable on $\mathbb{R}$;

(6)
$$2)\quad \left(\frac{2}{\pi}\right)^{n/2} \tilde{V}_N = \frac{1}{\pi^n} \int V_N(\boldsymbol{t}) e^{-i\boldsymbol{x}\boldsymbol{t}} d\boldsymbol{t} = 1 \text{ on } \Delta_N,$$

$$\Delta_N = \{|\boldsymbol{x}_j| \leqq N;\, j = 1, \ldots, n\},$$

(7)
$$3)\quad \frac{1}{\pi^n} \int V_N(\boldsymbol{t})\, d\boldsymbol{t} = 1,$$

(8)
$$4)\quad \frac{1}{\pi^n} \int |V_N(\boldsymbol{t})|\, d\boldsymbol{t} \leqq M \quad (N \geqq 1).$$

Property 1) is established without difficulty. Property 3) follows from the equation

$$\frac{1}{\pi} \int \frac{\sin \nu t}{t}\, dt = 1 \quad (\nu > 0),$$

where the improper Riemann integral converges uniformly on the interval $N \leqq \nu \leqq 2N$. As a consequence it is legitimate to carry out an integration of this integral relative to the parameter ν under the integral sign:

$$\frac{1}{\pi^n} \int V_N(\boldsymbol{t}) d\boldsymbol{t} = \frac{1}{(\pi N)^n} \prod_{j=1}^{n} \int dt_j \int\limits_N^{2N} \frac{\sin \nu t_j}{t_j}\, d\nu$$

$$= \frac{1}{N^n} \left(\int\limits_N^{2N} d\nu\, \frac{1}{\pi} \int \frac{\sin \nu t}{t}\, dt \right)^n = 1.$$

Property 4) is obvious:

$$\frac{1}{N}\int_{-\infty}^{\infty}\frac{|\cos Nt - \cos 2Nt|}{t^2}\,dt = \frac{2}{N}\int_0^{\infty}\frac{\left|\sin\frac{N}{2}t\sin\frac{3}{2}Nt\right|dt}{t^2}$$

$$= 2\int_0^{\infty}\frac{\left|\sin\frac{u}{2}\sin\frac{3}{2}u\right|}{u^2}\,du < \infty.$$

Consider the function

$$D_\lambda(\boldsymbol{t}) = \prod_{j=1}^{n}\frac{\sin\lambda_j t_j}{t_j},$$

which is the analogue of the Dirichlet kernel in the n-dimensional case. Its Fourier transform (see 1.5.7 (10))

$$\widetilde{D_\lambda(\boldsymbol{t})} = \widetilde{\prod_{j=1}^{n}\frac{\sin\lambda_j t_j}{t_j}} = \left(\sqrt{\frac{\pi}{2}}\right)^n (1)_{\Delta_\lambda},$$

where $(1)_{\Delta_\lambda}$ is a function equal to unity on $\Delta_\lambda = \{|x_j| < \lambda_j, j = 1, \ldots, n\}$ and zero outside Δ_λ. It is thus bounded along with its Fourier transform, so that equation (4) can be applied to it with $a = N$. Accordingly, recalling that

$$(1)_{\Delta_\lambda}(\boldsymbol{x}) = \prod_{j=1}^{n}(1)_{\lambda_j}(x_j),$$

where $(1)_{\lambda_j}$ is a function of one variable x_j equal to unity on the interval $|x_j| < \lambda_j$ and zero for the remaining x_j, we obtain

(9) $$\tilde{V}_N = \frac{1}{N^n}\int_{\Omega_N}\widetilde{\prod_{j=1}^{n}\frac{\sin\lambda_j t_j}{t_j}}\,d\lambda = \left(\frac{1}{N}\sqrt{\frac{\pi}{2}}\right)^n\int_{\Omega_N}(1)_{\Delta_\lambda}(\boldsymbol{x})\,d\lambda$$

$$= \prod_{j=1}^{n}\frac{1}{N}\sqrt{\frac{\pi}{2}}\int_N^{2N}(1)_{\lambda_j}(x_j)\,d\lambda_j = \prod_{j=1}^{n}\mu(x_j),$$

where

(10) $$\mu(\xi) = \sqrt{\frac{\pi}{2}}\begin{cases} 1 & (|\xi| < N), \\ \frac{1}{N}(2N - \xi) & (N < |\xi| \leqq 2N), \\ 0 & (2N < |x|). \end{cases}$$

We have obtained an effective formula for $\tilde{V}_N$. (6) is a direct consequence.

Suppose that $f \in L_p (1 \leqq p \leqq \infty)$. Then

$$(11) \qquad \sigma_N(f, \boldsymbol{x}) = \left(\frac{2}{\pi}\right)^{n/2} (V_N * f) = \frac{1}{\pi^n} \int V_N(\boldsymbol{x} - \boldsymbol{u}) f(\boldsymbol{u}) d\boldsymbol{u}$$

is a function in L_p differing from the convolution $V_N * f$ only by a constant factor. This function is the analogue of the periodic de la Vallée Poussin sum of order N. Since $V_N \in \mathfrak{M}_{2N}$ (entire of exponential type $2N$ relative to all the x_j lying in L), then $\sigma_N(f, \boldsymbol{x}) \in \mathfrak{M}_{2N,p}$ (see 3.6.2) for all $f \in L_p$. Moreover, if $\omega_N \in \mathfrak{M}_{N,p}$, then we have the identity

$$(12) \qquad \sigma_N(\omega_N, \boldsymbol{x}) = \omega_N(\boldsymbol{x}).$$

Indeed, $V_N \in L$, so that $\tilde{V}_N$ is a multiplicator. Moreover, in view of (9) and (10) we have $\tilde{V}_N = \left(\frac{\pi}{2}\right)^n$ on Δ_N, so that, in view of Lemma 8.5.2,

$$\sigma_N(\omega_{N_0}, \boldsymbol{x}) = \omega_{N_0}(\boldsymbol{x}) \quad (N_0 < N).$$

From the resulting equation, (12) follows as $N_0 \to N$. The legitimacy of the passage to the limit is easy to deduce from the effective formula (5) for V_N.

If $f \in L_p$ and $\omega_N \in L_p$ is an entire function of exponential type N, then in view of (12)

$$\sigma_N(f, \boldsymbol{x}) - f(\boldsymbol{x}) = \sigma_N(f - \omega_N, \boldsymbol{x}) + \omega_N(\boldsymbol{x}) - f(\boldsymbol{x}),$$

so that

$$(13) \qquad \|\sigma_N(f, \boldsymbol{x}) - f(\boldsymbol{x})\|_p \leqq (1 + M) E_N(f)_p,$$

i.e. the approximation of f by means of $\sigma_N(f)$ has the order of the best approximation of f by means of functions of exponential type N.

If p is finite, then the right side of (13) tends to zero as $N \to \infty$ (see 5.5.1). Hence it follows that

$$(14) \qquad \sigma_N(f) \to f \text{ weakly as } N \to \infty.$$

For $p = \infty$ the quantity $E_N(f)$ already does not tend to zero, but property (14) holds all the same. Indeed, on the basis of 8.3 (1) $(0 < \alpha \leqq 1)$

$$\int |\Delta^2_{hx_j} G_\alpha(|\boldsymbol{u}|)| d\boldsymbol{u} \leqq M|h|^\alpha.$$

Therefore

$$|\Delta^2_{hx_j} \int G_\alpha(|\boldsymbol{x}\boldsymbol{u}|)\, f(\boldsymbol{u})\, d\boldsymbol{u}|$$

$$= \int |\Delta^2_{hx_j} G_\alpha(\boldsymbol{u})\, f(\boldsymbol{x} - \boldsymbol{u}) d\boldsymbol{u}| \leqq \|f\|_{L_\infty} M|h|^\alpha \quad (j = 1, \ldots, n),$$

$$\|f\|_{L_\infty} = \sup_{\boldsymbol{x}\in\mathbb{R}} \text{vrai}\, |f(\boldsymbol{x})|.$$

We see that the function $F(\boldsymbol{x}) = I_\alpha(f)$ satisfies the condition

$$|\Delta^2_{hx_j} F(\boldsymbol{x})| \leqq c|h|^\alpha \quad (j = 1, \ldots, n).$$

Since moreover it is bounded, then it belongs to $H^\alpha_\infty(\mathbb{R})$ and is accordingly uniformly continuous on $\mathbb{R}$. Hence it belongs to C.

But then

$$E_N(I_\alpha f)_\infty = E_N(F)_\infty \to 0 \quad (N \to \infty).$$

This shows that

$$\sigma_N(f) \to F \text{ weakly},$$

so that

$$(15) \quad (\sigma_N(f), \varphi) = \left(\frac{2}{\pi}\right)^{n/2} (V_N * f, \varphi) = \left(\frac{2}{\pi}\right)^{n/2} (I_{-\alpha}(V_N * I_\alpha f), \varphi)$$

$$= (\sigma_N(I_\alpha f), I^*_{-\alpha}\varphi) \to (I_\alpha f, I^*_{-\alpha}\varphi) = (f, \varphi), \quad N \to \infty,$$

and we have proved (14) in the case $p = \infty$.

Thus, (14) holds for any $f \in L_p$ and any p $(1 \leqq p \leqq \infty)$. It is important that this property is preserved for any function f regular in the sense of L_p. In order to verify this, it suffices to perform on f the calculation (15) presented above.

Finally, we note the following inequalities holding for $f \in L_p$, which are important for us:

$$(16) \qquad \|I_r(\sigma_N(f) - \sigma_{2N}(f)]\|_p \leqq \gamma_r N^{-r} \|\sigma_N(f) - \sigma_{2N}(f)\|_p,$$

$$(17) \qquad \|I_r \sigma_1(f)\|_p \leqq \gamma_r \|\sigma_1(f)\|_p,$$

where r is any real number and γ_r does not depend on N and f.

For $r > 0$ inequality (17) follows from the fact that the operation I_r has a kernel belonging to L (see 8.1 (13) and 1.5.1 (5)), and inequality (16) from the fact that (see 8.5.1 and 8.6 (6))

$$\widetilde{\sigma_N}(f) - \widetilde{\sigma_{2N}}(f) = 0 \quad \text{on } \Delta_N.$$

Now for r negative inequalities (16) and (17) follow from the inequality proved in the following section, if we recall that

$$\sigma_N(f) - \sigma_{2N}(f) \in \mathfrak{M}_{\mu N,p}.$$

In § 8.8 it will be proved that inequalities (16) and (17) remain valid for any (generalized) function regular in the sense of L_p.

8.7. An Inequality for the Operation $I_{-r}(r > 0)$ over Functions of Exponential Type

Suppose that $g = g_\nu \in \mathfrak{M}_{\nu p}(\mathbb{R}_n) = \mathfrak{M}_{\nu p}$, i.e. g is a function of exponential type ν relative to each variable x_j, and belongs to $L_p = L_p(\mathbb{R}_n)$. We apply to it the operation (see 1.5.9)

$$I_{-r}g = \widehat{(1 + |x|^2)^{r/2}\tilde{g}}. \tag{1}$$

The basic objective of this section is to show that we have the inequality

$$\|I_{-r}g\|_{L_p(\mathbb{R}_n)} \leqq \varkappa_r(1 + \nu)^r \|g\|_{L_p(\mathbb{R}_n)} \tag{2}$$

$$(r, \nu > 0, 1 \leqq p \leqq \infty),$$

where $\varkappa_r$ is a constant not depending on ν. Put $\omega(\boldsymbol{x}) = (1 + |\boldsymbol{x}|^2)^{r/2}$. For any $\nu > 0$ we introduce a function $\omega_\nu(\boldsymbol{x})$ of period 2ν (relative to each variable x_j), defined by the equation

$$\omega_\nu(\boldsymbol{x}) = (1 + |\boldsymbol{x}|^2)^{r/2} \quad \{|x_j| \leqq \nu, \ j = 1, \ldots, n\}. \tag{3}$$

Let

$$\omega_\nu(\boldsymbol{x}) = \sum_{\boldsymbol{k}} c_{\boldsymbol{k}}^\nu e^{i\frac{\pi}{\nu}\boldsymbol{k}\boldsymbol{x}} \tag{4}$$

be its Fourier series. We shall show that for $r > r_0$, where r is sufficiently large, the following inequality holds:

$$\sum_{\boldsymbol{k}} |c_{\boldsymbol{k}}^\nu| \leqq \varkappa_r(1 + \nu^2)^{r/2}, \tag{5}$$

where $\varkappa_r$ does not depend on ν. Hence as a consequence of Theorem 3.2.1 it follows that the interpolation formula

$$I_{-r}g = \sum_{\boldsymbol{k}} c_{\boldsymbol{k}}^\nu g\left(\boldsymbol{x} + \frac{\boldsymbol{k}\pi}{\nu}\right) \tag{6}$$

holds, from which it directly follows that inequality (2) holds:

$$\|I_{-r}g\|_p = \sum_k |c_k^\nu| \, \|g\|_p \leqq \varkappa_r(1 + \nu^2)^{r/2}\|g\|_p, \tag{7}$$

as well as the fact that $I_{-r}g$ is an entire function of exponential type ν (see 3.5).

For small r the considerations connected with the estimation of the sum $\Sigma|c_k^\nu|$ become more complicated. But from the fact that inequality (2) is true for large r, we deduce from general considerations that it is true with the appropriate constant $\varkappa_r$ and for any $r > 0$.

We restrict ourselves to the consideration of the two-dimensional case. For $n \geqq 2$ the considerations are more complicated, but analogous. We have $\left(\sum_k' = \sum_{k \neq 0}\right)$

$$\begin{aligned} \sum |c_{kl}^\nu| &= |c_{00}^\nu| + \sum_k' |c_{k0}^\nu| + \sum_l' |c_{0l}^\nu| + \sum_k' \sum_l' |c_{kl}^\nu| \\ &= J_0 + J_1 + J_2 + J_3, \end{aligned}$$

$$J_0 = \frac{1}{\nu^2} \int_{-\nu}^{\nu} \int_{-\nu}^{\nu} (1 + u^2 + v^2)^{r/2}\, du\, dv \leqq (1 + 2\nu^2)^{r/2} \leqq c_1(1 + \nu^2)^{r/2}.$$

$$\begin{aligned} J_1 &= \frac{1}{\nu^2} \sum_k' \left| \int_{-\nu}^{\nu} \int_{-\nu}^{\nu} \omega(u, v) e^{-\frac{ik\pi}{\nu}u}\, du\, dv \right| \\ &\leqq \max_{|v| \leqq \nu} \frac{2}{\nu} \sum_k' \left| \int_{-\nu}^{\nu} \omega(u, v) e^{-\frac{ik\pi}{\nu}u}\, du \right| \\ &= \max_v c \sum_k' \frac{1}{k} \left| \int_{-\nu}^{\nu} \frac{\partial \omega}{\partial u} e^{-\frac{ik\pi u}{\nu}}\, du \right| \\ &\leqq \max_v c_2 \left\{ \sum_k' \left| \int_{-\nu}^{\nu} \frac{\partial \omega}{\partial u} e^{-\frac{ik\pi u}{\nu}}\, du \right|^2 \right\}^{1/2} \\ &\leqq \max_v c_3 \left\{ \nu \int_{-\nu}^{\nu} \left(\frac{\partial \omega}{\partial u}\right)^2 du \right\}^{1/2} \end{aligned}$$

$$\leqq \max_v c_4 \left\{ \nu \int_{-0}^{\nu} (1 + u^2 + v^2)^{r-2} u^2 du \right\}^{1/2} \leqq c_5(1 + \nu^2)^{r/2} \ (r \geqq 2).$$

Here we have integrated by parts and applied Parseval's inequality relative to the variable u. Analogously

$$J_3 \leqq c_5(1+\nu^2)^{r/2} \quad (r \geqq 2).$$

Finally, integrating by parts and using Parseval's inequality relative to both variables u and v, we get

$$J_3 = \sum_k{}' \sum_l{}' \frac{1}{kl\pi^2} \left| \int_{-\nu}^{\nu}\int_{-\nu}^{\nu} \frac{\partial^2 \omega}{\partial u\,\partial v} e^{-\frac{i\pi}{\nu}(ku+lv)} du\,dv \right|$$

$$\leqq c_6 \left\{ \sum{}' \sum{}' \left| \int_{-\nu}^{\nu}\int_{-\nu}^{\nu} \frac{\partial^2 \omega}{\partial u\,\partial v} e^{-\frac{i\pi}{\nu}(ku+lv)} du\,dv \right|^2 \right\}^{1/2}$$

$$\leqq c_7 \left\{ \nu^2 \int_0^{\nu}\int_0^{\nu} u^2 v^2 (1+u^2+v^2)^{r-4} du\,dv \right\}^{1/2} \leqq c_8 (1+\nu^2)^{r/2}$$

$$(r \geqq 4).$$

We have proved (5) for $r \geqq 4$.

Now suppose that r is any positive number. At first we suppose that $g = \mathfrak{M}_{\nu p}$. We choose a natural number s so that

$$2^{s-1} < 1 + \nu \leqq 2^s,$$

and represent g in the form (see 8.6 (11), (12))

$$g(\boldsymbol{x}) = \sigma_{2^s}(g, \boldsymbol{x}) = \sum_{j=0}^{s} q_j,$$

where

$$q_0 = \sigma_{2^0}(g, \boldsymbol{x}),\ q_j = \sigma_{2^j}(g, \boldsymbol{x}) - \sigma_{2^{j-1}}(g, \boldsymbol{x}) \ (j = 1, \ldots, s).$$

Suppose that the number $\varrho > r$ is sufficiently large that inequality (2) is satisfied. Then we have

$$I_{-r} g = \sum_0^s J_{\varrho - r} J_{-\varrho} q_j$$

and (explanation below)

(8) $$\|I_{-r} g\|_p \ll \sum_{j=0}^{2} \frac{1}{2^{(\varrho - r)j}} \|I_{-\varrho} q_j\|_p \ll \sum_{j=0}^{s} \frac{1}{2^{(\varrho - r)j}} 2^{\varrho j}$$

$$\ll \sum_{j=0}^{s} 2^{rj} \ll 2^{sj} \ll (1+\nu)^r.$$

The first relation in this chain holds on the basis of the inequalities (8.6 (16), (17)) already established ($\varrho - r > 0$). The second relation follows from the fact that ϱ is a number such that inequality (2) holds for it with r replaced by ϱ.

Thus inequality (2) is proved for any r. Of course, the considerations just presented yield a rough constant $\varkappa_r$. But cases are known when it is possible to obtain a precise (smallest) value for it. Consider for example the case $n = 1$.

In view of the evenness of $\omega(t) = (1 + t^2)^{r/2}$ $\left(\theta_j = \left(j + \frac{1}{2}\right)\frac{\nu}{k}\right)$

$$(9)\quad c^{\nu}_{-k} = c^{\nu}_k = \frac{1}{\nu}\int\limits_0^{\nu} \omega(t)\cos\frac{k\pi}{\nu}t\,dt = \frac{1}{\nu}\sum_{j=0}^{k-1}\int\limits_{\theta_j - \frac{\nu}{2k}}^{\theta_j + \frac{\nu}{2k}} \omega(t)\cos\frac{k\pi}{\nu}t\,dt$$

$$= \frac{1}{\nu}\sum_0^{k-1}(-1)^{j-1}\int\limits_0^{\frac{\nu}{2k}}[\omega(\theta_j + t) - \omega(\theta_j - t)]\sin\frac{k\pi}{\nu}t\,dt.$$

If $r \geqq 1$, then calculations show that

$$\omega''(t) > 0.$$

Hence the difference inside the square brackets inside the integral in the right side of (9) increases monotonically with j. It follows that the terms in the sum in the right side of (9) increase in absolute value, successively changing sign, and the sign of c_k coincides with the sign of the last term in this sum, corresponding to $j = k - 1$. Thus we have proved that

$$(10)\qquad (-1)^k c^{\nu}_k \geqq 0 \quad (k = 0, \pm 1, \pm 2, \ldots; r \geqq 1).$$

It follows from the evenness of $\omega_\nu(t)$ that

$$\omega_\nu(t) = c^{\nu}_0 + 2\sum_1^{\infty} c^{\nu}_k \cos\frac{k\pi}{\nu}t.$$

Therefore, in view of (10), we have the remarkable equation[1]

$$(1 + \nu^2)^{r/2} = \omega_\nu(\nu) = c^{\nu}_0 + 2\sum_1^{\infty}(-1)^k c^{\nu}_k = \sum_{-\infty}^{\infty}|c^{\nu}_k|.$$

[1] P. I. Lizorkin [8].

Thus we have proved that for $r \geqq 1$ and $n = 1$ one may take $\varkappa_r = 1$ in inequality (7). In this form that constant is unimprovable[1]. But we shall not dwell on the proof of this fact.

8.8. Decomposition of a Regular Function into Series Relative to de la Vallée Poussin Sums

If f is a generalized function which is regular in the sense of L_p, then it is natural to put (see 1.5 (10))

$$(1) \qquad \sigma_N(f) = \left(\frac{2}{\pi}\right)^{n/2} (V_N * f) = \left(\frac{2}{\pi}\right)^{n/2} I_{-\varrho}(V_N * I_\varrho f),$$

where ϱ is a positive constant sufficiently large that $I_\varrho f \in L_p$. But $V_N \in L$, so that $V_N * I_\varrho f$ belongs to L_p. In view of the fact that V_N is of exponential type 2N (see 3.6.2), $V_N * I_\varrho f \in \mathfrak{M}_{2N,p}$. Application of the operation $I_{-\varrho}$ to this last function (see 8.7) does not lead out of the class $\mathfrak{M}_{2N,p}$.

Thus,

$$(2) \qquad \sigma_N(f) \in \mathfrak{M}_{2N,p},$$

for any function f regular in the sense of L_p.

Further, for any real λ,

$$(3) \qquad I_\lambda \sigma_N(f) = \sigma_N(I_\lambda f)$$

(see 1.5.10 (5)). Since, for a regular function f, $\sigma_N(f) \in \mathfrak{M}_{2N,p}$, then for $r > 0$ inequalities 8.6 (16) and (17) obviously are satisfied for it. For $r < 0$ these inequalities also hold for any regular function f, because for it $\sigma_N(f) - \sigma_{2N}(f) \in L_p$ and $\tilde{\sigma}_N(f) - \tilde{\sigma}_{2N}(f) = 0$ on Δ_N (see 8.5.1 and 8.6 (6)).

It will be convenient for us to associate with each regular function the following series:

$$(4) \qquad f = \sigma_{2^0}(f) + \sum_{k=1}^{\infty} (\sigma_{2^k}(f) - \sigma_{2^{k-1}}(f)],$$

which, as we have explained, converges weakly to f. We call this series the *decomposition of the regular function f relative to de la Vallée Poussin sums.*

---

[1] See note to page 311.

For any real r it is legitimate to apply to it term by term the operation

$$I_r f = I_r \sigma_{2^0}(f) + \sum_{k=1}^{\infty} I_r[\sigma_{2^k}(f) - \sigma_{2^{k-1}}(f)]$$

$$= \sigma_{2^0}(I_r f) + \sum_{k=1}^{\infty} [\sigma_{2^k}(I_r f) - \sigma_{2^{k-1}}(I_r f)],$$

because, if f is a regular function, then $I_r f$ is as well, and so $I_r f$ decomposes into the form of a Vallée Poussin series converging to it weakly —the second series in (5). The terms of the second and first series are respectively equal in view of (3).

8.9. Representation of Functions of the Classes $B^r_{p\theta}$ in Terms of de la Vallée Poussin Series. Null Classes ($1 \leqq p \leqq \infty$)

We suppose that $r > 0$, $1 \leqq p \leqq \infty$, $1 \leqq \theta \leqq \infty$, $B^r_{p\theta}(\mathbb{R}_n) = B^r_{p\theta}$ ($B^r_{p\infty} = H^r_p$). We will start from the following definition of the class $B^r_{p\theta}$ (5.6 (5)): the function f belongs to $B^r_{p\theta}$ if the following norm is finite for it:

(1) $$^5\|f\| = \|f\|_{B^r_{p\theta}} = \|f\|_p + \left(\sum_{s=0}^{\infty} 2^{s\theta r} E_{2^s}(f)_p^\theta\right)^{1/\theta}.$$

The other definition equivalent to this (5.6 (6)) consists in saying that $f \in B^r_{p\theta}$ if it is possible to represent f in the form of a series converging to it in the sense of L_p:

$$f = \sum_{s=0}^{\infty} Q_s,$$

in which the Q_s are functions of entire exponential type of spherical degree 2^s on $\mathbb{R}_n$ such that the norm

(2) $$^6\|f\| = \|f\|_{B^r_{p\theta}} = \left(\sum_{s=0}^{\infty} 2^{s\theta r} \|Q_s\|_p^\theta\right)^{1/\theta}$$

is finite.

We shall show that the following definition is equivalent to those just given:

The function $f \in B^r_{p\theta}$ if it is a generalized function regular in the sense of L_p and if the de la Vallée Poussin series corresponding to it (converging weakly to it)

(3) $$f = \sum_{s=0}^{\infty} q_s,$$

(4) $$q_0 = \sigma_{2^0}(f),\ q_s = \sigma_{2^s}(f) - \sigma_{2^{s-1}}(f) \quad (s = 1, 2, \ldots),$$

is such that

$$(5)\qquad {}^{7}\|f\| = \|f\|_{B^r_{p\theta}} = \left(\sum_{s=0}^{\infty} 2^{sr\theta}\|q_s\|_p^\theta\right)^{1/\theta} < \infty .$$

Indeed, suppose that f is a function regular in the sense of L_p, for which (3)—(5) hold. Then q_s is an entire function of exponential type 2^{s+1} relative to all the variables. But then it is exponential of spherical type $\sqrt{n}\, 2^{s+1}$ (see 3.2.6) and accordingly of type 2^{s+l}, where we are supposing that l is a natural number such that $2\sqrt{n} \leqq 2^l$. Putting $q_s^* = 0$ ($s = 0, 1, \ldots, l$) and $q_{s+l}^* = q_s$ ($s = 0, 1, \ldots$), we find that $f = \sum_{s=0}^{\infty} q_s^*$, where q_s^*, is of spherical type 2^s and

$$\left(\sum_{s=0}^{\infty} 2^{sr\theta}\|q_s^*\|_p^\theta\right)^{1/\theta} < \infty ,$$

i.e. ${}^{6}\|f\| < \infty$ and $f \in B^r_{p\theta}$.

Now suppose that ${}^{5}\|f\| < \infty$. Then $f \in L_p$ and (see 8.6 (13))

$$(6)\qquad \|q_s\|_p \leqq \|\sigma_{2^s}(f) - f\|_p + \|\sigma_{2^{s-1}}(f) - f\|_p\,)_p$$
$$\leqq 2ME_{2^{s-1}}(f) \quad (s = 1, 2, \ldots),$$
$$\|q_0\| \leqq M\|f\|_p .$$

Therefore ${}^{7}\|f\| \ll {}^{5}\|f\|$, and we have proved the equivalence of the norm (5) with the norms (1) and (2).

Remark. The equivalence is preserved, if the de la Vallée Poussin sums $\sigma_{2^s}(f)$ are replaced respectively by the Dirichlet sums $s_{2^s}(f)$ (see further 8.10), however under the condition that $1 < p < \infty$. For $p = 1, \infty$, the constant M in inequalities (6) depends on s and moreover is not bounded as $s \to \infty$.

We introduce the zero class $B^0_{p\theta}$ of generalized functions.

By definition, *the function* $f \in B^0_{p\theta}$ if it *is regular in the sense of* L_p *and if its de la Vallée Poussin series is such that*

$$(7)\qquad \|f\|_{B^0_{p\theta}} = \left(\sum_{s=0}^{\infty} \|q_s\|_p^\theta\right)^{1/\theta} < \infty .$$

In particular,

$$(8)\qquad \|f\|_{H^0_p} = \sup_s \|q_s\|_p < \infty .$$

The definitions (7) and (8) presented here for the null classes have the advantage that they do not depend on the definitions of the corresponding classes for positive values of r. But it is possible to make the following definition of $B^0_{p\theta}$: this is the class of functions f of the type $I_{-r\omega}$, where $\varphi \in B^r_{p\theta}$, and r is an arbitrary positive number. We note further that the apparatus with which the original definitions of the H- and B-classes were given for positive r is impractical for direct generalizations to the case $r \leqq 0$.

8.9.1. Isomorphism Classes $B^r_{p\theta}$ for Different r

Theorem. *The operation $I^r (r > 0)$ realizes the isomorphism*

$$I_r(B^0_{p\theta}) = B^r_{p\theta} \quad (1 \leqq p, \theta \leqq \infty, B^r_{p\infty} = H^r_p). \tag{1}$$

The equation (1) *yields a representation of the functions of the class $B^r_{p\theta}$ in terms of a convolution of the Bessel-MacDonald kernel G_r with functions of the class $B^0_{p\theta}$, in general generalized.*

Proof. Suppose that $f \in B^0_{p\theta}$. Then f is regular in the sense of L_p and decomposes into the series

$$f = \sum_{s=0}^{\infty} q_s, \; q_0 = \sigma_{2^0}(f), \; q_s = \sigma_{2^s}(f) - \sigma_{2^{s-1}}(f) \quad (s = 1, 2, \ldots), \tag{2}$$

where

$$\|f\|_{B^0_{p\theta}} = \left(\sum_{s=0}^{\infty} \|q_s\|^\theta\right)^{1/\theta} < \infty.$$

But

$$F = I_r f$$

is also a regular function, decomposing into the series

$$F = \sum_{s=0}^{\infty} Q_s, \tag{3}$$

$$Q_0 = \sigma_{2^0}(F), \; Q_s = \sigma_{2^s}(F) - \sigma_{2^{s-1}}(F) \quad (s = 1, 2, \ldots).$$

In addition

$$\|Q_s\|_p = \|I_r q_s\|_p \leqq c \cdot 2^{-rs} \|q_s\|_p.$$

Accordingly,

$$\|F\|_{B^r_{p\theta}} = \left(\sum_{s=0}^{\infty} 2^{s\theta r} \|Q_s\|^\theta_p\right)^{1/\theta} \leqq c \left(\sum_{s=0}^{\infty} \|q_s\|^\theta_p\right)^{1/\theta} = \|f\|_{B^0_{p\theta}}. \tag{4}$$

Conversely, if $F \in B^p_{r\theta}$, then for F the decomposition (3) holds, and

$$\|F\|_{B^r_{p\theta}} = \left(\sum_{s=0}^{\infty} 2^{sr\theta} \|Q_s\|^\theta_p\right)^{1/\theta} < \infty.$$

For $f = I_{-r}F$ one has the decomposition (2) (see 8.6 (16)) and

$$\|q_s\|_p = \|I_{-r}Q_s\|_p \leqq c \cdot 2^{sr}\|Q_s\| \ (s = 0, 1, \ldots),$$

so that

$$\|f\|_{B^0_{p\theta}} = \left(\sum_{s=0}^{\infty} \|q_s\|_p^\theta\right)^{1/\theta} \leqq c\|F\|_{B^0_{p\theta}}.$$

The theorem is proved.

8.9.2. The classes $B^r_{p\theta}$ for $r < 0$. The concept of a regular function and its expansion into a de la Vallée Poussin series yields the possibility of extending the classes $B^r_{p\theta}$ to negative r. It is natural to suppose that *the function f is in $B^r_{p\theta}$, where r is an arbitrary real number, if it is regular in the sense of L_p and its decomposition into the de la Vallée Poussin series*

$$f = \sum_{s=0}^{\infty} q_s = \sigma_{2^0}(f) + \sum_{s=1}^{\infty} [\sigma_{2^s}(f) - \sigma_{2^{s-1}}(f)] \tag{1}$$

is such that

$$\|f\|_{B^r_{p\theta}} = \left(\sum_{s=0}^{\infty} 2^{sr\theta}\|q_s\|_p^\theta\right)^{1/\theta} < \infty. \tag{2}$$

It is easy to see, reasoning as in the preceding section, that for arbitrary real r and r_1 the operation I_{r_1-r} realizes the isomorphism

$$I_{r_1-r}(B^r_{p\theta}) = B^{r_1}_{p\theta} \quad (1 \leqq p, \theta \leqq \infty;\ B^r_{p\infty} = H^r_p). \tag{3}$$

8.10. Series Relative to Dirichlet Sums ($1 < p < \infty$)

If p satisfies the inequalities $1 < p < \infty$, then the theorem presented above may be developed on the basis of the Dirichlet kernels

$$D_N(\mathbf{t}) = \prod_{j=1}^{n} \frac{\sin Nt_j}{t_j}, \tag{1}$$

which are the analogues of the periodic sums of Dirichlet.

The kernel $D_N(\mathbf{t})$ has the following properties:

I) It is an entire function of exponential type N relative to each variable z_j ($j = 1, \ldots, n$), belonging to L_p, where $1 < p \leqq \infty$.

$$\text{II)}\quad \left(\frac{2}{\pi}\right)^{n/2} \tilde{D}_N = (1)_{\Delta_N} = \begin{cases} 1 \text{ on } \Delta_N = \{|x_j| < N\}, \\ 0 \text{ outside } \Delta_N \end{cases} \tag{2}$$

(see 1.5.7 (10)).

(3) $$\text{III)}\ \frac{1}{\pi^n}\int D_N(\boldsymbol{t})\,d\boldsymbol{t} = 1 \quad (N > 0).$$

IV) The convolution

$$S_N(f, \boldsymbol{x}) = D_N * f = \frac{1}{(2\pi)^n}\int D_N(\boldsymbol{x} - \boldsymbol{t})\, f(\boldsymbol{t})\,d\boldsymbol{t}$$

for $f \in L_p$ ($1 < p < \infty$) is an entire function of exponential type N relative to each variable (see 3.6.2), belonging to L_p (see 1.5.7 (9), (13); 3.6.2; $S_N(f) \in \mathfrak{M}_{Np}$). In addition

(4) $$\|D_N * f\|_p \leqq \varkappa_p \|f\|_p,$$

where $\varkappa_p$ depends only on p. For $p = 1, \infty$ this is no longer the case.

V) If $\omega_N \in \mathfrak{M}_{Np}$,

then

(5) $$S_N(\omega_N) = \omega_N.$$

The fact that

(6) $$D_{N_0} * \omega_N = \frac{1}{(2\pi)^n}\int D_{N_0}(\boldsymbol{x} - \boldsymbol{t})\,\omega_N(\boldsymbol{t})\,d\boldsymbol{t} = \omega_N(x) \quad (N < N_0),$$

follows from 8.5.2. Further

$$\left|\frac{\sin N_0 t_j}{t_j}\right| \leqq \varphi(t_j) = \begin{cases} N_1 & |t_j| \leqq 1, \\ \dfrac{1}{|t_j|} & |t_j| > 1 \end{cases} \qquad (N < N_0 < N_1),$$

so that

$$|D_{N_0}(\boldsymbol{x} - \boldsymbol{t})\,\omega_N(\boldsymbol{t})| \leqq \varphi(\boldsymbol{x} - \boldsymbol{t})\,|\omega_N(\boldsymbol{t})| \in L(R_n),\ \varphi(\boldsymbol{t}) = \prod_{j=1}^{n}\varphi(t_j)$$

$\left(\omega_N \in L_p(\mathbb{R}_n), \varphi \in L_q(\mathbb{R}_n), \dfrac{1}{p} + \dfrac{1}{q} = 1\right)$ Moreover $D_{N_0}(\boldsymbol{x} - \boldsymbol{t}) \to D_N(\boldsymbol{x} - \boldsymbol{t})$ ($N_0 \to N$) for all $\boldsymbol{t}$. Accordingly, by the Lebesgue theorem one may replace N_0 by N in (6).

VI) $\widetilde{D_N * f} = \tilde{f}$ on Δ_N (see (2) and 8.5.1).

It follows from (4) and (5) that if $f \in L_p$ $(1 < p < \infty)$ and ω_N is a best function in the class $\mathfrak{M}_{Np}$ approximating it in the sense of L_p, then

$$\|f - S_N(f)\|_p \leqq \|f - \omega_N\|_p + \|S_N(\omega_N) - f\|_p \leqq (1 + \varkappa_p)\, E_N(f)_p \to 0 \quad (N \to \infty). \tag{7}$$

In particular, thus

$$S_N(f) \to f \ (N \to \infty) \text{ weakly}. \tag{8}$$

Reasoning as in the proof of 8.6 (14) (see 8.6 (15)), where we need to replace V_N by D_N, we find that property (8) is preserved for any function regular in the sense of L_p.

In such a case a function f regular in the sense of L_p may be expanded into the series

$$f = S_{2^0}(f) + \sum_{k=1}^{\infty} [S_{2^k}(f) - S_{2^{k-1}}(f)], \tag{9}$$

converging to it weakly (see 8.8 (4)). It is legitimate to apply the operation I_ϱ, where ϱ is any real number, to this series term by term. It is Important to note that the kth term of the series (9) is an ordinary function of the class $\mathfrak{M}_{2^k p}$. Moreover it is important that

$$\widetilde{S_{2^k}(f)} - \widetilde{S_{2^{k-1}}(f)} = 0 \quad \text{on } \Delta_{2^{k-1}} \quad (k = 1, 2, \ldots)$$

and that the inequalities

$$\|I_r[S_N(f) - S_{2N}(f)]\|_p \leqq \lambda_r N^{-r} \|S_N(f) - S_{2N}(f)\|_p, \tag{10}$$

$$\|I_r S_1(f)\| \leqq \lambda_r \|S_1(f)\|_p \tag{11}$$

hold for any real r and function f regular in the sense of L_p.

Reasoning as in 8.9—8.9.2, where we need to replace $\varrho_l(f)$ throughout by $S_l(f)$, we may obtain the following theorem on the basis of what has been said.

8.10.1. Theorem. *Suppose that* $1 < p < \infty$, $1 \leqq \theta \leqq \infty$ *and* r *is any real number. Then* $f \in B^r_{p\theta}$ $(B^r_{p\infty} = H^r_p)$ *if and only if* f *is regular in the sense of* L_p, *and its series (converging to it weakly)*

$$f = \sum_{k=0}^{\infty} \beta_s,$$

$$\beta_0 = S_{2^0}(f), \quad \beta_s = S_{2^s}(f) - S_{2^{s-1}}(f) \quad (s = 1, 2, \ldots),$$

is such that

$$\|f\|_{B^r_{p\theta}} = \left(\sum_{s=0}^{\infty} 2^{sr\theta}\|\beta_s\|_p^\theta\right)^{1/\theta} < \infty,$$

$$\|f\|_{H^r_p} = \|f\|_{B^r_{p\infty}} = \sup_s 2^{sr}\|\beta_s\|_p < \infty$$

(see the Remark in 8.9).

We shall prove a lemma complementing the results of 1.5.6 (in the same notations).

8.10.2. Lemma. *Suppose that f is a generalized function regular in the sense of $L_p(1 < p < \infty)$ for which*

(1) $$\left\|\left\{\sum_{\boldsymbol{k}} |\delta_{\boldsymbol{k}}(f)|\right\}^{1/2}\right\|_p < \infty.$$

Then $f \in L_p$.

Proof. For any number $N > 0$ we define a set of integer-valued vectors $\boldsymbol{k}$:

$$\Omega_N = \{\boldsymbol{k}\colon |k_j| < N\}$$

and the sum

$$f_N = \sum_{\boldsymbol{k}\in\Omega_N} \delta_{\boldsymbol{k}}, \ \delta_{\boldsymbol{k}} = \delta_{\boldsymbol{k}}(f).$$

Since f is regular, then $f_N \to f$ weakly as $N \to \infty$. On the other hand, in view of (1), f_N converges in the sense of L_p: as $N, N' \to \infty$ with $N < N'$, we get

$$\|f_{N'} - f_N\|_p < c\left\|\left(\sum_{\boldsymbol{k}\in\Omega_{N'}-\Omega_N} |\delta_{\boldsymbol{k}}|^2\right)^{1/2}\right\|_p \to 0.$$

Therefore f_N tends to f in the sense of L_p, and $f \in L_p$.

Analogously one proves the following theorem (for notations see 1.5.6.1).

8.10.3. Theorem. *If the generalized function $f(x)$ of one variable is regular in the sense of $L_p(-\infty, \infty)$, $1 < p < \infty$, and if*

(1) $$\left\|\left\{\sum_{l>0} \beta_l(f)^2\right\}^{1/2}\right\|_p < \infty,$$

then it belongs to L_p.

8.10.4. *Example.* Below we present an example of a function $g(x) \in L_p(-\infty, \infty) = L_p(2 < p < \infty)$ of entire exponential type 1, whose Fourier transform is a generalized function which is not summable over any interval $(a, b) \subset (-1,1)$.

Put

(1) $$\psi_k(x) = \begin{cases} a_k \left(2^k - \frac{1}{2} < |x| < 2^k + \frac{1}{2}\right), \\ 0 \quad \text{for the remaining } k = 1, 2, \ldots, \end{cases}$$

where the numbers $a_k > 0$ are such that

(2) $$\sum_1^\infty a_k^2 = \infty, \sum_1^\infty a_k^p < \infty \quad (2 < p < \infty).$$

Further put $f_N = \sum_1^N \psi_k$ and

(3) $$f(x) = \sum_1^\infty \psi_k(x).$$

The series (3) obviously converges in the sense of $L_p = L_p(-\infty, \infty)$, and accordingly in the sense of S' as well, and also $f \in L_p \subset S'$ and $f \in L_p \subset S'$ and $\|f\|_p = (2 \sum |a_k|^p)^{1/p} < \infty$. Suppose further that

$$\lambda_k(x) = \alpha_k \sqrt{\frac{2}{\pi}} \int_{2^k - \frac{1}{2}}^{2^k + \frac{1}{2}} \frac{\sin xy}{y} \, dy \quad (k = 1, 2, \ldots).$$

It is easy to verify (see 1.5.7 (10)), that

$$\lambda_k'(x) = 2\alpha_k \sqrt{\frac{2}{\pi}} \cos 2^k x \, \frac{\sin \frac{x}{2}}{x}$$

$$= \alpha_k \sqrt{\frac{2}{\pi}} \left(e^{i2^k x} + e^{-i2^k x}\right) \frac{\sin \frac{x}{2}}{x} = \tilde{\psi}_k(x).$$

Put further

$$F_N(x) = \sum_1^N \lambda_k(x) = \frac{1}{2} \sqrt{\frac{2}{\pi}} \int_{-\infty}^{\infty} \sum_1^N \psi_k(y) \frac{\sin xy}{y} \, dy,$$

$$F(x) = \sum_1^\infty \lambda_k(x).$$

Then

$$(4) \qquad |F_N(x)| \ll \int_{-\infty}^{\infty} f(y) \left|\frac{\sin xy}{y}\right| dy$$

$$\ll \|f\|_p \left(\int_{-\infty}^{\infty} \left|\frac{\sin xy}{y}\right|^q dy \right)^{1/q} \ll \|f\|_p |x|^{1-\frac{1}{q}},$$

where the constants entering into these inequalities do not depend on N, $\|f\|_p$ and x. Therefore

$$(5) \qquad |F(x) - F_N(x) \leq c \left\| \sum_N^{\infty} \psi_k \right\|_p |x|^{1-\frac{1}{q}}$$

and $F_N(x) \to F(x)$ uniformly on any finite segment. But then $F(x)$ is a continuous function. From (4) and (5) it easily follows also that $F \in S'$ and $F_N \to F(S')$.

In such a case

$$F' = \sum_1^{\infty} \lambda_k' = \sum_1^{\infty} \tilde{\psi}_k = \tilde{f},$$

where all the operations (differentiation, summation and Fourier transformation) are understood in the sense of S'.

It is possible to prove (for the proof see the book of Zygmund [2], Volume II, Ch. XV, at the end of subsection 3.14) that the function $F(x)$ (considered as an ordinary function) does not have a derivative almost everywhere. But then the generalized derivative F' on any interval (a, b) is not a summable function. In other words, for any interval (a, b) there does not exist any function $\alpha(x)$ summable on that interval and such that

$$(6) \qquad (F', \varphi) = \int_a^b \alpha(x)\, \varphi(x) dx$$

for all functions $\varphi \in S$ with support on (a, b). Indeed, if there were such a function, then, integrating the right side of (6) by parts, we would obtain the equation

$$\int_a^b F(x)\, \varphi'(x)\, dx = \int_a^b \int_a^x \alpha(t)\, dt \varphi'(x)\, dx,$$

i.e.

$$\int_a^b \psi(x) \varphi'(x)\, dx = 0, \quad \psi(x) = F(x) - \int_a^x \alpha(t)\, dt,$$

for any function $\varphi \in S$ with support on (a, b). But then $\psi(x) \equiv C$ is a constant and F would be differentiable almost everywhere on (a, b). The fact that ψ is a constant function may be proved as follows. If this were not so, then it would be possible to choose a constant c_1 such that the function $\lambda(x) = \psi(x) + c_1$ would take on values of different signs at some two points. Suppose for definiteness that $a < x_1 < x_2 < b$ and $\lambda(x_1) < 0 < \lambda(x_2)$. For the functions φ in question obviously

$$\int_a^b \lambda(x)\, \varphi'(x) dx = 0,$$

because $\int_a^b \varphi'(x) dx = 0$. We select $\delta > 0$ sufficiently small that $\lambda(x) < 0$ on $(x_1 - \delta, x_1 + \delta)$ and $\lambda(x) > 0$ on $(x_2 - \delta, x_2 + \delta)$. We suppose that $\omega(x)$ is a function continuous on (a, b) equal to zero for $x < x_1 - \frac{\delta}{2}$ and $x > x_2 + \frac{\delta}{2}$ and such that $\omega'(x) = -1$ on $\left(x_1 - \frac{\delta}{2}, x_1 + \frac{\delta}{2}\right)$, $\omega'(x) = 0$ on $\left(x_1 + \frac{\delta}{2}, x_2 - \frac{\delta}{2}\right)$, and $\omega'(x) = 1$ on $\left(x_2 - \frac{\delta}{2}, x_2 + \frac{\delta}{2}\right)$. Its ε-average $\omega_\varepsilon(x)$ belongs, obviously, to the class of functions φ considered here, and therefore

$$0 = \int_a^b \lambda(x)\, \omega_\varepsilon'(x)\, dx \to \int_a^b \lambda(x)\, \omega'(x)\, dx > 0 \quad (\varepsilon \to 0).$$

We have arrived at a contradiction.

Put

$$g(x) = \widehat{(1)_\Delta \tilde{f}} = \frac{1}{\pi} \int D_1(x - t)\, f(t)\, dt, \quad \Delta = \{|x| < 1\}.$$

Since $f \in L_p$, then the function $g \in L_p$ is also an entire function of exponential type 1. Its Fourier transform

$$\tilde{g} = (1)_\Delta \tilde{f}$$

is a generalized function equal to $\tilde{f}$ on Δ, i. e. $(\tilde{g}, \varphi) = (\tilde{f}, \varphi)$ for all $\varphi \in S$ with support in Δ. But then $\tilde{f}$, and accordingly $\tilde{g}$ as well, are not representable on any interval $(a, b) \subset \Delta$ by a summable function.

Chapter 9

The Liouville Classes L

9.1. Introduction

We shall denote the Liouville classes by $L_p^r(\mathbb{R}_n)$ ($r \geqq 0$, $L_p^0(\mathbb{R}_n) = L_p(\mathbb{R}_n)$) in the isotropic case and by $L_p^{\boldsymbol{r}}(\mathbb{R}_n)$ in the anisotropic case. For integer r, $\boldsymbol{r}$ they coincide with the Sobolev classes W:

$$W_p^r(\mathbb{R}_n) = L_p^r(\mathbb{R}_n) \quad (r = 0, 1, \ldots),$$

$$W_p^{\boldsymbol{r}}(\mathbb{R}_n) = L_p^{\boldsymbol{r}}(\mathbb{R}_n) \quad (r_j = 0, 1, 2, \ldots; j = 1, \ldots, n).$$

Generalizations to the case when p has a vectorial character are possible.

The classes L^r, $L^{\boldsymbol{r}}$ for fractional r, $\boldsymbol{r}$ are the most natural extensions of the classes W^r, $W^{\boldsymbol{r}}$.

For orientation we will already note here their basic properties.

Functions $F \in L_p^r(\mathbb{R}_n)$ are defined in the form of the integrals already well-known to the reader:

$$(1) \qquad F(\boldsymbol{x}) = I_r f = \frac{1}{(2\pi)^{n/2}} \int G_r(|\boldsymbol{x} - \boldsymbol{u}|)\, f(\boldsymbol{u})\, d\boldsymbol{u},\ f \in L_p(\mathbb{R}_n),$$

where (see 8.1) G_r is the Bessel-MacDonald kernel. If r is a natural number, then (see 8.2) F runs through the class $W_p^r(\mathbb{R}_n)$ when f runs through $L_p(\mathbb{R}_n)$. We have in addition the isomorphism $I_r L_p(\mathbb{R}_n) = W_p^r(\mathbb{R}_n)$. For fractional r equation (1) is taken as the definition of the class $L_p^r(\mathbb{R}_n)$ (at least in this book; see 9.2.3), i.e. we assume that $F \in L_p^r(\mathbb{R}_n)$ if and only if $F = I_r f$, where $f \in L_p(\mathbb{R}_n)$, and we put

$$\|F\|_{L_p^r(\mathbb{R}_n)} = \|f\|_{L_p(\mathbb{R}_n)}.$$

For any $r > 0$, $L_p^r \subset H_p^r$. Moreover, "up to arbitrarily small ε", the classes L_p^r, as well as the B_p^r, coincide with the H_p^r, and indeed

$$H_p^{r+\varepsilon} \to L_p^r \to H_p^r.$$

From these imbeddings it follows that in any case, "up to arbitrarily small ε", the same imbedding theorems are valid as are valid for the classes H_p^r, since for example

$$L_p^r(\mathbb{R}_n) \to H_p^r(\mathbb{R}_n) \to H_q^{r-\frac{n}{p}+\frac{m}{q}}(\mathbb{R}_m) \to L_q^{r-\frac{n}{p}+\frac{m}{q}-\varepsilon}(\mathbb{R}_m).$$

The following imbedding holds (see 9.3, $B_p^r = B_{pp}^r$)

$$B_p^r \to L_p^r \quad (1 \leqq p \leqq 2), L_p^r \to B_p^r \quad (2 \leqq p \leqq \infty). \tag{2}$$

The classes B and L coincide only if $p = 2$ ($B_2^r = L_2^r$). If $p \neq 2$ they differ essentially from one another.

The classes L_p^r are unified by a single integral representation in terms of functions $f \in L_p$. The classes B_p^r are unified by the same representation, but in terms of functions $f \in B_p^0$, where B_p^0 is a class essentially distinct from L_p. For $p > 2$ it contains the generalized functions (see 8.9.1).

The family of classes L_p^r is remarkable further for the fact that it is closed with relation to those imbedding theorems which carry out a transition from one metric to another. Thus, there is the following imbeding of different metrics:

$$L_p^r(\mathbb{R}_n) \to L_q^\varrho(\mathbb{R}_n) \tag{3}$$

$$\left(\varrho = r - n\left(\frac{1}{p} - \frac{1}{q}\right) \geqq 0, \quad 1 < p < q < \infty\right).$$

Another more general case is the imbedding

$$L_p^r(\mathbb{R}_n) \to L_q^\varrho(\mathbb{R}_m) \tag{4}$$

$$\left(\varrho = r - \frac{n}{p} + \frac{m}{q} \geqq 0, \quad 1 < p < q < \infty\right),$$

where along with the change of dimension one carries out a passage from one metric to another. As to imbedding theorems, where the number of dimensions changes without a change in the metric, then the coore-sponding direct theorem may be expressed as follows $\left(1 \leqq m < n,\ 1 < p < \infty,\ \varrho = r - \frac{n-m}{p} > 0\right)$:

$$L_p^r(\mathbb{R}_n) \to B_p^\varrho(\mathbb{R}_m), \tag{5}$$

and the inverse as follows:

$$B_p^\varrho \mathbb{R}(_m) \to L_p^r(\mathbb{R}_n). \tag{6}$$

On the basis of what was said above B may be replaced by L for $p = 2$ in (5) and (6), and moreover in (5), if $1 < p < 2$, and in (6), if $2 < p < \infty$. In the remaining cases such a replacement is not valid. Thus, the theorem for imbedding of different dimensions is not in general closed relative to the classes L.

We have the following situation:

$$B_p^r(\mathbb{R}_n) \to L_p^r(\mathbb{R}_n) \to B_p^\varrho(\mathbb{R}_m) \to B_p^r(\mathbb{R}_n) \quad (1 < p \leqq 2),$$

$$L_p^r(\mathbb{R}_n) \to B_p^r(\mathbb{R}_n) \to B_p^\varrho(\mathbb{R}_m) \to L_p^r(\mathbb{R}_n) \quad (2 \leqq p < \infty),$$

which shows that the two distinct classes $B_p^r(\mathbb{R}_n)$ and $L_p^r(\mathbb{R}_n)$ of functions, defined on $\mathbb{R}_n$, yield one and the same set of traces on $\mathbb{R}_m$(the class $B_p^\varrho(\mathbb{R}_m)$).

One should note that the imbedding

$$B_p^r(\mathbb{R}_n) \rightleftharpoons B_p^\varrho(\mathbb{R}_m)$$

for the indicated m, n, p, r, ϱ may be obtained as a consequence of the imbeddings (5) and (6). Indeed,

$$B_p^r(\mathbb{R}_n) \rightleftharpoons L_p^{r+\frac{1}{p}}(\mathbb{R}_{n+1}) \rightleftharpoons B_p^\varrho(\mathbb{R}_m).$$

The facts which we have presented here are characteristic. In the anisotropic case the similar facts hold. They will also be proved in this chapter.

9.2. Definitions and Basic Properties of the Classes L_p^r and $L_p^{\boldsymbol{r}}$

Suppose that $1 \leqq m \leqq n$, $\boldsymbol{x} = (\boldsymbol{u}, \boldsymbol{y})$, $\boldsymbol{u} = (x_1, \ldots, x_m) \in \mathbb{R}_m$, $\boldsymbol{y} = (x_{m+1}, \ldots, x_n) \in \mathbb{R}_{n-m}$. For functions $\varphi(\boldsymbol{x}) = \varphi(\boldsymbol{u}, \boldsymbol{y})$ of the basic class S we shall denote their Fourier transforms (direct and inverse) relative to the variable $\boldsymbol{u}$ by $\tilde{\varphi}^{\boldsymbol{u}}$, $\hat{\varphi}^{\boldsymbol{u}}$($\tilde{\varphi}^{\boldsymbol{u}} = \tilde{\varphi}$, $\hat{\varphi}^{\boldsymbol{u}} = \hat{\varphi}$ for $m = n$). For example,

$$(1) \qquad \tilde{\varphi}^{\boldsymbol{u}}(\boldsymbol{u}, \boldsymbol{y}) = \frac{1}{(2\pi)^{m/2}} \int \varphi(\boldsymbol{t}, \boldsymbol{y}) e^{-i\boldsymbol{t}\boldsymbol{u}}\, d\boldsymbol{t}.$$

The operations $\tilde{\varphi}^{\boldsymbol{u}}$, $\hat{\varphi}^{\boldsymbol{u}}$ map S into S and are weakly continuous (see further 9.2.1), so that for arbitrary generalized functions $f \in S'$, defined on $\mathbb{R}_n$, the corresponding Fourier transforms $\tilde{f}^{\boldsymbol{u}}$, $\hat{f}^{\boldsymbol{u}}$ are correctly defined by the functionals

$$(\tilde{f}^{\boldsymbol{u}}, \varphi) = (f, \tilde{\varphi}^{\boldsymbol{u}}),$$

$$(\hat{f}^{\boldsymbol{u}}, \varphi) = (f, \hat{\varphi}^{\boldsymbol{u}}).$$

If $\lambda(\boldsymbol{u})$ is a C^∞ function of polynomial growth, depending only on $\boldsymbol{u}$, then for $f \in S'$

$$\widehat{\lambda \tilde{f}^{\boldsymbol{u}\boldsymbol{u}}} = \widehat{\lambda \tilde{f}}, \quad \widetilde{\lambda \tilde{f}^{\boldsymbol{u}\boldsymbol{u}}} = \widetilde{\lambda f},$$

which follows directly from the validity of these equations for $\varphi \in S$.

We introduce the operation

$$F = I_{\boldsymbol{u}r} f = \widehat{(1 + |\boldsymbol{u}|^2)^{-r/2} \tilde{f}^{\boldsymbol{u}\boldsymbol{u}}} = \widehat{(1 + |\boldsymbol{u}|^2)^{-r/2} \tilde{f}} \tag{2}$$

($|\boldsymbol{u}|^2 = \sum_1^m u_j^2$, $I_{\boldsymbol{u}r} = I_r$ for $m = n$, $f \in S'$), corresponding to a real number r, mapping S' onto S' in a $1-1$ way. For $m = 1$, when $\mathbb{R}_m = \mathbb{R}_{x_j}$ is the x_j coordinate axis, we will denote it further by $I_{x_j r}$.

For functions $f \in L_p = L_p(\mathbb{R}_n)$ $(1 \leqq p \leqq \infty)$ this operation for $r > 0$ reduces to the convolution

$$F = I_{\boldsymbol{u}r} f = \frac{1}{(2\pi)^{m/2}} \int_{\mathbb{R}_m} G_r(|\boldsymbol{u} - \boldsymbol{t}|_m)\, f(\boldsymbol{t}, \boldsymbol{y})\, d\boldsymbol{t} \left(|\boldsymbol{t}|_m^2 = \sum_1^m t_j^2 \right), \tag{3}$$

where G_r is the Bessel-MacDonald kernel. This is proved as follows.

For $f \in S'$ one has the equations

$$\tilde{f}(\boldsymbol{x}) = \hat{f}(-\boldsymbol{x}) = \widehat{f(-\boldsymbol{x})},$$

which follow using the usual "transportations" from the fact that they are evidently true for any $\varphi \in S$. Further, if $\tilde{\Lambda} \in L$ and $f \in L_p$, then

$$\widetilde{\Lambda f} = \widehat{\Lambda \tilde{f}(-\boldsymbol{u})}\,(-\boldsymbol{x}) = \frac{1}{(2\pi)^{n/2}} \int \hat{\Lambda}(-\boldsymbol{x} - \boldsymbol{u})\, f(-\boldsymbol{u})\, d\boldsymbol{u}.$$

In particular if $\hat{\Lambda}(\boldsymbol{u}) = \hat{\Lambda}(-\boldsymbol{u})$, then $\widehat{\Lambda \tilde{f}} = \widehat{\Lambda \hat{f}}$. Therefore for $f \in L_p$, $\varphi \in S$, recalling that $G_r(|\boldsymbol{u}|_m) = G_r(|-\boldsymbol{u}|_m \in L_p(\mathbb{R}_m)$, we get

$$(I_{\boldsymbol{u}r} f, \varphi) = \left(f, \widetilde{(1 + |\boldsymbol{u}|^2)^{-r/2} \hat{\varphi}}\right)$$

$$= \left(f, \widehat{(1 + |\boldsymbol{u}|^2)^{-r/2} \tilde{\varphi}}\right) = \left(f, \widehat{(1 + |\boldsymbol{u}|^2)^{-r/2} \tilde{\varphi}^{\boldsymbol{u}}}\right)$$

$$= \frac{1}{(2\pi)^{m/2}} \int f(\boldsymbol{u}, \boldsymbol{y})\, d\boldsymbol{u}\, d\boldsymbol{y} \int G_r(|\boldsymbol{u} - \boldsymbol{t}|_m)\, \varphi(\boldsymbol{t}, \boldsymbol{y})\, d\boldsymbol{t}$$

$$= \frac{1}{(2\pi)^{m/2}} \int \varphi(\boldsymbol{x})\, d\boldsymbol{x} \int G_r(|\boldsymbol{t} - \boldsymbol{u}|_m)\, f(\boldsymbol{u}, \boldsymbol{y})\, d\boldsymbol{u} \quad (d\boldsymbol{x} = d\boldsymbol{t}\, d\boldsymbol{y}),$$

which proves (3).

Now we introduce the functional classes $L^r_{up} = L^r_{up}(\mathbb{R}_n)$,

$$L^r_{x_ip} = L^r_{x_ip}(\mathbb{R}_n),\ L^{\boldsymbol{r}}_{up} = L^{\boldsymbol{r}}_{up}(\mathbb{R}_n),\quad \boldsymbol{r} = (r_1, \ldots, r_m).$$

By definition *the function $F \in S'$ belongs to*

$$L^r_{up} = L^r_{up}(\mathbb{R}_n),\ L^r_{x_ip} = L^r_{x_ip}(\mathbb{R}_n),\ 1 \leqq p \leqq \infty,\ -\infty < r < \infty,$$

if it is representable respectively in the form

$$F = I_{ur}f,\quad F = I_{x_ir}f,$$

where $f \in L_p$. In addition we introduce norms $\|F\|_{L^r_{up}} = \|f\|_p$, in particular $\|F\|_{L^r_{x_ip}} = \|f\|_p$, by carring over the norms through the isomorphisms

$$I_{ur}(L_p) = L^r_{up},\quad I_{x_ir}(L_p) = L^r_{x_ip},$$

realized by the operations I_{ur}, I_{x_ir}. The class $L^{\boldsymbol{r}}_{up} = L^{\boldsymbol{r}}_{up}(\mathbb{R}_n)$ corresponding to any real vector $\boldsymbol{r}$ is defined as the intersection

$$L^{\boldsymbol{r}}_{up} = \bigcap_{j=1}^{m} L^{r_j}_{x_jp}$$

with the norm

$$\|f\|_{L^{\boldsymbol{r}}_{up}} = \sum_{j=1}^{m} \|f\|_{L^{r_j}_{x_jp}}.$$

In view of what was indicated above the class $L^{\boldsymbol{r}}_{up}$ may be defined further as the class of functions representable for almost all $\boldsymbol{y}$ in the form of the integral (3), where $f(\boldsymbol{x}) = f(\boldsymbol{u}, \boldsymbol{y}) \in L_p(\mathbb{R}_n)$.

Under the condition $1 < p < \infty$

$$L^r_{up} = L^{r,\ldots,r}_{up} \quad (r \geqq 0), \tag{4}$$

$$L^r_{up} = W^r_{up} = W^{r,\ldots,r}_{up} \quad (r = 1, 2, \ldots), \tag{5}$$

$$L^{\boldsymbol{r}}_{up} = W^{\boldsymbol{r}}_{up}\ (\boldsymbol{r} = (r_1, \ldots, r_m)), \tag{6}$$

the r_i being nonnegative integers. Equations (5) and (6) show that the classes L^r_{up}, $L^{\boldsymbol{r}}_{up}$ may be considered as the extension to arbitrary real r, $\boldsymbol{r}$ of the Sobolev classes W^r_{up}, $W^{\boldsymbol{r}}_{up}$. But we need to keep in mind that functions of the classes L^r_{up}, $L^{\boldsymbol{r}}_{up}$ have been defined by us on the entire space $\mathbb{R}_n$, while the functions of the Sobolev classes may be given on arbitrary open regions $g \subset \mathbb{R}_n$.

The first equation (5) for $m = n$ was proved in 8.2. If now $m < n$, suppose for the time being that $f \in S$ (the class of basic functions).

Then $f \in W_p^r(\mathbb{R}_m)$ for any $\boldsymbol{y}$. It is also obvious that $f \in L_p^r(\mathbb{R}_m)$ for any $\boldsymbol{y}$. After all, the function $f(\boldsymbol{u}, \boldsymbol{y})$ relative to $\boldsymbol{u}$ belongs to $S = S(\mathbb{R}_m)$, and the operation I_r (relative to $\boldsymbol{u}$) maps it into a function of the class $S(\mathbb{R}_m)$ and thus into $L_p(\mathbb{R}_m)$. Therefore, in view of what was already proved in 8.2,

$$c_1\|f\|_{W_p^r(\mathbb{R}_m)} \leqq \|I_{\boldsymbol{u}(-r)}f\|_{L_p(\mathbb{R}_m)} \leqq c_2\|f\|_{W_p^r(\mathbb{R}_m)} \tag{7}$$

where c_1 and c_2 do not depend on f and $\boldsymbol{y}$.

Raising these inequalities to the pth power, applying elementary inequalities[1], integrating with respect to $\boldsymbol{y}$ and applying again elementary inequalities[1], we arrive at the inequalities

$$c'\|f\|_{W_{\boldsymbol{u}p}^r} \leqq \|I_{\boldsymbol{u}(-r)}f\|_{L_p} \leqq c''\|f\|_{W_{\boldsymbol{u}p}^r}, \tag{8}$$

so far for functions $f \in S$.

If now $f \in W_p^r(\mathbb{R}_n)$, then we define a sequence of functions f_l, $l = 1, 2, \ldots$, such that $\|f_l - f\|_{W_p^r(\mathbb{R}_n)} \to 0$ $(l \to \infty)$.

It follows from (8) that

$$c'\|f_k - f_l\|_{W_{\boldsymbol{u}p}^r} \leqq \|\varphi_k - \varphi_l\|_{L_p}$$
$$\leqq c''\|f_k - f_l\|_{W_{\boldsymbol{u}p}^r} \to 0 \; (k, l \to \infty) \quad (\varphi_k = I_{\boldsymbol{u}(-r)}f_k),$$

and then in view of the completeness of $W_{\boldsymbol{u}p}^r$ and the fact (see (3)) that

$$f_l(\boldsymbol{u}, \boldsymbol{y}) = \frac{1}{(2\pi)^{m/2}} \int G_r(|\boldsymbol{u} - \boldsymbol{t}|_m)\, \varphi_l(\boldsymbol{t}, \boldsymbol{y}) d\boldsymbol{t},$$

where $G_r(|\boldsymbol{u}|_m) \in L_p(\mathbb{R}_m)$, the second inequality (8) holds, where

$$\|I_{\boldsymbol{u}(-r)}\|_{L_p} = \|f\|_{L_{\boldsymbol{u}p}^r}.$$

If now $f \in L_{\boldsymbol{u}p}^r$ and $I_{\boldsymbol{u}(-r)}f = \varphi$, then we may choose a sequence of functions φ_l such that $\|\varphi_l - \varphi\|_{L_p} \to 0$, so that in view of the first inequality in (8) and the completeness of $W_{\boldsymbol{u}p}^r$ we obtain the first inequality in (8). Thus the first equation in (5) is proved.

From the first equation in (5), applied to each axis $\mathbb{R}_{x_j}$, $j = 1, \ldots, n$, (6) follows obviously.

We turn to the proof of (4). Suppose that $F \in L_{\boldsymbol{u}p}^{r,\ldots,r}$ $(r > 0)$. Then

$$\psi = \overbrace{\sum_{j=1}^{m} (1 + u_j^2)^{r/2}\, \tilde{F}} \in L_p, \; \|\psi\|_p \ll \|F\|_{L_{\boldsymbol{u}p}^{r,\ldots,r}}$$

---

[1] We have in mind the inequalities

$$c\,|\textstyle\sum a_k|^p \leqq \sum a_k^p \leqq c_1 |\sum a_k|^p,$$

where the numbers $a_k > 0$ and c, c_1 depend only on p and the (finite) number of terms within the summation sign.

and, since the function

$$(1 + |\boldsymbol{u}|^2)^{r/2} \left(\sum_{j=1}^{m} (1 + u_j^2)^{r/2} \right)^{-1}$$

is a Marcinkiewicz multiplier (see 1.5.5, Example 9 for $r = r_j$, $\sigma = 1$, and also take into account here and below the Remark in 1.5.4.1), then

$$I_{\boldsymbol{u}(-r)} F = \widehat{(1 + |\boldsymbol{u}|^2)^{r/2} \tilde{F}} \in L_p,$$

$$\|F\|_{L^r_{\boldsymbol{u}p}} = \|I_{\boldsymbol{u}(-r)} F\|_p \ll \|\psi\|_p \ll \|F\|_{L^{r,\ldots,r}_{\boldsymbol{u}p}},$$

from which it follows that $L^{r,\ldots,r}_{\boldsymbol{u}p} \to L^r_{\boldsymbol{u}p}$. Conversely, if $F \in L^r_{\boldsymbol{u}p}$, then

$$f = \widehat{(1 + |\boldsymbol{u}|^2)^{r/2} \tilde{F}} \in L_p, \ \|f\|_p = \|F\|_{L^r_{\boldsymbol{u}p}},$$

and, since the function

$$(1 + |\boldsymbol{u}|^2)^{-r/2} (1 + u_j^2)^{r/2} \ (r \geqq 0)$$

is a Marcinkiewicz multiplier (see 1.5.5, Example 3), then

$$f_j = \widehat{(1 + u_j^2)^{r/2} \tilde{F}} \in L_p, \quad \|f_j\|_p \ll \|f\|_p = \|F\|_{L^r_{\boldsymbol{u}p}},$$

so that $L^r_{\boldsymbol{u}p} \to L^{r,\ldots,r}_{\boldsymbol{u}p}$, and (6) is proved.

From what has been said it follows that

$$W^r_{\boldsymbol{u}p} \leftrightarrows L^r_{\boldsymbol{u}p} \leftrightarrows L^{r,\ldots,r}_{\boldsymbol{u}p} \leftrightarrows W^{r,\ldots,r}_{\boldsymbol{u}p} \qquad (r = 0, 1, \ldots),$$

which implies the second equation (5). The nontrivial part of it is the imbedding $W^{r,\ldots,r}_{\boldsymbol{u}p} \to W^r_{\boldsymbol{u}p}$ $(1 < p < \infty)$, expressing the fact that if the function $f \in L_p$ and has nonmixed derivatives of order r relative to the variables $x_1, \ldots, x_m$ separately, belonging to L_p, then it has also arbitrary mixed derivatives of order r relative to the indicated variables, also belonging to L_p. There exist examples showing that this imbedding does not hold for $p = 1$ and $p = \infty$ (B. S. Mitjagin [1]).

9.2.1. The weak continuity of the operation $\tilde{\varphi}^{\boldsymbol{u}} (\varphi \in S)$ follows from the following considerations. We will write $\tilde{\varphi}$ instead of $\tilde{\varphi}^{\boldsymbol{u}}$. Suppose given a natural number l and an integer-valued nonnegative vector $\boldsymbol{k} = \boldsymbol{s} + \boldsymbol{\varrho}$, where $\boldsymbol{s} = (k_1, \ldots, k_m, 0, \ldots, 0)$, $\boldsymbol{\varrho} = (0, \ldots, 0, k_{m+1}, \ldots, k_n)$. Then obviously the derivative

$$D^{\boldsymbol{k}} \tilde{\varphi} = D^{\boldsymbol{s}} \widetilde{\varphi^{(\rho)}}.$$

We have (explanation below)

$$
\begin{aligned}
(1 + |\boldsymbol{x}|^2)^l \, |D^{(\boldsymbol{k})}\tilde{\varphi}| &\leqq (1 + |\boldsymbol{y}|^2)^l \, (1 + |\boldsymbol{u}|^2)^l \, |D^{\boldsymbol{s}}\widetilde{\varphi^{(\varrho)}}| \\
&\leqq c(1 + |\boldsymbol{y}|^2)^l \sum_{(l',\boldsymbol{s}')\in\mathscr{E}_{l,\boldsymbol{s}}} \max_{\boldsymbol{u}} \, (1 + |\boldsymbol{u}|^2)^{l'} \, |D^{(\boldsymbol{s}')}\varphi^{(\varrho)}(\boldsymbol{u}, \boldsymbol{y})| \\
&= c(1 + |\boldsymbol{y}|^2)^l \sum (1 + |\boldsymbol{u}_0|^2)^{l'} \, |D^{(\boldsymbol{s}'+\varrho)}\varphi(\boldsymbol{u}_0, \boldsymbol{y})| \\
&\leqq c \sum (1 + |\boldsymbol{y}|^2 + |\boldsymbol{u}_0|^2)^{2l} \, |D^{(\boldsymbol{s}'+\varrho)}\varphi(\boldsymbol{u}_0, \boldsymbol{y})| \leqq c \sum \varkappa(2l, \boldsymbol{s}' + \varrho, \varphi).
\end{aligned} \tag{1}
$$

The second inequality follows from 1.5 (4). In the third term the constant c depends on l and $\boldsymbol{k}$, but not on φ and $\boldsymbol{y}$. The sum in the third and subsequent terms is extended over some finite set $\mathscr{E}_{l,\boldsymbol{s}}$, depending on l and $\boldsymbol{s}$, of pairs $(l', \boldsymbol{s}')$ of natural numbers l' and nonnegative $\boldsymbol{s}'$. In the fourth term $\boldsymbol{u}_0 \in \mathbb{R}_m$ indeed depends on the corresponding term in the sum and on $\boldsymbol{y}$; $\boldsymbol{u}_0$ is the point of maximum (relative to $\boldsymbol{u}$) of the corresponding term in the sum for fixed $\boldsymbol{y}$. The weak continuity of $\tilde{\varphi}^{\boldsymbol{u}}$ follows from the resulting inequalities (1).

9.2.2. Theorem on derivatives. *Suppose that* $1 < p < \infty$, $F \in L_p^{\boldsymbol{r}} = L_p^{\boldsymbol{r}}(\mathbb{R})$, $\mathbb{R} = \mathbb{R}_n$, $\boldsymbol{r} = (r_1, \ldots, r_n) > \boldsymbol{0}$ $(r_j > 0)$ *and that* $\boldsymbol{l} = (l_1, \ldots, l_n) \geqq 0$ *is a vector with integer entries, for which*

$$\varkappa = 1 - \sum_{j=1}^{n} \frac{l_j}{r_j} \geqq 0. \tag{1}$$

Suppose further that

$$\varrho = \varkappa \boldsymbol{r}. \tag{2}$$

Then the derivative

$$F^{(\boldsymbol{l})} = \widehat{(i\boldsymbol{x})^{\boldsymbol{l}}\hat{F}} \in L_p^{\varrho} \tag{3}$$

and

$$\|F^{(\boldsymbol{l})}\|_{L_p^{\varrho}} \leqq c\|F\|_{L_p^{\boldsymbol{r}}}. \tag{4}$$

Proof. By hypothesis $F \in L_p^{\boldsymbol{r}}$, so that

$$\psi = \widehat{\Lambda \hat{F}} \in L_p, \quad \Lambda = \sum_{j=1}^{n} (1 + x_j^2)^{r_j/2}, \quad \|\psi\|_p \leqq c\|F\|_{L_p^{\boldsymbol{r}}}. \tag{5}$$

In order to prove (3), (4), we have to establish that for any $s = 1, \ldots, n$

$$\widehat{(1 + x_s^2)^{\frac{\varkappa r_s}{2}} \widehat{F^{(\boldsymbol{l})}}} = \widehat{(1 + x_s^2)^{\varkappa r_s} (i\boldsymbol{x})^{\boldsymbol{l}} \hat{F}} \in L_p.$$

But this follows from (5), if one takes into account the fact that the function

$$(1 + x_s^2)^{\frac{\varkappa r_s}{2}} (i\boldsymbol{x})^{\boldsymbol{l}} \Lambda^{-1}$$

is a Marcinkiewicz multiplier (see 1.5.5, Example 6).

Theorem 9.2.2 just proved is in a certain sense analogous to Theorem 5.6.3 for B-classes. However Theorem 9.2.2 is true for $1 < p < \infty$ and $\varkappa \geqq 0$, while Theorem 5.6.3 is true for $1 \leqq p \leqq \infty$, but for $\varkappa > 0$.

Example. Suppose that f is a function defined on the disk $\sigma = \{\varrho^2 = x^2 + y^2 \leqq 1\}$ by the equations

$$(1) \qquad f = xy \ln \varrho^2 \ (\varrho > 0), \quad f = 0 \ (\varrho = 0)$$

and extended to the entire $\mathbb{R}_2$-plane in such a way that it along with its partial derivatives to the second order inclusive are bounded and continuous on the region $\varrho > {}^1/_2$ (see Theorem 3 in the remarks to 4.3.6).

It is easy to verify that f, $\frac{\partial f}{\partial x}$, $\frac{\partial f}{\partial y}$ are continuous and bounded, and $\frac{\partial^2 f}{\partial x^2}$ and $\frac{\partial^2 f}{\partial y^2}$ are bounded on $\mathbb{R}_2$, while $\frac{\partial^2 f}{\partial x \partial y}$ is continuous for $\varrho > 0$ but unbounded in the neighborhood of the origin.

This example shows that for $p = \infty$ Theorem 9.2.2 is in general not true.

9.2.3. Remark on derivatives of fractional order. In this book we operate with expressions of the form

$$(1) \qquad \widehat{(i\boldsymbol{x})^{\boldsymbol{\alpha}} \hat{f}} = f^{(\boldsymbol{\alpha})} \ (f \in S')$$

only in the case of integer vectors $\boldsymbol{\alpha}$. If $\boldsymbol{\alpha} \geqq \boldsymbol{0}$, then $f^{(\boldsymbol{\alpha})}$ is the derivative of $f \in S$ of order $\boldsymbol{\alpha}$. The function $(i\boldsymbol{x})^{\boldsymbol{\alpha}}$ for integer $\boldsymbol{\alpha}$ is C^∞ and of polynomial growth, so that the expression (1) correctly defines $f^{(\boldsymbol{\alpha})} \in S'$.

If the real number α is not an integer, then the function $(it)^\alpha$ $(-\infty < t < \infty)$ is not single-valued. However one may agree to understand by this expression a single-valued branch of this function

$$(it)^\alpha = |t|^\alpha \exp\left\{\frac{1}{2} \pi i \alpha \operatorname{sign} t\right\},$$

and then for a natural number α

$$(it)^\alpha = \underbrace{(it) \ldots (it)}_{\alpha \text{ times}}.$$

If further α, β are natural numbers, then

$$(it)^{\alpha+\beta} = (it)^{\alpha}(it)^{\beta}.$$

If now $\boldsymbol{\alpha} = (\alpha_1, \ldots, \alpha_n)$, $\boldsymbol{\beta} = (\beta_1, \ldots, \beta_n)$ are real vectors and $\boldsymbol{x} = (x_1, \ldots, x_n)$ is a real vector variable, then we put

$$(i\boldsymbol{x})^{\boldsymbol{\alpha}} = (ix_1)^{\alpha_1} \ldots (ix_n)^{\alpha_n},$$

and then the following equation is obviously satisfied:

$$(i\boldsymbol{x})^{\boldsymbol{\alpha}}(i\boldsymbol{x})^{\boldsymbol{\beta}} = (i\boldsymbol{x})^{\boldsymbol{\alpha}+\boldsymbol{\beta}}.$$

Now it is natural to define the derivative $f^{(\boldsymbol{\alpha})}$ of order $\boldsymbol{\alpha}$ for arbitrary, not necessarily integer, vectors $\boldsymbol{\alpha}$, using the expression (1). However here there arises a difficulty consisting in that for fractional $\boldsymbol{\alpha}$ the function $(i\boldsymbol{x})^{\boldsymbol{\alpha}}$ is not C^{∞}. It also is not a Marcinkiewicz multiplier, and thus is not applicable in the sense of the definitions considered in this book, even for $f \in L_p$. The way out of this situation is found in considering instead of S another class Ω of basic functions, consisting of functions orthogonal to polynomials (P. I. Lizorkin [5]).

Over Ω one defines a class of functionals (generalized functions) Ω'. In the terms of Ω' it makes sense to consider the expression $\widehat{(i\boldsymbol{x})^{\alpha}\tilde{f}}$ for fractional vectors $\boldsymbol{\alpha}$. In addition one proves that the class $L^r_{x_jp}$, where $r > 0$ is in general fractional, may be defined as consisting of the functions f for which the norm

$$\|f\|_{L_p} + \|\widehat{(ix_j)^r\tilde{f}}\|_{L_p}$$

is finite. This norm turns out to be equivalent to the norm $\|f\|_{L^r_{x_jp}}$ introduced by us earlier.

9.3. Interrelationships among Liouville and other Classes

We will suppose that $\mathbb{R} = \mathbb{R}_n$, $L = L(\mathbb{R})$, $H = H(\mathbb{R})$, ..., and

$$B^r_{pp} = B^r_p, \quad B^{\boldsymbol{r}}_{pp} = B^{\boldsymbol{r}}_p.$$

The following imbeddings are valid ($r \geqq 0$, $\boldsymbol{r} \geqq 0$):

(1) $$L^r_p \to H^r_p, \quad L^{\boldsymbol{r}}_p \to H^{\boldsymbol{r}}_p \quad (1 \leqq p \leqq \infty),$$

(2) $$B^r_p \to L^r_p, \quad B^{\boldsymbol{r}}_p \to L^{\boldsymbol{r}}_p \quad (1 \leqq p \leqq 2)^1,$$

(3) $$L^r_p \to B^r_p, \quad L^{\boldsymbol{r}}_p \to B^{\boldsymbol{r}}_p \; (2 \leqq p \leqq \infty, \; B^{\boldsymbol{r}}_{\infty} = B^{\boldsymbol{r}}_{\infty\infty} = H^{\boldsymbol{r}}_{\infty})^1.$$

[1] O. V. Besov [5] for integer r, $\boldsymbol{r}$, $1 < p < \infty$; P. I. Lizorkin [6] the general case, $1 < p < \infty$. See further the Remark to 9.3.

It follows from (2) and (3) that

(4) $$B_2^r = L_2^r, \quad B_2^{\mathbf{r}} = L_2^{\mathbf{r}}$$

and, in particular,

(5) $$B_2^r = W_2^r, \quad B_2^{\mathbf{r}} = W_2^{\mathbf{r}} \quad (r, r_j = 0, 1, \ldots).$$

Thus the value of the parameter $p = 2$ is exceptional—for it the corresponding B and L classes, and for natural r, $\mathbf{r}$, the class W as well, coincide.

We present proofs for (1)—(3), for the present for the case $n = 1$. In this case the imbeddings entering into each of the pairs (1), (2), or (3) respectively, coincide.

Suppose that $f \in L_p = L_p(\mathbb{R}_1)$ and that $\sigma_N(f)$ is its de la Vallée Poussin sum. Then

$$\|\sigma_{2^0}(f)_p \leqq M\|f\|_p,$$

$$\|\sigma_{2^k}(f) - \sigma_{2^{k-1}}(f)\|_p \leqq 2M\|f\|_p,$$

and this shows that (see 8.9.1 (1))

$$\|f\|_{B_p^0} = \sup_k \{\|\sigma_2^0(f)\|_p, \quad \|\sigma_2^k(f) - \sigma_2^{k-1}(f)\|_p\} \leqq 2M\|f\|_p,$$

i.e. $L_p \to H_p^0$. But since the operation I_r realizes the isomorphisms

$$I_r(L_p) = L_p^r, \quad I_r(H_p^0) = H_p^r,$$

then

$$L_p^r \to H_p^r.$$

Let us prove (3). The case $p = \infty$ has already been considered. Suppose that $2 \leqq p < \infty$ and that $f \in L_p = L_p(-\infty, \infty)$, so that it is regular in the sense of L_p.

Then, taking into account the fact that $\beta_k(f)$ has the same meaning as in 8.10.1, we obtain (explanation below)

$$\|f\|_p^p \gg \left\| \left\{ \sum_0^\infty \beta_k(f)^2 \right\}^{1/2} \right\|_p^p \geqq \left\| \left\{ \sum_0^\infty |\beta_k(f)|^p \right\}^{1/p} \right\|_p^p$$

$$= \int_{-\infty}^{\infty} \sum_0^\infty |\beta_k(f)|^p dx = \sum_0^\infty \|\beta_k(f)\|_p^p = \|f\|_{B_p^0}.$$

The first inequality follows from 1.5.6.1, the second from 3.3.3, and, finally, the last from Theorem 8.10.1. Accordingly, $L_p \to B_p^0$.

Now suppose under the same notations that $f \in B_p^0$, $1 < p \leqq 2$. Then (explanation below)

$$\|f\|_{B_p^0}^p = \sum_0^\infty \|\beta_k(f)\|_p^p = \int_{-\infty}^\infty \sum |\beta_k(f)|^p\,dx \gg \int \{\sum \beta_k(f)^2\}^{p/2} dx \gg \|f\|_p^p.$$

The first relation follows from 8.10.1, the third from 3.3.3, and the last from 1.5.6.1 Accordingly, $B_p^0 \to L_p$.

For $p = 1$ the reasoning is different. Suppose that $f \in B_1^0$. Then it is regular in the sense of L and is represented in the form of a de la Vallée Poussin sum 8.9 (3) weakly converging to it in the sense of S':

$$f = \sum_0^\infty q_s,$$

where $\sum_0^\infty \|q_s\|_L < \infty$. But then obviously $f \in L$ and

$$\|f\|_L \leqq \|f\|_{B_1^0}.$$

We have proved (1)—(3) for $n = 1$. But then the following imbeddings are also true:

$$L_{x_j p}^r(\mathbb{R}_n) \to H_{x_j p}^r(\mathbb{R}_n), \tag{6}$$

$$B_{x_j p}^r(\mathbb{R}_n) \to L_{x_j p}^r(\mathbb{R}_n) \quad (1 \leqq p \leqq 2), \tag{7}$$

$$L_{x_j p}^r(\mathbb{R}_n) \to B_{x_j p}^r(\mathbb{R}_n) \quad (2 \leqq p \leqq \infty). \tag{8}$$

Indeed, it is clear directly from the definitions of the H- and B-classes that if the function $F(\boldsymbol{x}) = F(x_1, \boldsymbol{y})$, $\boldsymbol{y} = (x_2, \ldots, x_n)$, belongs to $H_{x_1 p}^r(\mathbb{R}_n)$, $B_{x_1 p}^r(\mathbb{R}_n)$, then for almost all $\boldsymbol{y}$ it belongs as a function of x_1 respectively to $H_p^r(\mathbb{R}_{x_1})$, $B_p^r(\mathbb{R}_{x_1})$, where $\mathbb{R}_{x_1}$ is the x_1-axis. Analogously, if $F(\boldsymbol{x}) \in L_{x_1 p}(\mathbb{R}_n)$, then for almost all $\boldsymbol{y}$ it belongs to $L_p^r(\mathbb{R}_{x_1})$. This follows from the integral representation 9.2 (3) of functions of the class $L_{x_1 p}^r(\mathbb{R}_{x_1})$.

In addition, the following equations are valid up to an equivalence:

$$\|F\|_{\Lambda_{x_1 p}^r(\mathbb{R}_n)} = \left(\int \|F(x_1, \boldsymbol{y})\|_{\Lambda_p^r(\mathbb{R}_{x_1})}^p d\boldsymbol{y}\right)^{1/p}, \tag{9}$$

$$\|F\|_{L_{x_1 p}^r(\mathbb{R}_n)} = \left(\int \|f(x_1, \boldsymbol{y})\|_{L_p(\mathbb{R}_{x_1})}^p d\boldsymbol{y}\right)^{1/p}, \tag{10}$$

where in place of Λ we have to substitute H or B, and in (10) F and f are connected by equation 9.2 (3) for $m = 1$. The inequalities defining the imbeddings (6)—(8) then follow from the inequalities, already proved in the one-dimensional case, relating the norms in the integrals in (9), (10).

From (6)—(8), where we may already replace x_1 by x_j ($j = 1, \ldots, n$), the imbeddings (1)—(3) follow if one recalls, in the proof of the first imbeddings (1)—(3), that

$$H_p^r = H_p^{r,\ldots,r}\ (1 \leqq p \leqq \infty),\ B_p^r = B_p^{r,\ldots,r}\ (1 \leqq p \leqq \infty),$$

$$L_p^r = L_p^{r,\ldots,r}\ (1 < p < \infty).$$

9.4. Integral Representation of Anisotropic Classes

In this section we take up the study of the operation

$$F = \widehat{\Lambda_{\boldsymbol{r}} \tilde{f}} = I_{\boldsymbol{r}} f,$$

(1)
$$\Lambda = \Lambda_{\boldsymbol{r}} = \left\{ \sum_{j=1}^{n} (1 + x_j^2)^{r_j \sigma/2} \right\}^{-\frac{1}{\sigma}}$$

(2)
$$(\sigma > 0, \boldsymbol{r} = (r_1, \ldots, r_n) > \boldsymbol{0}),$$

depending on the positive vector $\boldsymbol{r}$ and the parameter σ. It is analogous to the operation I_r already studied for the case when r is a number, and in the one-dimensional case, they coincide for $r = r_1$ and $\sigma = 1$. In case $n > 1$, $r_1 = \ldots = r_n = r$, the operations I_r and $I_{\boldsymbol{r}}$ do not coincide even for $\sigma = 1$. However they have analogous properties, which, for example, is clear from the fact that the function

(3)
$$(1 + |\boldsymbol{x}|^2)^{r/2} \Lambda_r(\boldsymbol{x})$$

and the quantity inverse to it for any $\sigma > 0$ is a Marcinkiewicz multiplier (for $1 < p < \infty$, see 1.5.5, Example 9.8).

We shall write further

(4)
$$I_{-\boldsymbol{r}} F = f.$$

Since the multiplier $\Lambda_{\boldsymbol{r}}$ is a C^∞ function of polynomial growth, the same as the quantity inverse to it, then $I_{\boldsymbol{r}}$ transforms S' onto S' in a 1—1 way.

The operation $I_{\boldsymbol{r}}$ is remarkable for the fact that it realizes the isomorphism

(5)
$$L_p^{\boldsymbol{r}} = I_{\boldsymbol{r}}(L_p) \quad (1 < p < \infty).$$

Indeed, if $f \in L_p$, then

$$\|I_{-r_i} F\|_p \ll \|f\|_p,$$

which follows from the fact that the function

$$(1 + x_i^2)^{r_i/2} \Lambda_r(\boldsymbol{x}) \ (i = 1, \ldots, n) \tag{6}$$

is a Marcinkiewicz multiplier (see 1.5.5, Example 10). Therefore $F \in L_p^{\boldsymbol{r}}$ and

$$\|F\|_{L_p^{\boldsymbol{r}}} \ll \|f\|_p.$$

Conversely, if $F \in L_p^{\boldsymbol{r}}$, then

$$\|f\|_p \ll \|F\|_{L_p^{\boldsymbol{r}}}.$$

This follows from the fact that the function (see 1.5.5, Example 11)

$$\Lambda_{\boldsymbol{r}}^{-1}(\boldsymbol{x}) \left\{ \sum_{j=1}^{n} (1 + x_j^2)^{r_j/2} \right\}^{-1} \tag{7}$$

is a Marcinkiewicz multiplier.

Suppose that $\boldsymbol{r} > \boldsymbol{0}$, $\lambda, \delta > 0$. Then, as we have proved, we have the isomorphism

$$I_{(\lambda+\delta)\boldsymbol{r}}(L_p) = L_p^{(\lambda+\delta)\boldsymbol{r}} \ (1 < p < \infty). \tag{8}$$

It is remarkable that although the operation

$$I_{\lambda\boldsymbol{r}} I_{\delta\boldsymbol{r}}$$

in general differs from the operation $I_{(\lambda+\delta)\boldsymbol{r}}$, it nevertheless is equivalent in the sense that along with (8) one has the isomorphism

$$I_{\lambda\boldsymbol{r}} I_{\delta\boldsymbol{r}}(L_p) = L_p^{(\lambda+\delta)\boldsymbol{r}} \quad (1 < p < \infty).$$

This follows from the fact that the functions μ, μ^{-1}, considered in Example 12 of 1.5.5, are Marcinkiewicz multipliers.

9.4.1. Estimates of anisotropic kernels. Suppose given $\boldsymbol{r} = (r_1, \ldots, r_n) > 0$ and $\boldsymbol{l} = (l_1, \ldots, l_n)$. Suppose that σ is positive and large enough so that the inequalities

$$\sum_1^n \frac{l_j}{r_j} < \sigma - \sum_1^n \frac{1}{r_j} + \frac{1}{r_i} \quad (i = 1, \ldots, n)$$

are satisfied. Our immediate goal is to show that in such a case the Fourier transform (see 9.4 (2))

$$\hat{\Lambda} = \hat{\Lambda}_{\boldsymbol{r}}(x) = G_{\boldsymbol{r}}(\boldsymbol{x}) \tag{1}$$

is an ordinary function, having the „derivative"[1]

$$I_{-l}G_r = \widehat{\prod_1^n (1 + x_j^2)^{l_j/2} \Lambda}$$

(an ordinary function), satisfying the inequalities

$$(2) \qquad |I_{-l}G_r(\boldsymbol{x})| \leqq c \begin{cases} \left\{\sum_1^n |x_j|^{r_j\varkappa}\right\}^{-1} & \left(\varkappa = \sum_1^n \frac{1 + l_j}{r_j} - 1 > 0\right), \\ \ln\left(\frac{1}{|\boldsymbol{x}|} + 1\right) & (\varkappa = 0), \\ 1 & (\varkappa < 0), \end{cases}$$

$$(3) \qquad |I_{-l}G_r(\boldsymbol{x})| \leqq c e^{-c_1|\boldsymbol{x}|} \; (|\boldsymbol{x}| > 1),$$

where c_1 is positive and sufficiently small and σ enters only into the constant c, from which it follows that $G_r \in L$. This in particular shows that

$$(4) \qquad I_r f = \int G_r(\boldsymbol{x} - \boldsymbol{u})\, f(\boldsymbol{u})\, d\boldsymbol{u}$$

is the ordinary convolution for $f \in L_p$ $(1 \leqq p \leqq \infty)$.

9.4.2. Suppose that Ω denotes the complex plane with the cut $-\infty < x \leqq 0$ and $\varrho = \lambda + \mu i$ a complex number. In what follows we will suppose without explanations that z^ϱ is a single-valued branch of the many-valued function z^ϱ, defined on Ω and equal to $x^\varrho = x^\lambda e^{i\mu \ln x}$ on the ray $0 < x < \infty$. In other words, if $z = x + iy \in \Omega$, then we suppose always that $z^\varrho = |z|^\varrho e^{i\varrho \arg z}$, where $|\arg z| < \pi$.

Lemma 1. *Suppose that* $0 < \alpha \leqq 1$. *Then*

$$(1) \qquad |z^\alpha - A^\alpha| \geqq M\, |z - A|^\alpha \quad (z \in \Omega, \quad A \geqq 0)$$

where M *does not depend on* z *and* A.

Proof. First we consider the single-valued analytic function

$$(2) \qquad f(z) = \frac{z^\alpha - 1}{(z - 1)^\alpha}$$

---

[1] It is possible to show that this assertion is preserved if one replaces in it I_{-l} for integer l by the differential operation $D^l G_r = \widehat{(i\boldsymbol{x})^l \Lambda}$ (P. I. Lizorkin [10]).

on the region Ω^* of the complex plane with two cuts $-\infty < x \leqq 0$, $1 \leqq x < \infty$, equal to

$$f(x) = \frac{x^\alpha - 1}{(x-1)^\alpha}$$

on the upper side of the cut $1 \leqq x < \infty$.

In order to construct such a function, we need to suppose that the function z^α in the numerator in (2) is defined by the formula $z^\alpha = \varrho^\alpha e^{i\alpha\theta}$ ($z = \varrho e^{i\theta}$, $\varrho > 0$, $-\pi < \theta < \pi$); i.e. that z^α (in the numerator) is understood as a single-valued branch of z^α, defined on Ω, equal to x^α for $0 < x < \infty$. As to the function $(z-1)^\alpha$ in the denominator, it is understood in the sense $(z-1)^\alpha = r^\alpha e^{i\alpha\varphi}$ ($z - 1 = re^{i\varphi}$, $r > 0$, $0 \leqq \varphi \leqq 2\pi$).

The function $f(z)$ thus defined has a limit $\lim_{z\to\infty} f(z) = 1$. Moreover, it is bounded on both sides of both cuts. Thus, it is bounded on the whole boundary of Ω^*, and, by the maximum principle, is bounded on Ω^*:

$$M \geqq \left|\frac{z^\alpha - 1}{(z-1)^\alpha}\right| = \frac{|z^\alpha - 1|}{|(z-1)^\alpha|} = \frac{|z^\alpha - 1|}{|z-1|^\alpha},$$

and we have proved inequality (1) for $A = 1$ and all $z \in \Omega^*$. But then we have proved it for $z \in \Omega$ as well, because for a real $z = x > 1$ the inequality

$$x^\alpha - 1 \leqq (x-1)^\alpha \quad (0 < \alpha \leqq 1)$$

is very well known.

If now A is any positive number, then for $z \in \Omega$

$$|z^\alpha - A^\alpha| = A^\alpha \left|\left(\frac{z}{A}\right)^\alpha - 1\right| \leqq MA^\alpha \left|\frac{z}{A} - 1\right|^\alpha = M|z - A|^\alpha,$$

and we have proved (1).

Lemma 2. *Suppose that $\alpha \geqq 1$. Then for $z \in \Omega$ and any $A > 0$ we have*

$$|z^\alpha - A^\alpha| \leqq M|z - A|\,(A^{\alpha-1} + |z|^{\alpha-1}), \tag{3}$$

where M does not depend on z and A.

Proof. Indeed, join the points A and z by a segment c:

$$\zeta = A + t\,(z - A) \quad (0 \leqq t \leqq 1),$$

lying, obviously, in Ω. Then

$$z^\alpha - A^\alpha = \alpha \int_C z^{\alpha-1} dz = \alpha \int_0^1 [A + t(z - A)]^{\alpha-1}(z - A)\, dt,$$

from which (3) follows ($(a + b)^\beta \leqq c(a^\beta + b^\beta)$, $\beta > 0$, c does not depend on a and b).

9.4.3. We introduce the notations

$$(1) \qquad V = \sum_1^n (1 + u_j^2)^{\frac{r_j \sigma}{2}}, \quad U = \sum_1^n (1 + u_j^2)^{\frac{r_i}{2 r_n}} \quad (r_j > 0)$$

and note the inequalities

$$(2) \qquad U^{r_n \sigma} \leqq cV,$$

where c does not depend on V,

$$(3) \qquad |u_n| \leqq (1 + u_n^2)^{1/2} = (1 + u_n^2)^{\frac{r_n}{2 r_n}} \leqq U.$$

Here (2) follows from the fact that for $\beta > 0$ and any $x_j > 0$

$$\left(\sum_1^n x_j^\beta \right)^{1/\beta} \leqq c_\beta \sum_1^n x_j,$$

and c_β does not depend on x_j.

In the plane of the complex variable $w_n = u_n + iv_n$ we introduce a curve $L_{\boldsymbol{u}'}$:

$$(4) \qquad u_n + ikU \ (0 < k < 1, -\infty < u_n < \infty),$$

depending on the constant k and some parameter $\boldsymbol{u}' = (u_1, \ldots, u_{n-1})$. We introduce moreover a curve $L^*_{\boldsymbol{u}'}$. Suppose that

$$(5) \qquad B = k \left(\sum_1^{n-1} (1 + u_j^2)^{\frac{r_j}{2 r_n}} + 1 \right),$$

where k is a constant. If $B \leqq 1$, we will suppose that $L^*_{\boldsymbol{u}'} = L_{\boldsymbol{u}'}$. If on the other hand $B > 1$, i.e. when $L_{\boldsymbol{u}'}$ lies above the point i, put

$$L^*_{\boldsymbol{u}'} = L_{\boldsymbol{u}'} + l_{\boldsymbol{u}'},$$

where $l_{\boldsymbol{u}'}$ is the segment $[i, iB]$ passed through twice. More precisely, we suppose that the oriented curve $L^*_{\boldsymbol{u}'}$ obtained as follows: first the point of $L^*_{\boldsymbol{u}'}$ runs through the left piece of $L_{\boldsymbol{u}'}$ corresponding to the increase of u_n on the interval $(-\infty, 0)$. Then it goes down along the segment $l_{\boldsymbol{u}'}$ to i, goes around i, rises up to $L_{\boldsymbol{u}'}$ and goes off to $+\infty$ along the right piece of $L_{\boldsymbol{u}'}$.

We denote by $E_{\boldsymbol{u}'}$ the set of points w_n filling out the portion of the complex w_n plane between the real u_n axis and the curve $L^*_{\boldsymbol{u}'}$.

Since $\Lambda = V^{-1/\sigma}$ is a C^∞ function of polynomial growth, then it makes sense to say that $\hat{\Lambda} \in S'$. For small r_j the integral

$$\hat{\Lambda}(\boldsymbol{x}) = \frac{1}{(2\pi)^{n/2}} \int e^{i\boldsymbol{x}\boldsymbol{u}} \Lambda(\boldsymbol{u})\, d\boldsymbol{u}$$

already may not converge absolutely. Therefore the study of the function $\hat{\Lambda}$ will be carried out by the indirect method of introducing the auxiliary function

$$V^{-\varrho/\sigma} = \Lambda_{\varrho,\boldsymbol{r},\sigma} \quad (\varrho = \lambda + i\mu, \lambda > 0,\ \Lambda_{1,\boldsymbol{r},\sigma} = \Lambda_{\boldsymbol{r}} = \Lambda) \tag{6}$$

with the complex parameter ϱ. For sufficiently large λ it makes sense to write out directly in terms of the Lebesgue (absolutely converging) integral[1]

$$I_{-\boldsymbol{l}}\hat{\Lambda}_{\varrho,\boldsymbol{r},\sigma} = \frac{1}{(2\pi)^{n/2}} \int \frac{\prod_1^n (1+u_j^2)^{l_j/2} e^{i\boldsymbol{x}\boldsymbol{u}}\, d\boldsymbol{u}}{V^{\varrho/\sigma}} \tag{7}$$

$$= \frac{1}{(2\pi)^{n/2}} \int e^{i\boldsymbol{x}'\boldsymbol{u}'} d\boldsymbol{u}' \int \frac{\prod_1^n (1+u_j^2)^{l_j/2} e^{ix_n u_n}}{V^{\varrho/\sigma}}\, du_n$$

$$\big(\boldsymbol{x}' = (x_1, \ldots, x_{n-1}),\ \boldsymbol{u}' = (u_1, \ldots, u_{n-1})\big).$$

Along with (7) we will consider further for $x_n > 0$ the function

$$\mu_\varrho^{(\boldsymbol{l})}(\boldsymbol{x}) = \frac{1}{(2\pi)^{n/2}} \int e^{i\boldsymbol{x}'\boldsymbol{u}'} d\boldsymbol{u}' \int_{L_u^*} \frac{\prod_1^n (1+u_j^2)^{l_j/2} e^{ix_n u_n}}{V^{\varrho/\sigma}}\, du_n. \tag{8}$$

If $x_n < 0$, then $\mu_\varrho^{(\boldsymbol{l})}(\boldsymbol{x})$ is defined analogously, but as $L^*_{\boldsymbol{u}'}$ one chooses a curve in the complex plane, symmetric to it relative to the real axis. The consideration for that second integral, which we shall also denote by $\mu_\varrho^{(\boldsymbol{l})}(\boldsymbol{x})$, leads to analogous results.

---

[1] In the consideration of the operation $D^{\boldsymbol{l}}\check{\Lambda}_\varrho$ in (7), (8), the product $\prod_1^n (1+x_j)^{l_j/2}$ is replaced by $(i\boldsymbol{x})^{\boldsymbol{l}}$.

For $x_n > 0$ and real $\varrho > \varrho_0$, where ϱ_0 is sufficiently large, the inside integrals in (7) and (8) are equal to one another. Indeed, in (8) in V there enters the complex term

$$(1+z^2)^{\frac{r_n\sigma}{2}} = [1+(x+iy)^2]^{\frac{r_n\sigma}{2}} = (\xi+i\eta)^{\frac{r_n\sigma}{2}} \quad (z \in E_{\boldsymbol{u}'}),$$

$$\xi = 1 + x^2 - y^2, \quad \eta = 2xy.$$

The number $\xi + i\eta$ can belong to the cut $-\infty < \xi \leqq 0$ if and only if $x = 0$, $y^2 \geqq 1$, i.e. if

$$z = iy, \quad y^2 \geqq 1. \tag{9}$$

But points of the form (9) do not belong to $E_{\boldsymbol{u}'}$, which shows that V is a single-valued analytic function of w_n on $E_{\boldsymbol{u}'}$. In what follows (see 9.4.6 (7)) it will be proved that in addition for sufficiently small k

$$-\pi < \arg V < \pi \tag{10}$$

(if it is supposed *a priori* that $|\arg V| \leqq \pi$), which shows that when w_n runs through $E_{\boldsymbol{u}'}$ the point V belongs to Ω (a plane with a cut $-\infty < x \leqq 0$). But then $V^{\varrho/\sigma}$ (for real σ and complex ϱ) is also a single-valued analytic function. Thus, the function in the inside integrals (7) and (8) is a single-valued function in the region $E_{\boldsymbol{u}'}$. That they are equal follows from the fact that the integral along the segment c_ξ of points $\xi + i\eta$ $(0 \leqq \eta \leqq kU)$ tends to zero as $|\xi| \to \infty$:

$$\left|\int_{c_\xi} \frac{(1+u_n^2)^{l_n/2} e^{ix_n u_n}}{V^{\varrho/\sigma}}\, du_n\right| \leqq \int_0^{kV} \frac{e^{-x_n\eta}\, d\eta}{|1+(\xi+i\eta)^2|^{\frac{r_n\varrho - l_n}{2}}}$$

$$\leqq \frac{1}{(|\xi|-1)^{\frac{r_n\varrho - l_n}{2}}} \int_0^\infty e^{-x_n\eta}\, d\eta \to 0.$$

Similarly one proves the equality of the inside integrals in (7) and in the expression corresponding to (8) for $x_n < 0$ as well. This shows that

$$I_l \hat{\Lambda}_{\varrho,\boldsymbol{r},\sigma}(\boldsymbol{x}) = \mu_\varrho^{(\boldsymbol{l})}(\boldsymbol{x}) \quad (x_n \neq 0,\ \varrho > \varrho_0), \tag{11}$$

if ϱ_0 is a sufficiently large positive number.

In 9.4.6 we will obtain estimates for the function $\mu_\varrho^{(\boldsymbol{l})}(\boldsymbol{x})$.

On the basis of these estimates and the analytic properties of $I_{-l}\hat{\Lambda}_{\varrho,r,\sigma}$ and $\mu_{\varrho}^{(l)}$ one succeeds in proving (see 9.4.7) that the equation (11) in fact holds for all complex $\varrho = \lambda + i\mu$, in particular for $\varrho = 1$. Thus the generalized function $\hat{\Lambda}$ is summable on $\mathbb{R}_n$: $\hat{\Lambda}(\boldsymbol{x}) = \mu_1(\boldsymbol{x}) = \mu_1^{(0)}$. The estimates which will be obtained for $\mu_1^{(l)}(\boldsymbol{x})$ carry over directly to $I^{-l}\hat{\Lambda}^{(l)}(\boldsymbol{x})$, which leads to inequalities 9.4.1 (1), (2).

9.4.4. We begin by estimating the n-fold integral ($r_j/s > 0$, $\boldsymbol{l} = (l_1, \ldots, l_n) \geqq \boldsymbol{0}$, explanation below)

(1)
$$\int \frac{\prod_1^n (1+u_j^2)^{l_j/2}}{V^{\frac{s}{\sigma}}}\, d\boldsymbol{u} = \int \frac{\prod_1^n (1+u_j^2)^{l_j/2}\, d\boldsymbol{u}}{\left\{\sum_1^n (1+u_j^2)^{\frac{r_j\sigma}{2}}\right\}^{\frac{s}{\sigma}}}$$

$$= \int\limits_{|\boldsymbol{u}|<1} + \int\limits_{|\boldsymbol{u}|>1} \leqslant 1 + \int\limits_{\substack{|\boldsymbol{u}|>1\\ u_j>0}} \frac{\prod_1^n (1+u_j^2)^{l_j/2}}{\left\{\sum_1^n u_j^{r_j\sigma}\right\}^{\frac{s}{\varrho}}}\, d\boldsymbol{u}$$

$$\leqslant 1 + \int\limits_{\substack{|\xi|>\beta>0\\ \xi_j>0}} \frac{\prod_{j=1}^n (1+\xi_j^{2/r_j})^{l_j/2}\, \xi_j^{\frac{1}{r_j}-1}}{\left(\sum_1^n \xi_j\right)^s}\, d\xi$$

$$\leqslant 1 + \int_\beta^\infty \frac{\varrho^{\sum_1^n \frac{1+l_j}{r_j}-n}\, \varrho^{n-1} d\varrho}{\varrho^s} = 1 + \int_\beta^\infty \frac{d\varrho}{\varrho^{1+\left(s-\sum_1^n \frac{1+l_j}{r_j}\right)}} < \infty,$$

if

(2)
$$s > \sum_1^n \frac{1+l_j}{r_j}.$$

In the third relation the estimate of the integral on $\{|\boldsymbol{u}| > 1\}$ reduces to an estimate on $\{|\boldsymbol{u}| > 1, u_j > 0;\ j = 1, \ldots, n\}$. In the fourth term we have made the replacement of variables $u_j^{r_j} = \xi_j$, with the Jacobian which appears in the numerator inside the integral in the fifth term.

In this calculation we have further used the inequality $\left(\sum_1^n \xi_j^\sigma\right)^{1/\sigma} \gg \sum_1^n \xi_j$ ($\sigma > 0$); here β is a positive number small enough that the ball $|\boldsymbol{u}| < 1$ contains the ball $|\xi| < \beta$. In the fifth relation we pass to polar coordinates. It follows from (1) and (2) that for $\varrho = \lambda + i\mu$, $\boldsymbol{l} = \mathbf{0}$ and

$$(3) \qquad \lambda > \sum_1^n \frac{1}{r_j}$$

the integral 9.4.3 (7) converges absolutely and may be written in the form

$$\int_R = \int_{R'} d\boldsymbol{u}' \int du_n, \quad \boldsymbol{u}' = (u_1, \ldots, u_{n-1}),$$

where the inside integral (relative to u_n) converges absolutely for any $\boldsymbol{u}'$($|V^\varrho| = |V|^\lambda > |u_n|^{\lambda r_n}$, $\lambda r_n > 1$, see (3)).

9.4.5. We shall show that for any $\varphi \in S$ the function

$$(1) \qquad \Phi(\varrho) = (I_{-\boldsymbol{l}} \hat{\Lambda}_{\varrho,\boldsymbol{r},\sigma}, \varphi) \quad (\varrho = \lambda + i\mu,\ \lambda > 0),$$

is analytic on $\{\lambda > 0\}$. We have, obviously,

$$(2) \qquad \Phi(\varrho) = (\Lambda_{\varrho,\boldsymbol{r},\sigma}, \psi) = \int \frac{\psi(\boldsymbol{u}) d\boldsymbol{u}}{V^{\varrho/\sigma}} \left(\psi = \prod_1^n (1 + u_j^2)^{l_j/2} \hat{\varphi} \in S\right).$$

The derivative of Φ formally has the form

$$(3) \qquad \Phi'(\varrho) = -\frac{1}{\sigma} \int \frac{\psi \ln V}{V^{\varrho/\sigma}} d\boldsymbol{u}.$$

The functions in the integrals in (2) and (3) are continuous ($V \geqq 1$), and, moreover,

$$\left|\frac{\psi(\boldsymbol{u})}{V^{\varrho/\sigma}}\right| = \frac{|\psi(\boldsymbol{u})|}{V^{\lambda/\sigma}} \leqq |\psi(\boldsymbol{u})| \in L,$$

$$\left|\frac{\ln V \psi(\boldsymbol{u})}{V^{\varrho/\sigma}}\right| = \frac{|\ln V \psi(\boldsymbol{u})|}{V^{\lambda/\sigma}} \ll |\psi(\boldsymbol{u})| \in L,$$

where the right sides do not depend on ϱ. This shows that the differentiation (3) is legitimate and that $\Phi'(\varrho)$ is continuous. Accordingly, Φ is analytic for $\lambda > 0$.

9.4.6. Below we shall prove that if the parameter σ is sufficiently large $\left(\text{more precisely, } r_n(\varkappa - \sigma) < 1,\ \varkappa = \sum_1^n \frac{1 + l_j}{r_j} - \lambda\right)$[1], then the

[1] Here $\varkappa = \varkappa(\lambda)$, but $\varkappa(1)$ turns into the quantity considered in 9.4.1 (2).

integral (see 9.4.3. (8))

(1) $$\mu_\varrho^{(l)}(\boldsymbol{x}) = \frac{1}{(2\pi)^{n/2}} \int e^{i\boldsymbol{x}'\boldsymbol{u}'}\, d\boldsymbol{u}' \int\limits_{L^*_{\boldsymbol{u}'}} \frac{\prod_1^n (1+u_j^2)^{l_j/2} e^{ix_n u_n}}{V^{\varrho/\sigma}}\, du_n$$

$$(\varrho = \lambda + i\mu,\ \lambda > 0,\ \text{moreover}^1\ |\mu| < 1).$$

is a function continuous relative to $(\varrho, \boldsymbol{x})$ on the set $\{\lambda > 0,\ |\mu| < 1,\ x_n \neq 0\}$, analytic relative to ϱ, and the following estimates hold:

(2) $$|\mu_\varrho^{(l)}(\boldsymbol{x})| \ll \begin{cases} |x_n^{-r_n\varkappa}| & (\varkappa > 0), \\ |\ln|x_n|| + 1 & (\varkappa = 0),\ (|x_n| < 1), \\ 1 & (\varkappa < 0), \end{cases}$$

(3) $$|\mu_\varrho^{(l)}(\boldsymbol{x})| \ll e^{-c|x_n|} \qquad (c > 0,\ |x_n| > 1).$$

Denote by V_* the result of replacing u_n in V by the complex variable $u_n + i\eta_n \in E_{u'}$, where $E_{u'}$ is the region between $L_{u'}$ and the axis $\eta_n = 0$.

Obviously,

$$V_* = V + \omega,$$

where $(\eta_n = \eta)$

$$\omega = \left(1 + (u_n + i\eta)^2\right)^{\frac{r_n\sigma}{2}} - (1 + u_n^2)^{\frac{r_n\sigma}{2}}$$

$$= (1 + u_n^2 + 2u_n\eta i - \eta^2)^{\frac{r_n\sigma}{2}} - (1 + u_n^2)^{\frac{r_n\sigma}{2}}.$$

We shall estimate ω from above. If $0 < r_n\sigma < 2$, then in view of 9.4.2 (1) (explanation below)

(4) $$\begin{aligned} |\omega| &\leqq M\,|2u_n\eta i - \eta^2|^{\frac{r_n\sigma}{2}} \ll ||u_n|\eta + \eta^2|^{\frac{r_n\sigma}{2}} \\ &\ll (|u_n|kU)^{\frac{r_n\sigma}{2}} + k^{r_n\sigma}U^{r_n\sigma} \ll k^{\frac{r_n\sigma}{2}}\, U^{r_n\sigma} \ll k^{\frac{r_n\sigma}{2}}\, V. \end{aligned}$$

The application of inequality 9.4.2 (1) is legitimate, since it was explained in 9.4.3 that the complex point in the first bracket defining ω belongs to Ω (the plane with a cut $-\infty < u_n \leqq 0$).

We suppose that the constants entering into the inequalities $\ll$ do not depend on k. The third inequality follows from the fact that $(x + y)^\alpha \ll x^\alpha + y^\alpha$ $(x, y > 0)$; the next-to-last from the fact that $|u_n| \leqq U$ (see 9.4.3 (3)). The last inequality follows from 9.4.3 (2).

---

[1] The restriction $|\mu| < 1$ is in fact inessential.

If now $r_n\sigma \geqq 2$, then (see 9.4.2 (3))

$$(5)\qquad |\omega| \ll |2u_n\eta i - \eta^2| \left[(1 + u_n^2)^{\frac{r_n\sigma}{2} - 1} + |2u_n\eta i - \eta^2|^{\frac{r_n\sigma}{2} - 1}\right] \ll kV^{r_n\sigma} \ll kV,$$

because (see 9.4.3 (3)) and 9.4.3 (4), and take into account that $0 \leqq \eta \leqq kU$)

$$|2u_n\eta i - \eta^2| \ll |u_nkU| + k^2U^2 \leqq kU^2$$

and

$$(1 + u_n^2)^{\frac{r_n\sigma}{2} - 1} \leqq U^{r_n\sigma - 2}.$$

It follows from (4) and (5) that for sufficiently small k it is possible to achieve that for all $u_n + i\eta \in E_{\boldsymbol{u}'}$

$$(6)\qquad |\omega| < \gamma V \qquad \left(\gamma < \frac{1}{2}\right),$$

where γ may be chosen as small as desired, so that

$$(7)\qquad (1 - \gamma)\, V \leqq |V_*| \leqq (1 + \gamma) V.$$

Inequalities (6), (7), are in particular satisfied on the curve $L_{\boldsymbol{u}'}^*$ (the upper boundary of $E_{\boldsymbol{u}'}$). For the differential of the length of the arc $L_{\boldsymbol{u}'}^*$ we have the estimate

$$(8)\qquad \text{on } L_{u'} : dL_{u'}^* = \sqrt{1 + \left(k\frac{\partial\eta}{\partial u_n}\right)^2}\, du_n = \sqrt{1 + (ku_n)^2\,(1 + u_n^2)^{-1}}\, du_n \leqq \sqrt{2}\, du_n,$$

$$\text{on } l_{u'} : dL_{u'}^* = |d\eta|.$$

The argument of V_* (i.e. of V on $E_{\boldsymbol{u}'}$), for sufficiently small k, is estimated as follows (supposing *a priori* that $|\arg V_*| \leqq \pi$):

$$(9)\qquad |\arg V_*| \ll \left|\frac{\operatorname{Im} V_*}{\operatorname{Re} V_*}\right| \leqq \frac{|\omega|}{V - |\omega|} \leqq \frac{\gamma V}{(1 - \gamma)\, V} < 1.$$

Thus we have proved inequality 9.4.3. (10), which we needed in order to show that $V^{\varrho/\sigma}$ is a single-valued analytic function of the complex variable $w_n \in E_{\boldsymbol{u}'}$. It follows also from (9) and (7) that

$$(10)\qquad |V_*^{\varrho/\sigma}| = \left|(|V_*|\, e^{i\arg V_*})^{\frac{\lambda + i\mu}{\sigma}}\right| = |V_*|^{\frac{\lambda}{\sigma}}\, e^{-\frac{\mu}{\sigma}\arg V_*} \geqq c^{-\frac{1}{\sigma}}\, |V_*|^{\frac{\lambda}{\sigma}} \geqq cV^{\frac{\lambda}{\sigma}} \qquad (|\mu| < 1),$$

where c thus does not depend on $\lambda > 0$ and $|\mu| < 1$.

We have

$$\mu_\varrho^{(l)}(\boldsymbol{x}) = \frac{1}{(2\pi)^{n/2}} \int e^{i\boldsymbol{x}'\boldsymbol{u}'}\, d\boldsymbol{u}'$$

$$(11) \quad \times \left(\int_{-\infty}^{\infty} \frac{\prod_1^n (1+u_j^2)^{l_j/2} e^{ix_n(u_n+ikU)} \sqrt{1+ku_n(1+u_n^2)^{-1}}\, du_n}{V_*^{\varrho/\sigma}} + \int_{l_{\boldsymbol{u}'}} \frac{\prod_1^n (1+u_j^2)^{l_j/2} e^{ix_n w_n}}{V^{\varrho/\sigma}}\, dw_n \right) = I_1^{(l)} + I_2^{(l)},$$

where V_* is here understood as V on $L_{\boldsymbol{u}'}$. The second integral arises, given only that

$$k\left(\sum_1^{n-1} (1+u_j^2)^{\frac{r_j}{2r_n}} + 1\right) > 1.$$

The modulus of the function in the integral in $I_1^{(l)}$ does not exceed

$$(12) \qquad c\, \frac{\prod_1^n (1+u_j^2)^{l_j/2} e^{-kx_nU}}{V^{\lambda/\sigma}} = \alpha(\lambda, x_n, \boldsymbol{u}),$$

where c does not depend on $\boldsymbol{u}$, x_n, $\varrho = \lambda + i\mu$ ($\lambda > 0$, $|\mu| < 1$). We shall also estimate the integral of (12) relative to $\boldsymbol{u} \in \mathbb{R}_n$. We shall see that it is finite for any $\boldsymbol{l} \geqq \boldsymbol{0}$, $x_n > 0$, $\lambda > 0$. Since $\alpha(\lambda, x_n, \boldsymbol{u})$ increases with decreasing x_n and λ ($V \geqq 1$), then we have

$$\alpha(\lambda, x_n, \boldsymbol{u}) \leqq \alpha(\lambda_0, x_n^0, \boldsymbol{u}) \in L(\mathbb{R}_n) = L$$
$$(\lambda \geqq \lambda_0 > 0,\ x_n \geqq x_n^0 > 0).$$

Moreover, the function in the integral in $I_1^{(l)}$ is continuous relative to $(\varrho, \boldsymbol{x}, \boldsymbol{u})$. In such a case, on the basis of the Weierstrass test, $I_1^{(l)} = I_1^{(l)}(\varrho, \boldsymbol{x})$ is a continuous function of $(\varrho, \boldsymbol{x})$. If the expression in the integral in $I_1^{(l)}$ is differentiated relative to the complex ϱ, then the modulus of the resulting derivative will be equal, up to a constant coefficient, to

$$(13) \qquad \frac{|\boldsymbol{u}^{\boldsymbol{l}}| e^{-kx_nU} |\ln V_*|}{|V_*^{\varrho/\sigma}|} \leqq c\, \frac{|\boldsymbol{u}^{\boldsymbol{l}}| e^{-kx_n^0 U}}{V^{\frac{\lambda_0-\varepsilon}{\sigma}}} = \alpha(\lambda_0 - \varepsilon, x_n^0, \boldsymbol{u})$$

$$\left(\lambda > \lambda_0 - \frac{\varepsilon}{2} > \lambda_0 - \varepsilon > 0;\ |\mu| < 1,\ x_n \geqq x_0 > 0\right).$$

Now, since the right side of (13) belongs to L relative to $\boldsymbol{u}$, then, as a consequence of the Weierstrass test of the convergence of an integral one may say that for the indicated ϱ and $\boldsymbol{x}$ there exists a derivative $\frac{\partial}{\partial \varrho} I_1^{(\boldsymbol{l})}$, continuous relative to $(\varrho, \boldsymbol{x})$. This shows that the function $I_1^{(\boldsymbol{l})}(\varrho, \boldsymbol{x})$ is analytic relative to $\varrho (\boldsymbol{l} \geqq 0, \lambda > 0, |\mu| < 1, x_n > 0)$.

We note that the constants in inequalities (12) and (13), just as in the inequalities following later in the estimation of $I_2^{(\boldsymbol{l})}$, depend continuously on ϱ.

For $0 < x_n < 1$ (the explanations are the same as in 9.4.4)

$$|I_1^{(\boldsymbol{l})}| \ll \int \frac{\prod_1^n (1 + u_j^2)^{l_j/2} e^{-k x_n U}}{V^{\lambda/\sigma}} d\boldsymbol{u}$$

$$\leqq \int\limits_{|\boldsymbol{u}|<1} + \int\limits_{|\boldsymbol{u}|>1} \frac{\prod_1^n (1 + u_j^2)^{l_j/2} e^{-k x_n \sum_1^n u_j^{r_j/r_n}}}{\left(\sum_1^n u_j^{r_j \sigma}\right)^{\lambda/\sigma}} d\boldsymbol{u} \tag{14}$$

$$\ll 1 + \int\limits_{\substack{|\xi|>\beta>0 \\ \xi_j>0}} \frac{\prod_1^n (1 + \xi_j^{2/r_j})^{l_j/2} \xi^{\frac{1}{r_j}-1} e^{-k x_n \sum_1^n \xi_j^{1/r_n}}}{|\sum \xi_j|^\lambda} d\xi$$

$$\ll 1 + \int_\beta^\infty \varrho^{\varkappa-1} e^{-c x_n \varrho^{\frac{1}{r_n}}} d\varrho \ll 1 + \int_{\gamma x_n}^\infty e^{-cz} z^{r_n \varkappa - 1} dz \frac{1}{x_n^{r_n \varkappa}}$$

$$\ll \begin{cases} x_n^{-r_n \varkappa} & (\varkappa > 0), \\ 1 & (\varkappa < 0), \quad (0 < x_n < 1). \\ |\ln x_n| + 1 & (\varkappa = 0) \end{cases}$$

We need to note that the integral in the next-to-last of the relations, taken on $(1, \infty)$ converges, and on $(\gamma x_n, 1)$ it does not exceed

$$c \int_{\gamma x_n}^1 z^{r_n \varkappa - 1} dz \ll \begin{cases} 1 & (\varkappa > 0), \\ x_n^{r_n \varkappa} & (\varkappa < 0), \quad \left(0 < x_n < \frac{1}{\gamma}\right). \\ |\ln x_n| + 1 & (\varkappa = 0) \end{cases} \tag{15}$$

Now for $x_n > 1$

$$(16)\qquad |I_1^{(l)}| \ll \int\limits_{|\boldsymbol{u}|<1} e^{-kx_n} d\boldsymbol{u} + \int\limits_{\substack{|\boldsymbol{u}|>1\\ u_j>0}} (1+u_j^2)^{\frac{l_j}{2}} e^{-kx_n \sum_1^n u_j^{r_j/r_n}} d\boldsymbol{u}$$

$$\ll e^{-kx_n} + \int\limits_{\substack{|\xi|>\beta\\ \xi_j>0}} \prod_{j=1}^n \left(1 + \xi_j^{\frac{2r_n}{r_j}}\right)^{l_j/2} \xi_j^{\frac{r_n}{r_j}-1} e^{-kx_n \sum_1^n \xi_j} d\xi$$

$$\ll e^{-kx_n} + \int\limits_\beta^\infty \varrho^{r_n(\varkappa+\lambda)-1} e^{-cx_n\varrho} d\varrho \ll e^{-kx_n} + \int\limits_\beta^\infty e^{-\frac{c}{2}x_n\varrho} \ll e^{-c_1 x_n} \quad (c_1 > 0).$$

Now we turn to the estimation of $I_2^{(l)}$. The inside integral is taken along the cut (i, iB), where

$$B = k\left(\sum_1^{n-1} (1+u_j^2)^{\frac{r_j}{2r_n}} + 1\right).$$

The number B depends on $\boldsymbol{u}'$. For $\boldsymbol{u}' = \boldsymbol{0}$ it is minimal and is equal to kn. If $kn > 1$, then in the calculation of $I_2^{(l)}$ the ouside integration is carried out relative to all $\boldsymbol{u}' \in \mathbb{R}_{n-1}$. However if $kn < 1$, then the integration relative to $\boldsymbol{u}'$ is carried out relative to the exterior of some bounded neighborhood of the point $\boldsymbol{u}' = \boldsymbol{0}$. On $l_{\boldsymbol{u}'} = (i, iB)$ we have $u_n = iy$ $(1 \leqq y \leqq B)$. Here we have to understand the term $(1+u_n^2)^{\frac{r_n\sigma}{2}}$ entering into V along one side of $l_{\boldsymbol{u}'}$ as $(y^2-1)^{\frac{r_n\sigma}{2}} e^{i\frac{r_n\sigma}{2}\pi}$ and on the other as $(y^2-1)^{\frac{r_n\sigma}{2}} \times e^{-i\frac{r_n\sigma}{2}\pi}$. The corresponding values of V on the different sides of $l_{\boldsymbol{u}'}$ are complex conjugate, so that their product is equal to the square of their modulus, and the inside integral in $I_2^{(l)}$ is equal to

$$(17)\qquad -\int\limits_1^B \prod_1^{n-1} (1+u_j^2)^{l_j/2} y^{l_n} e^{-x_n y}$$

$$\times \left\{\frac{1}{\left\{A + [(y^2-1)e^{-i\pi}]^{\frac{r_n\sigma}{2}}\right\}^{\varrho/\sigma}} - \frac{1}{\left\{A + [(y^2-1)e^{i\pi}]^{\frac{r_n\sigma}{2}}\right\}^{\varrho/\sigma}}\right\} dy,$$

$$A = \sum_1^{n-1} (1+u_j^2)^{\frac{r_j\sigma}{2}} \qquad (\varrho = \lambda + i\mu,\ \lambda > 0).$$

The modulus of the expression in curly brackets does not exceed (explanation below)

$$\frac{\left|\left(A+[(y^2-1)\,e^{i\pi}]^{\frac{r_n\sigma}{2}}\right)^{\varrho/\sigma}-\left(A+[(y^2-1)\,e^{-i\pi}]^{\frac{r_n\sigma}{2}}\right)^{\varrho/\sigma}\right|}{\left|A-(y^2-1)^{\frac{r_n\sigma}{2}}\right|^{2\lambda/\sigma}}$$

$$=A^{-\frac{\lambda}{\sigma}}\frac{\left|\left\{1+\left(\tau e^{i\frac{\pi}{2}}\right)^{r_n\sigma}\right\}^{\varrho/\sigma}-\left\{1+\left(\tau e^{-i\frac{\pi}{2}}\right)^{r_n\sigma}\right\}^{\varrho/\sigma}\right|}{|1-\tau^{r_n\sigma}|^{2\lambda/\sigma}}$$

$$\ll a^{-r_n\lambda}\tau^{r_n\sigma}\left|\sin\frac{r_n\varrho\pi}{2}\right|\ll a^{-r_n\lambda}\tau^{r_n\sigma},$$

where the constants in the inequalities in every case may be regarded as not depending locally on $\varrho=\lambda+i\mu$. Here

(18) $$\tau=a^{-1}\sqrt{y^2-1},\qquad a=A^{\frac{1}{r_n\sigma}},$$

and since $1<y<B$, then

$$0<\tau<a^{-1}\sqrt{B^2-1}\ll a^{-1}B$$

(19) $$=\frac{k\left\{\sum_1^{n-1}(1+u_j^2)^{\frac{r_j}{2r_n}}+1\right\}}{\left\{\sum_1^{n-1}(1+u_j^2)^{\frac{r_j\sigma}{2}}\right\}^{\frac{1}{r_n\sigma}}}\leq ck=\omega<1,$$

where ω may be supposed less than unity, choosing a sufficiently small k. On this basis we have dropped the denominator in the second term, bounded below by a positive constant. The function in the modulus sign in the numerator is analytic on the interval $|\tau|<1$, and equal to zero for $\tau=0$. The law of the mean is applicable to it. Thus, the expression in the integral in $I_2^{(l)}$ does not exceed in modulus

(20) $$c\prod_1^{n-1}(1+u_j^2)^{l_j/2}\,y^{l_n}e^{-x_ny}a^{-r_n\lambda}\tau^{r_n\sigma},$$

where c does not depend on $\boldsymbol{u}'$, y, $x_n>0$ and in any case locally on $\lambda>0$.

We shall show that the function (20) is summable over the region $(\boldsymbol{u}', \boldsymbol{y})$ of definition of the integral $I_2^{(l)}$ for arbitrary $\boldsymbol{x}$, ϱ as indicated above. Moreover, it is directly evident that it increases with decreasing

x_n and λ. This leads to the fact that $I_2^{(\boldsymbol{l})}(\varrho, \boldsymbol{x})$ is continuous relative to the indicated $(\varrho, \boldsymbol{x})$ and is indeed the derivative (relative to $\boldsymbol{x}$) of order $\boldsymbol{l}$ of I_2. Finally, if one differentiates the expression in the integral in (17) relative to ϱ, one obtains

$$\prod_1^{n-1} (1 + u_j^2)^{l_j/2} y^{l_n} e^{-x_n y} \frac{1}{\sigma} \left\{ \frac{\ln \left(A + [(y^2 - 1) e^{-i\pi}]^{\frac{r_n\sigma}{2}}\right)}{\left(A + [(y^2 - 1)\, e^{-i\pi}]^{\frac{r_n\sigma}{2}}\right)^{\varrho/\sigma}} \right.$$

$$\left. - \frac{\ln \left(A + [(y^2 - 1)\, e^{i\pi}]^{\frac{r_n\sigma}{2}}\right)^{\varrho/\sigma}}{\left(A + [(y^2 - 1) e^{i\pi}]^{\frac{r_n\sigma}{2}}\right)^{\varrho/\sigma}} \right\}.$$

If the expression in curly brackets is reduced to a common denominator, the modulus taken, and A then carried throughout across the brackets then, as we know, in the estimation from above one may drop the denominator, as bounded below by a positive constant. As to the numerator, it is evidently estimated from above as follows:

$$\ln A \left| \left\{1 + \left(\tau e^{+i\frac{\pi}{2}}\right)^{r_n\sigma}\right\}^{\varrho/\sigma} - \left\{1 + \left(\tau e^{-\frac{i\pi}{2}}\right)^{r_n\sigma}\right\}^{\varrho/\sigma} \right|$$

$$+ \left| \left\{1 + \left(\tau e^{\frac{i\pi}{2}}\right)^{r_n\sigma}\right\}^{\varrho/\sigma} \ln \left\{1 + \left(\tau e^{-\frac{i\pi}{2}}\right)^{r_n\sigma}\right\} \right.$$

$$\left. - \left\{1 + \left(\tau e^{-\frac{i\pi}{2}}\right)^{r_n\sigma}\right\}^{\varrho/\sigma} \ln \left\{1 + \left(\tau e^{\frac{i\pi}{2}}\right)^{r_n\sigma}\right\} \right| \ll (\ln A + 1)\, \tau^{r_n\sigma}$$

$$= (\ln a^{r_n\sigma} + 1)\, \tau^{r_n\sigma} \ll a^{\varepsilon} \tau^{r_n\sigma} \qquad (\varepsilon > 0,\ a \geqq 1).$$

The constant in the right side depends on the arbitrarily small ε. But it is possible to take it so that it does not depend on ϱ from some small neighborhood of ϱ_0. As a result we find that the integrand expression in (17), differentiated with respect to ϱ (continuous with respect to $(\boldsymbol{u}', y, \varrho, x_n)$, $x_n > 0, \lambda > 0$), does not exceed in modulus a function, analogous to (20),

$$c \prod_1^{n-1} (1 + u_j^2)^{l_j/2} y^{l_n} e^{-x_n y} a^{\varepsilon - r_n\lambda} \tau^{r_n\sigma} \in L. \tag{20'}$$

This shows that the function $I_2^{(\boldsymbol{l})}(\varrho, \boldsymbol{x})$ is analytic in ϱ.

Thus (explanation below)

(21) $$|I_2^{(l)}| \ll \int \prod_1^{n-1} (1+u_j^2)^{l_j/2}\, a^{-r_n\lambda} \int_1^B y^{l_n} e^{-x_n y} \tau^{r_n\sigma}\, dy$$

$$\ll \int \prod_1^{n-1} (1+u_j^2)^{l_j/2}\, a^{1-r_n\lambda}\, d\boldsymbol{u}' \int_0^\omega (1+a^2\tau^2)^{l_n/2}$$

$$\times e^{-x_n\sqrt{1+a^2\tau^2}}\, \tau^{r_n\sigma}\, d\tau \ll 1 + \int_{\substack{|\xi|>\beta>0\\ \xi_j>0}} \prod_1^{n-1} \left(1+\xi_j^{\frac{2r_n}{r_j}}\right)^{\frac{l_j}{2}}$$

$$\times \xi_j^{\frac{r_n}{r_j}-1}\, a^{1-r_n\lambda}\, d\xi \int_0^\omega (1+a^2\tau^2)^{l_n/2} e^{-x_n\sqrt{1+a^2\tau^2}}\, \tau^{r_n\sigma}\, d\tau$$

$$\ll 1 + \int_0^\omega \tau^{r_n\sigma}\, d\tau \int_\beta^\infty \varrho^{\,l_n+1-r_n\lambda+r_n\sum_1^{n-1}\frac{1+l_j}{r_j}-1} e^{-x_n\sqrt{1+c^2\varrho^2\tau^2}}\, d\varrho$$

$$= 1 + \int_0^\omega \tau^{r_n\sigma}\, d\tau \int_\beta^\infty \varrho^{r_n\varkappa-1} e^{-x_n\sqrt{1+c^2\varrho^2\tau^2}}\, d\varrho$$

$$\ll 1 + \frac{1}{x_n^{r_n\varkappa}} \int_0^\omega \frac{d\tau}{\tau^{r_n(\varkappa-\sigma)}} \int_{\beta x_n\tau}^\omega \zeta^{r_n\varkappa-1} e^{-c\zeta}\, d\zeta \quad (c>0).$$

In the first relation we have made use of the estimate (20). In the second we have replaced y by τ using formula (18), taking account also of inequality (19). In the third relation the integral in $\boldsymbol{u}'$ is decomposed into two integrals: relative to $|\boldsymbol{u}'| < 1$ and $|\boldsymbol{u}'| > 1$, of which the first is evidently bounded. Moreover, we have taken account of the symmetric properties relative to $\boldsymbol{u}'$ of the integrands. The question reduces to integration relative to $u_j > 0$. Here we have carried out the change of variables

$$\xi_j = u_j^{r_j/r_n}, \quad du_j = \frac{r_n}{r_j}\, \xi_j^{\frac{r_n}{r_j}-1}\, d\xi_j \quad (j = 1, \ldots, n-1).$$

In the fourth relation we have changed the order of integration, introduced polar coordinates ($|\boldsymbol{\xi}| = \varrho$) in the space of the $\boldsymbol{\xi}$, used the fact that the variables a and ϱ have one hand the same order:

$$\varrho = \left(\sum_1^{n-1} u_j^{2r_j/r_n}\right)^{1/2} \ll \left\{\sum_1^{n-1} (1+u_j^2)^{\frac{r_j\sigma}{2}}\right\}^{\frac{1}{r_n\sigma}} = a \ll \varrho,$$

and the inequality $(1+\varrho^2\tau^2)^{l_n/2} \ll \varrho^{l_n}$ ($\varrho > \beta$, $0 < \tau < \omega$) is applicable.

Finally, in the last relation we have used the inequality $\sqrt{1 + c^2\varrho^2\tau^2} > c\varrho\tau$ and introduced the substitution $x_n\varrho\tau = \xi$, $x_n\tau d\varrho = d\xi$.

Suppose that $\varkappa > 0$ and the parameter σ is chosen so that $r_n(\varkappa - \sigma) < 1$. Then the integral relative to ζ in the right hand side of (21) over the interval $(0, \infty)$ is finite, just as is the integral relative to τ. Hence

$$|I_2^{(l)}| \ll x_n^{-r_n\varkappa}.$$

If $\varkappa = 0$, then the integral relative to ζ has for small x_n the order $\ln(x_n\tau)$. Therefore for $\sigma > 0$ we will have

$$|I_2^{(l)}| \ll |\ln x_n| + 1.$$

Finally, for $\varkappa < 0$ the integral relative to ζ has for small x_n order $x_n^{r_n\varkappa}$, so that for $\sigma > 0$

$$|I_2^{(l)}| \ll 1.$$

We have proved that for an appropriate σ

$$|I_2^{(l)}| \ll \begin{Bmatrix} x_n^{-r_n\varkappa}, & \varkappa > 0, \\ 1, & \varkappa < 0, \\ \ln \dfrac{1}{x_n}, & \varkappa = 0 \end{Bmatrix} \quad (0 < x_n < 1).$$

Finally, in order to obtain an estimate of $I_2^{(l)}$ for large x_n, we decompose the integral $I_2^{(l)}$ into two integrals: relative to $|\boldsymbol{u}'| < 1$ and relative to $|\boldsymbol{u}'| > 1$. The first integral (see the third term in formula (21)) has order e^{-x_n} $(x_n > 1)$. In order to estimate the second, we use the next-to-last integral in (21). Then we get

$$|I_2^{(l)}| \ll e^{-x_n} + \int_0^{\omega} \tau^{r_n\sigma} d\tau \int_{\beta}^{\infty} \varrho^{r_n\varkappa - 1} e^{-x_n\sqrt{1+c^2\varrho^2\tau^2}}\, d\varrho$$

$$= e^{-x_n} + \int_0^{\omega} \frac{d\tau}{\tau^{r_n(\varkappa-\sigma)}} \int_{\beta\tau}^{\infty} \xi^{r_n\varkappa - 1} e^{-x_n\sqrt{1+c^2\xi^2}}\, d\xi$$

$$\leqq e^{-x_n} + \int_0^{\omega} \frac{d\tau}{\tau^{r_n(\varkappa-\sigma)}} \left(\int_{\beta\tau}^{\beta\omega} \xi^{r_n\varkappa - 1} d\xi\, e^{-x_n} + \int_{\beta\omega}^{\infty} e^{-x_n c_1 \xi}\, d\xi \right) \ll e^{-c_2 x_n}$$

$$(r_n(\varkappa - \sigma) < 1;\ c_1, c_2, \sigma > 0).$$

In the case $x_n < 0$ the curve $L_{\boldsymbol{u}'}^*$ (see (1)) in the complex u_n plane is taken symmetrically relative to the $u_n = 0$ axis. The proof in this case is analogous.

Thus, we have proved inequalities (2) and (3), in which, as we have already noted, the constants depend continuously on ϱ.

9.4.7. In the definition of the function $\mu_\varrho(\boldsymbol{x}) = \mu_\varrho^{(0)}$ by formula 9.4.3. (8) the role of the variable u_n was distinguished. Having this in view, put $\mu_\varrho(\boldsymbol{x}) = u_{\varrho n}(\boldsymbol{x})$. One may equally well introduce the functions $\mu_{\varrho j}(\boldsymbol{x})$ ($j = 1, \ldots, n$), where the role of u_n is played by u_j. If $\boldsymbol{x} = (x_1, \ldots, x_n)$ is a point for which $x_i \neq 0$, $x_j \neq 0$, then

$$\mu_{\varrho i}(\boldsymbol{x}) = \mu_{\varrho j}(\boldsymbol{x}),$$

because this equation holds in every case for large real ϱ, and therefore for arbitrary complex $\varrho = \lambda + i\mu$ ($\lambda > 0$), as a consequence of the analyticity of both functions relative to ϱ for fixed $\boldsymbol{x}$. For a given j the function $\mu_{\varrho j}(\boldsymbol{x})$ is defined and continuous relative to $\boldsymbol{x}$ at any point $\boldsymbol{x}$ having the coordinate $x_j \neq 0$. From what has been said it is clear that $\mu_{\varrho j}(\boldsymbol{x})$ may be extended by continuity to any point $\boldsymbol{x} \neq \boldsymbol{0}$, and then

$$(1) \qquad \mu_\varrho(\boldsymbol{x}) = \mu_{\varrho 1}(\boldsymbol{x}) = \ldots = \mu_{\varrho 11}(\boldsymbol{x}) \quad (\boldsymbol{x} \neq 0).$$

In addition for a vector $\boldsymbol{l} \geq \boldsymbol{0}$ there exists a $\sigma_0 > 0$ such that for $\sigma > \sigma_0$ the function $\mu_\varrho^{(\boldsymbol{l})}(\boldsymbol{x})$ is continuous and

$$(2) \qquad |\mu_\varrho^{(\boldsymbol{l})}(\boldsymbol{x})| \ll \begin{cases} |x_j|^{-r_j\varkappa} & (\varkappa > 0), \\ |\ln|x_j|| + 1 & (\varkappa = 0), \\ 1 & (\varkappa < 0) \end{cases}$$

$$\left(\varkappa = \sum_1^n \frac{1 + l_j}{r_j} - \lambda, \ |x_j| < 1, \ \varrho = \lambda + i\mu, \ \lambda > 0\right),$$

$$|\mu_\varrho^{(\boldsymbol{l})}(\boldsymbol{x})| \ll e^{-c|x_j|} \ (|x_j| > 1, c > 0).$$

Hence there follow immediately the estimates

$$(3) \qquad |\mu_\varrho^{(\boldsymbol{l})}(\boldsymbol{x}) \leq c \begin{cases} \left\{\sum_1^n |x_j|^{r_j\varkappa}\right\}^{-1} & (\varkappa > 0), \\ |\ln|\boldsymbol{x}|| + 1 & (\varkappa = 0), \quad (|\boldsymbol{x}| < 1), \\ 1 & (\varkappa < 0), \end{cases}$$

$$(4) \qquad |\mu_\varrho^{(\boldsymbol{l})}(\boldsymbol{x})| \leq ce^{-c'|\boldsymbol{x}|} \quad (|\boldsymbol{x}| > 1, \ c > 0),$$

where the constants c and c' entering into the inequalities depend continuously on ϱ.

For $\boldsymbol{l} = \boldsymbol{0}$ it follows from these estimates that

$$\mu_\varrho(\boldsymbol{x}) = \mu_\varrho^{(0)}(\boldsymbol{x}) \in L = L(\mathbb{R}_n).$$

Indeed, if $\varkappa \leqq 0$ this is obvious. If now $\varkappa > 0$, then in view of the fact that $\lambda > 0$

$$\frac{1}{\varkappa}\sum_1^n \frac{1}{r_j} - 1 = \frac{\lambda}{\varkappa} > 0$$

and, accordingly (explanation as in 9.4.4)

$$\int |\mu_\varrho(\boldsymbol{x})|\,d\boldsymbol{x} = \int_{|\boldsymbol{x}|<1} \left\{\sum_1^n |x_j|^{r_j\varkappa}\right\}^{-1} d\boldsymbol{x}$$

$$+ \int_{|\boldsymbol{x}|>1} e^{-c|\boldsymbol{x}|}\,d\boldsymbol{x} \ll \int_{\substack{|\xi|>\beta\\ \xi_j<0}} \left(\sum_1^n \xi_j\right)^{-1} \prod_1^n \xi^{\frac{1}{r_j\varkappa}-1}\,d\xi + 1 \ll \int_0^\beta \varrho^{\frac{1}{\varkappa}\sum_1^n\frac{1}{r_j}-2}\,d\varrho + 1 \ll 1.$$

Now suppose that $\varphi \in S$. Then, since $\mu_\varrho \in L$ the following has meaning:

$$(\mu_\varrho, \varphi) = \int \mu_\varrho(\boldsymbol{x})\,\varphi(\boldsymbol{x})\,d\boldsymbol{x} \quad (\lambda > 0). \tag{5}$$

Let us denote by $\mu_{\varrho*}^{(\boldsymbol{l})}(\boldsymbol{x})$ the right side of (3) without the factor $c = c(\varrho)$. It is easy to see, in view of the monotonic properties of the function $\varkappa$ (relative to λ) and the continuity of $c(\varrho)$ relative to ϱ that for any $\varrho_0 = \lambda_0 + i\mu_0$ $(\lambda_0 > 0, |\mu_0| < 1)$ there exists a $\delta > 0$ such that if $|\varrho - \varrho_0| < \delta$ $(\varrho = \lambda + i\mu, \lambda > 0, |\mu| < 1)$, then

$$\begin{aligned} &|\mu_\varrho^{(\boldsymbol{l})}(\boldsymbol{x})| \leqq c\mu_{\lambda_0-\delta*}^{(\boldsymbol{l})}(\boldsymbol{x}) \in L(|\boldsymbol{x}| < 1),\\ &|\mu_\varrho^{(\boldsymbol{l})}(\boldsymbol{x})| \leqq ce^{-c_1|\boldsymbol{x}|} \in L(|\boldsymbol{x}| > 1), \end{aligned} \tag{6}$$

where c and c_1 do not depend on the indicated ϱ. Therefore the Weierstrass test for uniform convergence (locally relative to ϱ) of the integral (5) is satisfied and (μ_ϱ, φ) depends continuously on the complex $\varrho(\lambda > 0)$. The derivative of $\varphi\mu_\varrho$ relative to ϱ is also continuous relative to $(\varrho, \boldsymbol{x})$, and for it there hold the same estimates as (6) (see 9.4.6 (13) and (20)). This shows that the function (μ_ϱ, λ) is differentiable and accordingly analytic relative to ϱ.

9.4.8. For arbitrary complex $\varrho = \lambda + i\mu(\lambda > 0)$, and accordingly for $\varrho = 1$ as well, one has for sufficiently large $\sigma(r_n(\varkappa - \sigma) < 1)$ the equation

$$\hat{\Lambda}_{\varrho,\boldsymbol{r},\sigma} = \mu_\varrho(\boldsymbol{x}). \tag{1}$$

In fact, the functions

$$(\hat{\Lambda}_{\varrho,\boldsymbol{r},\sigma,\varphi}) \text{ and } (\mu_\varrho, \varphi) \quad (\varphi \in S)$$

are analytic in $\varrho(\lambda > 0)$ and coincide for real sufficiently large ϱ. Hence they coincide for arbitrary ϱ. But then they coincide as well for arbitrary $\varphi \in S$, which implies (1).

9.4.9. Other estimates of the anisotropic kernel. In the following lemma one estimates the differences of the kernel G_r in the metric of $L_p(\mathbb{R}_{n-1})$. These estimates will be used in the proof of the imbedding theorems. We introduce the notations

$$\boldsymbol{x} = (\eta, \zeta),\ \eta = (x_1, \ldots, x_{n-1}),\ x_n = \zeta.$$

Lemma[1]. *Suppose that* $\boldsymbol{r} = (r_1, \ldots, r_n) > 0$, $1 < p < \infty$, *and suppose given a noninteger positive number L such that for some* j, $j = 1, \ldots, n-1$, *the inequalities*

(1) $$1 - \frac{1}{r_n} < \frac{L}{r_j} < \min_i \left\{\sigma - \sum_1^n \frac{1}{r_k} - \frac{1}{r_i}\right\}$$

are satisfied. Then

(2) $$\int_{\mathbb{R}_{n-1}} |\Delta_{x_j h}^{s+1} G_r(\eta, \zeta)\, d\eta \leqq c|h|^L |\zeta|^{-r_n\left(\frac{1}{r_n} + \frac{L}{r_j} - 1\right)},$$

where s is the integer part of L, i.e. $L = s + l$, $0 \leqq l < 1$, *s an integer.*

Proof. Without loss of generality we may suppose $h > 0$, $j = 1$, $G_r = G$ and introduce the kernel

$$K_\nu(t) = \widehat{(1 + t^2)^{-\nu/2}}\ (-\infty < t < \infty).$$

We shall write

$$G(t) = G(t, x_2, \ldots, x_n),$$

$$I_{-L} G(t) = \psi(t).$$

We shall understand the sth difference with step h of the function $\varphi(\zeta)$ in the sense

$$\Delta_h \varphi = \varphi(\zeta + h) - \varphi(\zeta),\ \Delta_h^{s+1} \varphi = \Delta_k \Delta_h^s \varphi.$$

From the further estimates it will be clear that ψ is a summable function of t in every case for almost all $x_2, \ldots, x_{n+1}$.

We have

$$G(t) = \int K_{s+l}(t - \xi) \psi(\xi)\, d\xi,$$

(3) $$\Delta_h^{s+1} G(t) = \left\{\int_{-\infty}^{t} + \int_{t}^{t+(s+1)h}\right.$$

$$\left. + \int_{t+(s+1)h}^{\infty}\right\} \Delta_h^{s+i} K_{s+l}(t - \xi)\, \psi(\xi)\, d\xi = I_1 + I_2 + I_3,$$

[1] P. I. Lizorkin [10].

$$\int |I_1| dt \leqq \int_{-\infty}^{\infty} |\psi(\xi)| d\xi \int_{\xi}^{\infty} |\Delta_h^{s+1} K_{s+l}(t-\xi)| \, dt$$

$$\leqq \|\psi\|_L \int_0^{\infty} |\Delta_h^{s+1} K_{s+l}(t)| dt, \quad L = L(-\infty, \infty).$$

But (explanation below)

$$\int_0^{\infty} |\Delta_h^{s+1} K_{s+1}(t)| dt$$

$$= \int_0^{\infty} dt \int_0^h \cdots \int_0^h \left|K_{s+l}^{(s+1)}\left(t + \sum_1^{s+1} t_k\right)\right| dt_1 \cdots dt_{s+1}$$

$$\ll \int_0^{\infty} dt \int_0^h \cdots \int_0^h \frac{dt_1 \cdots dt_{s+1}}{\left(t + \sum_1^{s+1} t_k\right)^{1-(s+l)+s+1}} \ll \int_0^{\infty} dt \int_0^{ch} \frac{\varrho^s d\varrho}{(t+\varrho)^{2-l}}$$

$$\ll \int_0^{ch} \varrho^{s+l-1} d\varrho \leqq h^{s+l}.$$

In the second relation (inequality) we have applied the (third) estimate 8.1 (7). In the third we have introduced polar coordinates in the space of the $t_1, \ldots, t_{s+1}$ and recalled that $\sum_1^{s+1} t_k \geqq \left(\sum_1^{s+1} t_k^2\right)^{1/2}$. In the fourth relation we have reversed the order of integration.

The integral I_3 is estimated analogously:

$$\int |I_j| dt \leqq ch^{s+l} \|\psi\|_L \quad (j = 1, 3). \tag{4}$$

Further

$$\int_{-\infty}^{\infty} |I_2| dt = \int_{-\infty}^{\infty} dt \left| \int_t^{t+(s+1)h} \Delta_h^{s+1} K_{s+l}(t-\xi) \, \psi(\xi) \, d\xi \right|$$

$$\leqq \int_{-\infty}^{\infty} |\psi(\xi)| d\xi \int_{\xi-(s+1)h}^{\xi} |\Delta_h^{s+1} K_{s+l}(t-\xi)| dt$$

$$= \|\psi\|_L \int_{-(s+1)h}^{0} |\Delta_h^{s+1} K_{s+l}(t)| dt$$

$$= \|\psi\|_L \int_{-(s+1)h}^{0} dt \left| \int_0^h \cdots \int_0^h \left\{ K_{s+l}^{(s)}\left(t + \sum_1^s t_k + h\right) \right.\right.$$

$$\left.\left. - K_{s+l}^{(s)}\left(t + \sum_1^s t_k\right)\right\} dt_1 \cdots dt_s \right|$$

$$\leqq 2\|\psi\|_L \int_{-(s+1)h}^{(s+1)h} dt \int_0^h \cdots \int_0^h \left| K_{s+l}^{(s)}\left(t + \sum_1^s t_k\right)\right| dt_1 \cdots dt_s$$

$$\ll \|\psi\|_L \int_{-(s+1)h}^{(s+1)h} dt \int_0^{ch} \frac{\varrho^{s-1} d\varrho}{|t+\varrho|^{1-l}} = \|\psi\|_L \int_0^{ch} \varrho^{s+l-1} d\varrho \int_{-(s+1)\frac{h}{\varrho}}^{\frac{s+1}{\varrho}h} \frac{du}{|1+u|^{1-l}}$$

$$\leqq 2\|\psi\|_L \int_0^{ch} \varrho^{s+l-1}\, d\varrho \int_0^{\frac{s+1}{\varrho}h+1} \frac{dv}{v^{1-l}} \ll \|\psi\|_L \int_0^{ch} \varrho^{s+l-1} \left[\left(\frac{h}{\varrho}\right)^l + 1\right] d\varrho \ll \|\psi\|_L h^{l+s}.$$

From (3)—(5), after some additional integration of the inequalities relative to $(x_1, \ldots, x_{n-1})$, it follows that

$$\int_{\mathbb{R}_{n-1}} |\Delta_{x_1 h}^s G_r(\eta, \xi) d\eta \ll h^{s+l_1} \int_{\mathbb{R}_{n-1}} |I_{x_1, -(s+l_1)} G_r| d\eta$$

$$(s = 0, 1, \ldots, j = 1, \ldots, n-1).$$

Using estimates 9.4.1 (2) $\left(\text{recall that } \varkappa = \dfrac{1+s+l_1}{r_1} + \sum_2^n \dfrac{1}{r_j} - 1 > 0\right)$ we get

$$\int_{\mathbb{R}_{n-1}} |I_{x_1, -(s+l_1)} G_r| d\eta \ll \int_{\xi_j > 0} \frac{d\eta}{\xi^{r_n \varkappa} + \sum_1^{n-1} x_j^{r_j \varkappa}}$$

$$\ll \int \frac{\prod_1^{n-1} \lambda_j^{\frac{1}{r_j \varkappa} - 1} d\lambda}{\xi^{r_n \varkappa} + \sum_1^{n-1} \lambda_j} \ll \int \frac{\varrho^{\sum_1^{n-1} \frac{1}{r_j \varkappa} - 1}}{\xi^{r_n \varkappa} + \varrho} d\varrho$$

$$= \frac{1}{\xi^{r_n \varkappa}} \int\limits_{\varrho < \xi^{r_n \varkappa}} \varrho^{\sum_1^{n-1} \frac{1}{r_j \varkappa} - 1} d\varrho + \int\limits_{\xi^{r_n \varkappa} < \varrho} \varrho^{\sum_1^{n-1} \frac{1}{r_j \varkappa} - 2} d\varrho .$$

$$\ll \frac{1}{\xi^{r_n \varkappa}} \xi^{r_n \sum_1^{n-1} \frac{1}{r_j}} + \xi^{r_n \left(\sum_1^{n-1} \frac{1}{r_j} - \varkappa \right)} = \xi^{-r_n \left(\frac{s+l_1}{r_1} + \frac{1}{r_n} - 1 \right)}.$$

The first inequality follows from the first inequality 9.4.1 (2) ($\varkappa > 0$) and the symmetry of its right hand side. In the second relation we have made the substitution $x_j^{r_j \varkappa} = \lambda_j$. In the third we have introduced polar coordinates. In the last one we need to recall that

$$\varkappa - \sum_1^{n-1} \frac{1}{r_j} = \frac{s + l_1}{r_1} + \frac{1}{r_n} - 1 > 0 .$$

9.5. Imbedding Theorems

9.5.1. Theorem of different dimensions. *The following imbedding holds:*

(1) $$L_p^{\boldsymbol{r}}(\mathbb{R}_n) \to B_p^{\boldsymbol{\varrho}}(\mathbb{R}_m) ,$$

(2) $$\boldsymbol{\varrho} = (\varrho_1, \ldots, \varrho_m), \quad \varrho_i = r_i \varkappa \ (i = 1, \ldots, m) ,$$
$$1 < p \leqq \infty, \quad B_\infty^{\boldsymbol{\varrho}} = H_\infty^{\boldsymbol{\varrho}}, \quad 1 \leqq m < n,$$
$$\varkappa = 1 - \frac{1}{p} \sum_{m+1}^{n} \frac{1}{r_j} > 0 .$$

Under the condition $2 \leqq p \leqq \infty$ we have $L_p^{\boldsymbol{r}} \to B_p^{\boldsymbol{r}}$ (see 9.3 (3)) and the theorem follows from the corresponding theorem for the B-classes (see 6.5). Therefore it is only essential to prove it for $1 \leqq p < 2$. However the proof presented below is good for all finite p.

Proof. It is sufficient to carry out the proof in the case $m = n - 1$, because if $m < n - 1$ then one may pass from n to $n - 1$ using the imbedding (1). The passage from $n - 1$ to m is realized using the corresponding theorem for B classes (see 6.5). This is possible because of the transitivity of the relations (2) (see 7.1).

Thus, it is necessary only to prove the imbedding

(3) $$L_p^{\boldsymbol{r}}(\mathbb{R}_n) \to B_p^{\boldsymbol{\varrho}}(\mathbb{R}_{n-1}) ,$$

(4) $$\boldsymbol{\varrho} = (\varrho_1, \ldots, \varrho_{n-1}), \quad \varrho_i = r_i \varkappa ,$$

(5) $$\varkappa = 1 - \frac{1}{p r_n} > 0 .$$

We have (see 9.3 (1))

$$L_p^r(\mathbb{R}_n) \to H_p^r(\mathbb{R}_n) \to H_p^\varrho(\mathbb{R}_{n-1}) \to L_p(\mathbb{R}_{n-1}).$$

This shows that an arbitrary function $f \in L_p^r(\mathbb{R}_n)$ has a trace $g(x) = f|_{\mathbb{R}_{n-1}}$ on $\mathbb{R}_{n-1}$, belonging to $L_p(\mathbb{R}_{n-1})$, and the inequality

(6) $$\|g\|_{L_p(\mathbb{R}_{n-1})} \leqq c\|f\|_{L_p^r(\mathbb{R}_n)}$$

is satisfied. We will suppose that $\boldsymbol{y} = (x_1, \ldots, x_{n-1}) \in \mathbb{R}_{n-1}$, $z = x_n$. Suppose as always that $\bar{\varrho}_1$ is the largest integer less than ϱ_1. Let $f \in L_p^r(\mathbb{R}_n)$. In view of Theorem 9.2.2

$$\frac{\partial^{\bar{\varrho}_1} f}{\partial x_1^{\bar{\varrho}_1}} \in L_p^{r'}(\mathbb{R}_n),$$

where

$$\boldsymbol{r}' = \varkappa' \boldsymbol{r}, \quad \varkappa' = 1 - \frac{\bar{\varrho}_1}{r_1} > 0$$

(because $\bar{\varrho}_1 < \varrho_1 < r_1$), and

$$\left\|\frac{\partial^{\bar{\varrho}_1} f}{\partial x_1^{\bar{\varrho}_1}}\right\|_{L_p^{r'}(\mathbb{R}_n)} \leqq c\|f\|_{L_p^r(\mathbb{R}_n)}.$$

Therefore we have the representation

(7) $$\frac{\partial^{\bar{\varrho}_1} f}{\partial x_1^{\bar{\varrho}_1}} = \int G_{r'}(\boldsymbol{y} - \boldsymbol{\eta}, z - \zeta)\, v(\boldsymbol{\eta}, \zeta)\, d\boldsymbol{\eta}\, d\zeta \left(v \in L_p(\mathbb{R}_n)\right)$$

and

(8) $$\|v\|_{L_p(\mathbb{R}_n)} = \left\|\frac{\partial^{\bar{\varrho}_1} f}{\partial x_1^{\bar{\varrho}_1}}\right\|_{L_p^{r'}(\mathbb{R}_n)} \leqq c\|f\|_{L_p^r(\mathbb{R}_n)}.$$

Put

(9) $$w(\boldsymbol{y}) = \int G_{r'}(\boldsymbol{y} - \boldsymbol{\eta}, \zeta)\, v(\boldsymbol{\eta}, \zeta)\, d\boldsymbol{\eta}\, d\zeta = \left.\frac{\partial^{\bar{\varrho}_1} f}{\partial x_1^{\bar{\varrho}_1}}\right|_{\mathbb{R}_{n-1}}$$

(recall the evenness of $G_{r'}$). The explanation of the fact that the formal substitution $z = 0$ in (7) leads to the trace $\dfrac{\partial^{\bar{\varrho}_1} f}{\partial x_1^{\bar{\varrho}_1}}$ on $\mathbb{R}_m$ will be given at the end of the proof.

Let

$$\Lambda(\boldsymbol{y}, z) = \Delta_{x_1 h}^2 G_{r'}(\boldsymbol{y}, z)$$

be the second difference of $G_{r'}$ with step h in the direction of the x_1 axis. Then

$$\Delta^2_{x_1 h} w = \int_{-\infty}^{\infty} \left(\int_{\mathbb{R}_{n-1}} \Lambda(\mathbf{y} - \boldsymbol{\eta}, \zeta) v(\boldsymbol{\eta}, \zeta) \, d\boldsymbol{\eta} \right) d\zeta$$

$$= \int_{-\infty}^{\infty} \left(\int_{\mathbb{R}_{n-1}} \Lambda(\boldsymbol{\eta}, \zeta) \, v(\mathbf{y} - \boldsymbol{\eta}, \zeta) \, d\boldsymbol{\eta} \right) d\zeta ,$$

so that, twice applying the Minkowski inequality, we get

$$(10) \quad \|\Delta^2_{x_1 h} w\|_{L_p(\mathbb{R}_{n-1})} \leqq \int_{-\infty}^{\infty} \left\{ \int_{\mathbb{R}_{n-1}} d\mathbf{y} \left| \int_{\mathbb{R}_{n-1}} \Lambda(\boldsymbol{\eta}, \zeta) \, v(\mathbf{y} - \boldsymbol{\eta}, \zeta) d\boldsymbol{\eta} \right|^p \right\}^{1/p} d\zeta$$

$$\leqq \int_{-\infty}^{\infty} d\zeta \int_{\mathbb{R}_{n-1}} |\Lambda(\boldsymbol{\eta}, \zeta)| \, d\boldsymbol{\eta} \left(\int |v(\mathbf{y} - \boldsymbol{\eta}, \zeta)|^p d\mathbf{y} \right)^{1/p} = \int_{-\infty}^{\infty} I(h, \zeta) \| \, v(\boldsymbol{\eta}, \zeta) \|_{L_p(\mathbb{R}_{n-1})} d\zeta ,$$

$$(11) \qquad I(h, \zeta) = \int_{\mathbb{R}_{n-1}} |\Delta^2_{x_1 h} G_{r'}(\boldsymbol{\eta}, \zeta)| \, d\boldsymbol{\eta} .$$

Put

$$\alpha_1 = \varrho_1 - \bar{\varrho}_1$$

and note that

$$\frac{1}{r_{n'}} + \frac{\alpha_1}{r_1'} = \frac{1}{1 - \dfrac{\bar{\varrho}_1}{r_1}} \left(\frac{1}{r_n} + \frac{\alpha_1}{r_1} \right) = \frac{\dfrac{1}{r_n} + \dfrac{\alpha_1}{r_1}}{\dfrac{1}{p r_n} + \dfrac{\alpha_1}{r_1}} > 1 \; (p > 1) .$$

Therefore it is possible to define l_1 so as to satisfy the inequality $0 < l_1 < \alpha_1$ and such that

$$\frac{1}{r_n'} + \frac{l_1}{r_1} > 1 .$$

In such a case, in view of the estimate 9.4.9 (2) for the kernel $G_{r'}$,

$$(12) \qquad I(h, \zeta) \ll |h|^{1 + l_1} |\zeta|^{\beta'} ,$$

$$(13) \qquad I(h, \eta) \ll |h|^{l_1} |\zeta|^{\beta} .$$

Here the absolute value of the second difference has been replaced by the sum of the absolute values of the first differences, which exceeds it, and

$$\beta = -r_n' \left(\frac{1}{r_n'} + \frac{l_1}{r_1'} - 1 \right) ,$$

$$\beta' = -r_n' \left(\frac{1}{r_n'} + \frac{l_1 + 1}{r_1'} - 1 \right) = \beta - \frac{r_n}{r_1} .$$

We introduce also the numbers

$$\alpha = \frac{r'_n}{r'_1} p(l_1 - \alpha_1) - 1 < -1,$$

$$\alpha' = \frac{r'_n}{r'_1} p(l_1 - \alpha_1 + 1) - 1 > -1, \quad \alpha' = \alpha + \frac{r_n}{r_1} p.$$

The numbers α, α', β, β' are connected by the following relations:

$$\alpha + p + p\beta = -1 - \frac{r'_n a_1 p}{r'_1} + r'_n p = -1 - \frac{r_n \alpha_1 p}{r_1} + r_n p \left(1 - \frac{\bar{\varrho}_1}{r_1}\right)$$

$$= -1 - \frac{r_n \alpha_1 p}{r_1} + r_n p \left(1 - \frac{\varrho_1 - \alpha_1}{r_1}\right) = -1 + r_n p \left(1 - \frac{\varrho_1}{r_1}\right)$$

$$= -1 + r_n p \left(1 - 1 + \frac{1}{r_n p}\right) = 0,$$

$$\alpha' + p + p\beta' = 0.$$

Below we shall make use of *Hardy's inequality*, wich holds for $\varphi(t) \geqq 0$:

$$\int_0^\infty t^\alpha \Big(\int_{|\zeta|<t} \varphi(\zeta) d\zeta\Big)^p dt = \left\{\int_0^\infty dt \Big(\int_{|u|<1} \varphi(tu) t^{\frac{\alpha}{p}+1} du\Big)^p\right\}^{1/p}$$

$$\leqq \int_{|u|<1} \Big(\int_0^\infty \varphi(tu)^p t^{\alpha+p} dt\Big)^{1/p} du = c \Big(\int_{-\infty}^\infty \varphi(\zeta)^p |\zeta|^{\alpha+p} d\zeta\Big)^{1/p}.$$

$$c = \int_0^1 \frac{du}{u^{1+\frac{\alpha+1}{p}}} < \infty, \text{ when } \alpha < -1, \quad 1 \leqq p < \infty.$$

Its proof thus reduces to the change of variable $\zeta = tu$ and thereupon to the application of the Minkowski inequality[1]. Analogously one proves the inequality

$$\int_0^\infty t^\alpha \Big(\int_{|\zeta|>t} \varphi(\zeta) d\zeta\Big)^p dt \leqq c_1 \int_{-\infty}^\infty \varphi(\zeta)^p |\zeta|^{\alpha+p} d\zeta,$$

$$c_1 = \int_1^\infty \frac{du}{u^{1+\frac{\alpha+1}{p}}} \quad (\alpha > -1, \ 1 \leqq p < \infty).$$

[1] See the book of Hardy, Littlewood and Polya [1].

Now we have (in the third relation writing $t = h^{r_1/r_n}$)

$$\|f\|^p_{b^{\varrho_1}_{x_1 p}(\mathbb{R}_{n-1})} = \int_0^\infty h^{-1-p\alpha_1} \|\Delta^2_{x_1 h}\omega\|^p_{L_p(\mathbb{R}_{n-1})} dh$$

$$= \int_0^\infty h^{-1-p\alpha_1} dh \left\{ \left[\int_{|\zeta|<h^{r_1/r_n}} + \int_{|\zeta|>h^{r_1/r_n}} \right] \times I(\eta, \zeta) \, \|v(\eta, \zeta)\|_{L_p(\mathbb{R}_{n-1})} d\zeta \right\}$$

$$\ll \int_0^\infty h^{-1-p\alpha_1} dh \left(\int_{|\zeta|<h^{r_1/r_n}} h^{l_1} \zeta^\beta \|v(\eta, \zeta\|)_{L_p(\mathbb{R}_{n-1})} d\zeta \right.$$

$$\left. + \int_{|\zeta|>h^{r_1/r_n}} h^{l_1+1} \zeta^{\beta'} \|v(\eta, \zeta)\|_{L_p(\mathbb{R}_{n-1})} d\zeta \right)^p$$

$$= \int_0^\infty t^\alpha dt \left(\int_{|\zeta|<t} \zeta^\beta \|v(\eta, \zeta\|_{L_p(\mathbb{R}_{n-1})} d\zeta \right)^p$$

$$+ \int_0^\infty t^{\alpha'} dt \left(\int_{|\zeta|>t} \zeta^{\beta'} \|v(\eta, \zeta)\|_{L_p(\mathbb{R}_{n-1})} d\zeta \right)^p \ll \int_{-\infty}^\infty \|v(\eta, \zeta)\|^p_{L_p(\mathbb{R}_{n-1})} d\zeta = \|v\|^p_{L_p(\mathbb{R}_n)}.$$

Since in the resulting inequality x_1 may be replaced by x_j, then we have proved (see also (8)) that

$$\|f\|_{b^{\varrho_i}_{x_i p}(\mathbb{R}_n)} \ll \|v\|_{L_p(\mathbb{R}_n)} \quad (i = 1, \ldots, n-1),$$

so that (see further (6)) we have for $z = 0$

$$\|f(\boldsymbol{y}, z)\|_{B^{\boldsymbol{\varrho}}_p(\mathbb{R}_{n-1})} \ll \|f\|_{L^{\boldsymbol{r}}_p(\mathbb{R}_n)}.$$

It is obviously true for any z, not necessarily equal to zero, as is proved analogously. This proves (3).

The function $f \in L^{\boldsymbol{r}}_p(\mathbb{R}_n)$ may be written out in the form

$$f(\boldsymbol{y}, z) = \int G_r(\boldsymbol{y} - \boldsymbol{\eta}, z - \zeta) \, \lambda(\boldsymbol{\eta}, \zeta) \, d\boldsymbol{\eta} \, d\zeta,$$

$$\|f\|_{L^{\boldsymbol{r}}_p(\mathbb{R}_n)} = \|\lambda\|_{L_p(\mathbb{R}_n)}.$$

Hence

$$f(\boldsymbol{y}, z + h) - f(\boldsymbol{y}, z)$$

$$= \int G_{\boldsymbol{r}}(\boldsymbol{y} - \boldsymbol{\eta}, z - \zeta) \, [\lambda(\boldsymbol{\eta}, \zeta + h) - \lambda(\boldsymbol{\eta}, \zeta)] \, d\boldsymbol{\eta} \, d\zeta$$

and, in view of (14),

$$\left\|\frac{\partial^{\bar{\varrho}_1}}{\partial x_1^{\bar{\varrho}_1}}[f(\boldsymbol{y}, z+h) - f(\boldsymbol{y}, z)]\right\|_{L_p(\mathbb{R}_{n-1})} \leqslant \|f(\boldsymbol{y}, z+h) - f(\boldsymbol{y}, z)\|_{B_p^{\boldsymbol{\varrho}}(\mathbb{R}_{n-1})}$$

$$\leqslant \|f(\boldsymbol{y}, z+h) - f(\boldsymbol{y}, z\|_{L_p^{\boldsymbol{r}}(\mathbb{R}_n)}$$

$$= \|\lambda(\eta, \zeta + h) - \lambda(\eta, \zeta)\|_{L_p(\mathbb{R}_n)} \to 0 \quad (h \to 0) \quad (1 \leqq p < \infty).$$

This shows that if one fixes z in $\dfrac{\partial^{\bar{\varrho}_1}}{\partial x_1^{\bar{\varrho}_1}} f(\boldsymbol{y}, z)$, the resulting function of $\boldsymbol{y} \in \mathbb{R}_{n-1}$ is the trace of $\dfrac{\partial^{\bar{\varrho}_1} f}{\partial x_1^{\bar{\varrho}_1}}$ on the subspace $x_n = z$.

9.5.2. Inverse theorem of different dimensions. *Suppose that* $1 \leqq p < \infty$ *and suppose given positive numbers* r_i, $i = 1, \ldots, n$, *and all possible vectors* $\boldsymbol{\lambda}$ *with nonnegative integer coordinates*

$$\boldsymbol{\lambda} = (\lambda_{m+1}, \ldots, \lambda_n),$$

for which

$$\varrho_i^{(\lambda)} = r_i \left(1 - \sum_{j=m+1}^{n} \frac{\lambda_j}{r_j} - \frac{1}{p} \sum_{j=m+1}^{n} \frac{1}{r_j}\right) = r_i \varkappa > 0. \tag{1}$$

Suppose further that to each vector $\boldsymbol{\lambda}$ *there has been assigned a function*

$$\varphi_{(\lambda)}(\boldsymbol{u}) = \varphi_{(\lambda)}(x_1, \ldots, x_m) \in B_p^{\varrho^{(\lambda)}}(\mathbb{R}_m).$$

Then it is possible to construct a function $f(\boldsymbol{x}) = f(x_1, \ldots, x_n)$ *of* n *variables, having the following properties:*

$$f \in L_p^{\boldsymbol{r}}(\mathbb{R}_n), \tag{2}$$

$$\|f\|_{L_p^{\boldsymbol{r}}(\mathbb{R}_n)} \leqq c \sum_{\lambda} \|\varphi_{(\lambda)}\|_{B_p^{\varrho^{(\lambda)}}(\mathbb{R}_m)}, \tag{3}$$

$$f^{(\lambda)}|_{\mathbb{R}_m} = \varphi_{(\lambda)}(\boldsymbol{u}). \tag{4}$$

Proof. We shall show that as f we make choose the function

$$f = \sum_{\lambda} f_{(\lambda)} \tag{5}$$

already defined in 6.8, the sum being extended over all possible admissible vectors $\boldsymbol{\lambda}$ and

$$f_{(\lambda)} = \sum_{s=0}^{\infty} \varphi_{(\lambda)}^{s} \prod_{j=m+1}^{n} b^{-\frac{s\lambda_j}{r_j\varkappa}} \Phi_{\lambda_j}\left(b^{\frac{s}{r_j\varkappa}} x_j\right). \tag{6}$$

It is essential to note that the functions

$$q_s = \overset{s}{\varphi}_{(\lambda)} = g_{b^{\frac{s}{r_1\varkappa}},\ldots,b^{\frac{s}{r_m\varkappa}}} \qquad (s = 0, 1, \ldots)$$

are entire of exponential type $b^{\frac{s}{r_j\varkappa}}$ relative to x_j $(j = 1, \ldots, m)$:

$$\varphi_{(\lambda)} = \sum_{s=0}^{\infty} \overset{s}{\varphi}_{(\lambda)}$$

and

$$\|\varphi_{(\lambda)}\|_{B_p^{\boldsymbol{\varrho}^{(\lambda)}}(\mathbb{R}_m)} = \left(\sum_{s=0}^{\infty} b^{sp}\|q_s\|_{L_p(\mathbb{R}_m)}^p\right)^{1/p} < \infty.$$

As to the functions $\Phi_{\lambda_j}(t)$, they may be regarded as equal to

$$\Phi_{\lambda_j}(t) = \frac{T_{\lambda_j}(A_j t)}{t^3}. \tag{7}$$

where the T_{λ_j} are appropriately chosen trigonometric polynomials and the A_j are numbers. In addition the Φ_{λ_j} are entire functions of exponential type 1. In distinction from 6.8, in the denominator of (7) we have t^3 (instead of t^2), which is inessential, and to take up for this we now have the functions Φ_{λ_j} in $L = L(-\infty, \infty)$ along with their derivatives of the first order.

The fact that $f \in B_p^{\mathbf{r}}(\mathbb{R}_n)$ and that the boundary properties (4) are satisfied was proved in Theorem 6.8. It remains to prove properties (2), (3).

We denote by $\mathbb{R}_{n-1}$ the subspace of points $(x_1, \ldots, x_{n-1})$. The following inequality holds (explanation below):

$$\|I_{x_i(-r_i)} f_{(\lambda)}(\boldsymbol{x})\|_{L_p(\mathbb{R}_{n-1})} \leqq \sum_s \|q_s\|_{L_p(\mathbb{R}_m)} b^{\frac{s}{\varkappa}\left(1 - \sum\limits_{m+1}^{n}\frac{\lambda_j}{r_j} - \frac{1}{p}\sum\limits_{m+1}^{n-1}\frac{1}{r_j}\right)} |\psi_s(x_n)| \tag{8}$$

$$= \sum_s \lambda_s a^{s/p} |\psi_s(x_n)|,$$

where

$$\lambda_s = \|q_s\|_{L_p(\mathbb{R}_m)} b^s, \quad a = b^{\frac{1}{r_n\varkappa}} \quad (a > 1), \tag{9}$$

$$\psi_s(t) = \Phi(a^s t) \text{ for } i = 1, \ldots, n-1, \tag{10}$$

$$\psi_s(t) = a^{-r_n s} I_{-r_n} \Phi(a^s t) \text{ for } i = n, \Phi = \Phi_{\lambda_n}. \tag{11}$$

The norm in the metric of $L_p(\mathbb{R}_{n-1})$ of each term of the series (6) is equal to the the product of the L_p-norms of the factors of which it is composed. Here we need to recall that

$$\|I_{x_i(-r_i)}q_s\|_{L_p(\mathbb{R}_m)} \leqslant b^{s/\varkappa}\|q_s\|_{L_p(\mathbb{R}_m)}$$

$$(i = 1, \ldots, m;\ s = 0, 1, \ldots)$$

(see 8.7),

$$\left\|I_{x_i(-x_i)}\Phi_{\lambda_i}\left(b^{\frac{s}{r_i\varkappa}}x_i\right)\right\|_{L_p(\mathbb{R}_{x_i})} \leqslant b^{\frac{s}{\varkappa}}\left\|\Phi_{\lambda_i}\left(b^{\frac{s}{r_i\varkappa}}x_i\right)\right\|_{L_p(\mathbb{R}_{x_i})},$$

$$\left\|\Phi_{\lambda_i}\left(b^{\frac{s}{r_i\varkappa}}x_i\right)\right\|_{L_p(\mathbb{R}_{x_i})} = c_i b^{-\frac{s}{pr_i\varkappa}} \quad (i = m+1, \ldots, n-1),$$

where $\mathbb{R}_{x_i}$ is the x_i-axis and c_i does not depend on $s = 0, 1, \ldots,$

From (8) it follows that (explanation below)

$$\|I_{x_i(-r_i)}f_{(\lambda)}(\boldsymbol{x})\|_{L_p(\mathbb{R}_{n-1})} \leqslant \left(\int_{-\infty}^{\infty}\left|\sum_s \lambda_s a^{s/p}\psi_s(y)\right|^p dy\right)^{1/p}$$

$$\leqslant \left(\sum_s \lambda_s^p\right)^{1/p} = \left(\sum_s b^{sp}\|q_s\|^p_{L_p(\mathbb{R}_m)}\right)^{1/p} = \|\varphi_{(\lambda)}\|_{B_p^{\varrho\,(\lambda)}(\mathbb{R}_m)},$$

which proves (2), (3). But in these relations we need to justify the second inequality.

We note the inequalities

(12) $$|\psi_s(t)| < A,$$

(13) $$|\psi_s(t)| < \frac{A}{a^s|t|} \quad (a^s|t| > 1),$$

(14) $$a^s\int_{-\infty}^{\infty}|\psi_s(t)|dt < A,$$

where the constant A does not depend on t and s. In the case (10) these inequalities follow immediately from the fact that $\Phi(t)$ is an entire function representable in the form (7). In the case (11), however, this requires explanations. The function $\Phi(t)$ is entire exponential of type 1, and belongs to L along with its derivatives, so that its Fourier transform $\sqrt{2\pi}\,\mu(x)$ has a continuous derivative and a compact support on $(-1, +1)$.

Thus, $\mu(1) = \mu(-1) = 0$, and accordingly ($r = r_n$)

(15) $$\Phi(t) = \int_{-1}^{+1}\mu(\lambda)e^{i\lambda t}\,d\lambda,$$

$$\Phi(a^s t) = \int_{-1}^{+1}\mu(\lambda)e^{i\lambda a^s t}d\lambda = a^{-s}\int_{-a^s}^{a^s}\mu(a^{-s}\,\xi)\,e^{i\xi t}d\xi,$$

$$|I_{t(-r)}\Phi(a^s t)| = a^{-s}\left|\int_{-a^s}^{a^s}(1+\xi^2)^{r/2}\mu(a^{-s}\xi)e^{i\xi t}d\xi\right|$$

$$= \left|\frac{a^{-s}}{t}\int_{-a^s}^{a^s}[\mu'(a^{-s}\xi)a^{-s}(1+\xi^2)^{r/2} + r(1+\xi^2)^{r/2-1}\,\xi\mu(a^{-s}\xi)]\,e^{i\xi t}d\xi\right|$$

$$\ll \frac{a^{-s}}{|t|}\,a^s(a^{-s}a^{rs} + a^{(r-2)s}a^s) = \frac{a^{(r-1)s}}{|t|}.$$

We have proved (13) (see (11)). Further, if we recall that $\Phi(a^s t)$ is entire of type a^s, we get

$$|\psi_s(t)| \leqq a^{-r_n s}a^{r_n s}\max|\Phi(a^s t)| < A,$$

$$\int|\psi_s(t)|dt \leqq a^{-r_n s}a^{r_n s}\int|\Phi(a^s t)|dt < Aa^{-s},$$

i.e. (12) and (14).

Now we have

$$\int_0^\infty\left|\sum_s \lambda_s a^{s/p}\psi_s(y)\right|^p dy \ll \Lambda_1 + \Lambda_2 + \Lambda_3, \tag{16}$$

where

$$\Lambda_1 = \sum_{m=0}^\infty\int_{a^{-m-1}}^{a^{-m}}\left|\sum_{s=0}^m \lambda_s a^{s/p}\psi_s(y)\right|^p dy,$$

$$\Lambda_2 = \sum_{m=0}^\infty\int_{a^{-m-1}}^{a^{-m}}\left|\sum_{s=m+1}^\infty \lambda_s a^{s/p}\psi_s(y)\right|^p dy,$$

$$\Lambda_3 = \int_1^\infty\left|\sum_{s=0}^\infty \lambda_s a^{s/p}\psi_s(y)\right|^p dy.$$

But $\left(\text{see (12)}, \frac{1}{p} + \frac{1}{q} = 1\right)$

$$\Lambda_1 \ll \sum_{m=0}^\infty a^{-m}\left(\sum_{s=0}^\infty \lambda_s a^{\frac{s(1-\varepsilon)}{p}}a^{\frac{s\varepsilon}{p}}\right)^p$$

$$\leqq \sum_{m=0}^{\infty} a^{-m} \sum_{s=0}^{m} \lambda_s^p a^{s(1-\varepsilon)} \left(\sum_{s=0}^{m} a^{\frac{s\varepsilon q}{p}} \right)^{p/q} \leqq \sum_{m=0}^{\infty} a^{-m} \sum_{s=0}^{m} \lambda_s^p a^{s(1-\varepsilon)} a^{m\varepsilon}$$

$$= \sum_{s=0}^{\infty} \lambda_s^p a^{s(1-\varepsilon)} \sum_{m=s}^{\infty} a^{-m(1-\varepsilon)} \ll \sum_{s=0}^{\infty} \lambda_s^p a^{s(1-\varepsilon)} a^{s(\varepsilon-1)} = \sum_{0}^{\infty} \lambda_s^p .$$

For $p = 1$ the third term in this chain may be dropped.

Further (see (14), explanation below)

$$\Lambda_2 \leqq \sum_{m=0}^{\infty} \int_{a^{-m-1}}^{a^{-m}} \sum_{s=m+1}^{\infty} \lambda_s^p a^s \, |\psi_s(y)| \sum_{s=m+1}^{\infty} |\psi_s(y)|^{p-1} dy$$

$$\ll \sum_{m=0}^{\infty} \sum_{s=m+1}^{\infty} \lambda_s^p a^s \int_{a^{-m-1}}^{a^{-m}} |\psi_s(y)| dy = \sum_{s=0}^{\infty} \lambda_s^p a^s \sum_{m=0}^{s} \int_{a^{-m-1}}^{a^{-m}} |\psi_s| dy$$

$$= \sum_{s=0}^{\infty} \lambda_s^p a^s \int_{0}^{1} |\psi_s| dy \ll \sum_{s=0}^{\infty} \lambda_s^p ,$$

because (see (13))

(17) $$\sum_{s=m+1}^{\infty} |\psi_s(y)| \ll \frac{1}{|y|} \sum_{s=m+1}^{\infty} a^{-s} \ll a^m a^{-m-1} \ll 1 \, (a^{-m-1} < y) .$$

Finally (see (14)),

$$\Lambda_3 \leqq \int_{1}^{\infty} \sum_{s=0}^{\infty} \lambda_s^p a^s |\psi_s(y)| \left(\sum_{s=0}^{\infty} |\psi_s(y)| \right)^{p-1} dy \ll \sum_{s=0}^{\infty} \lambda_s^p a^s \int_{1}^{\infty} |\psi_s| dy \ll \sum_{s} \lambda_s^p ,$$

because

$$\sum_{s=0}^{\infty} |\psi_s(y)| = |\psi_0(y)| + \sum_{1}^{\infty} |\psi_s(y)| \ll 1 \quad (1 \leqq y < \infty)$$

(see (12) and (17) for $m = 0$).

We have proved that the integral (16) does not exceed $\Sigma \lambda_s^p$. The similar fact is proved analogously for the same integral, extended over $\{-\infty < y < 0\}$.

9.5.3. If we suppose, as we have agreed to do, that $B_\infty^\varrho = H_\infty^\varrho$, then Theorem 9.5.2 for $p = \infty$ is no longer valid. Indeed, an arbitrary function

$$f(x, y) \in W_\infty^{1,1}(\mathbb{R}_2) = L_\infty^{1,1}(\mathbb{R}_2) \subset H_\infty^{1,1}(\mathrm{R}_2) \subset H_\infty^{\alpha,\alpha}(\mathbb{R}_2) ,$$

$$0 < \alpha < 1 ,$$

is uniformly continuous (after appropriate alterations on a set of plane measure zero). It satisfies on $\mathbb{R}_2$, and therefore on the $\mathbb{R}_1$ axis, a Lipschitz condition. However on the $\mathbb{R}_1$ axis one may define a function $\varphi(x_1) \in H^1_\infty(\mathbb{R}_1)$, not satisfying a Lipschitz condition (in fact nowhere differentiable; see the Remark to 5.6.2—5.6.3). Accordingly, there does not exist a function $f(x_1, x_2) \in W^{1,1}_\infty(\mathbb{R}_2)$ which can extend φ from $\mathbb{R}_1$ to $\mathbb{R}_2$.

9.5.4[1]**.** As a consequence of Theorems 9.5.1 and 9.5.2 one can obtain the imbedding theorems

$$B^{\boldsymbol{r}}_p(\mathbb{R}_n) \rightleftarrows B^{\boldsymbol{\varrho}}_p(\mathbb{R}_m), \tag{1}$$

$\boldsymbol{\varrho} = \varkappa \boldsymbol{r}$, $\varkappa = 1 - \frac{1}{p} \sum\limits_{m+1}^{n} \frac{1}{r_j} > 0, 1 < p < \infty$. Indeed (explanation below)

$$B^{\boldsymbol{r}}_p(\mathbb{R}_n) \to L_p^{\frac{r_1}{\varkappa_1}, \ldots, \frac{r_n}{\varkappa_1}, r_{n+1}}(\mathbb{R}_{n+1}) \to B_p^{\frac{\varkappa_2}{\varkappa_1} r_1, \ldots, \frac{\varkappa_2}{\varkappa_1} r_m}(\mathbb{R}_m) = B^{\boldsymbol{\varrho}}_p(\mathbb{R}_m), \tag{2}$$

where

$$\varkappa_1 = 1 - \frac{1}{p r_{n+1}} > 0, \quad \varkappa_2 = 1 - \frac{1}{p} \sum_{m+1}^{m} \frac{\varkappa_1}{r_j} - \frac{1}{p r_{n+1}} = \varkappa_1 \varkappa;$$

$$B^{\boldsymbol{\varrho}}_p(\mathbb{R}_m) \to L_p^{\frac{\varrho_1}{\varkappa_2}, \ldots, \frac{\varrho_m}{\varkappa_2}, \frac{r_{m+1}}{\varkappa_1}, \ldots, \frac{r_n}{\varkappa_1}, r_{n+1}}(\mathbb{R}_{n+1}) \to B^{\boldsymbol{r}}_p(\mathbb{R}_n). \tag{3}$$

The first imbedding in (2), as well as in (3), follows from Theorem 9.5.2, and the second in (2) and (3) from Theorem 9.5.1.

9.6. Imbedding Theorem with a Limiting Exponent

9.6.1. Lemma. *Suppose that* $g \in L_{q'}(\mathbb{R}_m)$, $f \in L_p(\mathbb{R}_n)$,

$$1 \leqq m \leqq n, \quad 1 < p < q < \infty, \quad \frac{1}{p} + \frac{1}{p'} = 1, \quad \frac{1}{q} + \frac{1}{q'} = 1, \tag{1}$$

$$\boldsymbol{x} = (x_1, \ldots, x_n) \in \mathbb{R}_n, \quad r(\boldsymbol{\xi}) = \left\{ \sum_1^n |\xi_j|^{2/\varkappa_j} \right\}^{1/2}, \quad \varkappa_j > 0,$$

$$\boldsymbol{\xi} = (\xi_1, \ldots, \xi_n), \quad \lambda = \frac{1}{p'} \sum_1^n \varkappa_j + \frac{1}{q} \sum_1^m \varkappa_j.$$

---

[1] This Remark is due to V. I. Burenkov.

Then the following inequality is valid[1]:

(2) $$\left| \int\limits_{\mathbb{R}_m} dx \int\limits_{\mathbb{R}_n} \frac{g(\boldsymbol{x})\, f(\boldsymbol{y}) d\boldsymbol{y}}{r^\lambda(\boldsymbol{x}-\boldsymbol{y})} \right| \leqq c \|f\|_{L_p(\mathbb{R}_n)} \|g\|_{L_{q'}(\mathbb{R}_m)}$$

where c does not depend on f, g and $(x_{m+1}, \ldots, x_n) \in \mathbb{R}_{n-m}$, *so that*

(3) $$\left\| \int\limits_{\mathbb{R}_n} \frac{f(\boldsymbol{y})\, d\boldsymbol{y}}{r^\lambda(\boldsymbol{x}-\boldsymbol{y})} \right\|_{L_q(\mathbb{R}_m)} \leqq c \|f\|_{L_p(\mathbb{R}_n)}.$$

Proof. In the one-dimensional case ($n = m = 1$, $\varkappa_1 = 1$), (2) is the Hardy-Littlewood inequality

(4) $$\left| \int\limits_{-\infty}^{\infty} \int\limits_{-\infty}^{\infty} \frac{g(\xi)\, f(\eta) d\xi\, d\eta}{|\xi-\eta|^{\frac{1}{p'}+\frac{1}{q}}} \right| \leqq c \|f\|_{L_p(\mathbb{R}_1)} \|g\|_{L_{q'}(\mathbb{R}_1)}.$$

We shall not prove this here[2]. The fact that (2) implies (3) is a theorem of F. Riesz (see Banach [1]), stating that *if a function F, measurable on* $\mathbb{R}_m$, *is such that the Lebesgue integral*

$$\int\limits_{\mathbb{R}_m} Fg\, dx$$

exists for any $g \in L_{q'}(\mathbb{R}_m)$, *then*

$$F \in L_q(\mathbb{R}_m)$$

and

$$\|F\|_{L_q(\mathbb{R}_m)} = \sup_{\|g\|_{L_{q'}(\mathbb{R}_m)} \leqq 1} \int\limits_{\mathbb{R}_m} Fg\, dx.$$

We write out the integral being estimated in the form

(5) $$I = \int\limits_{\mathbb{R}_m} g(\boldsymbol{x})\, d\boldsymbol{x} \int\limits_{\mathbb{R}_m} d\boldsymbol{y}' \left[\int\limits_{\mathbb{R}_{n-m}} \frac{f(\boldsymbol{y})\, d\boldsymbol{y}''}{r^\lambda} \right],$$

---

[1] Hardy and Littlewood [1] for the case $n = m = 1$; V. I. Il'in [8] for the general case.

[2] The proof will be found in the book of Hardy, Littlewood, and Polya [1].

$\mathbf{y}' = (y_1, \ldots, y_m)$, $\mathbf{y}'' = (y_{m+1}, \ldots, y_n)$. To the integral in brackets we apply the Hölder inequality

$$|[\cdot]| \leqq \left(\int\limits_{\mathbb{R}_{n-m}} |f(\mathbf{y})|^p d\mathbf{y}'' \right)^{1/p} \left(\int\limits_{\mathbb{R}_{n-m}} \frac{d\mathbf{y}''}{r^{\lambda p'}} \right)^{1/p'} = P(\mathbf{y}')\, Q(\mathbf{y}') .$$

But (explanation below)

$$Q^{p'} = \int\limits_{\mathbb{R}_{n-m}} \frac{dy_{m+1} \cdots dy_n}{\left\{ H^2 + \sum\limits_{m+1}^{n} |y_j|^{2/\varkappa_j} \right\}^{1/2\left(\sum\limits_{m+1}^{n} \varkappa_i + \varepsilon \right)}}$$

$$= \frac{1}{H^\varepsilon} \int\limits_{\mathbb{R}_{n-m}} \frac{du_{m+1} \cdots du_n}{\left\{ 1 + \sum\limits_{m+1}^{n} |u_j|^{2/\varkappa_j} \right\}^{1/2\left(\sum\limits_{m+1}^{n} \varkappa_j + \varepsilon \right)}} = \frac{c}{H^\varepsilon}$$

$$= \frac{c}{\left\{ \sum\limits_{1}^{m} |x_j - y_j|^{2/\varkappa_i} \right\}^{\frac{p'}{2} \sum\limits_{1}^{m} \varkappa_j \left(\frac{1}{p'} + \frac{1}{q} \right)}} \leqq \frac{c}{\left\{ \prod\limits_{j=1}^{m} |x_j - y_j| \right\}^{p'\left(\frac{1}{p'} + \frac{1}{q} \right)}} .$$

Above we have used the following notations:

$$H^2 = \left\{ \sum_{1}^{m} |x_j - y_j|^{2/\varkappa_j} \right\},$$

$$\varepsilon = \lambda p' - \sum_{m+1}^{n} \varkappa_j = \sum_{1}^{m} \varkappa_j + \frac{p'}{q} \sum_{1}^{m} \varkappa_j > 0 .$$

In the second equation we have introduced the substitution

$$u_j = \frac{y_j}{H^{\varkappa_j}} \quad (j = 1, \ldots, m-1).$$

The integral in the third term is denoted by c. Its finiteness on the unit sphere in $\mathbb{R}_{n-m}$ is obvious. Outside it, if one puts $u_j^{2/\varkappa_j} = \xi_j$, restricting the u_j to be positive, and introduces polar coordinates for $\boldsymbol{\xi} = (\xi_{m+1}, \ldots, \xi_n)$, then the corresponding integral is estimated as follows:

$$\int\limits_{|\xi|>\beta>0} \frac{\prod\limits_{m+1}^{n} \xi^{\frac{\varkappa_j}{2}-1} d\xi}{\left(\sum\limits_{m+1}^{n} \xi_j\right)^{\frac{1}{2}\left(\sum\limits_{m+1}^{n} \varkappa_j + \varepsilon\right)}} \ll \int\limits_{\beta}^{\infty} \varrho^{-1-\frac{\varepsilon}{2}} d\varrho < \infty. \quad (\varepsilon > 0)$$

The last inequality is obtained from the obvious inequalities ($\xi_j = |x_j - y_j|$)

$$\xi_j = \xi_j^{\frac{2}{\varkappa_j}\frac{\varkappa_j}{2}} \leqq \left(\sum_{s=1}^{n} \xi_s^{2/\varkappa_s}\right)^{\varkappa_j/2} \quad (j = 1, \ldots, m),$$

which remain to be multiplied out and raised to the degree $p'\left(\frac{1}{p'} + \frac{1}{q}\right)$.

Accordingly,

$$|I| \ll \int\limits_{-\infty}^{\infty}\int\limits_{-\infty}^{\infty} \frac{dx_1\, dy_1}{|x_1 - y_1'|^{\frac{1}{p'}+\frac{1}{q}}} \cdots \int\limits_{-\infty}^{\infty}\int\limits_{-\infty}^{\infty} \frac{|g(\boldsymbol{x})|\, P(\boldsymbol{y}')}{|x_m - y_m|^{\frac{1}{p'}+\frac{1}{q}}}\, dx_m\, dy_m,$$

from which (2) follows on m-fold application of the onedimensional inequality (4).

9.6.2. Generalization of the Sobolev imbedding theorem[1].

Theorem. *Under the hypotheses* $1 < p < q < \infty$, $1 \leqq m \leqq n$,

(1) $\boldsymbol{r} = (r_1, \ldots, r_n) \geqq \boldsymbol{0}, \quad \varkappa = 1 - \left(\frac{1}{p} - \frac{1}{q}\right) \sum_{1}^{m} \frac{1}{r_j} - \frac{1}{p} \sum_{m+1}^{n} \frac{1}{r_j} \geqq 0$

one has the imbedding

(2) $$L_p^{\boldsymbol{r}}(\mathbb{R}_n) \to L_q^{\boldsymbol{\varrho}}(\mathbb{R}_m),$$

$$\boldsymbol{\varrho} = (\varrho_1, \ldots, \varrho_m),\ \varrho_j = \varkappa r_j \quad (j = 1, \ldots, m).$$

Proof. We will denote by $I_{-\boldsymbol{r}}$ the operation inverse to $I_{\boldsymbol{r}}$($\boldsymbol{r} \geqq \boldsymbol{0}$, $I_{\boldsymbol{0}}$ is the identity operator), and recall that the operations $I_{\boldsymbol{r}}, I_{\boldsymbol{r}'}, I_{-\boldsymbol{r}}, I_{-\boldsymbol{r}'}$ are commutative. Suppose that $f \in L_p^{\boldsymbol{r}}(\mathbb{R}_n)$ ($\boldsymbol{r} > \boldsymbol{0}$). Then

$$f = I_{\boldsymbol{r}} g \ \left(g \in L_p(\mathbb{R}_n)\right)$$

and accordingly

$$f = I_{\boldsymbol{\varrho}} I_{\boldsymbol{r}(1-\varkappa)} h,$$

[1] See the Remarks to 6.1 and 9.6.2.

where

$$h = I_{-\varrho} I_{-\boldsymbol{r}(1-\varkappa)} I_{\boldsymbol{r}} g, \; \|h\|_{L_p(\mathbb{R}_n)} \leqq c\|g\|_{L_p(\mathbb{R}_n)},$$

because the function

$$\left\{\sum_1^m (1 + u_j^2)^{\frac{r_j \varkappa}{2\sigma}}\right\}^\sigma \left\{\sum_1^n (1 + u_j^2)^{\frac{r_j(1-\varkappa)}{2\sigma}}\right\}^\sigma \left\{\sum_1^n (1 + u_j^2)^{r_j/2\sigma}\right\}^{-\sigma}$$

is a Marcinkiewicz multiplier (see 1.5.5, Example 12 and the Remark at the end of 1.5.5). So,

$$f = I_\varrho u, \tag{3}$$

$$u = \int G_{(1-\varkappa)\boldsymbol{r}}(\boldsymbol{x} - \boldsymbol{y})\, h(\boldsymbol{y})\, d\boldsymbol{y} \tag{4}$$

and for sufficiently large values of the parameter σ one has the inequalities

$$|G_{(1-\varkappa)\boldsymbol{r}}(\boldsymbol{x})| \ll \left\{\sum_1^n |x_j|^{r_j(1-\varkappa)\left(\sum\limits_{s=1}^n \frac{1}{r_s(1-\varkappa)} - 1\right)}\right\}^{-1} \tag{5}$$

$$= \left\{\sum_1^n |x_j|^{r_j\left(\sum\limits_{s=1}^n \frac{1}{r_s} - 1 + \varkappa\right)}\right\}^{-1} \leqq \left\{\sum_1^n |x_j|^{2r_j}\right\}^{-1/2\left(\sum\limits_{s=1}^n \frac{1}{r_s} - 1 + \varkappa\right)} = r(\boldsymbol{x})^{-\lambda}.$$

Here we have to keep in mind that $\varkappa < 1$ and

$$\lambda = \sum_{s=1}^n \frac{1}{r_s} - 1 + \varkappa = \frac{1}{p'} \sum_1^n \frac{1}{r_j} + \frac{1}{q} \sum_1^m \frac{1}{r_j}, \tag{6}$$

and it is possible to apply the first estimate in 9.4.1 (2). In the last inequality we have applied the usual estimate $\left(\sum\limits_1^n \xi_j^\beta\right)^{1/\beta} \leqq c \sum\limits_1^n \xi_j$, where $c = c_n$ is a constant.

From (4), (5), (6), in view of Lemma 9.6.1 (see formula 9.6.1 (3), where we need to suppose $\varkappa_j = 1/r_j$), we obtain

$$\|u\|_{L_q(\mathbb{R}_m)} \leqq c\|h\|_{L_p(\mathbb{R}_n)}.$$

But then it follows from (3) that $f \in L_q(\mathbb{R}_m)$ for arbitrary fixed $x_{m+1}, \ldots, x_n$ and

$$\|f\|_{L_q^{\boldsymbol{\varrho}}(\mathbb{R}_m)} = \|u\|_{L_q(\mathbb{R}_m)} \leqq c_2\|h\|_{L_p(\mathbb{R}_n)} \leqq c_3\|g\|_{L_p(\mathbb{R}_n)} = c_3\|f\|_{L_p^{\boldsymbol{r}}(\mathbb{R}_n)};$$

and the theorem is proved.

9.6.3[1]. From Theorem 9.6.2, if one takes into account Theorem 9.5.1 and 9.5.2, it is possible as a corollary to obtain the analogous theorem for the spaces B:

$$B_p^{\boldsymbol{r}}(\mathbb{R}_n) \to B_q^{\boldsymbol{\varrho}}(\mathbb{R}_m), \tag{1}$$

under the hypotheses $1 < p < q < \infty$, $1 \leqq m \leqq n$, $\boldsymbol{r} > \mathbf{0}$, $\boldsymbol{\varrho} = \varkappa \boldsymbol{r}$, $\varkappa > 0$ (for $\varkappa$ see 9.6.2 (1)). Indeed (explanation below),

$$B_p^{\boldsymbol{r}}(\mathbb{R}_n) \to L_p^{\frac{r_1}{\varkappa_1}, \dots, \frac{r_n}{\varkappa_1}, r_{n+1}, r_{n+2}}(\mathbb{R}_{n+2}) \tag{2}$$

$$\to L_q^{\frac{\varkappa_2}{\varkappa_1}r_1, \dots, \frac{\varkappa_2}{\varkappa_1}r_n, \varkappa_2 r_{n+1}}(\mathbb{R}_{n+1}) \to B_q^{\frac{\varkappa_3\varkappa_2}{\varkappa_1}r_1, \dots, \frac{\varkappa_3\varkappa_2}{\varkappa_1}r_m}(\mathbb{R}_m) = B_q^{\boldsymbol{\varrho}}(\mathbb{R}_m),$$

where $r_{n+1}, r_{n+2} > 0$ are chosen sufficiently large that

$$\varkappa_1 = 1 - \frac{1}{p}\left(\frac{1}{r_{n+1}} - \frac{1}{r_{n+2}}\right) > 0,$$

$$\varkappa_2 = 1 - \left(\frac{1}{p} - \frac{1}{q}\right)\left(\sum_1^n \frac{\varkappa_1}{r_j} + \frac{1}{r_{n+1}}\right) - \frac{1}{p r_{n+2}} > 0.$$

In addition

$$\varkappa_3 = 1 - \frac{1}{q}\left(\sum_{m+1}^n \frac{\varkappa_1}{\varkappa_2 r_j} + \frac{1}{\varkappa_2 r_{n+1}}\right) = \frac{\varkappa\varkappa_1}{\varkappa_2}.$$

The imbeddings (2) follow directly from 9.5.2, 9.6.2, and 9.5.1.

9.7. Nonequivalence of the Classes $B_p^{\boldsymbol{r}}$ and $L_p^{\boldsymbol{r}}$

In conclusion we shall show that the classes $B_p^{\boldsymbol{r}}$ and $L_p^{\boldsymbol{r}}$ for $1 \leqq p < \infty$; $p \neq 2$, are not equivalent, i.e. they are essentially distinct. We restrict ourselves to the consideration of the one-dimensional case.

Suppose first that $1 < p < \infty$. Consider the sequence of functions

$$\Phi_N(t) = \sum_0^N \varphi_k(t) = \varkappa \sum_0^N \cos\left[(2^k + 1)t\right] \frac{\sin t}{t} \quad (N = 1, 2, \dots),$$

$$\varphi_k(t) = \begin{cases} 1 & (2^k < |t| < 2^k + 2), \\ 0 & \text{for the remaining } t \end{cases}$$

(see 1.5.7 (10)).

---

[1] This remark is due to V. I. Burenkov.

We note that for $1 < p < \infty$

$$(1)\qquad \int_{-\infty}^{\infty} |\hat{\varphi}_k(t)|^p\, dt \ll \int_{-\infty}^{\infty} \left|\frac{\sin t}{t}\right|^p dt = A < \infty,$$

$$(2)\qquad \int_{-\infty}^{\infty} |\hat{\varphi}_k(t)|^p\, dt \gg \int_{\pi/3}^{\pi/2} |\cos (2^k + 1)\, t|^p\, dt > B > 0$$

(on $\left(\frac{\pi}{3}, \frac{\pi}{2}\right)$ the function $|t^{-1} \sin t|$ is bounded below by a positive constant), where B does not depend on k.

Obviously (see 1.5.6.1)

$$\beta_k(\Phi_N) = \widehat{(1)_{\Delta_k}\tilde{\Phi}_N} = \widehat{\varphi_k(t)}, \quad \Delta_k = \{2^k \leqq |t| \leqq 2^{k+1}\}$$

and we have

$$(3)\qquad \|\Phi_N\|_p \ll \left\|\left(\sum_0^N \hat{\varphi}_k^2\right)^{1/2}\right\|_p \ll \|\Phi_N\|_p,$$

in which the constants entering into the inequality here and below do not depend on N.

From the first inequality (3) it follows that

$$(4)\qquad \|\Phi_N\|_p \ll \left(\int_{-\infty}^{\infty} N^{p/2} \left|\frac{\sin t}{t}\right|^p dt\right)^{1/p} \ll N^{1/2}.$$

Further, from the second inequality (3) for $2 \leqq p < \infty$ (see 3.3.3) it follows that

$$(5)\quad \|\Phi_N\|_p \gg \left\|\left(\sum_0^N |\hat{\varphi}_k|^p\right)^{1/p}\right\|_p = \left(\int_{-\infty}^{\infty} \sum_0^N |\hat{\varphi}_k|^p\, dt\right)^{1/p} > (NB)^{1/p} \gg N^{1/p}$$

and for $1 < p \leqq 2$, using the generalized Minkowski inequality 1.3.2 (1) with exponent $\alpha = \frac{2}{p} \geqq 1$,

$$(6)\qquad \|\Phi_N\|_p^p \gg \left\|\left(\sum_0^N \hat{\varphi}_k^2\right)^{1/2}\right\|_p^p = \int \left(\sum_0^N \hat{\varphi}_k^2\right)^{p/2} dt$$

$$\geqq \left\{\sum_0^N \left(\int |\hat{\varphi}_k|^p\, dt\right)^{2/p}\right\}^{p/2} \gg \left(\sum_0^N B^{2/p}\right)^{p/2} = BN^{p/2}.$$

From (4), (5), and (6) it follows that

$$N^{1/2} \ll \|\Phi_N\|_p \ll N^{1/2} \quad (1 < p < \infty). \tag{7}$$

On the other hand (see 8.9 (5)), in view of (1), (2) the quantity

$$\|\Phi_N\|_{B_p^0} = \left(\sum_{k=0}^{N} \|\hat{\varphi}_k\|_p^p\right)^{1/p} \approx N^{1/p}, \tag{8}$$

i.e. it has the strict order $N^{1/p}$.

We see that the orders of the magnitudes of (7) and (8) for $p \neq 2$ are distinct. This shows that the null classes $L_p^0 = L_p$ and B_p^0 and accordingly, the classes L_p^r and B_p^r for arbitrary r are not equivalent.

By using the functions Φ_N one proves analogously that also for any $\theta \neq 2$ the class $B_{p\theta}^0$ is not equivalent to L_p (see O. V. Besov [5], to whom the considerations presented above are due). For $\theta = 2$, $p \neq 2$, non-equivalence also holds, however it is proved differently (see K. K. Golovkin [1)].

We turn to the case $p = 1$. The one-dimensional de la Vallée Poussin kernel (see 8.6 (5), (10), (11))

$$V_N(t) = \frac{1}{N} \frac{\cos Nt - \cos 2Nt}{t^2}$$

has the Fourier transform $\tilde{V}_N = \sqrt{\frac{\pi}{2}}\, \mu_N^*(t)$, where

$$\mu_N^*(t) = \sqrt{\frac{\pi}{2}} \begin{cases} 1 & (|x| < N), \\ \frac{1}{N}(2N - x) & (N < |x| < 2N), \\ 0 & (2N < |x|). \end{cases}$$

If k and N are natural numbers and $k \leqq N$, then

$$\mu_{2^k}^*(t)\, \mu_{2^N}^*(t) = \mu_{2^k}^*(t).$$

Therefore the kth de la Vallée Poussin sum for the function V_{2^N} is equal to

$$\sigma_{2^k}(V_{2^N}, x) = \frac{2}{\pi}(V_{2^k} * V_{2^N}) = \frac{2}{\pi} \widehat{\tilde{V}_{2^k} \tilde{V}_{2^N}} = V_{2^k}(x)$$

and, accordingly, the decomposition of V_{2^N} into a series of de la Vallée Poussin sums has the form

$$V_{2^N} = V_{2^0} + \sum_{1}^{N} (V_{2^k} - V_{2^{k-1}}).$$

Hence

$$\|V_{2^N}\|_{B_1^0} = \|V_{2^0}\|_L + \sum_{k=1}^{N} \|V_{2^k} - V_{2^{k-1}}\|_L \to \infty, \quad (N \to \infty),$$

because (after the change of variables $u = 2^{k-1}t$)

$$\|V_{2^k} - V_{2^{k-1}}\|_L = \int \left| \frac{\cos 2u - \cos 4u}{2u^2} - \frac{\cos u - \cos 2u}{u^2} \right| du = c > 0,$$

where c does not depend on $k = 1, 2, \ldots$. On the other hand, the norm of V_{2^N} in the metric of L,

$$\|V_{2^N}\|_L = \int \frac{|\cos u - \cos 2u|}{u^2} du = c_1 < \infty$$

is bounded. This shows that the imbedding $B_1^0 \to L$ is noninversible.

Remarks

To Chapter 1

1.1.—1.4. We present here, usually without proof, known facts from the theory of functions of a real variable and the theory of Banach spaces, so that further on we may refer to them and so that the reader may become acquainted with the notations adopted. These facts may be found in the books: P. S. Aleksandrov and A. N. Kolmogorov [1], A. N. Kolmogorov and S. V. Fomin [1], S. Banach [1, 2], L. A. Ljusternik and V. I. Sobolev [1], I. P. Natanson [1], V. I. Smirnov [2], S. L. Sobolev [3].

1.5. Here we present with proofs elementary information, only that needed for the book at hand, from the theory of generalized functions over the class S, in the way in which this class was defined by L. Schwartz [1]. We note the papers of S. L. Sobolev [1, 2], where the concept of generalized function was introduced, and the books in Russian on the theory of generalized functions of Halperin [1], V. S. Vladimirov [1], I. M. Gel'fand and G. E. Šilov [2][1].

We note also the book of Hörmander [1], where far-reaching results on multiplicators are obtained. The multiplier μ earlier could not be defined as a bounded measurable function, thus allowing μ to be considered as an element of S' and as having the property that $\|\widehat{\mu \hat{f}}\|_p \leq c_p \|f\|_p$ for all $f \in S$. Hörmander proved that such a generalized function μ naturally represents a bounded measurable function $\mu(\boldsymbol{x})$.

1.5.2. Inequality (6) was proved in the papers of Littlewood and Paley [1]. The theorems presented are to be found in the one-dimensional periodic case in Ch. XV of Zygmund's book [2], and in the two-dimensional periodic case in Marcinkiewicz [1].

1.5.3. The theorem of Marcinkiewicz in the two-dimensional periodic case was proved in his paper [1] mentioned above. The passage to the nonperiodic case was carried out by S. G. Mihlin [1]. For further development see P. I. Lizorkin [5]. The condition introduced in 1.5.4,

---

[1] Translator's note: In this edition we refer where possible to English or German versions of these books.

that $D^k\lambda$ should be continuous in each coordinate closed angle at any point x with $x_i \neq 0$, $i \in e_k$, is employed for example in Example 5 of 1.1.5, which is used in the further theory.

1.5.9. The operation I_r was studied in a series of papers, among them L. Schwartz [1], Caldersn [1], Aronszajn and Smith [1, 2] Aronszajn, Mulla, Szeptycki [1], P. I. Lizorkin [1, 8], Nikol'skiĭ, Lions and Lizorkin [1], Taibleson [1].

1.5.10. For the concept of a generalized function regular in the sense of L_p see S. M. Nikol'skiĭ [17, 18].

To Chapter 2

The information presented in 2.1—2.5 is well-known and here is of an auxiliary nature. In particular, on interpolation see for example the books of V. L. Gončarov [1] and A. F. Timan [1]. We note further the books of N. I. Ahiezer [1], A. Zygmund [2] and I. P. Natanson [2] where, as in the books noted above, there is detailed information on trigonometric polynomials of one variable.

To Chapter 3

3.1. For entire functions of one variable of exponential type, bounded on the real axis, see the book of Ahiezer [1]. In particular, that book presents criteria 3.1 (5), (6) for entire functions of exponential type and a complete proof of the facts relating to the theory of the Borel integral, which we have omitted in our exposition.

3.2. In the derivation of the interpolation formula (4) for functions of exponential type we have followed the schema presented in the paper of Civin [1]. But in the arguments in 3.2.1 we have, as was done in the paper [8] of P. I. Lizorkin, enlisted the aid of generalized functions. In doing this we have made some improvements.

3.2.2. The interpolation formula (2), for the derivative of a function $f(z)$ of exponential type, bounded on the real axis, is found in the problem book [1] of Polya and Szegö under the hypothesis that it is already known that $|f(z)| \leqq Ae^{\sigma|y|}$ $(z = x + iy)$. A complete proof is given in the book of Ahiezer [1], § 84.

3.2.3. The approach for obtaining inequalities of the type of Bernšteĭn's inequality in the case of general norms is indicated in Ahiezer's book [1], (§ 81, Theorem 3). We have added to the conditions 1), 2) enumerated there another condition, condition 3).

3.3.—3.5. For trigonometric polynomials inequalities 3.4.3 (2), (3), (4) were obtained by S. M. Nikol'skiĭ [3] along with the analogous inequalities for entire functions of exponential type (3.3.2 (2), 3.3.5 (1), 3.4.2 (1)). For trigonometric polynomials the case 3.4.3 (3) for $n = 1$, $p' = \infty$ was known already to Jackson [2]. Inequalities 3.4.3 (2) for trigonometric polynomials in the case $n = p = 1$ follow from the results of S. M. Lozinskiĭ [1].

In the inequalities which we have presented the constants are exact in the sense of order, but they are not absolutely exact. In certain cases more exact, or absolutely exact, values of these constants are known. In connection with this we refer to the book of I. I. Ibragimov [1], and also to N. K. Bari [1].

Inequality 3.4.1 (1), under the hypothesis that $g_\nu(\boldsymbol{x}) = g_\nu(\boldsymbol{u}, \boldsymbol{y})$ is entire of exponential type $\boldsymbol{\nu}$ relative to all $x_1, \ldots, x_n$, was treated in S. M. Nikol'skiĭ [3]. Here we consider the more general case when g is entire only relative to $\boldsymbol{u}$.

3.3.7. My attention was directed to the inequality $\|f\|_{L_{(p_1, \ldots, p_n)}} \leq \|f\|_{L_{(q_1, \ldots, q_n)}}$, where $q_1, \ldots, q_n$ is a permutation of $p_1, \ldots, p_n$ ordered in nondecreasing order, by V. I. Burenkov.

To Chapter 4

4.1. Aside from the papers of Beppo Levi [1] and S. L. Sobolev [1—5] already noted in the text, the generalized derivative was studied in the papers of Tonelli [1], Evans [1], Nikodym [1], Calkin [1], Morrey [1], S. M. Nikol'skiĭ [3, 5]. Deny and Lions [1, 2], where there are further indications as to the literature on this question.

4.2. For formula (2) and inequality (6) for periodic functions of one variable, see S. B. Stečkin [1].

4.3. The fractional classes $W_p^r(\Omega) = B_p^r(\Omega)$, for $\boldsymbol{r}$ fractional, arose in a natural way as classes of traces of functions of integer classes $W_p^l(g)$ on manifolds $\Omega \subset g$ or a boundary of g of dimension m less than the dimension of the region g. This problem on traces was first solved for $p = 2$ in the papers of Aronszajn [1], V. M. Babič and L. N. Slobodeckiĭ [1], and then for $m = n - 1$ by Gagliardo [1], Slobodeckiĭ [1], and for arbitrary m and l by O. V. Besov [2]. In the last case not only fractional B-classes were required, but integer classes as well.

Zygmund [1] drew attention to the fact that from certain points of view, for example from the point of view of the problem on the order of the best approximation of functions by trigonometric polynomials, the class of measurable functions of period 2π of one variable satisfying the condition

$$(1) \qquad \left(\int_0^{2\pi} |\Delta_h^k f(x)|^p \, dx\right)^{1/p} \leqq M|h|,$$

where $k > 1$, more naturally supplements the class of functions of x of period 2π satisfying the condition

$$(2) \qquad \left(\int_0^{2\pi} |f(x+h) - f(x)|^p \, dx\right)^{1/p} \leqq M|h|^\alpha \quad (0 < \alpha < 1),$$

than the class of functions for which (1) is satisfied for $\alpha = 1$.

The theory of imbeddings of H- and B-classes yields a number of new examples which emphasize this fact.

4.3.4. At the present time a number of equivalent methods are known for defining the spaces $B_{p\theta}^r$. A number of these are collected in §2 of the survey [3] of V. I. Burenkov.

4.3.6. Let us introduce the concept of an open set $g \subset \mathbb{R}_n$ with a Lipschitzian boundary. If the set g is bounded, then its boundary Γ is said to be *Lipschitzian*, if for any point $\boldsymbol{x}^0 \in \Gamma$ there exists a rectangular system of coordinates $\boldsymbol{\xi} = (\xi_1, \ldots, \xi_n)$ with origin at $\boldsymbol{x}^0$, for which the box

$$(1) \qquad \Delta = \{|\xi_j| < \eta_j; j = 1, \ldots, n\},$$

excises from Γ a subset $\gamma = \Gamma\Delta$ described by the equation

$$(2) \qquad \xi_n = \psi(\lambda), \lambda = (\xi_1, \ldots, \xi_{n-1}),$$

$$\lambda \in \Delta' = \{|\xi_j| \leqq \eta_j; j = 1, \ldots, n-1\},$$

where $\psi(\lambda)$ satisfies a Lipschitz condition on Δ', i.e. there exists a constant M such that

$$(3) \qquad |\psi(\lambda') - \psi(\lambda)| \leqq M|\lambda' - \lambda|,$$

$$\lambda, \lambda' \in \Delta'.$$

If the set g is unbounded, then its boundary Γ is said to be *Lipschitzian* if there exist numbers η and M, not depending on $x^0 \in \Gamma$, and a finite set e of rectilinear coordinate systems, obtained by rotation of the given rectilinear system $(x_1, \ldots, x_n)$, such that there exists for any $\boldsymbol{x}_b \in \Gamma$ a rectilinear system of coordinates $\boldsymbol{\xi} = (\xi_1, \ldots, \xi_n)$ with origin at x^0 parallel to one of the system of the set e, and a cube (1) excising from Γ a subset $\gamma = \Gamma\Delta$ described by the equations (2), where $\psi(\lambda)$ satisfies the Lipschitz condition (3) on Δ'.

Theorem 1. *Suppose that the open set $\Omega \subset \mathbb{R}_n$ has a Lipschitzian boundary. Then any of the classes $W_p^l(\Omega)$ (l an integer, $1 < p < \infty$),*

$$H_p^r(\Omega)\ (r > 0,\ 1 \leqq p \leqq \infty),\ B_{p\theta}^r(\Omega)\ (r > 0,\ \ 1 \leqq p \leqq \infty,\ \ 1 \leqq \theta \leqq \infty)$$

can be extended linearly beyond the boundaries of Ω to $\mathbb{R}_n$ with preservation of the norm.

The regions Ω beyond whose limits it is possible to extend functions of anisotropic classes depend essentially on the defining class of the vector $\boldsymbol{r}$, or, more precisely, on the propertion in which its components lie.

Suppose given a positive vector $\boldsymbol{r} > \boldsymbol{0}$ ($r_j > 0, j = 1, \ldots, n$), $\delta > 0$ and another positive vector $\boldsymbol{a} > \boldsymbol{0}$. Denote by $P(\boldsymbol{r}) = P(\boldsymbol{r}, \varrho, \boldsymbol{a}, \delta)$ the set (*horn* with vertex at the origin) of points $\boldsymbol{x} = (x_1, \ldots, x_n) \in \mathbb{R}_n$ which satisfy the condition

$$a_j h < x_j^{r_j} < (a_j + \delta)\, h \quad (j = 1, \ldots, n), \qquad 0 < h < \varrho \tag{4}$$

or any set obtained from (4) by mirror reflections (possible several) relative to the $(n-1)$-dimensional coordinate planes. Thus, any horn $P(\boldsymbol{r})$ may be desribed by the inequalities

$$a_j h < |x_j|^{r_j} < (a_j + \delta) h,\ \operatorname{sign} x_i = \text{const}, \qquad 0 < h < \varrho. \tag{5}$$

We denote by the symbol

$$g_1 + g_2$$

the *vector sum of the sets* $g_1, g_2 \subset \mathbb{R}_n$, that is, the set of all possible sums $\boldsymbol{x} + \boldsymbol{y}$, where $\boldsymbol{x} \in g_1$, $\boldsymbol{y} \in g_2$.

We shall say that the open set $\Omega \in A_\varepsilon(\boldsymbol{r})$ ($\varepsilon > 0$), if: 1) it may be represented in the form of two sums

$$\Omega = \bigcup_1^N U^k = \bigcup_1^N U_\varepsilon^k, \tag{6}$$

where U_ε^k is the set of points $\boldsymbol{x} \in U^k$ distant by more than ε from the boundary of $\Omega - U^k$, and 2) there exist $\varrho, \boldsymbol{a}, \delta$ such that to each k

there may be assigned a horn $p^k = P^k(\boldsymbol{r}, \varrho, \boldsymbol{a}, \delta)$ such that

$$(7) \qquad U^k + P^k \subset \Omega \quad (k = 1, \ldots, N).$$

The relation (7) expresses the fact that for any point $\boldsymbol{x} \in U^k$, if the horn P^k is shifted parallel to itself in such a way that its vertex coincides with $\boldsymbol{x}$, then the resulting shifted horn lies in Ω.

We note that in the case when $r_1 = \cdots = r_n = r$, the horn P is a cone, resting on some manifold, with vertex at the origin. It is possible to show that in this case the concept of a region having a Lipschitzian boundary coincides with the concept of a region of the class $A_\varepsilon(r, \ldots, r)$.

Theorem 2. *Suppose given a region $\Omega \subset A_\varepsilon(\boldsymbol{r})$ and classes with the norms*

$$\|f\|_{W_p^{\boldsymbol{l}}(\Omega)} = \|f\|_{L_p(\Omega)} + \sum_{j=1}^{n} \left\| \frac{\partial^{l_j} f}{\partial x_j^{l_j}} \right\|_{L_p(\Omega)}$$

(l_j integers, $1 < p < \infty$),

$$(9) \qquad \|f\|_{B_{p\theta}^{\boldsymbol{r}}(\Omega)} = \|f\|_{L_p(\Omega)} + \sum_{i=1}^{n} \left(\int_0^H \left\| \Delta_{x_j h}^{k_j} \frac{\partial^{\varrho_j} f}{\partial x_j^{\varrho_j}} \right\|_{L_p(\Omega_{k_j} h)}^{\theta} \frac{dh}{h^{1+\theta(r_j - \varrho_j)}} \right)^{1/\theta}$$

$$(k_j > r_j - \varrho_j > 0, \quad 1 \leqq \theta < \infty, 1 \leqq p \leqq \infty),$$

$$(10) \qquad \|f\|_{H_p^{\boldsymbol{r}}(\Omega)} = \|f\|_{L_p(\Omega)} + \sum_{j=1}^{n} \sup_h \frac{\left\| \Delta_{x_j h}^{k_j} \dfrac{\partial^{\varrho_j} f}{\partial x_j^{\varrho_j}} \right\|_{L_p(\Omega_{k_j} h)}}{h^{r_j - \varrho_j}}$$

$$(1 \leqq p \leqq \infty).$$

Any of these classes may be extended beyond Ω to $\mathbb{R}_n$ in a linear way with preservation of norm.

For $W_p^l(\Omega)$ with $l_1 = \cdots = l_n = l$ this theorem was proved by Smith [1], strengthening a result of Calderon [2], who proved it for the stronger norm

$$\|f\|_{L_p(\Omega)} + \sum_{|\alpha| \leqq l} \|f^{(\alpha)}\|_{L_p(\Omega)} = \|f\|_{*W_p^l(\Omega)}.$$

Here the following imbeddings hold:

$$(11) \qquad W_p^l(\Omega) \to W_p^{l, \ldots, l}(\Omega) \to W_p^{l, \ldots, l}(\mathbb{R}_n) \to W_p^l(\mathbb{R}_n)$$

$$\to {}^*W_p^l(\mathbb{R}_n) \to {}^*W_p^l(\Omega) \to W_p^l(\Omega).$$

The first and the two last above imbeddings are trivial. The second is the result of Smith just mentioned, and the third is proved in 9.2. (11) implies Theorem 1 for the spaces $W_p^l(\Omega)$.

For unequal integers l_j this theorem was proved for $W_p^l(\Omega)$ simultaneously and independently by O. V. Besov [9, 10] and V. P. Il'in [6].

For the norms $B_{p\theta}^{\boldsymbol{r}}(1 \leqq \theta \leqq \infty)$ this theorem (and thus Theorem 1 as well) was proved by O. V. Besov [9, 10]. See also O. V. Besov and V. P. Il'in [1].

V. I. Burenkov [4, 5] showed that for each $\boldsymbol{r} > \mathbf{0}$ it is possible to indicate a region not lying in $A_\varepsilon(\boldsymbol{r})$ for which Theorem 2 on extension is already not satisfied. For example, any horn $P(\boldsymbol{k})$, where $\boldsymbol{k} \neq c\boldsymbol{r}$, is such a region.

In the same place it was proved that if Theorem 2 holds for the spaces $W_{\boldsymbol{p}}^{\boldsymbol{r}}(\Omega)$, $B_{\boldsymbol{p}}^{\boldsymbol{r}}(\Omega)$ for a class of regions of the form $A_\varepsilon(\boldsymbol{r})$, then only in the case, when in place of the horn $P(\boldsymbol{r})$ one considers the horn $P(\boldsymbol{r}, \boldsymbol{p})$ $P(\boldsymbol{r}, \boldsymbol{p}, \delta, \boldsymbol{a}, \delta)$, defined as the set of points $\boldsymbol{x} \in \mathbb{R}_n$ subjected to the inequalities

$$a_i h < x_i^{\varrho_i} < (a_i + \delta) h \quad (i = 1, \ldots, n), \quad 0 < h < \varrho,$$

where

$$\varrho_i = \frac{r_i}{\varkappa_i}, \quad \varkappa_i = 1 - \sum_{j=1}^{n} \frac{1}{r_j} \left(\frac{1}{p_j} - \frac{1}{p_i} \right).$$

We note a further theorem following from Theorems 1 and 2.

Theorem 3. *If $g \subset \mathbb{R}_n$ is an arbitrary open set and $g_1 \subset \bar{g}_1 \subset g$ another bounded open set, then functions of any of the classes mentioned in Theorems 1 and 2, which we shall denote by $\Lambda(g)$, can be extended from g_1 to $\mathbb{R}_n$ in a linear way with preservation of the norm (relative to g). This must be understood in the sense that to each function $f \in \Lambda(g)$ one can assign an extension of it, $\bar{f} \in \Lambda(\mathbb{R}_n)$, from g_1 (not g) such that*

$$\|\bar{f}\|_{\Lambda(\mathbb{R}_n)} \leqq c \|f\|_{\Lambda(g)},$$

where c does not depend on f, and the dependence of $\bar{f}$ on f is linear.

Indeed, suppose given a rectangular net, dividing $\mathbb{R}$ into cubes, and suppose that Ω is the set consisting of those cubes of the net which contain points of $\bar{g}_1$. The boundary of Ω obviously satisfies the condition $A_\varepsilon(\boldsymbol{r})$ for any $\boldsymbol{r}$. If the net is sufficiently dense, $g_1 \subset \Omega \subset g$. We extend functions $f \in \Lambda(g) \subset \Lambda(\Omega)$ by means of the indicated theorem with preservation of the norm from Ω to $\mathbb{R}$: for the corresponding extensions $\bar{f}$ one has

$$\|\bar{f}\|_{\Lambda(\mathbb{R}_n)} \ll \|f\|_{\Lambda(\Omega)} \ll \|f\|_{\Lambda(g)}.$$

Theorem 3 may be proved in a simpler way, putting $\bar{f} = f\varphi$, where φ is a "cap", i.e. a function infinitely differentiable on $\mathbb{R}_n$, equal to unity on g_1 and to zero outside g (for the classes $H_p^r(g)$ see S. M. Nikol'skiĭ [5]).

Special cases of Theorems 1 and 2, relating to extension from a region with a sufficiently smooth boundary and beyond the limits of a segment, were considered in the earlier papers of S. M. Nikol'skiĭ [4, 7], V. K. Dzjadik [1], O. V. Besov [4]. See further V. M. Babič [1].

4.4.1—4.4.3. For investigations of this sort see S. M. Nikol'skiĭ [11]. For inequality 4.4.3 (4) see the book [4] of S. L. Sobolev.

4.4.5. If the derivative $\frac{\partial f}{\partial x_1}$ is understood in the sense of Sobolev, then this lemma is proved immediately. Indeed, suppose that we are given on g two sequences of functions f_k and $\lambda_k \in L_p(g)$ such that

$$\int f_k \frac{\partial \varphi}{\partial x_1}\, d\boldsymbol{x} = -\int \lambda_k \varphi\, d\boldsymbol{x} \quad (k = 1, 2, \ldots) \tag{1}$$

for all continuously differentiable finite functions φ on g. If in addition $f_k \to f$, $\lambda_k \to \lambda$ in the sense of $L_p(g)$, then it follows obviously from (1) that

$$\int f \frac{\partial \varphi}{\partial x_1}\, d\boldsymbol{x} = -\int \lambda \varphi\, d\boldsymbol{x}, \quad f, \ \lambda \in L_p(g)$$

for all the indicated φ, i.e. λ is the derivative of f with respect to x_1 on g in the sense of Sobolev, Sobolev used this lemma extensively [4]. Here it is proved starting from the fundamental definition of generalized derivative adopted in this book (see the beginning of § 4.1).

4.4.9. On this theorem see S. M. Nikol'skiĭ [5].

4.8. In the periodic case this theorem was proved by Hardy and Littlewood [1]. It was formulated without proof by A. A. Dezin [1]. The proof for $p = 2$ was presented in the dissertation [1] of A. S. Foht.

To Chapter 5

5.2. The approximation method 5.2.1 (4) was studied by S. N. Bernšteĭn [2], pages 421—432. There he proved inequality 5.2.1 (7) for $p = \infty$, $m = 1$. The case $m = 1$, $1 \leqq p \leqq \infty$ is treated in S. M. Nikol'skiĭ [3]. Here we study the more general case $m \leqq n$ of approximation by entire functions of spherical exponential type.

The inequality 5.2.1 (7) by itself was obtained for $m = n = 1$, $1 \leqq p < \infty$ by another method by N. I. Ahiezer [1].

The periodic inequalities 5.3.2 (2) were first obtained ($n = k = 1$, $p = \infty$) by Jackson [1]. The case $n = 1$, $1 \leqq p \leqq \infty$ was treated in the investigations of Quade [1] and Ahiezer [1]. The representations 5.3.1 (11) (analogies of 5.2.1 (4)) are found in S. B. Stečkin [1]. In the case of functions satisfying a Lipschitz condition for the Fejér sums ($p = \infty$), inequality 5.3.2 (5) was treated by A. Zygmund [3], § 4.7.9, and in the general case ($p = \infty$) by Stečkin [1].

For the approximation theorem 5.2.4 and its periodic analogue presented in 5.3.3 for the case $p = p_1 = \ldots = p_n = \infty$ see S. N. Bernšteĭn [2], pages 421—432. If the numbers $p_1, \ldots, p_n$ are in general distinct, see S. M. Nikol'skiĭ [10]. For inequality 5.3.2 (6) for power moduli of continuity see S. M. Nikol'skiĭ [6].

5.4. The inverse theorem of S. M. Bernšteĭn on approximation by algebraic and trigonometric polynomials ($p = \infty$, $n = 1$) was proved in his Collected Works, vol. 1 [1], pages 11—104. It was made more precise in the periodic case (for noninteger r) by de la Vallée Poussin [1] and (for integer r) by Zygmund [1].

5.4.4. Ja. S. Bugrov [3, 4] also found similar inequalities for polyharmonic functions in the disk and halfplane and applied them in the study of the differential properties of functions up to the boundary.

5.5.3. The equivalence of the norms $\|\cdot\|_H$ for various admissible pairs (k, ϱ) may be proved by a direct method, without recourse to approximation methods. For the one-dimensional method see Marchoud [1]. For a more general study in this direction see K. K. Golovkin [1, 2]. It was proved by Zygmund [1] in the periodic one-dimensional case by the method of approximations. He fixed attention on the equivalence for integer r of the norms of the norms ${}^1\|\cdot\|_{H^*}$ and ${}^5\|\cdot\|_{H^*}$, expressed through best approximations. For the nonperiodic case see S. N. Bernšteĭn [2], pages 421—432. Here we consider that problem for approximations by entire functions of exponential spherical type.

5.5.4. In the periodic one-dimensional case for $p = \infty$ this is a classical theorem, proved in Bernšteĭn [1], pages 11—104, Jackson [1], de la Vallée Poussin [1], Zygmund [1]. For $1 \leqq p < \infty$ see Quade [1], Zygmund [1]. In the nonperiodic one-dimensional case for $1 \leqq p \leqq \infty$ see Ahiezer [1]. Here we give a generalization to the case of approximations by entire functions of exponential spherical type.

5.5.8. In the periodic case there are many results related to this question.

5.6.—5.6.1. In the exposition of these sections we made essential use of the paper [5] of Besov, and in the case of 5.6.1 also the paper [3] of T. I. Amanov. Besov placed at my disposal a new variant, presented in the text, of the proof of the imbedding ${}^4B' \to {}^5B$. This method has the advantage that carries over freely to more general cases of theorems of this kind, for which see K. K. Golovkin [2].

Among the various equivalent norms $\|\cdot\|_B$ (in particular, $\|\cdot\|_H$), we have introduced the norms ${}^2\|\cdot\|_B$ and ${}^4\|\cdot\|_{B_1}$, expressed in terms of directional derivatives. In the isotropic case they have at any rate the technical advantage that in place of a sum of integrals corresponding to all possible partial derivatives of orders $\boldsymbol{s}$ with $|\boldsymbol{s}| = \varrho$ one chooses only one integral. In the case $\varrho = 0$ such norms have not infrequently been used in the literature.

The equivalence of the classes ${}^1B^{\boldsymbol{r}}_{p\theta}(\mathbb{R}_n)$ and ${}^5B^{\boldsymbol{r}}_{p\theta}(\mathbb{R}_n)$ $(1 \leqq \theta < \infty)$ was proved by Besov [3, 5], and in the periodic one-dimensional case by A. A. Konjuškov [1] and P. L. Ul'janov [1]. The equivalence of ${}^1B^{\boldsymbol{r}}_{p\theta}(\mathbb{R}_n)$ and ${}^3B^{\boldsymbol{r}}_{p\theta}(\mathbb{R}_n)$ was proved in the same papers of Besov, and for $\theta = p$, r_i integers in the paper of S. V. Uspenskiĭ [3].

In the special cases $\theta = p = 2$ (for the admissible pairs $(r, 1)$, $(r, 2)$), the norms ${}^3\|\cdot\|$ were introduced and studied in earlier papers of Aronszajn [1], Babič and Slobodeckiĭ [1], and for $p = \theta \neq 2$ for noninteger by Gagliardo [1] and Slobodeckiĭ [1].

The expansion of the functions f in the form of the series 5.6 (7) with the norm ${}^6\|\cdot\|_B$ was treated in Besov [3]. An explicit application of the norm ${}^6\|\cdot\|_B$ is found in the paper of Amanov [3].

5.6.2—5.6.3. Suppose that the function $f(x)$ has derivatives $\dfrac{\partial^{\bar{r}_j} f}{\partial x_j^{\bar{r}_j}}$, satisfying relative to x_j on the bounded region Ω a Hölder condition of exponent $\alpha_j (0 < \alpha_j \leqq 1, r_j = \bar{r}_j + \alpha_j)$ uniformly relative to the other variables, and suppose that $\varrho = (\varrho_1, \ldots, \varrho_n)$ is an (integer) vector, for which $\sum_1^n \dfrac{\varrho_j}{r_j} < 1$. In 1911 Bernšteĭn (see [1], pages 96—104), for $r = r_1 = \cdots = r_n$, and in 1918 Montel (see [1]) for arbitrary positive r_j, proved that in such a case there exists on Ω a mixed continuous derivative $f^{(\varrho)}$, bounded on any region $\Omega_1 \subset \bar{\Omega}_2 \subset \Omega$. Bernšteĭn proved further that the derivatives $f^{(\varrho)}(|\varrho| = \bar{r} = r - \alpha, r = r_j)$ satisfy on Ω_1 a Hölder condition of exponent $\alpha' < \alpha$.

Theorem 5.6.3 strengthens these results, indicating the exact class to which $f^{(\varrho)}$ belongs, and extends (for $\Omega = \mathbb{R}_n$) to the case of the metric $L_p (1 \leqq p \leqq \infty)$ and to the B-classes. For H-classes, for $p = \infty$ and $r = r_1 = \cdots = r_n$ this theorem was proved by Bernšteĭn ([2], 426—432), and simultaneously and independently, for arbitrary $r_j (j = 1, \ldots, n)$, by S. M. Nikol'skiĭ [2]. In the same paper he also proved its unimprovability in the terms of H-classes, and in [5] the same thing in the metric L_p. For the classes B it was proved by Besov [5].

The question of the extension of Theorem 5.6.3 to the case of regions $\Omega \subset \mathbb{R}_n$ is solved using the extension theorem (see 4.3.6). In the paper referred to, Montel in fact considered Ω to be a square with sides parallel to the coordinate axes: ϱ is not necessarily an integer vector, and then $f^{(\varrho)}$ is a mixed partial derivative in the sense of Liouville.

Theorem 5.6.2 on the equivalence

$$B_{p\theta}^{r,\ldots,r}(\mathbb{R}_n) = B_{p\theta}^{r}(\mathbb{R}_n),$$

follows for r noninteger from Theorem 5.6.3. In the general case its proof was given, by other methods, by V. A. Solonnikov [1]. For H-classes see further S. M. Nikol'skiĭ, J. Lions and P. I. Lizorkin [1]. If $\mathbb{R}_n$ is replaced in (1) by Ω, property 5.6.2 (1) of course is preserved for the regions $\Omega \subset \mathbb{R}_n$ for which the theorem on extension of the classes $B_{p\theta}^{r,\ldots,r}(\Omega)$ is valid. In this connection see 4.3.6 and the Remark to 4.3.6 Burenkov [2] investigated regions for which the equivalence 5.6.2 (1) does not hold.

Suppose that $B_p^{\boldsymbol{r}} = B_{pp}^{\boldsymbol{r}}(\mathbb{R}_n)$, $H_p^{\boldsymbol{r}} = B_{p\infty}^{\boldsymbol{r}}(\mathbb{R}_n)$, $L_p = L_p(\mathbb{R}_n)$. The hypothesis of theorem 5.6.3 for $\varkappa = 0$ and $1 \leqq p \leqq 2$ $(\theta = p)$ implies that $f^{(\boldsymbol{l})} \in L_p$. This follows from 9.2.2 and 9.3 (2). For $2 \leqq p < \infty$ this is already not so: if, for example, $f \in B_p^{\boldsymbol{l}}(\mathbb{R}_1)$, it therefore does not in general follow that $f^{(\boldsymbol{l})}$ exists and lies in L_p (see 9.7).

For the classes $H_p^{\boldsymbol{r}} = B_{p\infty}^{\boldsymbol{r}}$ Theorem 5.6.3 in the case $\varkappa = 0$ is again false. Indeed, the function

$$f(\theta) = \sum_{s=1}^{\infty} b^{-s} \cos b^s\theta \quad (b > 1)$$

does not have a derivative anywhere (Weierstrass, Hardy, see Zygmund [2], Ch. II, 4.8—4.11), and at the same time it lies in the periodic class $H_{p}^{1}{}^{*}$. This last assertion is proved as follows. It is easily verified that

$$\|\cos b^s x\|_{L_p^*} = \|\cos b^s x\|_{L_p(0,2\pi)} \leqq K,$$

where K does not depend on $s = 1, 2, \ldots$. Choose an h satisfying $0 < h < 1$ and select an integer N such that

$$b^{-(N+1)} < h \leqq b^{-N}.$$

Then

$$\|\Delta_h^2 f\|_{L_p^*} \ll \sum_{s=1}^{N} b^{-s} \|\Delta_h^2 \cos b^s\theta\|_{L_p^*} + \sum_{N}^{\infty} b^{-s}$$

$$\ll h^2 \sum_{s=1}^{N-1} b^{-s} b^{2s} + b^{-N} \ll h^2 b^N + b^{-N} \ll h,$$

where we have applied inequality 4.4.4 (3) for trigonometric polynomials.

5.6.4. This example was given by Ju. S. Nikol'skiĭ in [1].

5.6.5. Properties 5.6.5. (1), (2) express continuity in the norm in the corresponding spaces W_p^l, $B_{p\theta}^r$. In the case of B-classes my attention was called to this property by P. I. Lizorkin.

To Chapter 6

6.1. The complement of V. I. Il'in [2] in the imbedding theorem (1) relates to the so-called limiting case (when $\varrho = 0, 1, 2, \ldots$) for $m < n$. This assertion in the case of the limiting exponent was known to V. I. Kondrašov [1], where he investigated also certain cases for $m < n$ and noninteger ϱ.

S. L. Sobolev proved further that in theorem (1), for $m = n$ one may take $p = 1$.

In the one-dimensional case, the question of traces does not arise, and one can speak only of a "pure" theorem of different metrics. It was proved by Hardy and Littlewood [1]. See further the book [1] of Hardy, Littlewood, and Polya.

6.4. Suppose that $\Gamma \subset \mathbb{R}_n$ is a sufficiently smooth surface of dimension $m < n$. In any case, under the condition $r - \dfrac{n-m}{p} > 0$ the trace $f|_\Gamma$ of the function $f \in H_p(\mathbb{R}_n)$ is correctly defined. Under that condition one has both direct and inverse imbeddings $H_p^r(\mathbb{R}_n) \rightleftarrows H_p^{r-\frac{n-m}{p}}(\Gamma)$ for which see S. M. Nikol'skiĭ [5]. For the corresponding generalization to B-classes see Besov [11].

6.7. Ja. S. Bugrov [4] showed that the imbedding

$$H_p^{r'}(\mathbb{R}_m) \to H_p^r(\mathbb{R}_n), \quad r' = r - \frac{n-m}{p},$$

is true not only for $r' > 0$ but also for $r' = 0$, if we replace its left side $H_p^0(\mathbb{R}_m)$ by $L_p(\mathbb{R}_m)$.

Various strengthenings of the extension theorems may be obtained by imposing additional conditions on the functions being extended.

L. D. Kudrjavcev [2] showed that in Theorem 6.6. the function $f \in H_p^r(\mathbb{R}_n)$, extending to $\mathbb{R}_n$ the function

$$\varphi \in H^{r-\frac{n-m}{p}}(\mathbb{R}_m) \quad \left(1 \leq m < n, r - \frac{n-m}{p} > 0\right),$$

may be constructed in such a way that it is infinitely differentiable on $\mathbb{R}_n - \mathbb{R}_m$ and so that the properties

$$(1) \qquad \int_{\mathbb{R}_n} \varrho^{(s-\alpha)+\varepsilon} \left| \frac{\partial^{\bar{r}+s} f}{\partial x_1^{s_1} \dots \partial x_n^{s_n}} \right|^p d\mathbb{R}_n < \infty,$$

$$\varepsilon > 0, \sum_1^n s_k = \bar{r} + s \; (1 \leqq p < \infty), \quad \varrho^2 = \sum_2^n x_j^2$$

are satisfied. These inequalities cease being valid for $\varepsilon = 0$. He obtained an analogous result for $p = \infty$. These facts reflect on the determination of the connection between the classes considered here and the so-called weight classes of functions, whose derivatives (or differences) are integrable to the pth power with a weight. If we start with B-classes instead of H-classes ($\theta < \infty$), then similar facts hold already for $\varepsilon = 0$ (S. V. Uspenskiĭ [3]).

A systematic investigation of the weight classes was begun in the book [2] of Kudrjavcev noted above. See also A. A. Vašarin [1], Uspenskiĭ [3], I. A. Kiprijanov [1] and A. Kufner [1]. For a bibliography on this question see Burenkov [3], Kudrjavcev [3], J. Nečas [1], S. M. Nikol'skiĭ [12].

Bugrov [1] proved for the unit disk σ in the terms of H-classes the following theorem, which is exact in the limit.

Suppose that the functions, of period 2π,

$$\varphi_k(\theta) \in H_{p^*}^{r+l-k-1} \; (k = 0, 1, \dots, l-1, 1 \leqq p \leqq \infty, r > 0).$$

Then the polyharmonic function $u(\varrho, \theta)$ of polar coordinates ($0 \leqq \varrho \leqq 1$, $0 \leqq \theta \leqq 2\pi$), *solving in the unit disk σ the boundary problem*

$$\Delta^l u(\varrho, \theta) = 0, \left. \frac{\partial^k u}{\partial \varrho_k} \right|_{\varrho=1} = \varphi_k(\theta) \quad (k = 0, 1, \dots, l-1),$$

where Δ is the Laplace operator, lies in the class $H_p^{r+l+\frac{1}{p}-1}(\sigma)$.

Bugrov [3] obtained an analogous result for the halfplane. These theorems were preceded by the exact results of Gjunter [1] and Kellogg [1] for a three-dimensional region with smooth boundary with $p = \infty$ and r noninteger, of Besov [1] for a halfspace with $1 \leqq p \leqq \infty$ and noninteger r, of Mozžerova [1] for a three-dimensional region with smooth boundary for $1 \leqq p < \infty$ and noninteger r, and of S. M. Nikol'skiĭ [4, 9] for the disk with $p = 2$ and r arbitrary. See further T. I. Amanov [2]. At the present time there are many results of this type with estimates of the solutions of various boundary value problems in terms of the classes considered here. See for example Solonnikov [1], Nečas [1, 2], and I. N. Vekua [1].

6.9. Suppose that $\boldsymbol{k} = (k_1, \ldots, k_n) \geqq 0$ (i.e. $k_j \geqq 0$ for all j) is an integer vector and $\boldsymbol{h} = (h_1, \ldots, h_n)$ any vector $(h_j \neq 0, j = 1, \ldots, n)$. By definition

$$\Delta_{\boldsymbol{h}}^{\boldsymbol{k}} f = \Delta_{h_1}^{k_1} \ldots \Delta_{h_n}^{k_n} f,$$

where $\Delta_{x_j h_j}^{k_j} f$ is the difference of f of order k_j with step h_j in the direction of $x_j (\Delta_{x_j h_j}^0 f = f)$. Take a vector $\boldsymbol{r} = (r_1, \ldots, r_n) \geqq \boldsymbol{0}$. Suppose that e is any subset of the set $e_n = \{1, \ldots, n\}$ of integers. Let $\boldsymbol{r}^e = (r_1^e, \ldots, r_n^e)$ be a vector whose components are subjected to the condition

$$r_j^e = \begin{cases} r_j, & j \in e, \\ 0, & j \notin e. \end{cases}$$

Put

$$\overline{r^e} = \left\{\overline{r_1^e}, \ldots, \overline{r_n^e}\right\},$$

$$\boldsymbol{\alpha}^e = r^e - \overline{r^e} = \{\alpha_1^e, \ldots, \alpha_n^e\},$$

where, if $r_j^e > 0$, then $\overline{r_j^e}$ is the largest integer less than r_j^e, and if $r_j^e = 0$, then $\overline{r_j^e} = 0$.

Introduce further the vector $\boldsymbol{\omega} = \{1, \ldots, 1\}$. *By definition the function $f(\boldsymbol{x}) = f(x_1, \ldots, x_n)$ lies in the class $S_p^{\boldsymbol{r}} H = S_p^{\boldsymbol{r}}$, if the norm*

$$\|f\|_{S_p^{\boldsymbol{r}} H} = \sum_e \sup_{\boldsymbol{h}} \left\| \frac{\Delta^{2\omega} f^{(\overline{\boldsymbol{r}^e})}(\boldsymbol{x})}{\boldsymbol{h}^{\alpha^e}} \right\|_{L_p(\mathbb{R}_n)}$$

is finite, where the sum is extended over all the subsets e, $\boldsymbol{h}^{\alpha^e} = h_1^{\alpha_1^e} \ldots h_n^{\alpha_n^e}$ and $f^{(\overline{\boldsymbol{r}^e})}$ is the partial derivative of order $\overline{\boldsymbol{r}}^e$. That sum contains a term $\|f\|_{L_p(\mathbb{R}_n)}$, corresponding to the empty set e.

Representation Theorem. *Suppose that $\boldsymbol{r} > \boldsymbol{0}$. For $f(\boldsymbol{x})$ to lie in $S_p^{\boldsymbol{r}}(H)$, it is necessary and sufficient that the representation*

(1) $$f(\boldsymbol{x}) = \sum_{\boldsymbol{k} \geqq \boldsymbol{0}} Q_{\boldsymbol{k}}(\boldsymbol{x}),$$

should hold, where the $Q_{\boldsymbol{k}}(\boldsymbol{x}) = Q_{2^{k_1 r_1}, \ldots, 2^{k_n r_n}}(\boldsymbol{x})$ are entire functions of the exponential type $2^{k_j r_j}$ relative to the x_j respectively, for which

(2) $$\|Q_{\boldsymbol{k}}\|_{L_p(\mathbb{R}_n)} \leqq c 2^{-\boldsymbol{kr}} \left(\boldsymbol{kr} = \sum_{j=1}^n k_j r_j\right),$$

and c does not depend on $\boldsymbol{k}$.

We have the imbeddings

(3) $$S_p^{\boldsymbol{r}}(\mathbb{R}_n) \to S_{p'}^{\varrho}(\mathbb{R}_n) \left(1 \leqq p < p' \leqq \infty, \varrho = \boldsymbol{r} - \left(\frac{1}{p} - \frac{1}{p'}\right)\omega > 0\right)$$

(4) $$S_p^{\boldsymbol{r}}(\mathbb{R}_n) \to S_p^{\overline{\boldsymbol{r}e_m}}(\mathbb{R}_m)\ (e_m = (1, \ldots, m);\ 1 \leqq m < n;$$

$$r_j - \frac{1}{p} > 0,\ j = m+1, \ldots, n).$$

Indeed, if $f \in S_p^{\boldsymbol{r}}(\mathbb{R}_n)$, then (1) and (2) hold. But then

$$\|Q_{\boldsymbol{k}}\|_{L_{p'}(\mathbb{R}_n)} \leqq c_1 2^{-\boldsymbol{k}\rho} \quad (\varrho > 0)$$

and $f \in S_{p'}^{\rho}(\mathbb{R}_n)$. Further,

$$\|Q_{\boldsymbol{k}}\|_{L_p(\mathbb{R}_m)} \leqq c_1 2^{-\boldsymbol{kr} + \frac{1}{p}\sum_{m+1}^{n} k_j r_j},$$

and, if one puts $x_{m+1} = \cdots = x_n = 0$ in (1), then for the trace of f on $\mathbb{R}_m$ we get the representation

$$\varphi(x_1, \ldots, x_m) = f(x_1, \ldots, x_m, 0, \ldots, 0) = \sum_{\boldsymbol{k}e_m \geqq 0} q_{\boldsymbol{k}e_m},$$

where the sum is extended over m-dimensional vectors $\boldsymbol{k}^{(m)} \geqq \boldsymbol{0}$ and

$$q_{\boldsymbol{k}e_m} = \sum_{\substack{k_j \geqq 0 \\ m+1 \leqq j \leqq n}} Q_{\boldsymbol{k}}(x_1, \ldots, x_m, 0, \ldots, 0),$$

while $\left(r_j - \frac{1}{p} > 0\right)$

$$\left\| q_{\boldsymbol{k}e_m} \right\| \ll 2^{-\sum_1^m k_j r_j} \sum_{\substack{k_j \geqq 0 \\ m+1 \leqq j \leqq n}} 2^{-\sum_{m+1}^{n} k_j \left(r_j - \frac{1}{p}\right)} \ll 2^{-\sum_1^m k_j r_j'},$$

which implies (4) in view of the inverse representation theorem.

For the vectors $\boldsymbol{r}^1, \ldots, \boldsymbol{r}^N$ we introduce the space

$$S^{\boldsymbol{r}^1, \ldots, \boldsymbol{r}^N} = S_p^{\boldsymbol{r}^1, \ldots, \boldsymbol{r}^N} = \bigcap_1^N S_p^{\boldsymbol{r}^j}$$

with the norm

$$\|f\|_{S_p^{\boldsymbol{r}^1, \ldots, \boldsymbol{r}^N}} = \sum_{j=1}^{N} \|f\|_{S_p^{\boldsymbol{r}^j}}.$$

The interpolation theorem holds:

$$(5)\qquad S^{\mathbf{r}^1,\ldots,\mathbf{r}^N} \to S^{\sum_1^N \lambda_k \mathbf{r}^k} \left(\lambda_k \geqq 0,\ \sum_1^N \lambda_k \leqq 1\right).$$

If $N = n$ and $\mathbf{r}^i = (0, \ldots, 0, r_i, 0, \ldots, 0)$, than

$$S^{\mathbf{r}^1,\ldots,\mathbf{r}^N} \equiv S^{\mathbf{r}^1,\ldots,\mathbf{r}^N} H \equiv H_p^{r_1,\ldots,r_n}.$$

These results were proved by S. M. Nikol'skiĭ in [15, 16].

We note the work of N. S. Bahvalov [1], where there is an independent proof of one side, the necessary side, of the theorem on the representation of functions of the periodic class $S_{*p}^{\mathbf{r}}H$: if $f \in S_{*p}^{\mathbf{r}}H$, then (1) and (2) hold. The extension of these theorems from H- to B-classes is due to Amanov [3], and, with another method, to Džabrailov [1].

6.10.2. This remark on mean functions was communicated to me by O. V. Besov.

To Chapter 7

7.2. The inequalities among the norms of partial derivatives with a parameter ε and the multiplicative inequalities will be found in the paper [7] of V. I. Il'in and in the papers by V. A. Solonnikov [1], K. K. Golovkin [1], Il'in and Solonnikov [1] and others.

Inequalities with ε are applied in the theory of differential equations when it is desired that one of the terms of the form

$$\varepsilon^{\alpha}\|f\| + \frac{1}{\varepsilon^{\beta}}\|f^{(k)}\|$$

should be less than some number given in advance.

It follows from the results of S. M. Nikol'skiĭ [11], relating to more general imbedding theorems, that the inequalities among polynorms

$$(1)\qquad \|f\|_{h_{p'}^{r'}(g)} \leqq c\|f\|_{w_p^r(g)},$$

$$(2)\qquad \|f\|_{w_{p'}^{l}(g)} \leqq c\|f\|_{w_p^r(g)}$$

$$\left(1 \leqq p < p' \leqq \infty,\ r' = r - \left(\frac{1}{p} - \frac{1}{p'}\right) n > 0,\ l < r'\right)$$

hold for any region $g \subset \mathbb{R}_n$, without the hypothesis that $\|f\|_{L_p(g)}$ is finite, given only that $\bar{r} \leqq r' < r$ in case (1), or $\bar{r} < l < r' < r$ in case (2).

The inequality

$$\|f\|_{w_p^{l-\frac{1}{p}}(\mathbb{R}_{n-1})} \leqq c\|f\|_{w_p^l(\mathbb{R}_n)} \quad (1 < p < \infty)$$

follows from the papers [4] of Kudrjavcev and [1] of Ju. S. Nikol'skiĭ on weight spaces, as well as the assertion on the possibility of extending functions $\varphi \in \omega_p^{l-\frac{1}{p}}(\mathbb{R}_{n-1})$ to $\mathbb{R}_n$, so that for functions extending them one has

$$\|f\|_{w_p^l(\mathbb{R}_n)} \leqq c\|\varphi\|_{w_p^{l-\frac{1}{p}}(\mathbb{R}_{n-1})} \quad (1 < p < \infty)$$

without the hypothesis that the norms $\|f\|_{L_p(\mathbb{R}_n)}$ and $\|\varphi\|_{L_p(\mathbb{R}_{n-1})}$ are finite. In these two theorems, as above, $0 < l - \frac{1}{p} < l$.

7.3. Boundary functions were introduced and studied in the papers [2, 3] of S. M. Nikol'skiĭ, [1] of T. I. Amanov, [1] of P. Pilika. Using these functions it was possible to establish the exactness (unimprovability) of the inequalities among H-norms presented here.

7.7. Many investigations have been devoted to the problem of compactness of classes of functions, starting with the work of Ascoli [1] and Arzelā [1]. The fundamental theorem of Arzelā on compactness relates to the class of continuous functions. What correspond to it in the metric of L_p are the theorem of Kolmogorov [1], for $p > 1$, and the theorem of Tulaĭkov [1] for $p = 1$. The question of compactness of classes of differentiable functions was treated in papers of Rellich [1], I. G. Petrovskiĭ and K. N. Smirnov [1], V. I. Kondrašov [1], M. Picone [1], C. Pucci [1], Kudrjavcev [1], Besov [12], Il'in [9] and others.

For the classes H_p^r, W_p^r presented here see S. M. Nikol'skiĭ [8]. In essence we are dealing here with weak compactness: from a sequence bounded in the metric of H_p^r or W_p^r it is possible to select a subsequence converging in the sense of $H_p^{r-\varepsilon}(\varepsilon > 0)$ to some function $f \in H_p^r$, W_p^r.

In Theorems 7.7.1—7.7.5 were dealing already with the compactness of a set in the metric of the space to which it belongs. In particular, it contains the theorem on compactness in L_p (see S. L. Sobolev [4], Ch. 1, § 4.3).

Theorems 7.7.2—7.7.5 were proved by P. I. Lizorkin and S. M. Nikol'skiĭ.

Besov [12] studied compactness questions for sets of functions f in H-classes, imposing supplementary conditions on f. For example, in the case $H_p^r(r < 1)$ it was supposed that

$$\|f(\boldsymbol{x} + \boldsymbol{h}) - f(\boldsymbol{x})\| \leqq \alpha(\boldsymbol{h})\ |\boldsymbol{h}|^r$$
$$(\alpha(\boldsymbol{h}) \to 0,\ |\boldsymbol{h}| \to 0).$$

To Chapter 8

8.1. The operation I_l corresponds to a certain extent to the operation of Weyl (see Zygmund [2], Ch. XII, 8)

$$f(x) = I_l^* \varphi = \frac{1}{\pi} \int_{-\pi}^{\pi} K_l(x - t)\, \varphi(t)\, dt, \tag{1}$$

$$K_l(t) = \sum_{\nu=1}^{\infty} \frac{\cos\left(\nu t + \frac{l\pi}{2}\right)}{\nu^l} \quad (l > 0),$$

$$\int_{-\pi}^{\pi} \varphi(t)\, dt = 0, \quad \varphi \in L.$$

It is closely connected with the (nonperiodic) operation

$$f(x) = \frac{1}{\Gamma(\alpha)} \int_a^x (x - t)^{\alpha-1}\, \varphi(t)\, dt$$

of Liouville. The kernels of I_l, I_l^* have for small t the same singularity $|t|^{\alpha-1}$. Here we have in view the one-dimensional case. Compare 8.1 (6), (13) and Zygmund [1], Ch. V, 2.1. The Liouville kernel $(x - t)^{\alpha-1}$ has the same singularity for $t = x$.

For estimates of the type 8.1 (7) for the partial derivatives of $G_r(|\boldsymbol{x}|)$ see Aroszajn and Smith [1].

8.2. The theorem on isomorphism of the classes $W_p^l(\mathbb{R}_n)$ was proved in the papers of Calderon [3] and Lions and Magenes [1].

8.3. For the estimates of the differences of the derivatives of $G_r(|\boldsymbol{x}|)$ see Aronszajn and Smith [1], Nikol'skiĭ, Lions and Lizorkin [1], Ch. I, S. M. Nikol'skiĭ (with a supplement by E. Nosilovskiĭ), [18].

8.4. For inequality (2) see Nikol'skiĭ, Lions and Lizorkin [1], Ch. I, and S. M. Nikol'skiĭ [18], Lemma 6.

In the periodic case for $n = 1, p = \infty$ it was known to I. P. Natanson [2], pages 119—120, if one takes into account the fact that $E_v^*(f)_\infty$ is the best approximation to the function f by trigonometric polynomials with mean value relative to the period equal to zero. It was known to S. B. Stečkin [2] with the usual understanding of $E_v^*(f)_\infty$. See also Sun' Jun-Šen [1].

Inequality (4) is an analogue to the corresponding one-dimensional inequality of Favard [1] in the periodic case. It is applied to obtain inequality 8.6 (16) ($r > 0$), and here the advice of my colleague S. A. Teljakovskiĭ was very useful.

8.6. N. I. Ahiezer and B. M. Levitan studied, with different objectives, a kernel more general than $V_N(t)$, corresponding to the more general trigonometric sums of de la Vallée Poussin $\frac{1}{p+1}(D_{N-p}^* + \cdots + D_N^*)$, where D_k^* is the Dirichlet kernel.

On expansions, regular in the sense of L_p, of functions into de la Vallée Poussin sums, see S. M. Nikol'skiĭ [17].

8.8—8.9.2. We present here facts relating to the concept of generalized function regular in the sense of $L_p (1 \leqq p \leqq \infty)$, and decompositions of them into weakly converging Valleé Poussin sums. For this see S. M. Nikol'skiĭ [17, 18]. The concepts themselves of the null classes $B_{p\theta}^0$, isomorphisms of the $B_{p\theta}^r$ for different r, and the integral representations of $B_{p\theta}^r$ in terms of null classes, and negative classes $B_{p\theta}^r$ were established from various considerations in the papers of Calderon [3], Aronszajn, Mulla and Szeptycki [1], Taibleson [1, 2], Nikol'skiĭ, Lions, and Lizorkin [1].

8.9. The collection $S_p' = S_p'(\mathbb{R}_n)$ of all generalized functions regular in the sense of L_p $(1 \leqq p \leqq \infty)$ (see 1.5.10) may be defined further as a sum

$$S_p' = \bigcup_k H_p^{r_k} \tag{1}$$

$(H_p^r = H_p^r(\mathbb{R}_n))$, where $\{r_k\}$ is any sequence of real numbers tending to $-\infty$. Indeed, if $f \in S_p'$, then for some $\varrho \geqq 0$ one has $I_\varrho f \in L_p$ (see 1.5.10), so that (see (8.2) $I_{\varrho+1} f \in W_p^1 \subset H_p^1$. But then $f \in H_p^{-\varrho} \to H_p^{r_k}$, if k is such that $r_k < -\varrho$. Conversely, if $f \in H_p^{r_k}$ for certain k, then $I_{-r_k+1} f \in H_p' \subset L_p$.

It is clear that H in (1) may be replaced by B or L (see 6.1).

We shall agree to say that the function $f \in S_p'$ has a spectrum in the region $G \subset \mathbb{R}_n$, if its Fourier transform $\bar{f}$ has a support on G, i.e. $\bar{f} = 0$ outside G.

From what has been said it follows that if the function $f \in S'_p$, then it belongs also to H^r_p for some r and decomposes into a series

$$f = \sum_0^\infty q_s \tag{2}$$

with the following properties: a) $q_s \in L_p$ and has a spectrum in $\Delta_{s+1} - \Delta_{s-1}$ $(s = 1, 2, \ldots)$, $\Delta_0 (s = 0)$, where $\Delta_s = \{|x_j| < a^s\}$, $a > 1$; the inequalities

$$\|q_s\|_{L_p} \leqq M a^{-sr} \quad (s = 0, 1, \ldots) \tag{3}$$

hold, where M does not depend on s.

Indeed, as the q_s we may choose the corresponding Vallée Poussin sums of the function f (see 8.9). In the case $1 < p < \infty$ property a) may be replaced by the following: a) $q_s \in L_p$ and has a spectrum in $\Delta_s - \Delta_{s-1}$ $(s = 1, 2, \ldots)$, $\Delta_0 (s = 0)$ (see 8.10.1).

If the function f is represented in the form of a series (2) converging weakly to it with the indicated properties a), b) for some real r, then we will say this series is a *regular expansion of* f.

Lemma. *Any formally constituted series*

$$\sum_0^\infty u_s, \tag{4}$$

whose terms satisfy the properties: a*) $u_s \in L_p$ *and has a spectrum* outside Δ_{n_s} ($n_s = \varkappa s$, $\varkappa > 0$ *a constant not depending on* s), b)

$$\|u_s\|_{L_p} \ll a^{-sr} \quad (s = 0, 1, \ldots), \tag{5}$$

converges slowly to some function $f \in S^r_p$. *The functions* u_s *themselves thus form a sequence which converges weakly to zero.*

Proof. Suppose that $\varkappa\varrho > -r$. Then (see 8.4 (4)) $\|I_\varrho u_s\|_{L_p} \ll a^{-s(\varkappa\varrho + r)}$ and therefore the series

$$\sum_0^\infty I_\varrho u_s = F$$

converges in the sense of L_p, and accordingly, weakly to some $F \in L_p$. But then the series (1) converges weakly to $f \in I_{-\varrho} F \in S'_p$.

We note that for $r > 0$ the series (1) converges in L_p under the hypothesis that condition b) holds (without a*)).

The following imbeddings are valid:

(6) $$L_p^r(\mathbb{R}_n) \to L_{p'}^{\varrho}(\mathbb{R}_n) \quad (1 < p < p' < \infty),$$

(7) $$B_{p\theta}^r(\mathbb{R}_n) \to B_{p'\theta}^{\varrho}(\mathbb{R}_n) \quad (1 \leqq p \leqq p' \leqq \infty; 1 \leqq \theta \leqq \infty, B_{r\infty}^r = H_p^r),$$

$$\varrho = r - n\left(\frac{1}{p} - \frac{1}{p'}\right),$$

where r is any real number.

Indeed, suppose that Λ_p^r denotes one of the classes figuring in the left sides of (6), (7). Suppose that k is a number such that $k + \varrho > 0$. Then (see 8.2, 8.7, 9.6.2, 6.2)

$$I_k(\Lambda_p^r) = \Lambda_p^{r+k} \to \Lambda_{p'}^{\varrho+k},$$

so that

$$\Lambda_p^r \to I_{-k}(\Lambda_{p'}^{\varrho+k}) = \Lambda_{p'}^{\varrho},$$

and we have proved (6) and (7).

Regarding imbedding theorems of different dimensions, the situation is more complicated, as will be clear in what follows.

Put $\boldsymbol{x} = (\boldsymbol{u}, \boldsymbol{v})$, $\boldsymbol{u} = (x_1, \ldots, x_m)$, $\boldsymbol{v} = (x_{m+1}, \ldots, x_n)$ $(1 \leqq m < n)$, and suppose that $\mathbb{R}_m(\boldsymbol{v}^0) = \mathbb{R}_m$ denotes a linear subspace of $\mathbb{R}_n$ of points $(\boldsymbol{u}, \boldsymbol{v}^0)$, where $\boldsymbol{v}^0$ is fixed and $\boldsymbol{u}$ arbitrary.

Definition. *Suppose that the function $f \in S_p' = S_p'(\mathbb{R}_n)$, $1 \leqq p \leqq \infty$, and that*

(8) $$f(\boldsymbol{u}, \boldsymbol{v}) = \sum_{s=0}^{\infty} q_s(\boldsymbol{u}, \boldsymbol{v})$$

is a regular expansion of it, having the property that for any s the spectrum of q_s lies in the spectrum of f. (We note that the terms of the Vallée Poussin series are subjected to this property.) Suppose further that for any of the regular expansions of f defined above the series

(9) $$f(\boldsymbol{u}, \boldsymbol{v}^0) = \sum_{s=0}^{\infty} q_s(\boldsymbol{u}, \boldsymbol{v}^0)$$

converges weakly relative to $\boldsymbol{u}$ (in the sense of $S(\mathbb{R}_m)$) to some function $f(\boldsymbol{u}, \boldsymbol{v}^0)$, not depending on the expansion of f.

Then that function (of $\boldsymbol{u}$) is called the trace of f on $\mathbb{R}_m$.

We note that if the function $f(\boldsymbol{u}, \boldsymbol{v})$ is entire of exponential type, then any of its regular expansions is a finite sum (8), and its trace on $\mathbb{R}_m$ is obviously $f(\boldsymbol{u}, \boldsymbol{v}^0)$.

Below we present some assertions without proof.

Theorem. *The traces of the function $f(\boldsymbol{u}, \boldsymbol{v})$ on $\mathbb{R}_m$ in the sense of the definition just made and in the sense of the definition of 6.3 coincide.*

Denote by $\mathfrak{M}_\lambda$ any set of points $\boldsymbol{x} = (\boldsymbol{u}, \boldsymbol{v})$ of the form

$$\mathfrak{M}_\lambda = A + \{|\boldsymbol{v}| < |\boldsymbol{u}|^\lambda\} \ (\lambda > 0),$$

where A is a bounded set in $\mathbb{R}_n$ lying in the cube $\Delta_M = \{|x_j| < a^M, a > 1\}$.

Theorem. *If the function $f(\boldsymbol{u}, \boldsymbol{v}) \in S'_p(\mathbb{R}_n)$ $(1 \leqq p \leqq \infty)$ lying in a set $\mathfrak{M}_\lambda$, then it has a trace $f(\boldsymbol{u}, \boldsymbol{v}^0) \in S'_p(\mathbb{R}_m)$.*

Moreover, for the classes of functions $H^r_p(\mathbb{R}_n)$ having spectra in $\mathfrak{M}_\lambda$, one has imbeddings with constants depending on M and λ:

$$(10)\quad H^r_p(\mathbb{R}_n) \to \begin{cases} H_p^{\left(r-\frac{n-m}{p}\right)\lambda}(\mathbb{R}_n) & \left(r - \dfrac{n-m}{p} < 0, \lambda > 1\right), \\ H_p^{-\varepsilon}(\mathbb{R}_m) & \left(r - \dfrac{n-m}{p} = 0, \lambda \geqq 1\right), \\ H_p^{\left(r-\frac{\lambda(n-m)}{p}\right)}(\mathbb{R}_m) & \Big(0 < \lambda \leqq 1, \textit{ except for the} \\ \qquad \textit{case} \quad r - \dfrac{n-m}{p} = 0, \lambda = 1\Big). \end{cases}$$

Inverse Theorem. *The function*

$$\varphi(\boldsymbol{u}) \in H_p^{\left(r-\frac{\lambda(n-m)}{p}\right)}(\mathbb{R}_m) \quad (0 < \lambda \leqq 1)$$

or

$$\varphi(\boldsymbol{u}) \in H_p^{\left(r-\frac{n-m}{p}\right)\lambda}(\mathbb{R}_m) \quad (\lambda > 1)$$

may be extended to $\mathbb{R}_n$ in such a way that the extended function $f(\boldsymbol{u}, \boldsymbol{v}) \in H^r_p(\mathbb{R}_n)$, has a spectrum lying in a set of the type $\mathfrak{M}_\lambda$, and its trace $f(\boldsymbol{u}, 0) = \varphi(\boldsymbol{u})$. In addition we have the following imbeddings, with the corresponding inequalities, for which see 6.0 (13):

$$H_p^{\left(r-\frac{\lambda(n-m)}{p}\right)}(\mathbb{R}_m) \to H^r_p(\mathbb{R}_n) \quad (0 < \lambda \leqq 1),$$

$$H_p^{\left(r-\frac{n-m}{p}\right)\lambda}(\mathbb{R}_m) \to H^r_p(\mathbb{R}_n) \quad (\lambda > 1).$$

The imbedding (11) with $\lambda = 1$ and $r - \dfrac{n-m}{p} > 0$ is already known (see 6.5), but here it is given a stronger formulation, inclusding the assertion on the character of the spectrum of the extended function. For $\lambda > 1$ and $r - \frac{n-m}{p} = 0$ the (inverse) imbedding (12) and the corresponding (direct) imbedding (10) are already not mutually inverse.

We emphasize that in the relations (11), (12) no restrictions whatever are imposed on the spectra of the functions of the original (imbedded) classes.

In the case $0 < \lambda \leqq 1$, using the hypothesis of the extension theorem, the function $f(u, v)$ may be defined in the form of a weakly converging series:

$$f(u, v) = \sum_{s=0}^{\infty} Q_s(u, v),$$

$$Q_s(u, v) = \varphi_s(u) \prod_{j=m+1}^{M} F(2^{(s-k)\lambda} v_j), \; F(t) = 4 \left(\frac{\sin \frac{t}{2}}{t} \right)^2,$$

where

$$\varphi(u) = \sum_{s=0}^{\infty} \varphi_s(u),$$

$$\varphi_0(u) = \sigma_{2^0} f, \; \varphi_s(u) = (\sigma_{2^s} - \sigma_{2^{s-1}}) \varphi$$

(see 8.9).

Now in the case $\lambda > 1$ the function $f(u, v)$ is defined by the weakly converging series

$$f(u, v) = \sum_{s=0}^{\infty} q_{n_s}(u, v),$$

$$q_{n_s}(u, v) = \varphi_s(u) \prod_{j=m+1}^{M} \alpha_s(v_j), \quad \alpha_s(\xi) = \cos 3 \cdot 2^{n_s - 1} \xi F(2^{n_s - 1} \xi),$$

where $n_s (s = 0, 1, \ldots)$ is an increasing sequence of integers such that $\frac{n_s}{s_\lambda} \to 1$ $(s \to \infty)$, and the functions φ_s have the preceding sense.

The function $\psi(x, y)$ of two variables, having the Fourier transform

$$\bar{\psi} = \begin{cases} u^{-1} v^{-1} & (u, v > 2), \\ 0 & \text{for the remaining } (u, v) \end{cases}$$

lies in $H_2^{1/2}(\mathbb{R}_2)$ and at the same time has no trace on the axis $v = 0$.

Proof. As a regular expansion for ψ we select the series

$$\psi = \sum_{s=1}^{\infty} q_s,$$

where

$$q_s(x, y) = \frac{1}{2\pi} \iint\limits_{\Delta_s - \Delta_{s-1}} u^{-1} v^{-1} e^{i(xu+yv)} \, du \, dv ,$$

and

$$\|q_s\|^2_{L_2(\mathbb{R}_2)} \leqq 2 \int\limits_{2^{s-1}}^{2^s} u^{-2} du \int\limits_{2}^{2^s} - v^{-2} \, dv \ll 2^{-s} ,$$

so that $\psi \in H_2^{1/2}(\mathbb{R}_2)$. Further

$$S_N(x) = \sum_1^N q_s(x, 0) = \frac{1}{2\pi} \iint\limits_{\Delta_N} u^{-1} v^{-1} e^{ixu} \, du \, dv$$

$$= \frac{1}{2\pi} \int\limits_2^{2^N} v^{-1} \, dv \int\limits_2^{2^N} u^{-1} e^{ixu} \, du = cN \int\limits_2^{2^N} u^{-1} e^{ixu} \, du .$$

The function $S_N(x)$ does not converge weakly, since, for example, for a function φ such that $\tilde{\varphi} = e^{-x^2} \in S'$ one has

$$(S_N, \varphi) = (\tilde{S}_N, \tilde{\varphi}) = cN \int\limits_2^{2^N} u^{-1} e^{-u^2} \, du \to \infty \quad (N \to \infty).$$

It is possible to construct an example showing that in (10) it is not possible to replace $\varepsilon > 0$ by $\varepsilon = 0$.

The assertions indicated above may be extended from the classes H_p^r to $B_{p\theta}^r$.

8.10 – 8.10.1. The facts presented here, relating to the expansion of functions of the classes $B_{p\theta}^r$ in series in Dirichlet sums in the case $1 < p < \infty$, are close to the results of P. I. Lizorkin [7], and also to the results of M. D. Ramazanov [1], who investigated from his point of view classes of functions somewhat different from $B_{p\theta}^r$.

To Chapter 9

9.1. Suppose that

(1) $$1 < p < q < \infty ,$$

(2) $$1 \leqq m \leqq n, \varrho = r - \frac{n}{p} + \frac{m}{q} \geqq 0$$

and that r is an integer. Then 9.1 (4) implies the imbedding

$$W_p^r(\mathbb{R}_n) = L_p^r(\mathbb{R}_n) \to L_q^\varrho(\mathbb{R}_m) \to W_q^{[\varrho]}(\mathbb{R}_m)$$

(S. L. Sobolev with complements due to V. I. Kondrašov and V. I. Il'in; see 6.1). For noninteger ϱ and $p = q$ one has (see (9.3 (1), 6.2 (4), 6.5 (1'))

$$L_p^r(\mathbb{R}_n) \to H_p^r(\mathbb{R}_n) \to H_p^\varrho(\mathbb{R}_m) \to W_p^{[\varrho]}(\mathbb{R}_m)\,.$$

The case $\varrho = 0$, $p = 1 < q < \infty$ is interesting. In this case for an integer $r(L_p^r = W_p^r)$ it was proved, by S. L. Sobolev [4] for $m = n$, and by E. Gagliardo [2] for $m < n$, that the imbedding (3) also remains valid.

9.2.2. The theorem on derivatives was proved for $p = 2$ by S. N. Bernšteĭn [1], p. 98, for $l_1 = l_2 = 2$ and by L. N. Slobodeckiĭ [3] in the general case. It was proved for $1 < p < \infty$ and integer $l = l_1 = \cdots = l_n$ by A. I. Košelev [1], and for arbitrary $\boldsymbol{l} > \boldsymbol{0}$ by P. I. Lizorkin [10]. For the periodic case and any $\boldsymbol{l}$ it was proved by Ju. L. Bessonov [1, 2].

9.4—9.6. The results presented here relating to isotropy classes L_p^r are basically due to Lizorkin, published without proof in his note [10]. He placed at my disposal a quantity of manuscript material, which lies at the basis of my exposition. I have reduced the question throughout to the operation I_r, while Lizorkin in corresponding cases operates using "pure" Liouville derivatives (see 9.2.3). The basic objective of these investigations is to obtain integral representations for functions of the anisotropy classes $L_p^{\boldsymbol{r}}$ for any $\boldsymbol{r} \geqq \boldsymbol{0}$ and on the basis of these representations to obtain a complete system of imbedding theorems for these classes. For the isotropy classes integral representations of this kind were obtained in the preceding chapter. The needed estimates arose there from the facts relating to the theory of the Bessel-MacDonald kernel. In the anisotropic case the corresponding kernel is more complicated. Of course, imbedding theorems in the isotropic case may be obtained from the corresponding anisotropic theory if one puts $r_1 = \cdots = r_n = r$. For integer r, $\boldsymbol{r}$ we obtain the corresponding results for the W-classes, in particular the imbedding theorems of S. L. Sobolev, with which historically this multidimensional theory started.

9.4.1. The estimates (2), (3) for $I_{-l}G_{\boldsymbol{r}}(x)$ are equivalent in the case $r_1 = \cdots = r_n = r$ to the isotropic estimates 8.1 (7).

9.5.1—9.5.2. Theorems 9.5.1 and 9.5.2 were obtained in their full generality by Lizorkin in [10]. They contain a number of the preceding results relating to the case of integer $\boldsymbol{r}(W_p^{\boldsymbol{r}} = L_p^{\boldsymbol{r}})$ and to arbitrary $\boldsymbol{r}$ for $p = 2$ of Aronszajn [1], Slobodeckiĭ [1] (see further Babič and

Slobodeckiĭ [1]), Gagliardo [1], Besov [2], Lizorkin [9], Uspenskiĭ [1]. For more details see the survey [12] by S. M. Nikol'skiĭ.

To these are related the corresponding results for the isotropy classes $L_p^r = L_p^{r,\ldots,r}$, due to Stein [1], Aronzajn, Mulla and Szeptycki [1] and Lizorkin [3]. For details see V. I. Burenkov [3].

These results were obtained by various methods.

In this book in extending functions from $\mathbb{R}_m$ to $\mathbb{R}_n$ we have applied the method of expanding them into series of entire functions of exponential type and then accumulating the terms by means of special functions (S. M. Nikol'skiĭ [5]). With these aims other authors have also applied another method, based on the averaging of functions in the sense of Steklov (see A. A. Dezin [1], Gagliardo [1]).

We note a sufficiently simple direct proof of the imbedding theorems for different dimensions in the anisotropic case for the integer classes $L_p^{\boldsymbol{r}} = W_p^{\boldsymbol{r}}$, due to Solonnikov [1].

9.6.2. The imbedding theorem of S. L. Sobolev (with complements due to Kondrašov and Il'in) (see 6.1 and the Remark to 6.1) reduces to Theorem 9.6.2 as a special case, and in the terms of the (integer) classes W_p^l it is exact in the limit.

In the isotropic case of fractional l, Theorem 9.6.2 was proved by Stein [1] and Lizorkin [5]. In the anisotropic case the proof, presented in the text, is due to Lizorkin [10].

During proof-reading we became acquainted with the paper of Sadosky and Cotlar [1], in which, for rational vectors $\boldsymbol{r} \geqq \boldsymbol{0}$ classes equivalent to the classes $L_p^{\boldsymbol{r}}$ are defined and some imbedding theorems proved for them.

Literature

Agmon, F.

1. The L_p approach to the Dirichlet problem. Annali della Scuola Normale Superiore di Pisa, **3**, 13, 4, 405—448 (1959).

Ahiezer, N. I.

1. Lectures on the theory of approximation, Moscow: "Nauka" 1965, 407 pages [in Russian].

Aleksandrov, P. S. (Alexandroff, P. S.)

1. Introduction to the general theory of sets and functions, State Publishing House for Technical Literature 1948, 411 pages [in Russian].

Aleksandrov, P. S. (Alexandroff, P. S.), Kolmogorov, A. N.

1. Introduction to the theory of functions of a real variable, 3rd. Ed., Moscow—Leningrad 1938, 268 pages [in Russian].

Amanov, T. I.

1. Boundary functions of the classes $H_p^{r_1,\ldots,r_n}$ and $\overset{r_1,\ldots,r_n}{H_p^*}$. Izv. Akad. Nauk SSSR, ser. matem. **19**, No. 1, 17—32 (1955) [in Russian].
2. On the solution of the biharmonic problem. Dokl. Akad. Nauk SSSR, **87**, 389—392 (1953) [in Russian].
3. Representation and imbedding theorems for the functional spaces $S^r_{p\theta}B(\mathbb{R}_n)$ and $\overset{r}{S^*_{p\theta}}B(\mathbb{R}_n)$ $(0 \leqq x_j \leqq 2\pi)$. Trudy Steklov Institute of the Akad. Nauk SSSR, Moscow, **77**, 5—34 (1965) [in Russian].
4. (Doctoral dissertation) Investigation of the properties of classes of functions with dominating mixed derivatives, representation, imbedding, extension and interpolation theorems, Novosibirsk, 1967 [in Russian].

Aronszajn, N.

1. Boundary values of functions with finite Dirichlet integral, Conference on Partial Differential Equations, Studies in Eigenvalue Problems, University of Kansas, 1965.

Aronszajn, N., Mulla, F., Szeptycki, P.

1. On spaces of potentials connected with L_p classes. Ann. Inst. Fourier **13**, No. 2, 211—306 (1963).

Aronszajn, N., Smith, K. T.

1. Functional spaces and functional completion. Ann. Inst. Fourier **6**, 125—185 (1956).
2. Theory of Bessel Potentials. I. Ann. Inst. Fourier **11**, 385—475 (1961).

Arzelà, C.

1. Esistenza degli integrali delle equazioni a derivate parziali. Acc. delle Scienze di Bologna, **6**, No. 3 (1936).

Ascoli, G.

1. Le curve limiti di una varietà data di curve. Memoria Acc. dei Lincei, **3**, No. 18, 521—586 (1884).

Babič, V. M.

1. On the question of the distribution of functions. Uspehi matem. nauk, **8**, issue 2 (54), 111—113 (1953) [in Russian].

Babič, V. M., Slobodecskiĭ, L. N.

1. On the boundedness of the Dirichlet integral. Dokl. Akad. Nauk SSSR, **106**, 604—606 (1956) [in Russian].

Bahvalov, N. S.

1. Imbedding theorems for classes of functions with several bounded derivatives. Vestnik Moscow State University, Matem., mehan. No. 3, 7—16 (1953) [in Russian].

Banach, S.

1. Théorie des opérations linéaires, Warszawa 1932, 254 pages.
2. Course in functional analysis, Soviet School, Kiev, 1948, 216 pages [in Ukranian].[1]

Bari, N. K.

1. A generalization of inequalities of S. N. Bernšteĭn and A. A. Markov. Izv. Akad. Nauk SSSR, ser. matem. **18**, 59—176 (1954) [in Russian].

Bernšteĭn, S. N.

1. Collected Works, Vol. I, Moscow: Publishing House of the Akad. Nauk SSSR, 1952, 581 pages [in Russian].
2. Collected Works, Vol II, Moscow: Publishing House of the Akad. Nauk SSSR 1954, 628 pages [in Russian].

Besov, O. V.

1. On certain properties of harmonic functions given on a halfplane. Izv. Akad. Nauk SSSR, ser. matem. **20**, 469—484 (1956) [in Russian].
2. On a certain family of functional spaces. Imbedding and extension theorems. Dokl. Akad. Nauk SSSR, **126**, 1163—1165 (1959) [in Russian].
3. On certain conditions for the derivatives of periodic functions to belong to L_p. Naučnye dokl. vysš. šk. No. 1, 12—17 (1959) [in Russian].
4. On extensions of functions preserving the properties of the integral modulus of smoothness of the second order. Matem. sb. **58**, 673—684 (1962) [in Russian].
5. Investigation of a family of functional spaces in connection with imbedding and extension theorems. Trudy Steklov Institute of the Akad. Nauk SSSR, **60**, 42—81 (1961) [in Russian].
6. On imbedding theorems for spaces of differentiable functions. Uspehi matem. nauk, **16**, No. 5, 207—208 (1961) [in Russian].
7. An example relating to the theory of imbedding theorems. Dokl. Akad. Nauk SSSR **143**, 1014—1016 (1962) [in Russian].
8. On the density of finite functions in $L^l_{p,\theta}$ and extension of functions. Trudy Steklov Institute of the Akad. Nauk. SSSR, **89**, 18—30 (1867) [in Russian].
9. Extension of functions from L^l_p and W^l_p. Trudy Steklov Institute of the Akad. Nauk SSSR, **89**, 5—17 (1967) [in Russian].
10. On the theory of imbedding and extension of classes of differentiable functions. Abstract of a doctoral dissertation, Matem. zametki, **1**, No. 2, 235—244 (1967) [in Russian].
11. On a family of functional spaces. Imbedding and extension theorems, Abstract of a candidate dissertation, 1960 [in Russian].
12. On certain properties of the spaces $H_p^{r_1,\ldots,r_n}$. Izv. vysš. uč. zaved., ser. matem. No. 1 (14), 16—23 (1960) [in Russian].

Besov, O. V., Il'in, V. P.

1. A natural extension of the class of domains in imbedding theorems. Matem. sb. **75** (117); **4**, 483—495 (1968) [in Russian].

[1] [2] is a translation of [1], with a few new notes.

Bessonov, Ju. L.

1. Approximation of periodic functions lying in the classes $W_{p\theta}^{r_1,r_2}$ by Fourier sums. Dokl. Akad. Nauk SSSR, **147**, 519–522 (1962) [in Russian].
2. On the existence of mixed derivatives of fractional order in L_p. Uspehi matem. nauk, **19**, issue 4 (118), 163–170 (1964) [in Russian].

Bochner, S.

1. Vorlesungen über Fouriersche Integrale, Akademie-Verlag 1932.

Bugrov, Ja. S.

1. The Dirichlet problem for the disk. Dokl. Akad. Nauk SSSR, **115**, 639–642 (1965) [in Russian].
2. Properties of polyharmonic functions. Izv. Akad. Nauk SSSR, ser. matem. **22**, 491–514 (1958) [in Russian].
3. Polyharmonic functions in the halfplane. Matem. sb. **60**, 486–498 (1963) [in Russian].
4. On imbedding theorems for the H-classes of S. M. Nikol'skiĭ. Sibirsk. matem. ž. **4**, 1012–1028 (1963) [in Russian].

Burenkov, V. I.

1. Local lemmas for certain classes of differentiable functions. Trudy Steklov Institute of the Akad. Nauk SSSR, **77**, 65–71 (1965) [in Russian].
2. Certain properties of the classes $W_p^{(r)}(\Omega)$ and $W_p^{(r,r)}(\Omega)$ for $0 < r < 1$. Trudy Steklov Institute of the Akad. Nauk SSSR, **77**, 72–88 (1965) [in Russian].
3. Imbedding and extension theorems for classes of differentiable functions of several variables given on the entire space. Itogi nauki, Matematičeskiĭ analiz, 1965, Moscow, Publishing House of the All-Union Institute for Scientific and Technical Information ("VINITI") of the Akad. Nauk SSSR, 71–155 (1966) [in Russian].
4. Certain properties of classes of differentiable functions and connections with imbedding and extension theorems. Abstract of a candidate dissertation, Moscow, 1966 [in Russian].
5. On imbedding theorems for the region $R_k = \{\alpha_i h < x_i^{k_i} < \beta_i h,\ 0 < h < 1\}$. Matem. sb., **75** (117); **4**, 496–501 (1968) [in Russian].

Calderon, A. P.

1. Singular integrals, Notes a course taught at the Massachussetts Institute of Technology, 1959.
2. Lebesgue spaces of differentiable functions, Conference on partial differential equations, University of California, 1960.

Calkin, J. W.

1. Functions of several variables and absolute continuity. I. Duke Math. J. **6**, 176–186 (1940).

Civin, P.

1. Inequalities for trigonometric integrals. Duke Math. J. **8**, 656–665 (1941).

Deny, J., Lions, J. L.

1. Espaces de Beppo Levi et applications. C. R. Acad. des Sc. **239**, 1174–1177 (1954).
2. Les espaces de type de Beppo Levi. Annales de l'Institut Fourier **5**, 1953–1954, 305–370 (1955).

Dezin, A. A.

1. On imbedding theorems and the extension problem. Dokl. Akad. Nauk SSSR, **88**, 741–743 (1953) [in Russian].

Džabrailov, A. D.

1. On certain functional spaces. Direct and inverse imbedding theorems. Dokl. Akad. Nauk SSSR, **159**, 254–257 (1964) [in Russian].

Džafarov, A. S.

1. On certain properties of functions of several variables. In the collection "Investigations on current problems of the theory of functions of a complex variable", Moscow: Physics and Mathematics Publishing House 1960, pp. 537—544 [in Russian].

Dzjadyk, V. K.

1. On extension of functions satisfying a Lipschitz condition in the metric of L_p. Matem. sb. **40** (82), 239—242 (1956) [in Russian].

Evans, G. C.

1. Note on a theorem of Bochner. Amer. J. of Math. **50**, 123—126 (1928).
2. Complements of potential theory. II. Amer. J. of Math. **55**, 29—49 (1933).

Favard, J.

1. Application de la formule sommatoire d'Euler à la démonstration de quelques propriétés extrémales des intégrales des fonctions périodiques ou presque périodiques. Matematisk Tidskrift (B), 81—94 (1936).

Foht, A. S.

1. Estimates of the solutions of equations of parabolic type close to the boundary of the region in which they are given, Abstract of a candidate dissertation, Moscow, 1963 [in Russian].
2. Certain inequalities for the solutions of equations of elliptic type and their derivatives close to the boundary of the region in the metric of L_2. Trudy Steklov Mathematical Institute of the Akad. Nauk SSSR, **77**, 168—191 (1965) [in Russian].

Gagliardo, E.

1. Caratterizazioni della trace sulla frontiera relative ad alcune classi di funzioni in n variabili. Rend. Semin. matem. Università di Padova, **27**, 284—305 (1957).

Gel'fand, I. M., Šilov, G. E.

1. Generalized functions, issue 1, 439 pages, issue 2, 337 pages, Moscow, 1958 [in Russian].

Gjunter, N. M.

1. Potential theory and its application to the fundamental problems of mathematical physics. State Publishing House for Technical Literature 1953, 416 pages [in Russian].

Golovkin, K. K.

1. On the impossibility of certain inequalities between functional norms. Trudy Steklov Institute of the Akad. Nauk SSSR, **70**, 5—25 (1964) [in Russian].
2. On equivalent normalizations of fractional spaces. Trudy Steklov Institute of the Akad. Nauk SSSR, **66**, 364—383 (1962) [in Russian].

Gončarov, V. L.

1. Theory of interpolation and approximation of functions, 2nd Ed., State Publishing House for Technical Literature 1954, 328 pages [in Russian].

Halperin, I.

1. Introduction to the theory of generalized functions. On the basis of lectures of Laurent Schwartz, Moscow: Publishing House for Foreign Literature 1954, 64 pages [in Russian].[1]

Hardy, G. H., Littlewood, J. E., Polya, G.

1. Inequalities, Cambridge University Press 1934.

[1] Translator's note. This is a translation. I do not however know the original reference.

Hardy, G. H., Littlewood, J. E.
1. Some properties of fractional integrals, I. Math. Zeit. **27**, 565—606 (1928).
2. A convergence criterion for Fourier series. Math. Zeit. **28**, 122—147 (1928).

Hausdorff, F.
1. Mengenlehre (3. Aufl.).
2. Theory of sets. ONTI, 1937, 304 pages [in Russian].
2. Zur Theorie der linearen Räume. Journ. f. reine und angew. Math. **167**, 294—311 (1932).

Hörmander, L.
1. Estimates for operators invariant relative to a displacement, Moscow: Publishing House for Foreign Literature 1962, 71 pages [in Russian].[1]

Ibragimov, I. I.
1. Extremal poperties of entire functions of finite order. Baku, 1962, 315 pages [in Russian].

Il'in, V. P.
1. On a theorem of G. H. Hardy and J. E. Littlewood. Trudy Steklov Institute of the Akad. Nauk SSSR, **53**, 128—144 (1959) [in Russian].
2. On an imbedding theorem for the limiting exponent. Dokl. Akad. Nauk SSSR, **96**, 905—908 (1954) [in Russian].
3. Certain inequalities for differentiable functions of several variables. Dokl. Akad. Nauk **135**, 778—782 (1960) [in Russian].
4. On the question of inequalities among the norms of the partial derivatives of functions of several variables. Dokl. Akad. Nauk SSSR, **150**, 975—977 (1963) [in Russian].
5. One inequalities among the norms of the partial derivations of functions of several variables. Trudy Steklov Institute of the Akad. Nauk SSSR, **84**, 144—173 (1965) [in Russian].
6. Integral representations of differentiable functions and their application to questions of the extension of functions of the classes $W_p^l(g)$. Sibirsk. matem. ž **7**, 573—586 (1967) [in Russian].
7. Estimates of functions having derivatives summable to a given power on hyperplanes of various dimensions. Dokl. Akad. Nauk SSSR, **78**, 633—636 (1951) [in Russian].
8. Certain integral inequalities and their application in the theory of differentiable functions of several variables. Matem. sb. **54**, 331—380 (1961) [in Russian].
9. On complete continuity of an imbedding operator for the case of an unbounded domain. Dokl. Akad. Nauk SSSR, **135**, 517—519 (1960) [in Russian].

Il'in, V. P., Solonnikov, V. A.
1. On certain properties of differentiable functions of several variables. Dokl. Akad. Nauk SSSR, **136**, 538—541 (1961) [in Russian].
2. On certain properties of differentiable functions of several variables. Trudy Steklov Institute of the Akad. Nauk SSSR, **66**, 205—226 (1962) [in Russian].

Jackson, D.
1. Über die Genauigkeit der Annäherung stetiger Funktionen durch ganze rationale Funktionen gegebenen Grades und trigonometrischen Summen gegebener Ordnung. Diss., Göttingen, 1911.
2. The theory of approximation. Amer. Math. Soc. Colloquium Publications, **11**, 1930.

[1] Translator's note. This is a translation, for which I do not have the original reference.

Jakovlev, G. N.
1. Boundary properties of functions of the class $W_p^{(l)}$ on regions with angular points. Dokl. Akad. Nauk SSSR, **140**, 73–76 (1961) [in Russian].
2. Boundary properties of a certain class of functions. Trudy Steklov Institute of the Akad. Nauk SSSR, **60**, 325–349 (1961) [in Russian].

Kellogg, O. D.
1. On the derivatives of harmonic functions on the boundary. Trans. Amer. Math. Soc. 33, No. 2 (1931).

Kiprijanov, I. A.
1. On a class of imbedding theorems with weights. Dokl. Akad. Nauk SSSR, **147**, 540–543 (1962) [in Russian].

Kolmogorov, A. N. (Kolmogoroff, A. N.)
1. Über Kompaktheit der Funktionenmengen bei der Konvergenz in Mittel. Götting. Nachr. 60–63 (1931).

Kolmogorov, A. N., Fomin, S. V.
1. Elements of the theory of functions and functional analysis, Moscow: "Nauka" 1968, 496 pages [in Russian].

Kondrašov, V. I.
1. On certain properties of functions of the space L_p. Dokl. Akad. Nauk SSSR, **48**, 563–566 (1945) [in Russian].
2. Behavior of functions on L_p^ν on manifolds of various dimensions. Dokl. Akad. Nauk SSSR, **6**, 1005–1012 (1950) [in Russian].

Konjuškov, A. A.
1. Best approximations by trigonometric polynomials and Fourier coefficients Matem. sb. **44** (86), 53–84 (1958) [in Russian].

Košelev, A. I.
1. Differentiability of the solutions of certain problems of potential theory. Matem. sb. **32** (74), 653–664 (1953) [in Russian].

Kudrjavcev, L. D.
1. On a generalization of a theorem of S. M. Nikol'skiĭ on the compactness of classes of differentiable functions. Uspehi matem. nauk **9**, No. 2 (59), 111–120 (1954) [in Russian].
2. Direct and inverse imbedding theorems. Applications to the solution by variational methods of elliptic equations. Trudy Steklov Institute of the Akad. Nauk SSSR, **55**, 1–181 (1959) [in Russian].
3. Imbedding theorems for weight spaces and their applications to the solution of the Dirichlet problem. Collection "Investigations on current problems of the constructive theory of functions". Akad. Nauk Azerb. SSSR, Baku 1965, pp. 493–501 [in Russian].
4. Imbedding theorems for classes of functions defined on the entire space or on a halfspace, II. Matem. sb. **70** (**112**); 1, 3–35 (1966) [in Russian].

Kufner, A.
1. Einige Eigenschaften der Sobolevschen Räume mit Belegungsfunktion. Čehoslov. matemat. žurnal **15**, 597–620 (1965).

Levi, B.
1. Sul principio di Dirichlet. Rend. Palermo **22**, 293–359 (1906).

Lions, J. L., Magenes, E.
1. Problèmes aux limites non homogènes (III). Annali della Scuola Norm. Sup. Pisa, **15**, 39–101 (1961).
2. Problèmes aux limites non homogènes et applications I. 372 pages, II, 251 pages, Paris: Dunod 1968.

Littlewood, J. E., Paley, R. E. A. C.

1. Theorems on Fourier series and power series. I. J. London Math. Soc. **6**, 230–233 (1931); II, Proc. London Math. Soc. **42**, 52–89 (1936); III, ibid., **43**, 105–126 (1937).

Lizorkin, P. I.

1. Imbedding theorems for functions of the space L_p. Dokl. Akad. Nauk SSSR, **143**, 1042–1045 (1962) [in Russian].
2. The spaces $L_p^r(\Omega)$. Extension and imbedding theorems. Dokl. Akad. Nauk SSSR, **145**, 527–530 (1962) [in Russian].
3. A characteristic of the boundary values of functions of $L_p^r(E_n)$ on hyperplanes. Dokl. Akad. SSSR, **150**, 984–986 (1963) [in Russian].
4. (L_p, L_q)-multipliers of Fourier integrals. Dokl. Akad. Nauk SSSR, **152**, 808–811 (1963) [in Russian].
5. Generalized Liouville differentiation and the functional spaces $L_p(\mathbb{R}_n)$. Imbedding theorems. Matem. sb. **60** (**102**), 325–363 (1963) [in Russian].
6. Formulas of Hirschman type and relations among the spaces $B_p^r(E_n)$ and $L_p^r(E_n)$. Matem. sb. **63**, 505–535 (1964) [in Russian].
7. On the Fourier transform and Besov spaces. The null scale $B_{p\theta}^0$. Dokl. Akad. Nauk SSSR, **163**, 1318–1321 (1965) [in Russian].
8. Estimates of trigonometric integrals and the Bernšteĭn inequality for fractional derivatives. Izv. Akad. Nauk SSSR, **29**, 109–126 (1965) [in Russian].
9. Boundary properties of functions from weight classes. Dokl. Akad. Nauk SSSR, **132**, 514–517 (1960) [in Russian].
10. Nonisotropic Bessel potentials. Imbedding theorems for the Sobolev spaces $L_p^{r_1,\dots,r_n}$ with fractional derivatives. Dokl. Akad. Nauk SSSR, **170**, 508–511 (1966) [in Russian].

Ljusternik, L. A., Sobolev, V. I.

1. Elements of functional analysis. Moscow: "Nauka" 1965, 520 pages [in Russian].

Lozinskiĭ, S. M.

1. On convergence and summability of Fourier series and interpolation processes. Matem. sb. **14** (56), 175–268 (1944) [in English].
2. Generalization of a theorem of S. N. Bernšteĭn on the derivative of a trigonometric polynomial. Dokl. Akad. Nauk SSSR, **55**, 9–12 (1947) [in Russian].

Marchoud, A.

1. Sur les dérivées et sur les différences des fonctions de variables réelles. J. Math. pures et appl. **6**, 337–425 (1927).

Marcienkiewicz, J.

1. Sur les multiplicateurs des séries de Fourier. Studia Math. **8**, 78–91 (1939).

Matveev, I. V., Nikol'skiĭ, S. M.

1. On the pasting-together of functions of the class $H_p^{(\alpha)}$. Uspehi matem. nauk, **18**, No. 5, 175–180 (1963) [in Russian].

Mihlin, S. G.

1. Fourier integrals and multiple singular integrals. Vestnik of the Leningrad State University, ser. No. 7, 143–155 (1957) [in Russian].

Mitjagin, B. S.

1. On certain properties of functions of two variables. Vestnik Moscow State University, ser. matem., meh., astr., fiz., him. No. 5, 137–152 (1959) [in Russian].

Montel, P.
1. Sur les polynomes d'approximation. Bull. Soc. Math. de France, 151–192 (1918).
Morrey C. B., Jr.
1. Functions of several variables and absolute continuity, II. Duke Math. J. **6**, No. 1, 187–215 (1940).
Mozžerova, N. I.
1. Boundary properties of harmonic functions in three-dimensional space. Dokl. Akad. Nauk SSSR, **118**, 636–638 (1958) [in Russian].
Natanson, I. P.
1. Theory of functions of a real variable. State Publishing House for Technical Literature, 1957, 552 pages [in Russian].[1]
2. Constructive theory of functions, Moscow–Leningrad, 1949, 658 pages [in Russian].
Nečas, J.
1. Sur les solutions des équations élliptiques aux derivées partielles du second ordre avec intégrale de Dirichlet non bornée. Czechoslov. Math. J. **10**, 283–298 (1960).
2. Les méthodes directes en théorie des equations elliptiques. Academia, éditions de l'Académie Tchécoslovaque ses Sciences, Prague, 1967, 351 pages.
Nikodym, O. M.
1. Sur une classe de fonctions considérées dans le problème de Dirichlet. Fund. Math. **21**, 129–150 (1933).
Nikol'skiĭ, Ju. S.
1. Boundary values of functions of weight classes. Dokl. Akad. Nauk SSSR, **164**, 503–506 (1965) [in Russian].
Nikol'skiĭ, S. M.
1. Theory of approximation of functions. 4. I, Functional analysis. Dnepropetrovskiĭ Universitet, 1947, 71 pages [in Russian].
2. Generalization of a proposition of S. N. Bernšteĭn on differentiable functions of several variables. Dokl. Akad. Nauk SSSR, **59**, 1533–1536 (1948) [in Russian].
3. Inequalities for entire functions of finite degree and their application to the theory of differentiable functions of several variables. Trudy Steklov Institute of the Akad. Nauk SSSR, **38**, 244–278 (1951) [in Russian].
4. On the question of the solution of the polyharmonic equation by variational methods. Dokl. Akad. Nauk SSSR, **33**, 409–411 (1953) [in Russian].
5. Properties of some classes of functions of several variables on differentiable manifolds. Matem. sb. **33** (75); **2**, 261–326 (1953) [in Russian].
6. On an inequality for periodic functions. Uspehi matem. nauk, **11**, No. 1(67), 219–222 (1956) [in Russian].
7. On extension of functions of several variables with preservation of differential properties. Matem. sb. **40** (82), 243–268 (1956) [in Russian].
8. Compactness of the classes $H_p^{r_1,\dots,r_n}$ of functions of several variables. Izv. Akad. Nauk SSSR, **20**, 611–622 (1956) [in Russian].
9. Boundary properties of functions defined on a region with angular points. Matem sb. I **40** (82), 303–318 (1956); II **44** (86), 127–144 (1957); III **45** (87), 181–194 (1958) [in Russian].

[1] *Translator's note.* There exists an English translation, at least of an earlier edition.

10. An imbedding theorem for functions with partial derivatives, considered in different metrics. Izv. Akad. Nauk SSSR, ser. matem., **22**, 321—336 (1958) [in Russian].
11. Certain properties of differentiable functions given on an n-dimensional open set. Izv. Akad. Nauk SSSR, ser. matem. **23**, 213—242 (1959) [in Russian].
12. On imbedding, extension, and approsimation theorems for differentiable functions of several variables. Uspehi matem. nauk, **16**, 5 (101), 63—114 (1961) [in Russian].
13. Correction to the paper "Properties of some classes of functions of several variables on differentiable manifolds". Matem. sb. **57** (99), 527 (1962) [in Russian].
14. On a problem of S. L. Sobolev. Sibirsk. matem. ž. **3**, 845—851 (1962) [in Russian].
15. Representation theorem for a class of differentiable functions of several variables in terms of entire functions of exponential type. Dokl. Akad. Nauk SSSR, **150**, 484—487 (1963) [in Russian].
16. Functions with dominating mixed derivatives satisfying a multiple Hölder condition. Sibirsk. matem. ž **4**, 1342—1364 (1963) [in Russian].
17. Constructive representation of null classes of differentiable functions of several variables. Dokl. Akad. Nauk SSSR, **170**, 542—545 (1966) [in Russian].
18. Integral representation and isomorphism of classes of differentiable functions of several variables. Third mathematical summer school on the constructive theory of functions, held in Kaceveli, June—July 1965, Kiev: Publishing House "Naukova dumka" (Ukranian: "Scientific thought") 1966, pp. 135—238 [Article is in Russian].

Nikol'skiĭ, S. M., Lions, J. L., Lizorkin, P. I.

1. Integral representation and isomorphism properties of some classes of functions. Annali Scuola Norm. Super. Pisa. Sci. fis., e mat., Ser. III, Vol XIX, No. 11, 127—178 (1965).

Paley, R. E. A. C., Wiener, N.

1. Fourier transforms in the complex domain. American Mathematical Society Colloquium Publications, 1934.

Petrovskiĭ, I. G., Smirnov, K. N.

1. On conditions for equicontinuity of families of functions. Bjull. Moscow State University, section A, issue 10, 1—15 (1938) [in Russian].

Picone, M.

1. Sulla derivazione parziale per serie. Bol. Un. Mat. It. III, No. 5, 24—33 (1950).

Pilika, P.

1. Boundary functions of the classes $B^{(r_1,\dots,r_n)}_{(p_1,\dots,p_n)}$. Dokl. Akad. Nauk SSSR, **128**, 677—679 (1959) [in Russian].

Plancherel, M., Polya, G.

1. Fonctions entières et intégrales de Fourier multiples. Comment. Math. Helvitici, I, 224—248 (1937); II, 110—163 (1938).

Pólya, G., Szegö, G.

1. Problems and theorems in analysis. I, 4th. Ed. (Die Grundlehren der mathematischen Wissenschaften, Bd. 193), Berlin—Heidelberg—New York: Springer 1970.

Pucci, C.

1. Compatezza di succesione di funzioni e derivabilità delle funzioni limita. Annali di Mathematica, **37**, 1—25 (1954).

Quade, E.

1. Trigonometric approximation in the mean. Duke Math. J. **3** (1937).

Ramazanov, M. D.

1. Apriori estimates of type L_p for solutions of parabolic equations. Dokl. Akad. Nauk, **161**, 530—533 (1961) [in Russian].

Rellich, F.

1. Ein Satz über mittlere Konvergenz. Gött. Nachr., 30—35 (1930).

Riesz, M.

1. Les fonctions conjuguées et les séries de Fourier. C. R. Acad. Sci. **178**, 1464—1476 (1924).
2. Sur les fonctions conjuguées. Math. Z. **27**, 218—244 (1927).

Sadosky, C., Cotlar, M.

1. On quasihomogeneous Bessel potential operators. Proceedings of a symposium in pure mathematics. Singular Integrals. Providence. Vol. X, 1957, pp. 275—287.

Šaginjan, A. L.

1. On best approximations by harmonic polynomials in a space. Dokl. Akad. Nauk SSSR, **90**, 141 (1953) [in Russian].

Schwartz, L.

1. Théorie des distributions. I, II, Paris 1957.

Slobodeckiĭ, L. N.

1. Spaces of S. L. Sobolev of fractional order and their application to boundary problems for partial differential equations. Dokl. Akad. Nauk SSSR, **118**, 243—246 (1958) [in Russian].
2. Generalized Sobolev spaces and their application to boundary problems for partial differential equations. Uč. zap. Leningr. ped. in-ta im A. I. Herzen **197**, 54—112 (1958) [in Russian].
3. Estimates of the solutions of elliptic and parabolic systems. Dokl. Akad. Nauk SSSR **120**, 468—471 (1959) [in Russian].

Smirnov, V. I.

1. Course in higher mathematics. Vol. V, State Publishing House for Technical Literature 1947, pp. 1—584 [in Russian].

Smith, K. T.

1. Inequalities for formally positive integrodifferential forms. Bull. Amer. Math. Soc. **67**, 368—370 (1961).

Sobolev, S. L.

1. The Cauchy problem in functional spaces. Dokl. Akad. Nauk SSSR, **3**, 291—294 (1935) [in Russian].
2. Méthode nouvelle à resoudre le problème de Cauchy pour les équations linéaires hyperboliques normales. Matem. sb. **1** (43), 39—72 (1936).
3. On a theorem of functional analysis. Matem. sb. **4** (46), 471—497 (1938) [in Russian].
4. Some applications of functional analysis in mathematical physics. Leningrad State University, 1950, pp. 1—255; Novosibirsk, 1962, pp. 1—255 [in Russian].
5. Some generalizations of imbedding theorems. Fund. Math. **47**, 277—324 (1959) [in Russian].

Solncev, Ju. K.
1. On an estimate of the mixed derivative in $L_p(g)$. Trudy Steklov Institute of the Akad. Nauk SSSR, **64**, 211—238 (1961) [in Russian].

Solonnikov, V. A.
1. Apriori estimates for second order parabolic equations. Trudy Steklov Institute of the Akad. Nauk SSSR, **70**, 132—212 (1964) [in Russian].

Sonin, N. Ja.
1. Recherches sur les fonctions cylindriques et le développement des fonctions continues en séries. Math. Ann. **16**, 1—80 (1880).

Stečkin, S. B.
1. On the order of best approximation of continuous functions. Izv. Akad. Nauk SSSR, ser. matem. **15**, 219—242 (1951) [in Russian].
2. On best approximation of conjugate functions by trigonometric polynomials. Izv. Akad. Nauk SSSR ser. matem. **20**, 197—206 (1956) [in Russian].

Stein, E. M.
1. The characterization of functions arising as potentials. Bull. Amer. Math. Soc. I **67**, 102—104 (1961); II **68**, 577—584 (1962).

Sun' Jun-Šen (Сунь Юн-Шен)
1. On best approximations of classes of functions represented in the form of convolutions. Dokl. Akad. Nauk SSSR **118**, 247—250 (1958) [in Russian].

Taibleson, M. N.
1. Lipschitz classes of functions and distributions in E_p. Bull. Amer. Math. Soc. **69**, 487—493 (1965).
2. On the theory of Lipschitz spaces of distributions on Euclidean n-space. I, Principal properties. J. Math. and Mech. **13**, No. 3, 407—479 (1964).

Timan, A. F.
1. Theory of approximation of functions of a real variable, Moscow: Physics and Mathematics Publishing House 1960, 624 pages [in Russian].

Titchmarch, E. C.
1. Introduction to the theory of the Fourier integral. Oxford University Press.

Tonelli, L.
1. Sulla quadratura della superficie. Rend. R. Accad. Lincei **6**, No. 3, 633—638 (1926).

Tulaĭkov, A. N.
1. Zur Kompaktheit in Raume L_p für $p = 1$. Gött. Nachr., 167—170 (1933).

Ul'janov, P. L.
1. On some equivalent conditions for the convergence of series and integrals. Uspehi matem. nauk, **8**, No. 6 (58), 138—141 (1953) [in Russian].

Uspenskiĭ, S. V.
1. Properties of the classes $W_p^{(r)}$ with fractional derivatives on differentiable manifolds. Dokl. Akad. Nauk SSSR, **132**, 60—62 (1960) [in Russian].
2. An imbedding theorem for the fractional order classes of S. L. Sobolev. Dokl. Akad. Nauk SSSR, **130**, 1960 [in Russian].
3. On imbedding theorems for weight classes. Trudy Steklov Institute of the Akad. Nauk SSSR, **50**, 282—303 (1961) [in Russian].

Vallée Poussin, Ch.-J. de la
1. Leçons sur l'approximation des fonctions d'une variable réelle, Paris 1919, 150 pages.

Vašarin, A. A.
1. Boundary properties for functions of the class $W_2^l(\alpha)$ and their application to the solution of a boundary problem of mathematical physics. Izv. Akad. Nauk SSSR, ser. matem. **23**, 421—454 (1959) [in Russian].

Vekua, I. N.

1. Generalized analytic functions, Moscow: Physics and Mathematics Publishing House; 1959, 628 pages [in Russian].

Vladimirov, V. S.

1. Methods of the theory of functions of several complex variables, Moscow: "Nauka" 1964, 411 pages [in Russian].

Watson, G. N.

1. Introduction to the theory of the Fourier integral. Theory of Bessel functions, published in Russian translation by the Foreign Languages Publishing House in 1949, 798 pages.

Whitney, H.

1. Analytic extensions of differentiable functions defined in closed sets. Trans. Amer. Math. Soc., I **36**, 63–89 (1943); II **40**, 309–317 (1946).

Zygmund, A.

1. Smooth functions, Duke Math. J. **12**, 47–76 (1945).
2. Trigonometrical Series, published in Russia by "Mir", 1965, Vol. 1, 615 pages, Vol. II, 537 pages.
3. Trigonometrical Series (old edition in Russian). GONTI (State Unified Scientific-Technical Publishing House) 1939, 323 pages.

Index of Names

Subject Index

DG 721/15/74

Die Grundlehren der mathematischen Wissenschaften in Einzeldarstellungen mit besonderer Berücksichtigung der Anwendungsgebiete

Eine Auswahl

---

23. Pasch: Vorlesungen über neuere Geometrie. In Vorbereitung
41. Steinitz: Vorlesungen über die Theorie der Polyeder. In Vorbereitung
45. Alexandroff/Hopf: Topologie. Band 1
46. Nevanlinna: Eindeutige analytische Funktionen
63. Eichler: Quadratische Formen und orthogonale Gruppen
102. Nevanlinna/Nevanlinna: Absolute Analysis
114. Mac Lane: Homology
127. Hermes: Enumerability, Decidability, Computability
144. Weil: Basic Number Theory
145. Butzer/Berens: Semi-Groups of Operators and Approximation.
146. Treves: Locally Convex Spaces and Linear Partial Differential Equations
147. Lamotke: Semisimpliziale algebraische Topologie
148. Chandrasekharan: Introduction to Analytic Number Theory
149. Sario/Oikawa: Capacity Functions
150. Iosifescu/Theodorescu: Random Processes and Learning. Cloth DM 75,—
151. Mandl: Analytical Treatment of One-dimensional Markov Processes
152. Hewitt/Ross: Abstract Harmonic Analysis. Vol. 2: Structure and Analysis for Compact Groups. Analysis on Locally Compact Abelian Groups
153. Federer: Geometric Measure Theory
154. Singer: Bases in Banach Spaces I
155. Müller: Foundations of the Mathematical Theory of Electromagnetic Waves
156. van der Waerden: Mathematical Statistics
157. Prohorov/Rozanov: Probability Theory
159. Köthe: Topological Vektor Spaces I
160. Agrest/Maksimov: Theory of Incomplete Cylindrical Functions and their Applications
161. Bhatia/Szegö: Stability Theory of Dynamical Systems
162. Nevanlinna: Analytic Functions
163. Stoer/Witzgall: Convexity and Optimization in Finite Dimensions I
164. Sario/Nakai: Classification Theory of Riemann Surfaces
165. Mitrinović/Vasić: Analytic Inequalities
166. Grothendieck/Dieudonné: Eléments de Géométrie Algébrique I
167. Chandrasekharan: Arithmetical Functions
168. Palamodov: Linear Differential Operators with Constant Coefficients
170. Lions: Optimal Control of Systems Governed by Partial Differential Equation
171. Singer: Best Approximation in Normed Linear Spaces by Elements of Linear Subspaces
172. Bühlmann: Mathematical Methods in Risk Theory
173. Maeda/Maeda: Theory of Symmetric Lattices
174. Stiefel/Scheifele: Linear and Regular Celestial Mechanics. Perturbed Two-body Motion—Numerical Methods—Canonical Theory
175. Larsen: An Introduction to the Theory of Multipliers
176. Grauert/Remmert: Analytische Stellenalgebren

177. Flügge: Practical Quantum Mechanics I
178. Flügge: Practical Quantum Mechanics II
179. Giraud: Cohomologie non abélienne
180. Landkof: Foundations of Modern Potential Theory
181. Lions/Magenes: Non-Homogeneous Boundary Value Problems and Applications I
182. Lions/Magenes: Non-Homogeneous Boundary Value Problems and Applications II
183. Lions/Magenes: Non-Homogeneous Boundary Value Problems and Applications III
184. Rosenblatt: Markov Processes. Structure and Asymptotic Behavior
185. Rubinowicz: Sommerfeldsche Polynommethode
186. Handbook for Automatic Computation. Vol. 2. Wilkinson/Reinsch: Linear Algebra
187. Siegel/Moser: Lectures on Celestial Mechanics
188. Warner: Harmonic Analysis on Semi-Simple Lie Groups I
189. Warner: Harmonic Analysis on Semi-Simple Lie Groups II
190. Faith: Algebra: Rings, Modules, and Categories I
192. Mal'cev: Algebraic Systems
193. Polya/Szegö: Problems and Theorems in Analysis I
194. Igusa: Theta Functions
195. Berberian: Baer*-Rings
196. Athreya/Ney: Branching Processes
197. Benz: Vorlesungen über Geometrie der Algebren
198. Gaal: Linear Analysis and Representation Theory
199. Nitsche: Vorlesungen über Minimalflächen
200. Dold: Lectures on Algebraic Topology
201. Beck: Continuous Flows in the Plane
202. Schmetterer: Introduction to Mathematical Statistics
203. Schoeneberg: Elliptic Modular Functions
204. Popov: Hyperstability of Control Systems
206. André: Homologie des algèbres commutatives
207. Donoghue: Monotone Matrix Functions and Analytic Continuation
208. Lacey: The Isometric Theory of Classical Banach Spaces
209. Ringel: Map Color Theorem
210. Gihman/Skorohod: The Theory of Stochastic Processes 1
211. Comfort/Negrepontis: The Theory of Ultrafilters
212. Switzer: Algebraic Topology. In Vorbereitung
213. Shafarevich: Basic Algebraic Geometry
214. van der Waerden: Group Theory and Quantum Mechanics. In Vorbereitung